Introduction to Partial Differential Equations with Applications

E. C. Zachmanoglou

Professor of Mathematics
Purdue University

Dale W. Thoe

Professor of Mathematics
Purdue University

Dover Publications, Inc., New York

Published in Canada by General Publishing Company, Ltd., 30 Lesmill Road, Don Mills, Toronto, Ontario.
Published in the United Kingdom by Constable and Company, Ltd.

This Dover edition, first published in 1986, is an unabridged, corrected republication of the work first published by The Williams & Wilkins Company, Baltimore, 1976.

Manufactured in the United States of America
Dover Publications, Inc., 31 East 2nd Street, Mineola, N.Y. 11501

Library of Congress Cataloging-in-Publication Data

Zachmanoglou, E. C.
Introduction to partial differential equations with applications.

Reprint. Originally published: Baltimore : Williams & Wilkins, c1976.
Bibliography: p.
Includes index.
1. Differential equations, Partial. I. Thoe, Dale W. II. Title.
[QA377.Z32 1986] 515.3′53 86-13604
ISBN 0-486-65251-3

PREFACE

In writing this introductory book on the old but still rapidly expanding field of Mathematics known as Partial Differential Equations, our objective has been to present an elementary treatment of the most important topics of the theory together with applications to problems from the physical sciences and engineering. The book should be accessible to students with a modest mathematical background and should be useful to those who will actually need to use partial differential equations in solving physical problems. At the same time we hope that the book will provide a good basis for those students who will pursue the study of more advanced topics including what is now known as the modern theory.

Throughout the book, the importance of the proper formulation of problems associated with partial differential equations is emphasized. Methods of solution of any particular problem for a given partial differential equation are discussed only after a large collection of elementary solutions of the equation has been constructed.

During the last five years, the book has been used in the form of lecture notes for a semester course at Purdue University. The students are advanced undergraduate or beginning graduate students in mathematics, engineering or one of the physical sciences. A course in Advanced Calculus or a strong course in Calculus with extensive treatment of functions of several variables, and a very elementary introduction to Ordinary Differential Equations constitute adequate preparation for the understanding of the book. In any case, the basic results of advanced calculus are recalled whenever needed.

The book begins with a short review of calculus and ordinary differential equations. A new elementary treatment of first order quasi-linear partial differential equations is then presented. The geometrical background necessary for the study of these equations is carefully developed. Several applications are discussed such as applications to problems in gas dynamics (the development of shocks), traffic flow, telephone networks, and biology (birth and death processes and control of disease). The method of probability generating functions in the study of stochastic processes is discussed and illustrated by many examples. In recent books the topic of first order equations is either omitted or treated inadequately. In older books the treatment of this topic is probably inaccessible to most students.

A brief discussion of series solutions in connection with one of the basic results of the theory, known as the Cauchy-Kovalevsky theorem, is included. The characteristics, classification and canonical forms of linear partial differential equations are carefully discussed.

For students with little or no background in physics, Chapter VI, "Equations of Mathematical Physics," should be helpful. In Chapters VII, VIII and IX where the equations of Laplace, wave and heat are studied, the physical problems associated with these equations are always used to motivate and illustrate the theory. The question of determining the well-posed problems associated with each equation is fundamental throughout the discussion.

The methods of separation of variables and Fourier series are introduced in the chapter on Laplace's equation and then used again in the chapters on the wave and heat equations. The method of finite differences coupled with the use of computers is illustrated with an application to the Dirichlet problem for Laplace's equation.

The last chapter is devoted to a brief treatment of hyperbolic systems of equations with emphasis on applications to electrical transmission lines and to gas dynamics.

The problems at the end of each section fall in three main groups. The first group consists of problems which ask the student to provide the details of derivation of some of the items in the text. The problems in the second group are either straightforward applications of the theory or ask the student to solve specific problems associated with partial differential equations. Finally, the problems in the third group introduce new important topics. For example, the treatment of nonhomogeneous equations is left primarily to these problems. The student is urged at least to read these problems.

The references cited in each chapter are listed at the end of that chapter. A guide to further study, a bibliography for further study and answers to some of the problems appear at the end of the book.

The book contains roughly twenty-five percent more material than can be covered in a one-semester course. This provides flexibility for planning either a more theoretical or a more applied course. For a more theoretical course, some of the sections on applications should be omitted. For a more applied version of the course, the instructor should only outline the results in the following sections: Chapter III, Section 4; Chapter IV, Sections 1 and 2; Chapter V Sections 5, 6, 7, 8, and 9 (the classification and characteristics of second order equations should be carefully discussed, however); Chapter VII, the proof of Theorem 10.1 and Section 11.

We are indebted to many of our colleagues and students for their comments concerning the manuscript, and extend our thanks for their help. Finally, to Judy Snider, we express our deep appreciation for her expert typing of the manuscript.

Purdue University
1975

E. C. ZACHMANOGLOU
DALE W. THOE

TABLE OF CONTENTS

CHAPTER VIII. THE WAVE EQUATION 261

CHAPTER IX. THE HEAT EQUATION 331

CHAPTER I

Some concepts from calculus and ordinary differential equations

In this chapter we review some basic definitions and theorems from Calculus and Ordinary Differential Equations. At the same time we introduce some of the notation which will be used in this book. Most topics may be quite familiar to the student and in this case the chapter may be covered quickly.

In Section 1 we review some concepts associated with sets, functions, limits, continuity and differentiability. In Section 2 we discuss surfaces and their normals and recall one of the most useful and important theorems of mathematics, the Implicit Function Theorem. In Section 3 we discuss two ways of representing curves in three or higher dimensional space and give the formulas for finding the tangent vectors for each of these representations. In Section 4 we review the basic existence and uniqueness theorem for the initial value problem for ordinary differential equations and systems.

1. Sets and Functions

We will denote by R^n the n-dimensional Euclidean space. A point in R^n has n coordinates $x_1, x_2, \ldots, x_n$ and its position vector will be denoted by x. Thus $x = (x_1, \ldots, x_n)$. For $n = 2$ or $n = 3$ we may also use different letters for the coordinates. For example, we may use (x, y) for the coordinates of a point in R^2 and (x, y, z) or (x, y, u) for the coordinates of a point in R^3.

The *distance* between two points $x = (x_1, \ldots, x_n)$ and $y = (y_1, \ldots, y_n)$ in R^n is given by

$$d(x, y) = \left[\sum_{i=1}^{n} (x_i - y_i)^2\right]^{1/2}.$$

An *open ball* with center $x^0 \in R^n$ and radius $\rho > 0$ is the set of points in R^n which are at a distance less than ρ from x^0,

(1.1) $$B(x^0, \rho) = \{x : x \in R^n, d(x, x^0) < \rho\}.$$

The corresponding *closed ball* is

(1.2) $$\bar{B}(x^0, \rho) = \{x : x \in R^n, d(x, x^0) \leqq \rho\}$$

and the surface of this ball, called a *sphere*, is

(1.3) $$S(x^0, \rho) = \{x : x \in R^n, d(x, x^0) = \rho\}.$$

A set A of points in R^n is called *open* if for every $x \in A$ there is a ball with center at x which is contained in A. A set in R^n is called *closed* if its complement is an open set. A point x is called a *boundary point* of a set A if every ball with center at x contains points of A and points of the complement of A. The set of all boundary points of A is called the *boundary* of A and is denoted by ∂A.

It should be easy for the student to show that the open ball $B(x^0, \rho)$ defined by equation (1.1) is an open set while the closed ball $\bar{B}(x^0, \rho)$ defined by (1.2) is a closed set. The boundary of both these sets is the sphere $S(x^0, \rho)$ defined by equation (1.3). As another example, in R^2 the set

$$\{(x_1, x_2) : x_2 > 0\}$$

is open, while the set

$$\{(x_1, x_2) : x_2 \geqq 0\}$$

is closed. The boundary of both of these sets is the x_1-axis. Also in R^2, the rectangle defined by the inequalities

$$a < x_1 < b, \qquad c < x_2 < d$$

is open, while the rectangle

$$a \leqq x_1 \leqq b, \qquad c \leqq x_2 \leqq d$$

is closed. However, the inequalities

$$a \leqq x_1 < b, \qquad c < x_2 < d$$

define a set which is neither open nor closed. All the above three sets have the same boundary which the student should describe by means of equalities and inequalities.

A *neighborhood* of a point in R^n is any open set containing the point. All open balls centered at a point x^0 are neighborhoods of x^0. Thus, a point has neighborhoods which are arbitrarily small. Note also that an open set is a neighborhood of each of its points.

An open set A in R^n is called *connected* if any two points of A can be connected by a polygonal path which is contained entirely in A. An open and connected set in R^n is called a *domain*. A domain will be usually denoted in this book by the letter Ω.

A set A in R^n is called *bounded* if it is contained in some ball of finite radius centered at the origin.

We consider now functions defined for all points in some set A in R^n and with values real numbers. Such functions are called real-valued functions. The set A is called the *domain* of f and the set of values of f is called its *range*. For simplicity in this discussion, we take $n = 2$. Thus we consider only functions of two independent variables. All the definitions and theorems that we mention below are valid for functions of more than two variables. The student should formulate the appropriate statements of the corresponding definitions and theorems for such functions.

Let f be a function defined for all points (x, y) in some set A in R^2. Let (x^0, y^0) be a fixed point in A or on ∂A. We say that f has *limit* L as (x, y) approaches (x^0, y^0) and we write

$$\lim_{(x,y)\to(x^0,y^0)} f(x, y) = L$$

if, given any $\epsilon > 0$, there is a $\delta > 0$ such that

$$|f(x, y) - L| < \epsilon$$

for all $(x, y) \neq (x^0, y^0)$ such that

$$(x, y) \in B((x^0, y^0), \delta) \cap A.$$

In other words, $f(x, y) \to L$ as $(x, y) \to (x^0, y^0)$ if given any positive number ϵ we can find a positive number δ such that the distance of $f(x, y)$ from L is less than ϵ for all points (x, y) of A which are at a distance less than δ from (x^0, y^0) excepting possibly the point (x^0, y^0) itself. The function f is said to be *continuous* at the point $(x^0, y^0) \in A$ if

$$\lim_{(x,y)\to(x^0,y^0)} f(x, y) = f(x^0, y^0).$$

f is called continuous in A if it is continuous at every point of A.

We now state an important theorem concerning the maximum and minimum values of a continuous function: Let A be a closed and bounded set in R^n and suppose that f is defined and continuous on A. Then f attains its maximum and minimum values in A; i.e., there is a point $a \in$ A such that

$$f(a) \geqq f(x), \quad \text{for every } x \in A$$

and a point $b \in$ A such that

$$f(b) \leqq f(x), \quad \text{for every } x \in A.$$

Next we recall the definition of the partial derivatives of f at a point (x^0, y^0) of its domain A. Since this involves the values of f at points in a neighborhood of (x^0, y^0), we must assume either that the domain A is open or that (x^0, y^0) is an interior point of A (see Problem 1.3). In either case the points $(x^0 + h, y^0)$ and $(x^0, y^0 + k)$ with sufficiently small $|h|$ and $|k|$ belong to A. We say that the partial derivative of f with respect to x exists at (x^0, y^0) if the limit

$$\lim_{h\to 0} \frac{f(x^0 + h, y^0) - f(x^0, y^0)}{h}$$

exists. Then the value of this limit is the value of the derivative,

$$\frac{\partial f}{\partial x}(x^0, y^0) = \lim_{h\to 0} \frac{f(x^0 + h, y^0) - f(x^0, y^0)}{h}.$$

Similarly,

$$\frac{\partial f}{\partial y}(x^0, y^0) = \lim_{k\to 0} \frac{f(x^0, y^0 + k) - f(x^0, y^0)}{k}.$$

If the partial derivatives of f with respect to x or y exist at every point in a subset B of A, then they are themselves functions with domain the set B. We will denote these functions by

$$D_1 f \quad \text{or} \quad \frac{\partial f}{\partial x} \quad \text{or} \quad f_x$$

and by

$$D_2 f \quad \text{or} \quad \frac{\partial f}{\partial y} \quad \text{or} \quad f_y.$$

In general, D_j will stand for the differentiation operator with respect to the jth variable in R^n,

$$D_j = \frac{\partial}{\partial x_j}; \qquad j = 1, \ldots, n.$$

If the partial derivatives themselves have partial derivatives, these are called the second order derivatives of the original function. Thus a function f of two variables may have four partial derivatives of the second order,

$$D_1^2 f = \frac{\partial^2 f}{\partial x^2}, \qquad D_2^2 f = \frac{\partial^2 f}{\partial y^2}$$

$$D_2 D_1 f = \frac{\partial^2 f}{\partial y \partial x}, \qquad D_1 D_2 f = \frac{\partial^2 f}{\partial x \partial y}$$

It is not hard to show that if $D_1 f$, $D_2 f$, $D_1 D_2 f$ and $D_2 D_1 f$ are continuous functions in an open set A then the mixed derivatives are equal in A,

$$D_2 D_1 f = D_1 D_2 f.$$

Partial derivatives of any order higher than the second may also exist and are defined in the obvious way.

Let f be a function defined in a set A and suppose that f and all its partial derivatives of order less than or equal to k are continuous in a subset B of A. Then f is said to be *of class C^k in B*. The collection of all functions of class C^k in B is denoted by $C^k(B)$. Thus, a short way of indicating that f is of class C^k in B is by writing $f \in C^k(B)$. $C^0(B)$ is the collection of all functions which are continuous in B and $C^\infty(B)$ is the collection of all functions which have continuous derivatives of all orders in B.

All polynomials of a single variable as well as the functions $\sin x$, $\cos x$, e^x are of class C^∞ in R^1. The function

$$f(x) = \begin{cases} 0 & \text{if} \quad x \leqq 0 \\ x & \text{if} \quad x > 0 \end{cases}$$

is of class C^0 in R^1 but f is not of class C^1 in R^1. The function

$$g(x) = \begin{cases} 1, & \text{if} \quad x \leqq 0 \\ 1 + x^2, & \text{if} \quad x > 0 \end{cases}$$

is of class C^1 in R^1 but g is not of class C^2 in R^1. These last two examples illustrate that we have the sequence of inclusions

$$C^0(B) \supset C^1(B) \supset C^2(B) \supset \ldots \supset C^\infty(B)$$

and that each class $C^j(B)$ is actually smaller than $C^{j-1}(B)$.

Problems

1.1 Prove that a closed set contains all of its boundary points while an open set contains none of its boundary points.

1.2 The closure $\bar{A}$ of a set A is defined to be the union of the set and of its boundary,

$$\bar{A} = A \cup \partial A.$$

Prove that if A is closed, then $\bar{A} = A$. Describe the closures of the sets given as examples in Section 1.1.

1.3 A point $x \in A$ is called an interior point of A if there is an open ball centered at x and contained in A. The set of interior points of A is called the interior of A and is denoted by $\mathring{A}$. Show that if A is open then $\mathring{A} = A$.

1.4 Give an example of a function defined in R^2 which is of class C^1 in R^2 but not of class C^2 in R^2.

1.5 Using the definition of limit of a function, show that

(a) $\displaystyle\lim_{(x,y)\to(0,0)} \frac{x^4 + y^4}{x^2 + y^2} = 0$ (b) $\displaystyle\lim_{(x,y)\to(0,0)} [xy \log (x^2 + y^2)] = 0$

1.6 Show that the mixed partial derivatives of the function

$$f(x, y) = \begin{cases} \dfrac{x^3y - xy^3}{x^2 + y^2} & \text{if} \quad (x, y) \neq (0,0) \\ 0 & \text{if} \quad (x, y) = (0, 0) \end{cases}$$

are not equal. Explain.

2. Surfaces and Their Normals. The Implicit Function Theorem

A useful way of visualizing a function f of one variable x is by drawing its graph in the (x, y)-plane. We write $y = f(x)$ and draw the locus of points of the form $(x, f(x))$ where x varies over the domain of f. Usually the graph of f is a curve in the (x, y)-plane. However not every curve in the (x, y)-plane is the graph of some function. For example a circle is not the graph of any function.

In principle, we should be able to draw the graph of a function of any number of variables but we are limited by our inability to draw figures in more than three dimensions. Consequently we can actually draw the graphs of functions of up to two variables only. If f is a function of x and y, we write $z = f(x, y)$ and draw in (x, y, z)-space the locus of points of the form $(x, y, f(x, y))$ where (x, y) varies over the domain of f. Although the graph of f is usually a surface, not every surface in three dimensions is the graph of some function of two variables. For example, a sphere is not the graph of any function. Since surfaces will play a central role in many of our discussions, we will describe now a class of surfaces more general than the surfaces obtained as graphs of functions. For simplicity we restrict the discussion to the case of three dimensions.

Let Ω be a domain in R^3 and let $F(x, y, z)$ be a function in the class $C^1(\Omega)$. The *gradient* of F, written grad F, is a vector valued function defined in Ω by the formula

$$\text{grad } F = \left(\frac{\partial F}{\partial x}, \frac{\partial F}{\partial y}, \frac{\partial F}{\partial z}\right). \tag{2.1}$$

The value of grad F at a point $(x, y, z) \in \Omega$ is a vector with components the values of the partial derivatives of F at that point. It is convenient to visualize grad F as a field of vectors (vector field), with one vector, grad $F(x, y, z)$, emanating from each point (x, y, z) in Ω.

We now make the assumption

$$\text{grad } F(x, y, z) \neq (0, 0, 0) \tag{2.2}$$

at every point of Ω. This means that the partial derivatives of F do not vanish simultaneously at any point of Ω. Under the assumption (2.2), the set of points (x, y, z) in Ω which satisfy the equation

$$F(x, y, z) = c, \tag{2.3}$$

for some appropriate value of the constant c, is a surface in Ω. This surface is called a *level surface* of F. The appropriate values of c are the values of the function F in Ω. For example, if (x_0, y_0, z_0) is a given point in Ω and if we take $c = F(x_0, y_0, z_0)$, the equation

$$F(x, y, z) = F(x_0, y_0, z_0)$$

represents a surface in Ω passing through the point (x_0, y_0, z_0). For different values of c, equation (2.3) represents different surfaces in Ω. Each point of Ω lies on exactly one level surface of F and any two points (x_0, y_0, z_0) and (x_1, y_1, z_1) of Ω lie on the same level surface if and only if

$$F(x_0, y_0, z_0) = F(x_1, y_1, z_1).$$

Thus, Ω can be visualized as being laminated by the level surfaces of F.

Quite often we consider c in equation (2.3) as a parameter and we say that equation (2.3) represents a *one-parameter family of surfaces* in Ω. Through each point in Ω passes a particular member of this family corresponding to a particular value of the parameter c.

As an example let

$$F(x, y, z) = x^2 + y^2 + z^2.$$

Then

$$\text{grad } F(x, y, z) = (2x, 2y, 2z)$$

and if Ω is the whole of R^3 except for the origin, condition (2.2) is satisfied at every point of Ω. The level surfaces of F are spheres with center the origin. As another example, let

$$F(x, y, z) = z.$$

Then grad $F(x, y, z) = (0, 0, 1)$ and condition (2.2) is satisfied at every point of R^3. The level surfaces are planes parallel to the (x, y)-plane.

Let us consider now a particular level surface S_c given by equation (2.3) for a fixed value of c. Under our assumptions on F, there is a tangent plane to S_c at each of its points. At the point of tangency the value of grad F is a vector normal to the tangent plane. For this reason we say that, at each point of S_c, the value of grad F is a vector normal to S_c.

Let us recall the equation of a plane in R^3. Since we are using the letters x, y, z for the coordinates we shall use $\mathbf{r}$ for the position vector of a point with coordinates (x, y, z). Thus $\mathbf{r} = (x, y, z)$. Suppose now that P is a plane passing through a fixed point $\mathbf{r}_0 = (x_0, y_0, z_0)$ and having $\mathbf{n} = (n_x, n_y, n_z)$ as

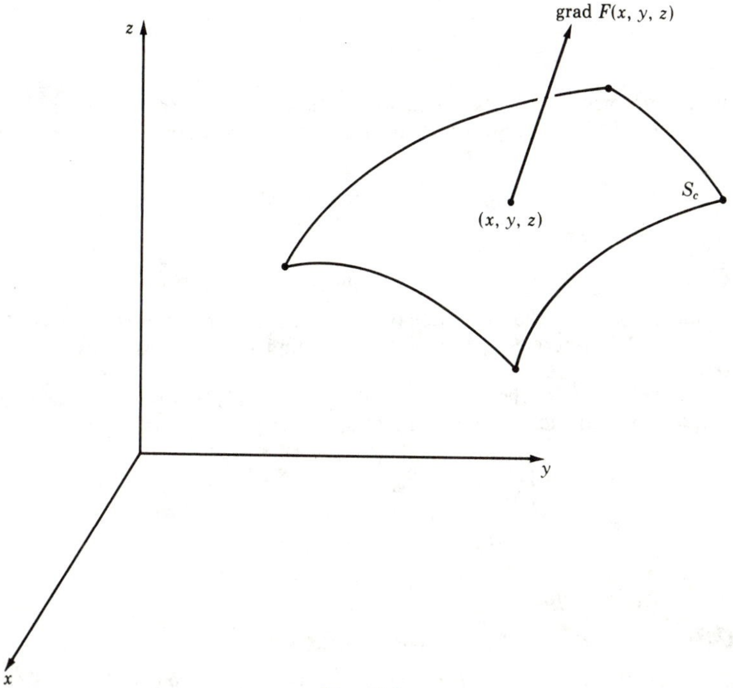

Fig. 2.1

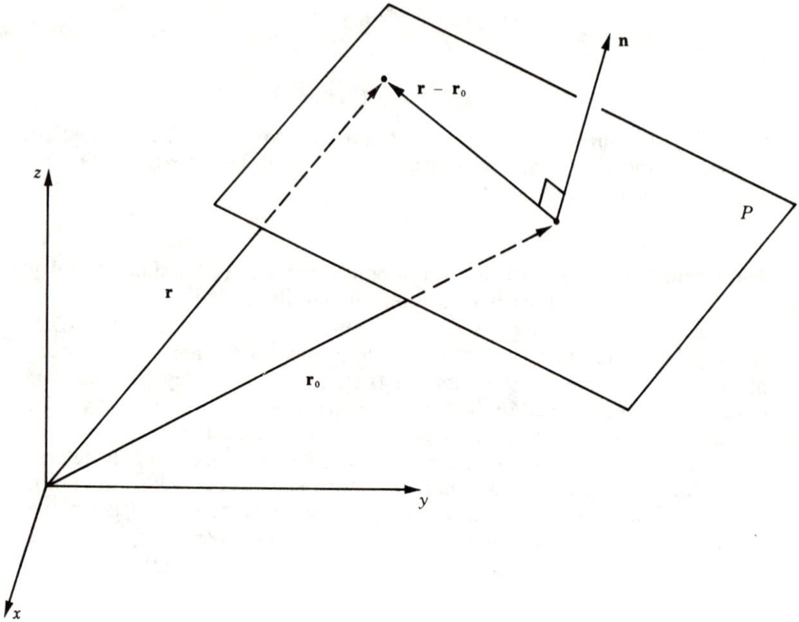

Fig. 2.2

its normal vector. If $\mathbf{r}$ is the position vector of an arbitrary point on P, then the vector $\mathbf{r} - \mathbf{r}_0$ lies on P and hence it must be normal to $\mathbf{n}$. It follows that

(2.4) $$(\mathbf{r} - \mathbf{r}_0)\cdot\mathbf{n} = 0$$

or, in terms of the coordinates,

(2.5) $$(x - x_0)n_x + (y - y_0)n_y + (z - z_0)n_z = 0.$$

Equation (2.5) is the equation of the plane P. Note that (2.5) has the form of (2.3) where $F(x, y, z)$ is the left hand side of (2.5) and grad $F = (n_x, n_y, n_z)$.

Returning now to the level surface S_c given by (2.3), it is easy to see that the equation of the tangent plane to S_c at the point (x_0, y_0, z_0) of S_c is

(2.6) $$(x - x_0)\left[\frac{\partial F}{\partial x}(x_0, y_0, z_0)\right] + (y - y_0)\left[\frac{\partial F}{\partial y}(x_0, y_0, z_0)\right] + (z - z_0)\left[\frac{\partial F}{\partial z}(x_0, y_0, z_0)\right] = 0$$

or, in vector form,

(2.7) $$(\mathbf{r} - \mathbf{r}_0)\cdot\text{grad } F(\mathbf{r}_0) = 0.$$

As an example, let us find the equation of the plane tangent to the sphere

$$x^2 + y^2 + z^2 = 6$$

at the point (1, 2, −1). We have

$$F(x, y, z) = x^2 + y^2 + z^2,$$

$$\text{grad } F(x, y, z) = (2x, 2y, 2z),$$

and

$$\text{grad } F(1, 2, -1) = (2, 4, -2).$$

The vector (2, 4, −2) is normal to the sphere at the point (1, 2, −1) and the equation of the tangent plane at (1, 2, −1) is

$$2(x - 1) + 4(y - 2) - 2(z + 1) = 0.$$

Again, let us consider the surface S_c given by equation (2.3) and suppose that the point (x_0, y_0, z_0) lies on this surface. We ask the following question: Is it possible to describe S_c by an equation of the form

(2.8) $$z = f(x, y),$$

so that S_c is the graph of f? This is equivalent to asking whether it is possible to solve equation (2.3) for z in terms of x and y. An answer to this question is given by the

Implicit Function Theorem

If $F_z(x_0, y_0, z_0) \neq 0$, then (2.3) can be solved for z in terms of x and y for (x, y, z) near the point (x_0, y_0, z_0). Moreover, the partial derivatives of z with respect to x and y, i.e., the partial derivatives of f in (2.8), can be obtained by implicit differentiation of (2.3) where z is considered as a function of x and y,

(2.9) $$\frac{\partial z}{\partial x} = -\frac{F_x}{F_z}, \qquad \frac{\partial z}{\partial y} = -\frac{F_y}{F_z}.$$

Of course, there is nothing special about the variable z. If F_x does not vanish at (x_0, y_0, z_0), then near (x_0, y_0, z_0) we can solve (2.3) for x and we can compute the derivatives of x with respect to y and z by implicit differentiation. In fact, if F satisfies condition (2.2) at every point of Ω, then the implicit function theorem asserts that, in a neighborhood of any point of Ω, we can always solve equation (2.3) for at least one of the variables in terms of the other two. A proof of the Implicit Function Theorem may be found in any book on Advanced Calculus. (See for example Taylor.[1])

Again, as an example let us consider the equation of the unit sphere

(2.10) $$x^2 + y^2 + z^2 = 1.$$

At the point (0, 0, 1) of this surface we have $F_z(0, 0, 1) = 2$. By the implicit function theorem, we can solve (2.10) for z near the point (0, 0, 1). In fact we have

(2.11) $$z = +\sqrt{1 - x^2 - y^2}, \qquad x^2 + y^2 < 1.$$

In the upper half space $z > 0$, equations (2.10) and (2.11) describe the same surface. The point $(0, 0, -1)$ is also on the surface (2.10) and $F_z(0, 0, -1) = -2$. Near $(0, 0, -1)$ we have

$$z = -\sqrt{1 - x^2 - y^2}, \qquad x^2 + y^2 < 1. \tag{2.12}$$

In the lower half space $z < 0$, equations (2.10) and (2.12) represent the same surface. On the other hand, at the point $(1, 0, 0)$ we have $F_z(1, 0, 0) = 0$ and it is easy to see that it is not possible to solve (2.10) for z in terms of x and y near this point. However, it is possible to solve for x in terms of y and z. Finally, near the point $(1/\sqrt{3}, 1/\sqrt{3}, 1/\sqrt{3})$ which satisfies (2.10) it is possible to solve for every one of the variables in terms of the other two.

Problems

2.1. Let $F(x, y, z) = z^2 - x^2 - y^2$

(a) Find grad F. What is the largest set in which grad F does not vanish?

(b) Sketch the level surfaces $F(x, y, z) = c$ with $c = 0, 1, -1$. (*Hint:* Set $r^2 = x^2 + y^2$).

(c) Find a vector normal to the surface

$$z^2 - x^2 - y^2 = 0$$

at the point $(1, 0, 1)$, and the equation of the plane tangent to the surface at that point. What happens at the point $(0, 0, 0)$ of the surface?

2.2. Sketch the surface described by the equation

$$(z - z_0)^2 - (x - x_0)^2 - (y - y_0)^2 = 0$$

where (x_0, y_0, z_0) is a fixed point. Show that if $\mathbf{n} = (n_x, n_y, n_z)$ is a vector normal to the surface then

$$n_z^2 - n_x^2 - n_y^2 = 0.$$

Show that $\mathbf{n}$ makes a 45° angle with the z-axis.

2.3. Find the equation of the plane tangent to the paraboloid

$$z = x^2 + y^2$$

at the point $(1, 1, 2)$.

2.4. In R^2, a level surface of a function F of two variables is a curve (level curve of F) and a tangent plane is a line. Find the equation of the line tangent to the curve

$$x^4 + x^2y^2 + y^4 = 21$$

at the point $(1, 2)$.

2.5. If possible, solve the equation

$$z^2 - x^2 - y^2 = 0.$$

for z in terms of x, y near the following points:
(a) $(1, 1, \sqrt{2})$; (b) $(1, 1, -\sqrt{2})$; (c) $(0, 0, 0)$.

2.6. By implicit differentiation derive formulas (2.9).
2.7. Prove the identities
(a) $\text{grad}(f + g) = \text{grad}\, f + \text{grad}\, g$
(b) $\text{grad}(fg) = f\,\text{grad}\, g + g\,\text{grad}\, f$.

3. Curves and Their Tangents

The most common way of describing a curve in R^3 is by means of a **parametric representation**. If $\mathbf{r}$ denotes the position vector of a point on a curve C, then C may be described by the vector equation

$$\mathbf{r} = \mathbf{F}(t), \qquad t \in I, \tag{3.1}$$

where I is some interval on the real axis and $\mathbf{F}(t) = (f_1(t), f_2(t), f_3(t))$ is a vector valued function of the parameter t. If we write $\mathbf{r} = (x, y, z)$, then the vector equation (3.1) is equivalent to the three equations

$$x = f_1(t), \qquad y = f_2(t), \qquad z = f_3(t);\ t \in I. \tag{3.2}$$

We will assume that the functions f_1, f_2 and f_3 belong to $C^1(I)$ and that their derivatives do not vanish simultaneously at any point of I,

$$\left(\frac{df_1(t)}{dt}, \frac{df_2(t)}{dt}, \frac{df_3(t)}{dt}\right) \neq (0, 0, 0), \qquad t \in I. \tag{3.3}$$

Then for each $t \in I$ the nonvanishing vector

$$\frac{d\mathbf{r}}{dt} = \frac{d\mathbf{F}(t)}{dt} = \left(\frac{df_1(t)}{dt}, \frac{df_2(t)}{dt}, \frac{df_3(t)}{dt}\right)$$

is tangent to the curve C at the point $(f_1(t), f_2(t), f_3(t))$ of C.

Example 3.1. It is easy to see that the equations

$$x = \cos t \qquad y = \sin t, \qquad z = t; \qquad t \in R^1 \tag{3.4}$$

represent a helix (see Fig. 3.1). We have

$$\left(\frac{dx}{dt}, \frac{dy}{dt}, \frac{dz}{dt}\right) = (-\sin t, \cos t, 1).$$

To $t = \pi/2$ corresponds the point $(0, 1, \pi/2)$ on the helix and at that point the vector $(-1, 0, 1)$ is tangent to the helix.

Another way of describing a curve in R^3 is by making use of the fact that the intersection of two surfaces is usually a curve. Let F_1 and F_2 be two real valued functions of class C^1 in some domain Ω in R^3 and suppose $\text{grad}\, F_1$ and $\text{grad}\, F_2$ do not vanish in Ω. Then, as we saw in Section 2, the set of points satisfying each of the equations

$$F_1(x, y, z) = c_1, \qquad F_2(x, y, z) = c_2 \tag{3.5}$$

is a surface, and hence the set of points satisfying both equations must lie on the intersection of these two surfaces. This intersection, if not empty, is in general a curve. In fact if we make the additional assumption that $\text{grad}\, F_1$ and $\text{grad}\, F_2$ are not collinear at any point of Ω then the intersection (if not empty) of the two surfaces given by each of the equations (3.5) is

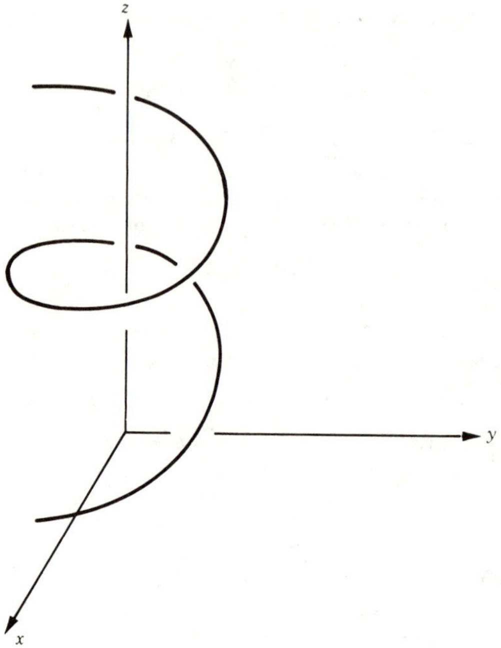

Fig. 3.1

always a curve. This assumption can be expressed in terms of the cross product,

$$\text{(3.6)} \qquad [\text{grad } F_1(x, y, z)] \times [\text{grad } F_2(x, y, z)] \neq \mathbf{0}, \; (x, y, z) \in \Omega.$$

We prove the above assertion by showing that, under the assumption (3.6), if (x_0, y_0, z_0) is any point in Ω satisfying equations (3.5), then near (x_0, y_0, z_0) the set of points (x, y, z) satisfying (3.5) can be described parametrically by equations of the form (3.2). Since

$$\text{(3.7)} \qquad \text{grad } F_1 \times \text{grad } F_2 = \left(\frac{\partial(F_1, F_2)}{\partial(y, z)}, \frac{\partial(F_1, F_2)}{\partial(z, x)}, \frac{\partial(F_1, F_2)}{\partial(x, y)}\right)$$

where, for example, $\partial(F_1, F_2)/\partial(y, z)$ is the Jacobian

$$\frac{\partial(F_1, F_2)}{\partial(y, z)} = \begin{vmatrix} \frac{\partial F_1}{\partial y} & \frac{\partial F_1}{\partial z} \\ \frac{\partial F_2}{\partial y} & \frac{\partial F_2}{\partial z} \end{vmatrix} = \frac{\partial F_1}{\partial y}\frac{\partial F_2}{\partial z} - \frac{\partial F_1}{\partial z}\frac{\partial F_2}{\partial y},$$

condition (3.6) means that at every point of Ω at least one of the Jacobians on the right side of (3.7) is different from zero. Suppose for example that

$$(3.8) \qquad \left.\frac{\partial(F_1, F_2)}{\partial(y, z)}\right|_{(x_0,\, y_0,\, z_0)} \neq 0.$$

Then a more general form of the Implicit Function Theorem (see Taylor,[1] Section 8.3) asserts that near (x_0, y_0, z_0) it is possible to solve the system of equations (3.5) for y and z in terms of x,

$$(3.9) \qquad y = g(x), \qquad z = h(x).$$

If we set $x = t$, equations (3.9) can be written in the parametric form

$$(3.10) \qquad x = t, \qquad y = g(t), \qquad z = h(t).$$

This shows that near (x_0, y_0, z_0) the set of points (x, y, z) satisfying (3.5) forms a curve with parametric representation given by (3.10). Note that in (3.10) the variable x is actually used as the parameter of the curve. In general, under the assumption (3.6), equations (3.5), with appropriate values of c_1 and c_2, represent a curve. Near any one of its points this curve can be represented parametrically, with one of the variables used as the parameter.

Example 3.2. Let

$$F_1(x, y, z) = x^2 + y^2 - z, \qquad F_2(x, y, z) = z.$$

We have grad $F_1 = (2x, 2y, -1)$, grad $F_2 = (0, 0, 1)$ and it is easy to see that

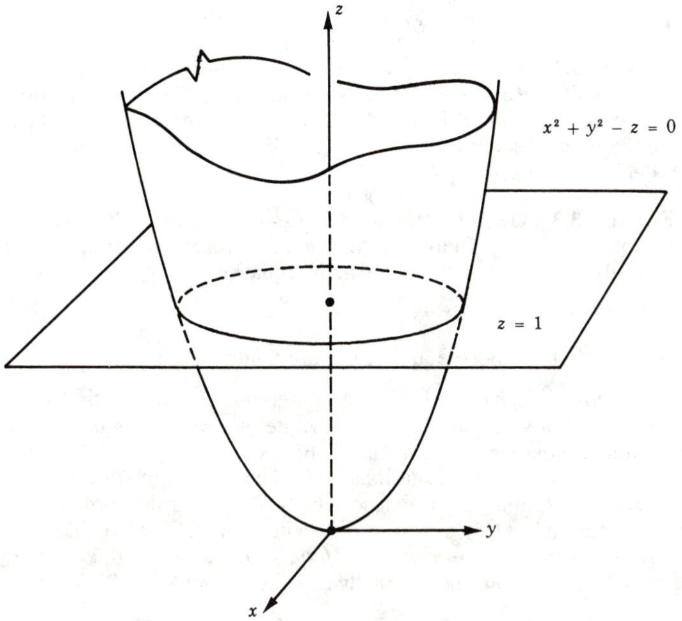

Fig. 3.2

if Ω is R^3 with the z-axis removed, then in Ω condition (3.6) is satisfied. The pair of equations

$$x^2 + y^2 - z = 0, \qquad z = 1$$

represents a circle which is the intersection of the paraboloidal surface represented by the first equation and the plane represented by the second equation. The point (0, 1, 1) lies on the circle, and near this point the circle has the parametric representation

$$x = t, \qquad y = +\sqrt{1 - t^2}, \qquad z = 1; \quad t \in (-1, 1).$$

Suppose now that a curve C is given parametrically by equations (3.2) where the functions f_1, f_2, f_3 satisfy condition (3.3). Is it possible to represent C by a pair of equations of the form (3.5)? Or, in geometric language, is C the intersection of two surfaces? Near any point of C, the answer to this question is yes. In fact let us consider the point (x_0, y_0, z_0) corresponding to $t = t_0$ and suppose that $f_1'(t_0) \neq 0$. Then, by the implicit function theorem, the equation

$$x - f_1(t) = 0$$

can be solved for t in terms of x for (x, t) near (x_0, t_0),

$$t = g(x). \tag{3.11}$$

Substituting (3.11) into the last two of equations (3.2) we obtain

$$y = f_2(g(x)), \qquad z = f_3(g(x)). \tag{3.12}$$

Equations (3.12) represent C near (x_0, y_0, z_0) as the intersection of two surfaces which are in fact cylindrical surfaces. The first of equations (3.12) describes a cylindrical surface with generators parallel to the z-axis while the second equation describes a cylindrical surface with generators parallel to the y-axis.

Example 3.3. The third equation in (3.4) is already solved for t and is valid for all z and t. Therefore, the helix represented parametrically by (3.4) is also represented by the pair of equations

$$x = \cos z, \qquad y = \sin z.$$

Note that each of these equations is a cylindrical surface.

It should be clear now that the two methods of representing a curve in R^3 are essentially equivalent. In the sequel we will use whichever representation is most suitable for our purposes.

Let us consider again equations (3.5). To each set of suitable values of c_1 and c_2 corresponds a curve described by (3.5). For different values of c_1 and c_2 equations (3.5) describe different curves. The totality of these curves is called a *two-parameter family of curves* and c_1 and c_2 are referred to as the parameters of this family.

Example 3.4. Let F_1, F_2 and Ω be as in Example 3.2. For suitable values of c_1 and c_2, the equations

$$x^2 + y^2 - z = c_1, \qquad z = c_2$$

represent a two-parameter family of curves which are circles in Ω. The suitable values of c_1 and c_2 are all pairs of real numbers satisfying the condition

$$c_1 + c_2 > 0.$$

We close this section with the following additional remarks about curves.

In the parametric representation (3.2) of a curve C, the coordinate functions f_1, f_2, f_3 were assumed to be in $C^1(I)$ and their derivatives were assumed to satisfy condition (3.3). These assumptions guarantee that the tangent vector $(f_1'(t), f_2'(t), f_3'(t))$ is always non-zero and varies continuously on C, and consequently that C has no corners. Such a curve C is sometimes called a *smooth* curve. In this book we will deal only with smooth curves even though we will not always use the word smooth.

A curve C has many different parametric representations. For example,

(3.13) $$x = t, \qquad y = t, \qquad z = t; \qquad t \in [0, \infty),$$

(3.14) $$x = t^2 - 1, \qquad y = t^2 - 1, \qquad z = t^2 - 1; \qquad t \in [1, \infty),$$

represent the same curve which is a straight radial line beginning at the origin and passing through the point (1, 1, 1). We often say that (3.13) and (3.14) are two parametrizations of this line. The student should be able to write down many other parametrizations of this line. More generally, suppose that a curve C is represented parametrically by equations (3.2). Let g be a function in $C^1(I)$ such that $dg(t)/dt > 0$ for $t \in I$, and set

$$t' = g(t), \qquad t \in I.$$

As t varies over the interval I, t' varies over some interval I'. Since g is a strictly increasing function, g has an inverse function h,

$$t = h(t'), \qquad t' \in I'.$$

A new parametric representation of C is given by the equations

$$x = f_1(h(t')), \qquad y = f_2(h(t')), \qquad z = f_3(h(t')); \qquad t' \in I'.$$

Representing a curve as the intersection of surfaces can also be done in many different ways. For example the curve represented parametrically by (3.13) or (3.14) can be represented by the equations

(3.15) $$y - x = 0, \qquad z - x = 0; \qquad x \geqq 0,$$

or by the equations

(3.16) $$x + y - 2z = 0, \qquad z - y = 0; \qquad x \geqq 0.$$

Further discussion of the topics of Sections 2 and 3 may be found in standard books on Advanced Calculus such as Taylor[1] or Kaplan.[2]

Problems

3.1. Let (x_0, y_0, z_0) be a point on the curve C described by equations (3.5). Show that the vector

$$[\text{grad } F_1(x_0, y_0, z_0)] \times [\text{grad } F_2(x_0, y_0, z_0)]$$

is tangent to C at (x_0, y_0, z_0).

3.2. Find a vector tangent to the space circle

$$x^2 + y^2 + z^2 = 1 \qquad x + y + z = 0$$

at the point $(1/\sqrt{14}, 2/\sqrt{14}, -3/\sqrt{14})$.

3.3. Let F be a real-valued function of class C^1 in a domain Ω of R^3 with grad $F \neq \mathbf{0}$ in Ω. Let S_c be the level surface

$$F(x, y, z) = c.$$

Let C be a C^1 curve in Ω given by

$$(x, y, z) = (f_1(t), f_2(t), f_3(t)), \qquad t \in I.$$

Suppose that the curve C lies on S_c. Then

$$F(f_1(t), f_2(t), f_3(t)) = c, \qquad t \in I.$$

Use the chain rule to show that at each point of C the vector tangent to C is orthogonal to the vector normal to S_c.

3.4. The curve C given by

$$(x, y, z) = (\sin t, \cos t, 2\cos 2t), \qquad t \in \left[-\frac{\pi}{2}, \frac{\pi}{2}\right]$$

lies on some level surface S_c of

$$F(x, y, z) = 2(x^2 - y^2 + 1) + z.$$

Find the level surface S_c. Verify that at each point of C the tangent vector to C is orthogonal to the normal vector to S_c.

3.5. Find a parametric representation for the space curve given by

$$x^2 + y^2 = 1, \qquad x + y + z = 0.$$

3.6. Describe the two-parameter family of curves

$$F_1(x, y, z) = c_1, \qquad F_2(x, y, z) = c_2$$

where

$$F_1(x, y, z) = \frac{z^2}{x^2 + y^2}$$

and F_2 is one of the following:

(a) $F_2(x, y, z) = z$

(b) $F_2(x, y, z) = x$

(c) $F_2(x, y, z) = z + x$.

4. The Initial Value Problem for Ordinary Differential Equations and Systems

The general first order ordinary differential equation is an equation of the form

$$F(t, x, x') = 0 \tag{4.1}$$

where F is a function of three variables and x' denotes the derivative dx/dt. Suppose that the point (t_0, x_0, x_0') in R^3 satisfies equation (4.1) and suppose further that $D_3F(t_0, x_0, x_0') \neq 0$. Then the implicit function theorem asserts that near the point (t_0, x_0, x_0'), equation (4.1) can be solved for x',

$$x' = f(t, x) \tag{4.2}$$

where f is a function of two variables defined in a neighborhood of the point (t_0, x_0) in R^2. In this section we consider only ordinary differential equations of the form (4.2).

A *solution* of the ordinary differential equation (4.2) on some interval I of the t-axis is a function $x(t)$ defined and continuously differentiable for $t \in I$ such that when $x(t)$ is substituted for x in (4.2), equation (4.2) becomes an identity in t for all $t \in I$. The *general solution* of (4.2) is the collection of all of its solutions.

If the function f in (4.2) depends only on t, the equation takes the elementary form

$$\frac{dx}{dt} = f(t), \tag{4.3}$$

and, if f is continuous in some interval I, the general solution of (4.3) in the interval I is given by

$$x(t) = \int_{t_0}^{t} f(\tau)d\tau + c \tag{4.4}$$

where t_0 is any fixed point in I and c is an arbitrary constant. Thus the general solution of (4.3) is a one-parameter family of curves given by (4.4) where c is the parameter of the family (see Fig. 4.1). If in addition to equation (4.3) we require that the solution satisfy the condition

$$x = x_0 \quad \text{when} \quad t = t_0, \tag{4.5}$$

then we must choose the parameter c in the general solution (4.4) to be equal to x_0. Condition (4.5) is called an *initial condition* and the problem of finding the solution of equation (4.3) satisfying the initial condition (4.5) is called an *initial value problem* or a *Cauchy problem*. The solution of the initial value problem (4.3), (4.5) is given by

$$x = \int_{t_0}^{t} f(\tau)d\tau + x_0. \tag{4.6}$$

It is often useful to indicate the dependence of the solution x given by (4.6) on the initial data t_0 and x_0. This is done by writing

$$x = x(t; t_0, x_0).$$

It is clear from equation (4.6) that the solution of the initial value problem (4.3), (4.5) depends continuously on t_0 and x_0; i.e., that $x(t; t_0, x_0)$ is a continuous function of t_0 and x_0. Of course the solution is a continuously differentiable function of t in the interval I.

Under some conditions on the function f, the general solution of

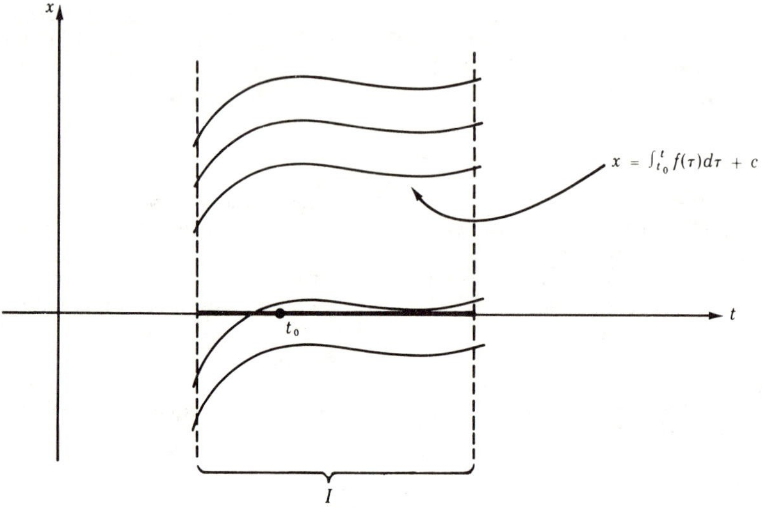

Fig. 4.1

equation (4.2) exists and also depends on one parameter, but it is usually difficult if not impossible to write a simple formula for it. It is easy to see, however, that if $x = x(t)$ is a solution of the initial value problem (4.2), (4.5) on I, then, for $t \in I$, $x(t)$ must satisfy the integral equation

(4.7) $$x(t) = \int_{t_0}^{t} f(\tau, x(\tau))d\tau + x_0.$$

Conversely, if $x(t)$ is a continuously differentiable function on I satisfying the integral equation (4.7), then $x(t)$ must be a solution of the initial value problem (4.2), (4.5). (Here we assume that $f(t, x)$ is a continuous function of (t, x).)

There is a fundamental theorem of ordinary differential equations which asserts the existence and uniqueness of solution of the initial value problem (4.2), (4.5). The proof of this theorem consists of showing that there is a unique solution of the integral equation (4.7) and consequently a unique solution of the initial value problem (4.2), (4.5). Before stating the theorem we must introduce some terminology and state the initial value problem in more precise language.

Let (t_0, x_0) be a point in R^2 and let f be a real valued function of two variables t and x defined for all points (t, x) of a rectangle Q centered at (t_0, x_0) and described by inequalities of the form

(4.8) $$|t - t_0| < a, \qquad |x - x_0| < b$$

where $2a$ and $2b$ are the lengths of the sides of Q (see Fig. 4.2). The initial value problem (4.2), (4.5) asks for a function $x(t)$ defined for t in some open interval I described by

(4.9) $$|t - t_0| < h, \qquad 0 < h < a$$

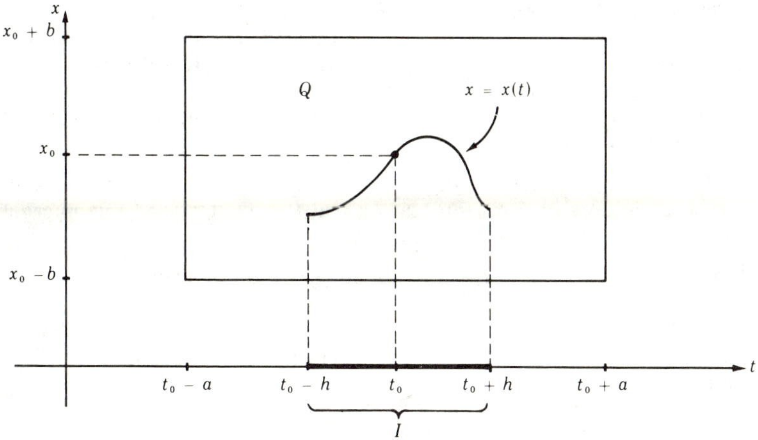

Fig. 4.2

such that $x = x(t)$ satisfies the differential equation

(4.2′)
$$\frac{dx(t)}{dt} = f(t, x(t)), \qquad \text{for} \quad t \in I$$

and the initial condition

(4.5′)
$$x(t_0) = x_0.$$

In geometrical language the problem is to find a curve $x = x(t)$ which passes through the given point (t_0, x_0) (as required by the initial condition (4.5)) and which at each of its points $(t, x(t))$ has slope given by the value of the function f at that point (as required by the differential equation (4.2)).

One of the conditions in the fundamental theorem that we state below requires that the function f satisfy a *Lipschitz condition* with respect to x in the rectangle Q. This means that there is a constant $A > 0$ such that

(4.10)
$$| f(t, x_1) - f(t, x_2) | \leqq A \, | x_1 - x_2 |$$

for every pair of points (t, x_1) and (t, x_2) in Q. Note that the function f will satisfy the Lipschitz condition (4.10) if $D_2 f$ exists and is bounded in Q, i.e., there is a constant $M > 0$ such that

(4.11)
$$| D_2 f(t, x) | \leqq M \qquad \text{for all} \quad (t, x) \in Q.$$

This follows from the mean value theorem,

$$f(t, x_1) - f(t, x_2) = D_2 f(t, \bar{x})(x_1 - x_2)$$

where $\bar{x}$ is some number between x_1 and x_2. Taking the absolute values of both sides,

$$| f(t, x_1) - f(t, x_2) | = | D_2 f(t, \bar{x}) | \, | x_1 - x_2 |$$

and since $(t, \bar{x}) \in Q$, (4.11) implies (4.10) with $A = M$.

Theorem 4.1. Suppose that in some rectangle Q given by inequalities of the form (4.8), the function $f(t, x)$ is continuous and satisfies a Lipschitz condition (4.10) with respect to x. Then there is an interval I given by an inequality of the form (4.9) and a unique function $x(t)$ defined for $t \in I$, such that $x(t) \in C^1(I)$ and $x = x(t)$ satisfies the differential equation (4.2′) and the initial condition (4.5′).

It should be emphasized that the theorem guarantees the existence of the solution only in some interval I about the point t_0, and this interval may be small. This is illustrated in the following example.

Example 4.1. Let $f(t, x) = x^2$ and let Q be the rectangle given by

$$|t| < 2, \qquad |x - 1| < 1$$

so that $(t_0, x_0) = (0, 1)$ and $a = 2$, $b = 1$. The function f is continuous in Q, and since

$$|D_2 f(t, x)| = |2x| < 4 \text{ in } Q,$$

f satisfies the Lipschitz condition (4.10) with $A = 4$. Now, it is easy to see that the unique solution to the initial value problem

$$\frac{dx}{dt} = x^2, \qquad x(0) = 1$$

is given by

$$x = \frac{1}{1 - t},$$

and this solution exists only in the subinterval I given by $|t| < 1$, of the original interval $|t| < 2$.

We consider next a system of n first order ordinary differential equations in n unknowns of the form

$$\text{(4.12)} \qquad \frac{dx_i}{dt} = f_i(t, x_1, \ldots, x_n); \qquad i = 1, \ldots, n.$$

Here t denotes the independent variable and $x_1, \ldots, x_n$ the unknowns. An initial value problem for the system (4.12) asks for n functions $x_1(t), \ldots, x_n(t)$ of the single variable t defined for t in some interval I centered at a point t_0 such that $x_1 = x_1(t), \ldots, x_n = x_n(t)$ satisfy the equations (4.12) and the initial conditions

$$\text{(4.13)} \qquad x_1(t_0) = x_1^0, \ldots, x_n(t_0) = x_n^0$$

where $x_1^0, \ldots, x_n^0$ are the given initial values. In geometrical language the problem is to find a curve C in R^n defined parametrically by the equations

$$x_1 = x_1(t), \ldots, x_n = x_n(t); \quad t \in I$$

such that C passes through a given point $(x_1^0, \ldots, x_n^0)$ (as required by the initial conditions (4.13)) and such that at each point of C a tangent vector to C is given by the vector $(f_1(t, x_1(t), \ldots, x_n(t)), \ldots, f_n(t, x_1(t), \ldots, x_n(t)))$

(as required by the differential equations (4.12)).

The fundamental theorem asserting the existence and uniqueness of solution of the initial value problem (4.12), (4.13) is essentially the same as Theorem 4.1. We need only extend the concepts of rectangle and Lipschitz condition to the case of several variables. Let Q be a box in R^{n+1} defined by the inequalities

$$|t - t_0| < a, \qquad |x_1 - x_1^0| < b_1, \ldots, |x_n - x_n^0| < b_n. \tag{4.14}$$

A function $f(t, x_1, \ldots, x_n)$ defined in Q is said to satisfy a Lipschitz condition in Q with respect to the variables $x_1, \ldots, x_n$, if there exist positive constants $A_1, A_2, \ldots, A_n$ such that

$$\begin{aligned} &|f(t, x_1^{(1)}, \ldots, x_n^{(1)}) - f(t, x_1^{(2)}, \ldots, x_n^{(2)})| \leq \\ &A_1 |x_1^{(1)} - x_1^{(2)}| + \cdots + A_n |x_n^{(1)} - x_n^{(2)}| \end{aligned} \tag{4.15}$$

for all pairs of points $(t, x_1^{(1)}, \ldots, x_n^{(1)})$ and $(t, x_1^{(2)}, \ldots, x_n^{(2)})$ in Q.

Theorem 4.2. Suppose that in some box Q defined by inequalities of the form (4.14), each of the functions $f_1, \ldots, f_n$ is continuous and satisfies a Lipschitz condition (4.15) with respect to the variables $x_1, \ldots, x_n$. Then there exists a unique set of functions $x_1(t), \ldots, x_n(t)$ defined for t in some interval I of the form (4.9) such that each function $x_i(t)$ belongs to $C^1(I)$ and

$$x_1 = x_1(t), \ldots, x_n = x_n(t)$$

satisfy the system of equations (4.12) for all $t \in I$ and the initial conditions (4.13).

The proofs of the theorems and further discussion of the topics of this section may be found in the books of Coddington[3] or Rabenstein.[4]

Problems

4.1. Find the general solution of the first order linear differential equation

$$\frac{dx}{dt} + p(t)x = q(t)$$

in some interval I where $p(t)$ and $q(t)$ are continuous functions of $t \in I$. (*Hint:* If $P(t)$ is an antiderivative of $p(t)$ in I, then $e^{P(t)}$ is an integrating factor).

4.2. Find the general solution of each of the following linear equations in the indicated intervals

(a) $$\frac{dx}{dt} + \frac{1}{t}x = 1, \qquad t > 0$$

(b) $$\frac{dx}{dt} + t^2 x = t^2, \qquad -\infty < t < \infty$$

(c) $$\frac{dx}{dt} - tx = 1, \qquad -\infty < t < \infty$$

(d) $$\frac{dx}{dt} + x = \sin t, \qquad -\infty < t < \infty.$$

4.3. For each of the equations in Problem 4.2 solve the initial value problem with initial condition $x(1) = 2$.

4.4. Find the general solution of each of the following equations

(a) $$\frac{dx}{dt} = x^3$$

(b) $$t\frac{dx}{dt} = \frac{x}{x^2 + 1}$$

(*Hint:* Use the method of separation of variables.)

4.5. Find the solution of the initial value problem

$$\frac{dx}{dt} = x^3, \qquad x(0) = 1.$$

Describe the largest interval I on which the solution is defined.

4.6 (a) Find the general solution of the system of equations,

$$\frac{dx_1}{dt} = x_2, \qquad \frac{dx_2}{dt} = -x_1.$$

(b) Find the solution of the above system satisfying the initial conditions

$$x_1(0) = 1, \qquad x_2(0) = 0.$$

Describe the largest interval I on which the solution is defined.

4.7. (a) Find the general solution of the system of equations

$$\frac{dx_1}{dt} = x_1x_2, \qquad \frac{dx_2}{dt} = -x_1x_2.$$

(b) Find the solution of the above system satisfying the initial conditions

$$x_1(0) = 1, \qquad x_2(0) = 1.$$

Describe the largest interval I on which the solution is defined.

4.8. Let $f(t, x_1, \ldots, x_n)$ be defined in a box Q given by the inequalities (4.14) and suppose that the derivatives $\partial f/\partial x_1, \ldots, \partial f/\partial x_n$ are bounded in Q. Show that f satisfies a Lipschitz condition (4.15) with respect to $x_1, \ldots, x_n$ in Q.

4.9. Show that the function

$$f(t, x) = \sqrt{|x|}$$

does not satisfy a Lipschitz condition with respect to x in any rectangle centered at the origin of the (t, x)-plane.

4.10. Consider the initial value problem

$$x' = 2\sqrt{|x|}$$
$$x(0) = 0.$$

Show that for any nonnegative value of the constant c, the function

$$x(t) = \begin{cases} 0, & \text{for } t < c \\ (t - c)^2 & \text{for } t \geq c \end{cases}$$

is a solution of the problem. Thus, in any interval centered at $t = 0$ there are infinitely many solutions of the problem. Does this contradict Theorem 4.1? Explain.

References for Chapter I

1. Taylor, A. E.: *Advanced Calculus*, Boston: Ginn and Co., 1955.
2. Kaplan, W.: *Advanced Calculus*, Ed. 2, Reading, Mass.: Addison Wesley Publishing Co., 1973.
3. Coddington, E. A.: *An Introduction to Ordinary Differential Equations*, Englewood Cliffs, N.J.: Prentice Hall, Inc., 1961.
4. Rabenstein, A. L.: *Elementary Differential Equations with Linear Algebra*, New York: Academic Press, 1970.

CHAPTER II

Integral curves and surfaces of vector fields

In this chapter we study methods for finding integral curves and surfaces of vector fields. The material of this chapter has important applications in many areas of mathematics, physics and engineering, but the main reason for including it in this book is the essential role that it plays in the study of quasi-linear first order partial differential equations presented in Chapter III.

In Sections 1 and 2 we discuss methods for finding the integral curves of a vector field by solving systems of ordinary differential equations. In Section 3 we find the general solution of the partial differential equation $Pu_x + Qu_y + Ru_z = 0$. In Section 4 we describe a method for finding an integral surface of a vector field containing a given curve. Finally, in Section 5 we apply the result of Section 3 to the study of solenoidal vector fields and obtain a theorem which is useful in many areas of physics and engineering. An application to plasma physics is also included in this section.

1. Integral Curves of Vector Fields

Let

(1.1) $$\mathbf{V}(x, y, z) = (P(x, y, z), Q(x, y, z), R(x, y, z))$$

be a vector field defined in some domain of R^3. In this book we will deal only with vector fields $\mathbf{V}$ in domains Ω in which the following two conditions are satisfied:

(a) $\mathbf{V}$ is nonvanishing in Ω; i.e., the component functions P, Q, R of $\mathbf{V}$ do not vanish simultaneously at any point of Ω,

(b) $P, Q, R \in C^1(\Omega)$.

For the "constant" vector field

$$\mathbf{V} = (1, 0, 0) \tag{1.2}$$

these conditions are satisfied in the domain Ω consisting of the whole of R^3. For the "radial" vector field

$$\mathbf{V} = (x, y, z) \tag{1.3}$$

the conditions are satisfied in the domain Ω consisting of R^3 minus the origin. For the vector field

$$\mathbf{V} = (y, -x, 0) \tag{1.4}$$

the conditions are satisfied in the domain Ω consisting of R^3 minus the z-axis. For the vector field

$$\mathbf{V} = \left(xyz, \frac{x + y}{z}, e^{x+y+z}\right) \tag{1.5}$$

the conditions are satisfied if Ω is either one of the half-spaces $z > 0$ or $z < 0$.

Definition 1.1. A curve C in Ω is an *integral curve of the vector field* $\mathbf{V}$ if $\mathbf{V}$ is tangent to C at each of its points.

Figures 1.1, 1.2 and 1.3 show some integral curves of the vector fields (1.2), (1.3) and (1.4), respectively. The integral curves of (1.2) are lines parallel to the x-axis. The integral curves of (1.3) are rays emanating from the origin. The integral curves of (1.4) are circles parallel to the (x,y)-plane and centered on the z-axis.

In physics, an integral curve of a vector field is often called a field line. If $\mathbf{V}$ is a force field, the integral curves of $\mathbf{V}$ are called lines of force. Under the influence of a force field due to a magnet, iron filings group themselves along the lines of force. If $\mathbf{V}$ is the velocity field of fluid flow, the integral curves of $\mathbf{V}$ are called lines of flow. These are the paths of motion of the fluid particles.

In the first two sections of this chapter we discuss methods for finding the integral curves of a vector field $\mathbf{V}$. We do this here because integral curves of vector fields play an essential role in the study of first order partial differential equations. Vector fields and their integral curves are also important in many other areas of mathematics, physics and engineering.

With the vector field $\mathbf{V} = (P, Q, R)$ we associate the system of ordinary differential equations,

$$\frac{dx}{dt} = P(x, y, z), \qquad \frac{dy}{dt} = Q(x, y, z), \qquad \frac{dz}{dt} = R(x, y, z). \tag{1.6}$$

A solution $(x(t), y(t), z(t))$ of (1.6), defined for t in some interval I, may be regarded as a curve in Ω. We will call this curve a *solution curve* of the system (1.6). Obviously, every solution curve of the system (1.6) is an integral curve of the vector field $\mathbf{V}$. Conversely, it can be shown (see Problem 1.9), that if C is an integral curve of $\mathbf{V}$, then there is a parametric representation

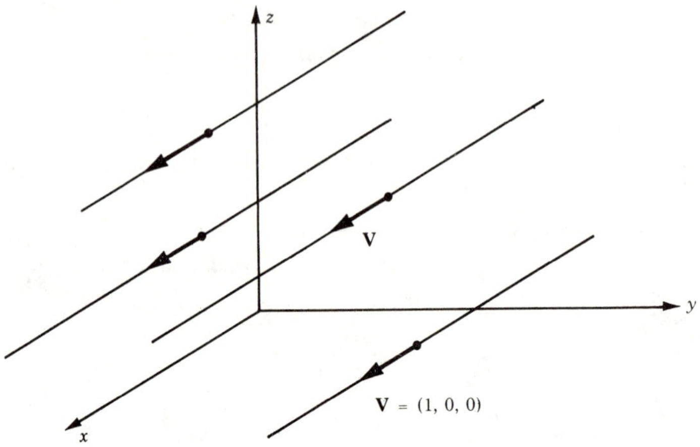

Fig. 1.1

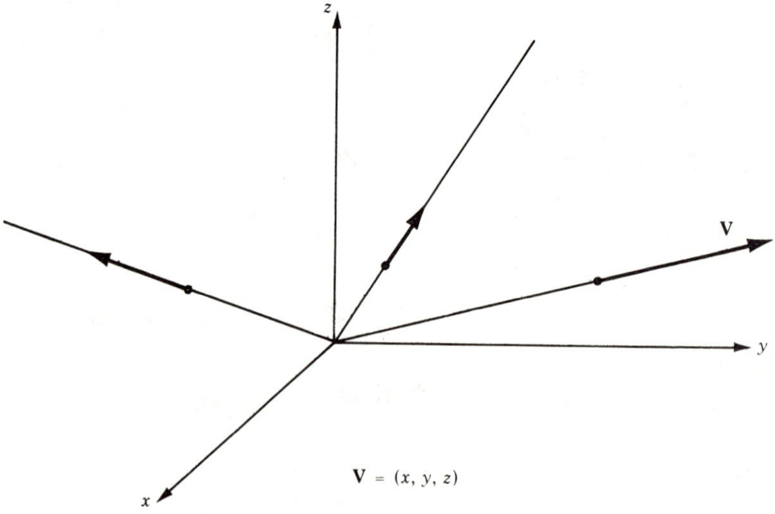

Fig. 1.2

$$x = x(t), \qquad y = y(t), \qquad z = z(t); \qquad t \in I,$$

of C, such that $(x(t), y(t), z(t))$ is a solution of the system of equations (1.6). Thus, every integral curve of **V**, if parametrized appropriately, is a solution curve of the associated system of equations (1.6).

The integral curves of simple vector fields, such as those given by equations (1.2) and (1.3), can sometimes be found by geometrical intuition. However, for more complicated vector fields this is not always

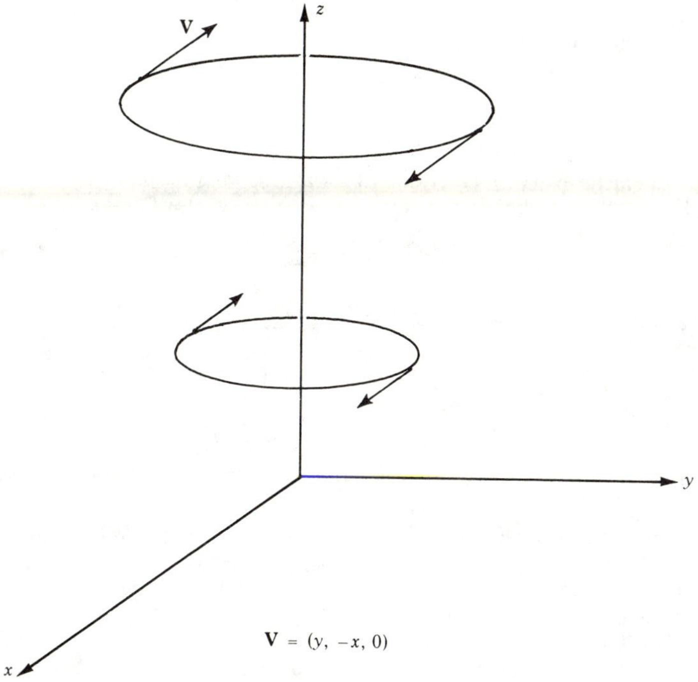

Fig. 1.3

possible. In any case, the integral curves of a vector field $\mathbf{V} = (P, Q, R)$ can be found by considering these curves as solution curves of the associated system of equations (1.6) and by solving this system.

Since the right hand sides of the system (1.6) do not depend on t, it is possible to eliminate t completely and consider any two of the variables x, y, z as functions of the third. If, for example, $P \neq 0$, then y and z may be considered as functions of the independent variable x, and the system (1.6) may be written in the form

$$\frac{dy}{dx} = \frac{Q}{P}, \qquad \frac{dz}{dx} = \frac{R}{P}. \tag{1.7}$$

Similarly, if $Q \neq 0$ or $R \neq 0$, the system (1.6) may be written in the form

$$\frac{dx}{dy} = \frac{P}{Q}, \qquad \frac{dz}{dy} = \frac{R}{Q} \tag{1.8}$$

or

$$\frac{dx}{dz} = \frac{P}{R}, \qquad \frac{dy}{dz} = \frac{Q}{R}, \tag{1.9}$$

respectively. In order to avoid distinguishing between dependent and independent variables, it is customary to write the equivalent systems (1.7)–(1.9) in the form

$$\frac{dx}{P} = \frac{dy}{Q} = \frac{dz}{R}. \tag{1.10}$$

Although we will usually write the system (1.6) in the form (1.10), the equivalent forms (1.7)–(1.9) should always be kept in mind.

The system of equations associated with the vector field (1.2) is

$$\frac{dx}{1} = \frac{dy}{0} = \frac{dz}{0}. \tag{1.11}$$

The systems associated with the vector fields (1.3) and (1.4), respectively, are

$$\frac{dx}{x} = \frac{dy}{y} = \frac{dz}{z}, \tag{1.12}$$

$$\frac{dx}{y} = \frac{dy}{-x} = \frac{dz}{0}. \tag{1.13}$$

The zeroes that appear in the denominators should not be disturbing. For example the zero appearing in the denominator in (1.13) simply means that $dz/dx = dz/dy = dz/dt = 0$, and therefore a solution curve of (1.13) must lie on a plane z = constant (see Fig. 1.3).

We will try to find the integral curves of a vector field $\mathbf{V}$ by considering them as intersections of surfaces and representing them in the form

$$u_1(x, y, z) = c_1, \qquad u_2(x, y, z) = c_2. \tag{1.14}$$

As we saw in Chapter I, Section 3, if the functions u_1 and u_2 satisfy the condition

$$\operatorname{grad} u_1(x, y, z) \times \operatorname{grad} u_2(x, y, z) \neq \mathbf{0}, \quad (x, y, z) \in \Omega, \tag{1.15}$$

then, for each pair of appropriate values of c_1 and c_2, equations (1.14) represent a curve in Ω. Geometrically, condition (1.15) means that grad u_1 and grad u_2 are not parallel at any point of Ω.

Definition 1.2. Two functions u_1 and u_2 in $C^1(\Omega)$ which satisfy condition (1.15) will be called *functionally independent* in Ω. (See Problem 1.10.)

Now, let C be a curve represented by equations (1.14). At each point of C, grad u_1 and grad u_2 are normal to C (see Fig. 1.4). In order for C to be an integral curve of $\mathbf{V}$, at each point of C, $\mathbf{V}$ must be tangent to C and, hence, normal to grad u_1 and grad u_2. It follows that in order for equations (1.14) to represent integral curves of the vector field $\mathbf{V}$ in Ω, the functions u_1 and u_2 must satisfy the conditions

$$P\frac{\partial u_1}{\partial x} + Q\frac{\partial u_1}{\partial y} + R\frac{\partial u_1}{\partial z} = 0, \qquad P\frac{\partial u_2}{\partial x} + Q\frac{\partial u_2}{\partial y} + R\frac{\partial u_2}{\partial z} = 0.$$

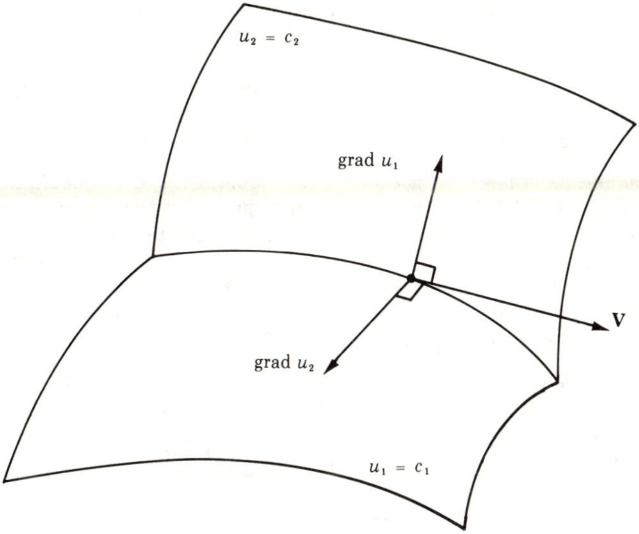

Fig. 1.4

Definition 1.3. A function u in $C^1(\Omega)$ is called a *first integral* of the vector field $\mathbf{V} = (P, Q, R)$ (or of its associated system $dx/P = dy/Q = dz/R$) in Ω, if at each point of Ω, $\mathbf{V}$ is orthogonal to grad u, i.e.

$$P\frac{\partial u}{\partial x} + Q\frac{\partial u}{\partial y} + R\frac{\partial u}{\partial z} = 0 \quad in \quad \Omega. \tag{1.16}$$

Equation (1.16) is a partial differential equation in the unknown function u of three independent variables x, y, z. According to Definition 1.3, any solution of the p.d.e. (1.16) is a first integral of $\mathbf{V}$.

We now can prove the following basic theorem which describes the integral curves of $\mathbf{V}$ in terms of two functionally independent first integrals of $\mathbf{V}$.

Theorem 1.1. Let u_1 and u_2 be any two functionally independent first integrals of $\mathbf{V}$ in Ω. Then the equations

$$u_1(x, y, z) = c_1 , \qquad u_2(x, y, z) = c_2 \tag{1.14}$$

describe the collection of all integral curves of $\mathbf{V}$ in Ω.

Proof. The assumptions of the theorem mean that u_1 and u_2 satisfy condition (1.15) in Ω and that each satisfies the p.d.e. (1.16) in Ω. We have already shown in Chapter I, Section 3, that condition (1.15) implies that for appropriate values of c_1 and c_2, equations (1.14) represent a two-parameter family of curves in Ω. Each member C of this family is an integral curve of $\mathbf{V}$ since $\mathbf{V}$ is tangent to C at each of its points (why?). Now let C be any integral curve of $\mathbf{V}$ in Ω. We must show that C is given by equations (1.14) for certain values of c_1 and c_2. Let (x_0, y_0, z_0) be any point of C and let C' be the curve given by (1.14) with $c_1 = u_1(x_0, y_0, z_0)$, $c_2 = u_2(x_0, y_0, z_0)$. Then C and C' are integral curves of **V** passing through the point (x_0, y_0, z_0), and, if parametrized appropriately, they are both solution curves of the system (1.6). By the uniqueness part of Theorem 4.2 of Chapter I we must have $C' = C$. The proof is complete.

It should be noted that, according to Theorem 1.1, the collection of all integral curves of **V** is a two-parameter family of curves. Moreover, the representation (1.14) of this two-parameter family of integral curves of **V** is not unique. The functions u_1 and u_2 may be replaced by any other pair of functionally independent first integrals of **V** in Ω. The following theorem shows that from any given first integrals of **V** we can obtain an infinite number of new first integrals.

Theorem 1.2. (a) If $f(u)$ is a C^1 function of a single variable u and if $u(x, y, z)$ is a first integral of **V** then $w(x, y, z) = f(u(x, y, z))$ is also a first integral of **V**.

(b) If $f(u, v)$ is a C^1 function of two variables u and v and if $u(x, y, z)$ and $v(x, y, z)$ are any two first integrals of **V** then $w(x, y, z) = f(u(x, y, z), v(x, y, z))$ is also a first integral of **V**.

We prove only (a) and leave (b) as an exercise. Using the chain rule, we obtain

$$P\frac{\partial w}{\partial x} + Q\frac{\partial w}{\partial y} + R\frac{\partial w}{\partial z} = Pf'\frac{\partial u}{\partial x} + Qf'\frac{\partial u}{\partial y} + Rf'\frac{\partial u}{\partial z}$$
$$= f'\left[P\frac{\partial u}{\partial x} + Q\frac{\partial u}{\partial y} + R\frac{\partial u}{\partial z}\right] = 0$$

where in the last equality we used the fact that u satisfies the p.d.e. (1.16). We have shown that w also satisfies (1.16) and hence it is a first integral of **V**.

According to Theorem 1.2, either of the functions u_1 or u_2 may be replaced by a function $f(u_1, u_2)$ in the equations (1.14) representing the integral curves of **V**. Of course we must make sure that in the new equations thus obtained, the functions involved are functionally independent.

Example 1.1. Let **V** be the vector field given by (1.2) and let $\Omega = R^3$. A first integral of **V** is a solution of the equation

(1.17) $$u_x = 0.$$

Any function of y and z only is a solution of this equation. For example,

$$u_1 = y, \qquad u_2 = z$$

are two solutions which are obviously functionally independent. The integral curves of **V** are described by the equations

(1.18) $$y = c_1, \qquad z = c_2,$$

and are straight lines parallel to the x-axis (see Fig. 1.1). The functions $y - z$ and e^{y+z} are also functionally independent first integrals, and the integral curves of **V** are also described by the equations

$$y - z = c_1 \qquad e^{y+z} = c_2. \tag{1.19}$$

Of course, different values of c_1 and c_2 must be used in (1.18) and (1.19) in order to get the same integral curve of **V**.

Example 1.2. Let **V** be the vector field given by equation (1.3) and let Ω be the octant $x > 0$, $y > 0$, $z > 0$. A first integral of **V** is a solution of the equation

$$xu_x + yu_y + zu_z = 0. \tag{1.20}$$

It is easy to verify that the functions

$$u_1(x, y, z) = \frac{y}{x}, \qquad u_2(x, y, z) = \frac{z}{x}$$

are first integrals of **V** in Ω. Moreover, they are functionally independent in Ω since they satisfy condition (1.15). Therefore, the integral curves of **V** in Ω are described by the equations

$$\frac{y}{x} = c_1, \qquad \frac{z}{x} = c_2. \tag{1.21}$$

They are rays emanating from the origin (see Fig. 1.2) and a parametric representation of them is

$$x = t, \qquad y = c_1 t, \qquad z = c_2 t; \qquad t > 0.$$

It is easy to check by direct computation that any function of u_1 and (or) u_2 is also a first integral of **V**. For example, if $f(u) = u^2$ then

$$u(x, y, z) = f(u_1(x, y, z)) = \left(\frac{y}{x}\right)^2$$

is a first integral of **V**. If $f(u_1, u_2) = u_1{}^2 - u_2{}^2$, then

$$u(x, y, z) = f(u_1(x, y, z), u_2(x, y, z)) = \frac{y^2 - z^2}{x^2}$$

is a first integral of **V**. Similarly

$$\frac{z}{y}, \frac{yz}{x}, \frac{x}{z}, \sin\left(\frac{y}{x}\right), \cos\left(\frac{z}{x}\right)^2,$$

are all first integrals of **V**.

Example 1.3. Let **V** be the vector field (1.4) and let Ω be R^3 minus the z-axis. A first integral of **V** is a solution of the equation

$$yu_x - xu_y = 0.$$

It is easy to verify that the functions

$$u_1(x, y, z) = x^2 + y^2, \qquad u_2(x, y, z) = z$$

are two functionally independent first integrals of **V**. Therefore, the integral curves of **V** in Ω are given by

$$x^2 + y^2 = c_1, \qquad z = c_2. \tag{1.22}$$

Equations (1.22) describe circles parallel to the (x, y)-plane and centered on the z-axis (see Fig. 1.3).

We usually want to find the integral curves of a vector field **V** in the largest possible domain Ω in which the component functions P, Q, R satisfy the basic assumptions (a) and (b) stated in the beginning of this section. If we can find two functionally independent first integrals u_1 and u_2 defined in the whole of Ω, then the collection of integral curves of **V** in Ω is given by equations (1.14). However, this may not always be possible. Frequently, we can find pairs of functionally independent first integrals of **V** which are defined only in subdomains of Ω. If u_1 and u_2 is such a pair defined in the subdomain Ω_0 of Ω then the integral curves of **V** in Ω_0 are given by equations (1.14). In order to find the integral curves of **V** in the whole of Ω we may have to express Ω as the union of (possibly overlapping) subdomains. In each of these subdomains the integral curves of **V** may be expressed by equations of the type (1.14) with different pairs of first integrals being used in different subdomains. To illustrate this let us consider the vector field **V** in Example 1.2. For this vector field, the largest domain Ω in which assumptions (a) and (b) are satisfied is R^3 minus the origin. In the subdomain of Ω consisting of the half-space $x > 0$, the integral curves of **V** are given by

$$\frac{y}{x} = c_1, \qquad \frac{z}{x} = c_2.$$

In the subdomain of Ω consisting of the half-space $z < 0$ the integral curves of **V** are given by

$$\frac{x}{z} = c_1, \qquad \frac{y}{z} = c_2.$$

The student should be able to write down the pair of equations describing the integral curves of (1.3) in any coordinate half-space.

Although Theorem 1.1 describes the integral curves of a vector field **V** $= (P, Q, R)$ in a domain Ω in terms of two functionally independent first integrals of **V** in Ω, the theorem does not tell us whether such first integrals always exist, or, if they do exist, how to find them. Since the integral curves of **V** are solution curves of the associated system of equations (1.10), it can be shown, using the existence theory for systems of ordinary differential equations, that in a neighborhood of any point of Ω there always exist two functionally independent first integrals of **V**. Of course this still does not tell us how to find these integrals. In the next section we describe some methods for finding functionally independent first integrals of **V** by solving the associated system of equations (1.10). Actually, these methods yield equations (1.14) directly as consequences of the differential equations (1.10). This is illustrated in the following example.

Example 1.4. Solve the system of equations

$$\frac{dx}{x} = \frac{dy}{y} = \frac{dz}{z} \tag{1.12}$$

which is associated with the vector field (1.3). The first of these equations is an ordinary differential equation in x and y and the general solution of this equation is

$$\frac{y}{x} = c_1. \tag{1.23}$$

The equation obtained by equating the first and last of the ratios in (1.12) is an ordinary differential equation in x and z, and the general solution of this equation is

$$\frac{z}{x} = c_2. \tag{1.24}$$

We have thus obtained equations (1.21) describing the integral curves of the vector field (1.3) (for $x > 0$) by solving the associated system of equations (1.12).

The left hand sides of equations (1.23) and (1.24) are first integrals of the vector field (1.3). In general, if the relation

$$u(x, y, z) = c \tag{1.25}$$

holds as a consequence of the system of equations (1.10), then the function $u(x, y, z)$ is a first integral of **V**. To see this, take the differential of (1.25) to obtain

$$\frac{\partial u}{\partial x}\, dx + \frac{\partial u}{\partial y}\, dy + \frac{\partial u}{\partial z}\, dz = 0.$$

Now, in view of equations (1.10), it follows that u satisfies equation (1.16).

Problems

1.1. Consider the "constant" vector field

$$\mathbf{V}(x, y, z) = (1, 1, 1).$$

(a) Describe the integral curves of **V** using "geometrical intuition."
(b) Write down equation (1.16) for this vector field and guess solutions of this equation to obtain two functionally independent first integrals of **V**. Write down equations (1.14) describing the integral curves of **V**.
(c) Write down the system of ordinary differential equations associated with **V** and obtain the equations of the integral curves by solving this system.

1.2. Follow the instructions of Problem 1.1 for each of the following vector fields

(a) $\mathbf{V} = (1, 1, 0)$, (b) $\mathbf{V} = (1, -1, 0)$.

1.3. Find the integral curves of the vector fields (1.2) and (1.4) by solving the systems of ordinary differential equations associated with the vector fields.

1.4. Find the integral curves of the following vector fields:

(a) $\mathbf{V} = (0, z, -y)$ (b) $\mathbf{V} = (y^3, -x^3, 0)$ (c) $\mathbf{V} = (1, 3x^2, 0)$.

1.5. Verify directly (without using the theory developed in this section) that equations (1.18) and (1.19) describe the same two-parameter family of curves in R^3.

1.6. If $\mathbf{V}$ and Ω are as in Example 1.2, show that yz/x^2 and $(y^2 - z^2)/x^2$ are functionally independent first integrals of $\mathbf{V}$ in Ω. Verify directly that the equations

$$\frac{yz}{x^2} = c_1, \qquad \frac{y^2 - z^2}{x^2} = c_2$$

and equations (1.21) describe the same two-parameter family of curves in Ω.

1.7. Let $u(x, y, z)$ be a first integral of $\mathbf{V}$ and let C be an integral curve of $\mathbf{V}$ given by

$$x = x(t), \qquad y = y(t), \qquad z = z(t); \qquad t \in I.$$

Show that C must lie on some level surface of u. [*Hint:* Compute $d/dt\,[u(x(t), y(t), z(t))]$].

1.8. Let $\mathbf{V}$ be a vector field defined in Ω and let f be a nonvanishing function in $C^1(\Omega)$. Show that the vector fields $\mathbf{V}$ and $f\mathbf{V}$ have the same integral curves in Ω.

1.9. Let C be an integral curve of the vector field $\mathbf{V} = (P, Q, R)$ and suppose that C is given parametrically by the equations,

$$x = x(t), \qquad y = y(t), \qquad z = z(t); \qquad t \in I$$

where the functions $x(t)$, $y(t)$, $z(t)$ are in $C^1(I)$ and the tangent vector

$$\mathbf{T}(t) = \left(\frac{dx(t)}{dt}, \frac{dy(t)}{dt}, \frac{dz(t)}{dt}\right)$$

never vanishes for $t \in I$.

(*a*) Show that there is a function $\mu(t)$ in $C^1(I)$ such that for every $t \in I$, $\mu(t) \neq 0$ and

$$\mathbf{V}(x(t), y(t), z(t)) = \mu(t)\mathbf{T}(t).$$

(b) Let $t = t(\tau)$ be a solution of the differential equation

$$\frac{dt}{d\tau} = \mu(t)$$

where τ varies over some interval I' as t varies over I. Set

$$\bar{x}(\tau) = x(t(\tau)),\ \bar{y}(\tau) = y(t(\tau)),\ \bar{z}(\tau) = z(t(\tau)).$$

Show that in terms of the new parametric representation

$$x = \bar{x}(\tau), \qquad y = \bar{y}(\tau), \qquad z = \bar{z}(\tau); \qquad \tau \in I',$$

the curve C is a solution curve of the system of equations associated with **V**,

$$\frac{dx}{d\tau} = P, \qquad \frac{dy}{d\tau} = Q, \qquad \frac{dz}{d\tau} = R.$$

1.10. The definition of functional independence in terms of condition (1.15) does not exactly coincide with the usual definition of functional independence given in books on advanced calculus.
   (a) Look up the definition of functional independence in an advanced calculus text.
   (b) Show that two functions that are functionally independent according to Definition 1.2 are also functionally independent according to the definition found in (a).

1.11. Prove part (b) of Theorem 1.2.

2. Methods of Solution of $dx/P = dy/Q = dz/R$

In this section we describe methods for finding the integral curves of a vector field $\mathbf{V} = (P, Q, R)$ in a domain Ω of R^3 by solving the associated system of equations

$$\frac{dx}{P} = \frac{dy}{Q} = \frac{dz}{R}. \tag{2.1}$$

We always assume that **V** is nonvanishing and C^1 in Ω. Since the integral curves of **V** and the solution curves of (2.1) are the same (if parametrized appropriately), we know from the theory developed in the previous section that the collection of solution curves of (2.1) in Ω is given by

$$u_1(x, y, z) = c_1, \qquad u_2(x, y, z) = c_2 \tag{2.2}$$

where u_1 and u_2 are any two functionally independent first integrals of (2.1) in Ω. For this reason we will refer to equations (2.2) as the *general solution* of the system (2.1). Recall that a function $u \in C^1(\Omega)$ is a first integral of (2.1) in Ω if it satisfies the p.d.e.

$$P\frac{\partial u}{\partial x} + Q\frac{\partial u}{\partial y} + R\frac{\partial u}{\partial z} = 0, \tag{2.3}$$

in Ω. Moreover, two functions u_1 and u_2 in $C^1(\Omega)$ are functionally independent in Ω if their gradients are never parallel in Ω, i.e.

$$\text{grad } u_1 \times \text{grad } u_2 \neq \mathbf{0}, \quad \text{in} \quad \Omega. \tag{2.4}$$

It is sometimes possible to find solutions of equation (2.3) by inspection. However, this is usually very difficult if not impossible. In practice we try to find the relations (2.2) as consequences of the differential equations (2.1). The difficulty with equations (2.1) is that, in general, each equation involves all three variables x, y, z. If one of the variables were missing from one of the equations then that equation would be an ordinary differential equation in two variables and we could try to solve this equation by a method such as the method of separation of variables.

Suppose for example that P and Q are functions of x and y only. Then the differential equation

$$\frac{dx}{P(x, y)} = \frac{dy}{Q(x, y)} \tag{2.5}$$

is an ordinary differential equation where either x or y may be considered as the independent variable. If we can find the general solution of (2.5) in the form

$$u_1(x, y) = c_1 \tag{2.6}$$

then it is easy to check that u_1 is a first integral of (2.1) (see Problem 2.1).

Example 2.1. Consider the system

$$\frac{dx}{x} = \frac{dy}{y} = \frac{dz}{xy(z^2 + 1)}. \tag{2.7}$$

The first equation does not involve z and can be solved immediately,

$$\frac{y}{x} = c_1. \tag{2.8}$$

The function $u_1(x, y, z) = y/x$ is a first integral of the system (2.7) since it satisfies the p.d.e.

$$xu_x + yu_y + xy(z^2 + 1)u_z = 0. \tag{2.9}$$

If one first integral u_1 of (2.1) is already known, then we can attempt to find another first integral by the following procedure: Using the relation $u_1(x, y, z) = c_1$ we try to eliminate one of the variables from one of the equations (2.1). We then try to solve the resulting ordinary differential equation in the remaining two variables.

Example 2.1 (continued). Using (2.8) we can eliminate y from the equation

$$\frac{dx}{x} = \frac{dz}{xy(z^2 + 1)}$$

to obtain

$$\frac{dx}{x} = \frac{dz}{xc_1x(z^2 + 1)}.$$

The solution of this equation is

$$c_1 \frac{x^2}{2} = \arctan z + c_2. \tag{2.10}$$

Substituting for c_1 from (2.8) we get

$$\frac{1}{2}xy - \arc\tan z = c_2. \tag{2.11}$$

It is easy to check that the function

$$u_2(x, y, z) = \frac{1}{2}xy - \text{arc tan } z$$

is a first integral of (2.7) since it satisfies the p.d.e. (2.9). Now,

$$\text{grad } u_1 = \left(-\frac{y}{x^2}, \frac{1}{x}, 0\right), \qquad \text{grad } u_2 = \left(\frac{y}{2}, \frac{x}{2}, -\frac{1}{1+z^2}\right),$$

which are never parallel. Hence, u_1 and u_2 are functionally independent first integrals of (2.7) in any domain where they are both defined. Since u_1 is defined for $x \neq 0$ and u_2 is defined everywhere, equations (2.8) and (2.11) describe the solution curves of (2.7) in the domain $x > 0$ and in the domain $x < 0$.

Example 2.2. Solve the system

$$\frac{ydx}{y^2 + z^2} = \frac{dy}{xz} = -\frac{dz}{xy}. \tag{2.12}$$

After canceling x, the last equation is an ordinary differential equation in y and z,

$$\frac{dy}{z} = -\frac{dz}{y}.$$

The general solution of this equation is

$$y^2 + z^2 = c_1. \tag{2.13}$$

The function $u_1 = y^2 + z^2$ is a first integral of (2.12) since it satisfies the p.d.e.

$$\frac{y^2 + z^2}{y} u_x + xzu_y - xyu_z = 0. \tag{2.14}$$

In order to find another first integral we use (2.13) to eliminate the variable y from the equation

$$\frac{y\,dx}{y^2 + z^2} = -\frac{dz}{xy}$$

to obtain

$$\frac{(c_1 - z^2)dx}{c_1} = -\frac{dz}{x}.$$

After separation of variables this equation becomes

$$x\,dx = \frac{c_1}{z^2 - c_1}\,dz.$$

We now use partial fraction decomposition on the right side,

$$x\,dx = \frac{\sqrt{c_1}}{2}\left(\frac{dz}{z - \sqrt{c_1}} - \frac{dz}{z + \sqrt{c_1}}\right)$$

and after integration of this equation we obtain

$$x^2 = \sqrt{c_1}\log\left(\frac{\sqrt{c_1} - z}{\sqrt{c_1} + z}\right) + c_2.$$

Again using equation (2.13), we can eliminate c_1 to get

$$x^2 - (y^2 + z^2)^{1/2}\log\left[\frac{(y^2 + z^2)^{1/2} - z}{(y^2 + z^2)^{1/2} + z}\right] = c_2. \tag{2.15}$$

It is left as an exercise to check that the function u_2 defined by the left hand side of this equation satisfies the p.d.e. (2.14) and that u_1 and u_2 are functionally independent in the domain $y > 0$.

Let us consider now the more difficult case in which none of the equations in (2.1) is an ordinary differential equation involving two variables only. In this case we try to introduce new variables and to derive from (2.1) a new differential equation involving only two of these new variables. In order to understand the procedure let us recall the parametric form of (2.1),

$$\frac{dx}{dt} = P, \quad \frac{dy}{dt} = Q, \quad \frac{dz}{dt} = R. \tag{2.16}$$

If a, b, c are any three constants, then as a consequence of (2.16) we also have the equation

$$\frac{d}{dt}(ax + by + cz) = aP + bQ + cR. \tag{2.17}$$

The nonparametric form of the four equations (2.16) and (2.17) is

$$\frac{dx}{P} = \frac{dy}{Q} = \frac{dz}{R} = \frac{d(ax + by + cz)}{aP + bQ + cR}. \tag{2.18}$$

Since a, b, c are arbitrary constants, (2.18) actually represents an infinite collection of equations which are all consequences of the original equations (2.1). For example we have

$$\frac{d(x + y + z)}{P + Q + R} = \frac{dx}{P}$$

and

$$\frac{d(x + y)}{P + Q} = \frac{d(x - z)}{P - R}.$$

Thus we can add or subtract numerators and denominators in the ratios of (2.1) and the resulting new ratio is also equal to the original ones.

Suppose now that we can find constants a, b, c and a', b', c' such that the equation

$$\frac{d(ax + by + cz)}{aP + bQ + cR} = \frac{d(a'x + b'y + c'z)}{a'P + b'Q + c'R}$$

is an ordinary differential equation involving only the new variables $X = ax + by + cz$, $X' = a'x + b'y + c'z$. Then we can try to find the general solution of this equation,

$$u(X, X') = c_1.$$

Consequently the function $u(ax + by + cz, a'x + b'y + c'z)$ would be a first integral of (2.1).

Example 2.3. Solve the system

$$\frac{dx}{y + z} = \frac{dy}{y} = \frac{dz}{x - y}. \tag{2.19}$$

Adding the numerators and denominators in the first and last ratio we obtain the ratio $d(x + z)/(x + z)$. Equating this to the middle ratio above we obtain the equation

$$\frac{d(x + z)}{x + z} = \frac{dy}{y}$$

which is an ordinary differential equation in the two variables $X = x + z$ and y. The general solution of this equation is

$$\frac{X}{y} = c_1.$$

Hence

$$\frac{x + z}{y} = c_1. \tag{2.20}$$

The function $u_1(x, y, z) = (x + z)/y$ is a first integral of (2.19) since it satisfies the p.d.e.

$$(y + z)u_x + yu_y + (x - y)u_z = 0. \tag{2.21}$$

It is left as an exercise to obtain another first integral using substitution from equation (2.20). Here we find another first integral as follows. Subtracting the numerators and denominators in the first two ratios in (2.19) and equating the result to the last ratio, we obtain the equation

$$\frac{d(x - y)}{z} = \frac{dz}{x - y}.$$

This is an ordinary differential equation in $Y = x - y$ and z, and its general solution is

$$Y^2 - z^2 = c_2.$$

Hence

(2.22) $$(x - y)^2 - z^2 = c_2.$$

The function $u_2 = (x - y)^2 - z^2$ is a first integral of (2.19) since it satisfies the p.d.e. (2.21). To check functional independence we compute

$$\text{grad } u_1 = \left(\frac{1}{y}, -\frac{x+z}{y^2}, \frac{1}{y}\right), \qquad \text{grad } u_2 = (2(x - y), -2(x - y), -2z)$$

and

$$\text{grad } u_1 \times \text{grad } u_2$$

$$= \left(\frac{2z(x+z)}{y^2} + \frac{2(x-y)}{y}, \frac{2(x-y+z)}{y}, \frac{2(x-y+z)(x-y)}{y^2}\right).$$

The second component is never zero if $x - y + z \neq 0$. Hence u_1 and u_2 are two functionally independent first integrals of (2.19) in the domain defined by the inequalities

$$x - y + z > 0, \qquad y > 0.$$

The student should sketch this domain.

Let us consider now the possibility of allowing the multipliers a, b, c in the above method to be functions of x, y, z. In this case equation (2.17) is not a consequence of equations (2.16) and it may not be correct (why?). Suppose, however, that we can find functions $a(x, y, z)$, $b(x, y, z)$, $c(x, y, z)$ and a function $X(x, y, z)$ such that

(2.23) $$dX = adx + bdy + cdz.$$

Then as a consequence of equations (2.16) we have the equation

(2.24) $$\frac{dX}{dt} = aP + bQ + cR.$$

The nonparametric form of equations (2.16) and (2.24) is

(2.25) $$\frac{dx}{P} = \frac{dy}{Q} = \frac{dz}{R} = \frac{dX}{aP + bQ + cR}.$$

If one of the equations in (2.25) involves only two of the variables x, y, z and X, we can proceed to solve it as an ordinary differential equation.

Example 2.4. Solve the system

(2.26) $$\frac{dx}{x(y-z)} = \frac{dy}{y(z-x)} = \frac{dz}{z(x-y)}.$$

Since

$$d \log (xyz) = \frac{1}{x} dx + \frac{1}{y} dy + \frac{1}{z} dz,$$

we can take

$$X = \log(xyz), \qquad a = \frac{1}{x}, \qquad b = \frac{1}{y}, \qquad c = \frac{1}{z}.$$

We then have $aP + bQ + cR = 0$ and

$$\frac{dx}{x(y - z)} = \frac{dy}{y(z - x)} = \frac{dz}{z(x - y)} = \frac{d \log(xyz)}{0}.$$

Hence

$$xyz = c_1$$

and the function $u_1 = xyz$ is a first integral of (2.26) as can be easily checked.

Problems

2.1. Show that if (2.6) gives the general solution of (2.5) then $u_1(x, y)$ is a first integral of (2.1).
2.2. Find the solution curve of the system (2.7) passing through the point (1, 1, 0).
2.3. Find the general solution of (2.7) in the domain $y > 0$.
2.4. Verify that the function u_2 defined by the left hand side of equation (2.15) satisfies the p.d.e. (2.14) and that the functions u_1 and u_2 of Example 2.2 are functionally independent in the domain $y > 0$.
2.5. Use (2.20) to obtain another first integral for the system (2.19).
2.6. Find another functionally independent first integral in the Example 2.4.
2.7. Find the integral curves of the following vector fields

(a) $\mathbf{V} = (\log(y + z), 1, -1)$ (b) $\mathbf{V} = (x^2, y^2, z(x + y))$.

2.8. Find the general solution of the following systems

(a) $\dfrac{dx}{y - z} = \dfrac{dy}{z - x} = \dfrac{dz}{x - y}$ (b) $\dfrac{dx}{yz} = \dfrac{dy}{-xz} = \dfrac{dz}{xy(x^2 + y^2)}$.

3. The General Solution of $Pu_x + Qu_y + Ru_z = 0$

In this section we find the general solution, i.e., the collection of all solutions, of the partial differential equation

$$P(x, y, z)\frac{\partial u}{\partial x} + Q(x, y, z)\frac{\partial u}{\partial y} + R(x, y, z)\frac{\partial u}{\partial z} = 0 \tag{3.1}$$

in some domain Ω of R^3. We assume that the functions P, Q, R belong to $C^1(\Omega)$ and do not vanish simultaneously at any point of Ω. A function $u(x, y, z)$ in $C^1(\Omega)$ is a solution of the p.d.e. (3.1) in Ω if substitution of $u(x, y, z)$ for u into (3.1) yields an equation in x, y, z which is satisfied identically in Ω.

In Section 1, any solution of equation (3.1) in Ω was called a first integral of the vector field $\mathbf{V} = (P, Q, R)$ or of the associated system of ordinary differential equations

$$\frac{dx}{P} = \frac{dy}{Q} = \frac{dz}{R}. \tag{3.2}$$

In Section 2 we described some methods for finding solutions of (3.1) by manipulation and integration of (3.2). The following theorem is a restatement of Theorem 1.2.

Theorem 3.1. Let u_1 and u_2 be any two solutions of (3.1) in Ω and let $F(u_1, u_2)$ be any C^1 function of two variables. Then

$$u(x, y, z) = F(u_1(x, y, z), u_2(x, y, z)) \tag{3.3}$$

is also a solution of (3.1) in Ω.

The following theorem asserts that if u_1 and u_2 are any two functionally independent solutions of (3.1) in Ω (i.e., u_1, u_2 satisfy condition (1.15)) then every solution of (3.1) has the form (3.3) near any point of Ω. More precisely,

Theorem 3.2. Let u_1 and u_2 be any two given functionally independent solutions of (3.1) in Ω and let u be any solution of (3.1) in Ω. Let (x_0, y_0, z_0) be any point of Ω. Then there is a neighborhood U of (x_0, y_0, z_0), $U \subset \Omega$, and a C^1 function $F(u_1, u_2)$ of two variables such that

$$u(x, y, z) = F(u_1(x, y, z), u_2(x, y, z)), \quad \text{for} \quad (x, y, z) \in U.$$

Recall that a neighborhood of (x_0, y_0, z_0) is any open set containing (x_0, y_0, z_0). In Theorem 3.2 the size of U is not specified. U may be very small or it may be the whole of the domain Ω.

Roughly, Theorems 3.1 and 3.2 say that the general solution of (3.1) is given by

$$u = F(u_1, u_2)$$

where u_1 and u_2 are any two fixed functionally independent solutions of (3.1) and F is an arbitrary C^1 function of two variables. The general solution of an ordinary differential equation of the first order involves an arbitrary constant. In contrast we see here that the general solution of the first order partial differential equation (3.1) involves an arbitrary function.

Example 3.1. Consider the p.d.e.

$$xu_x + yu_y + zu_z = 0$$

in the domain $x > 0$. In this domain the functions $u_1 = y/x$ and $u_2 = z/x$ are functionally independent solutions of this equation (see Example 1.2). The general solution of the equation is

$$u(x, y, z) = F\left(\frac{y}{x}, \frac{z}{x}\right)$$

where F is an arbitrary C^1 function of two variables.

Proof of Theorem 3.2. Since u, u_1, u_2 are solutions of (3.1) in Ω we must have at every point of Ω,

$$P\frac{\partial u}{\partial x} + Q\frac{\partial u}{\partial y} + R\frac{\partial u}{\partial z} = 0$$

$$P\frac{\partial u_1}{\partial x} + Q\frac{\partial u_1}{\partial y} + R\frac{\partial u_1}{\partial z} = 0$$

$$P\frac{\partial u_2}{\partial x} + Q\frac{\partial u_2}{\partial y} + R\frac{\partial u_2}{\partial z} = 0.$$

This set of equations forms a linear system of homogeneous equations in P, Q, R. Since $(P, Q, R) \neq (0, 0, 0)$ at every point of Ω the determinant of the coefficients of the system must vanish at every point of Ω,

$$\begin{vmatrix} \dfrac{\partial u}{\partial x} & \dfrac{\partial u}{\partial y} & \dfrac{\partial u}{\partial z} \\ \dfrac{\partial u_1}{\partial x} & \dfrac{\partial u_1}{\partial y} & \dfrac{\partial u_1}{\partial z} \\ \dfrac{\partial u_2}{\partial x} & \dfrac{\partial u_2}{\partial y} & \dfrac{\partial u_2}{\partial z} \end{vmatrix} = 0, \quad \text{in} \quad \Omega.$$

This determinant is of course the Jacobian of u, u_1, u_2 with respect to x, y, z. Hence

$$\frac{\partial(u, u_1, u_2)}{\partial(x, y, z)} = 0 \quad \text{in} \quad \Omega.$$

Now, the assumption that u_1 and u_2 are functionally independent in Ω means that at every point of Ω at least one of the Jacobians

$$\frac{\partial(u_1, u_2)}{\partial(x, y)}, \qquad \frac{\partial(u_1, u_2)}{\partial(x, z)}, \qquad \frac{\partial(u_1, u_2)}{\partial(y, z)}$$

is different from zero. According to a theorem of advanced calculus which is based on the Implicit Function Theorem (see Taylor,[1] Section 9.6, Theorem V and its generalization to $n = 3$ at the end of the section), the last two conditions on the Jacobians imply the conclusion of the theorem. The proof is complete.

Problems

3.1. Find the general solution of each of the following equations. Refer to previous examples and problems.

(a) $u_x = 0$

(b) $yu_x - xu_y = 0$

(c) $u_y + 3y^2u_z = 0$

(d) $x^2u_x + y^2u_y + z(x + y)u_z = 0.$

3.2. Find the general solution of each of the following equations. Refer to the examples of Section 2.

(a) $xu_x + yu_y + xy(z^2 + 1)u_z = 0$

(b) $\frac{1}{y}(y^2 + z^2)u_x + xzu_y - xyu_z = 0$

(c) $(y + z)u_x + yu_y + (x - y)u_z = 0$

(d) $x(y - z)u_x + y(z - x)u_y + z(x - y)u_z = 0.$

3.3. *Extension to n dimensions*. Let Ω be a domain in R^n with $x = (x_1, \ldots, x_n)$ denoting a point of Ω, and let

$$\mathbf{V}(x) = (P_1(x), \ldots, P_n(x)), \qquad x \in \Omega \tag{3.4}$$

be a nonvanishing C^1 vector field in Ω. A curve C is an integral curve of $\mathbf{V}$, if $\mathbf{V}$ is tangent to C at each of its points. The integral curves of $\mathbf{V}$, if parametrized appropriately, are the solution curves of the associated system of ordinary differential equations

$$\frac{dx_1}{P_1(x)} = \frac{dx_2}{P_2(x)} = \cdots = \frac{dx_n}{P_n(x)}. \tag{3.5}$$

A function $u \in C^1(\Omega)$ is called a first integral of $\mathbf{V}$ (or of the system (3.5)) in Ω if u is a solution of the p.d.e.

$$P_1(x)u_{x_1} + \cdots + P_n(x)u_{x_n} = 0, \quad \text{in} \quad \Omega. \tag{3.6}$$

The first integrals $u_1(x), \ldots, u_{n-1}(x)$ of $\mathbf{V}$ are said to be functionally independent in Ω if grad $u_1(x), \ldots,$ grad $u_{n-1}(x)$ are linearly independent at every point $x \in \Omega$. $(n-1)$ functionally independent first integrals of $\mathbf{V}$ can be found by solving the system of o.d.e.'s (3.5).

(a) Prove the extension of Theorem 1.1: The $(n-1)$-parameter family of integral curves of $\mathbf{V}$ in Ω is described by the equations

$$u_1(x) = c_1, \ldots, u_{n-1}(x) = c_{n-1} \tag{3.7}$$

where $u_1, \ldots, u_{n-1}$ are any $(n - 1)$ functionally independent first integrals of $\mathbf{V}$ in Ω.

(b) State and prove the extension of Theorem 3.2 which briefly says that the general solution of the p.d.e. (3.6) is given by

$$u(x) = F(u_1(x), \ldots, u_{n-1}(x))$$

where F is a C^1 function of $(n - 1)$ variables and $u_1, \ldots, u_{n-1}$ are any fixed $(n-1)$ functionally independent first integrals of $\mathbf{V}$.

3.4. Find the general solution of each of the equations

(a) $u_{x_1} = 0$ $\quad$ (n arbitrary).

(b) $x_1u_{x_1} + x_1x_2u_{x_2} + x_1x_3u_{x_3} + x_4u_{x_4} = 0$ $\quad$ $(n = 4)$.

4. Construction of an Integral Surface of a Vector Field Containing a Given Curve

An integral surface of a vector field $\mathbf{V} = (P, Q, R)$ is a surface S such that $\mathbf{V}$ is tangent to S at each of its points. We will use the following more precise definition.

Definition 4.1. A surface S in a domain Ω of R^3 is an *integral surface of the vector field* $\mathbf{V}$ if S is a level surface of a first integral of $\mathbf{V}$; i.e., S is described by an equation of the form

$$u(x, y, z) = c \tag{4.1}$$

where u is a solution of the equation

$$Pu_x + Qu_y + Ru_z = 0 \tag{4.2}$$

in Ω such that grad $u \neq \mathbf{0}$ in Ω.

Equation (4.2) means that at each point of the surface S given by equation (4.1), $\mathbf{V}$ is tangent to S. Note that according to Definition 4.1, an integral surface of $\mathbf{V}$ is always a member of a one-parameter family of integral surfaces of $\mathbf{V}$ given by equation (4.1) with c being considered a parameter. We will also call an integral surface of $\mathbf{V}$ a *solution surface of equation* (*4.2*) since it is a level surface of a solution of this equation.

The plane described by the equation $y/x = c$ is an integral surface of the vector field $\mathbf{V} = (x, y, z)$. The cylinder $x^2 + y^2 = c$ is an integral surface of $\mathbf{V} = (y, -x, 0)$.

In this section we will show that the integral surfaces of $\mathbf{V}$ are generated by the integral curves of $\mathbf{V}$. We will then describe a method which uses this fact for constructing an integral surface of $\mathbf{V}$ passing through a given curve. This method will be used in Chapter III to solve the initial value problem for a first order quasi-linear partial differential equation.

We first prove that since $\mathbf{V}$ is tangent to its integral surfaces and integral curves, the integral surfaces of $\mathbf{V}$ are generated by its integral curves, or, equivalently, the solution surfaces of equation (4.2) are generated by the solution curves of the associated system

$$\frac{dx}{P} = \frac{dy}{Q} = \frac{dz}{R}. \tag{4.3}$$

This last statement summarizes the content of the following theorem.

Theorem 4.1. If S is a solution surface of (4.2) in Ω, then for every point of S the solution curve of (4.3) passing through that point lies on S. Conversely, if u is a C^1 function in Ω, and if at each point $(x_0, y_0, z_0) \in \Omega$, the solution curve of (4.3) passing through (x_0, y_0, z_0) lies on the level surface of u passing through (x_0, y_0, z_0), then u is a solution of (4.2) in Ω (and hence the level surfaces of u are solution surfaces of (4.2)).

Proof. For the first part of the theorem, let S be given by the equation

$$u(x, y, z) = c$$

where u is a solution of (4.2) in Ω, and let (x_0, y_0, z_0) be a point of S. Then

$$u(x_0, y_0, z_0) = c.$$

Let C be a solution curve of (4.3) given parametrically by the equations

$$x = x(t), \qquad y = y(t), \qquad z = z(t); \qquad t \in I$$

and passing through the point (x_0, y_0, z_0). Then for some $t_0 \in I$,

$$x_0 = x(t_0), \qquad y_0 = y(t_0), \qquad z_0 = z(t_0)$$

and hence

$$u(x(t_0), y(t_0), z(t_0)) = c.$$

Now,

$$\begin{aligned}\frac{d}{dt} u(x(t), y(t), z(t)) &= \frac{\partial u}{\partial x}\frac{dx}{dt} + \frac{\partial u}{\partial y}\frac{dy}{dt} + \frac{\partial u}{\partial z}\frac{dz}{dt} \\ &= \frac{\partial u}{\partial x} P + \frac{\partial u}{\partial y} Q + \frac{\partial u}{\partial z} R = 0.\end{aligned}$$

In the first equality we used the chain rule, in the second equality we used the fact that C is a solution curve of (4.3) and in the last equality we used the fact that u is a solution of (4.2). It follows that the function $u(x(t), y(t), z(t))$ is constant, and since its value at $t = t_0$ is c, we must have

$$u(x(t), y(t), z(t)) = c, \qquad t \in I.$$

This means that C lies on S. For the second part of the theorem, the assumptions on u mean that at each point $(x_0, y_0, z_0) \in \Omega$, the vector **V** which is tangent to the solution curve of (4.3) passing through (x_0, y_0, z_0) is also tangent to the level surface of u passing through (x_0, y_0, z_0). Hence, at every point of Ω, grad u must be orthogonal to **V**, or

$$P\frac{\partial u}{\partial x} + Q\frac{\partial u}{\partial y} + R\frac{\partial u}{\partial z} = 0.$$

This means that u is a solution of (4.2) in Ω. The proof of the theorem is complete.

Let us now consider the problem of constructing an integral surface S of **V** passing through a given curve C in Ω. Since every integral surface of **V** is generated by integral curves of **V** we can try to construct S as follows. Suppose first that **V** is nowhere tangent to C. Then through each point of C construct the unique integral curve of **V** passing through that point. The totality of these curves forms the desired integral surface S (see Fig. 4.1).

Suppose on the other hand that **V** is tangent to C at every one of its points; in other words, suppose that C is an integral curve of **V**. Then every integral curve of **V** passing through any point of C coincides with C. Hence, in this case the integral curves of **V** passing through points of C generate nothing more than the curve C itself (see Fig. 4.2(a)). Nevertheless, in this case we can construct infinitely many integral surfaces of **V** containing the curve C. In fact, let C' be any curve intersecting C and such that **V** is nowhere tangent to C'. Let S' be the integral surface of **V** generated by the integral curves of **V** passing through points of C'. Since S' contains the intersection of C with C', S' must contain the whole of C (why?). Since we can draw infinitely many curves C' intersecting C, this procedure yields infinitely many integral surfaces of **V** all of which contain the curve C (see Fig. 4.2(b)).

Obviously, the above arguments, which are based on geometric intuition, need to be made precise. We do this in Theorems 4.2 and 4.3 below.

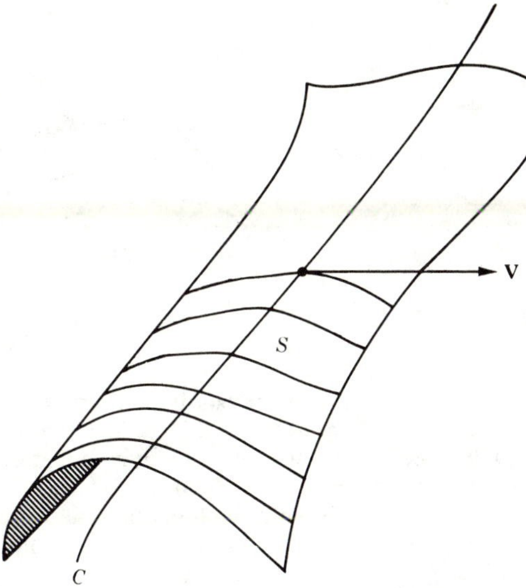

Fig. 4.1

In the proofs of these theorems we actually describe methods for finding the equation(s) of the integral surface(s) of **V** containing a given curve. Before stating and proving these theorems, we discuss a particularly simple example.

Example 4.1. Consider the vector field **V** = (1, 0, 0). The corresponding equation (4.2) is

(4.4) $$u_x = 0$$

and the associated system of equations is

(4.5) $$\frac{dx}{1} = \frac{dy}{0} = \frac{dz}{0}.$$

The integral curves of **V** are given by

$$y = c_1, \qquad z = c_2$$

and they are lines parallel to the x-axis. Suppose first that the curve C lies on the $x = 0$ plane and is given by the equations

(4.6) $$f(y, z) = 0, \qquad x = 0.$$

Then the cylindrical surface S given by

(4.7) $$f(y, z) = 0$$

is the integral surface of **V** containing the curve C. Next, suppose that C is an integral curve of **V** given by

(4.8) $$y = y_0, \qquad z = z_0.$$

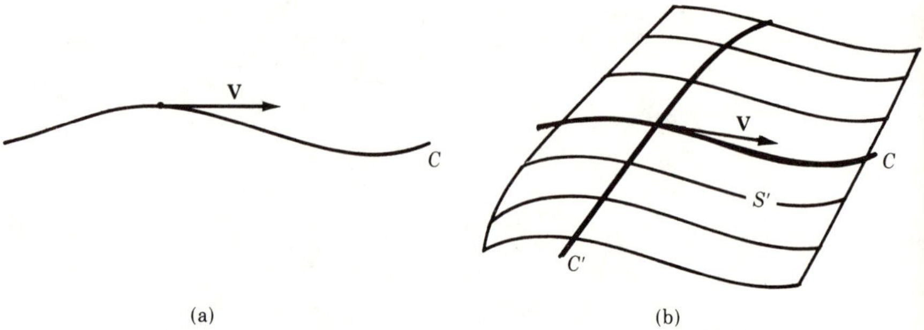

Fig. 4.2

Let C' be any curve on the $x = 0$ plane passing through the point $(0, y_0, z_0)$. Then C' is given by equations of the form (4.6) with the condition $f(y_0, z_0) = 0$. It is easy to see that the surface S' given by equation (4.7) is an integral surface of **V** containing the curve (4.8). In fact, any surface given by an equation of the form (4.7), with the function f subject only to the condition $f(y_0, z_0) = 0$, is an integral surface of **V** containing the curve (4.8).

We now formulate and prove the theorems. Let the given curve C in Ω be described parametrically by the equations

$$(4.9) \qquad x = x_0(t), \qquad y = y_0(t), \qquad z = z_0(t); \qquad t \in I$$

where the functions $x_0(t)$, $y_0(t)$, $z_0(t)$ belong to $C^1(I)$ and their derivatives do not vanish simultaneously on I. Let (x_0, y_0, z_0) be a point of C corresponding to $t = t_0 \in I$.

Theorem 4.2. Suppose that $\mathbf{V} = (P, Q, R)$ is not tangent to C at (x_0, y_0, z_0). Then, in some neighborhood Ω_0 of (x_0, y_0, z_0), there is a unique integral surface of **V** containing the part of C in Ω_0.

Note that this is a "local" theorem since it asserts the existence of the desired integral surface only in a neighborhood Ω_0 of (x_0, y_0, z_0). The theorem does not specify the size of the neighborhood Ω_0. It may be a very small open set containing (x_0, y_0, z_0) or it may be the whole domain Ω. Note also that the assumption of the theorem that **V** is not tangent to C at the point (x_0, y_0, z_0) implies that **V** is not tangent to C at points of C near (x_0, y_0, z_0) (i.e., at every point of C contained in some neighborhood of (x_0, y_0, z_0)). This follows from our assumption that both **V** and the tangent vector to C vary continuously.

Proof of Theorem 4.2. Let u_1 and u_2 be two functionally independent first integrals of **V** defined in some neighborhood of (x_0, y_0, z_0). Let U_1 and U_2 be functions of t defined for t near t_0 by the equations

$$(4.10) \qquad U_1 = u_1(x_0(t), y_0(t), z_0(t)), \qquad U_2 = u_2(x_0(t), y_0(t), z_0(t)).$$

The assumption that **V** is not tangent to C at (x_0, y_0, z_0) implies that near $(t_0, U_1(t_0), U_2(t_0))$ we can eliminate t between equations (4.10) and obtain a relation between U_1 and U_2. To see this let us compute the derivatives of U_1 and U_2 at $t = t_0$,

$$\left.\frac{dU_1}{dt}\right|_{t=t_0} = \text{grad } u_1(x_0, y_0, z_0)\cdot\left.\left(\frac{dx_0}{dt}, \frac{dy_0}{dt}, \frac{dz_0}{dt}\right)\right|_{t=t_0}$$

$$\left.\frac{dU_2}{dt}\right|_{t=t_0} = \text{grad } u_2(x_0, y_0, z_0)\cdot\left.\left(\frac{dx_0}{dt}, \frac{dy_0}{dt}, \frac{dz_0}{dt}\right)\right|_{t=t_0}$$

If both of these derivatives were zero, then the tangent to C at (x_0, y_0, z_0) would be parallel to the (non-zero) vector grad $u_1(x_0, y_0, z_0) \times$ grad $u_2(x_0, y_0, z_0)$. Since $\mathbf{V}(x_0, y_0, z_0)$ is also parallel to this vector, it would follow that $\mathbf{V}$ is tangent to C at (x_0, y_0, z_0), contradicting the assumption of the theorem. Hence, at least one of the derivatives dU_1/dt or dU_2/dt is not equal to zero at $t = t_0$. Suppose for example that $dU_1/dt|_{t=t_0} \neq 0$. Then we can solve the first of equations (4.10) for t and substitute in the second to obtain a relation of the form

$$U_2 - f(U_1) = 0 \tag{4.11}$$

where f is a C^1 function defined near $U_1(t_0)$. Now the equation

$$u_2(x, y, z) - f(u_1(x, y, z)) = 0 \tag{4.12}$$

describes the equation of the desired integral surface. In fact, the left-hand side of (4.12) is defined in a neighborhood Ω_0 of (x_0, y_0, z_0) and is a solution of (4.2) in Ω_0. Moreover, it follows from (4.10) and (4.11) that the surface (4.12) contains the part of C in Ω_0.

We do not give here a precise proof of the uniqueness part of Theorem 4.2. This is intuitively obvious from the fact that every integral curve passing through a point of C must lie on every integral surface of $\mathbf{V}$ containing C. Since $\mathbf{V}$ is not tangent to C near (x_0, y_0, z_0) the integral curves of $\mathbf{V}$ passing through C generate a surface near (x_0, y_0, z_0). Any two integral surfaces of $\mathbf{V}$ containing C must also contain this surface and hence they must coincide (at least in a neighborhood of (x_0, y_0, z_0)).

Example 4.2. Find the integral surface of $\mathbf{V} = (x, y, z)$ containing the curve C given by

$$y = x + 1, \qquad z = x^2; \qquad x > 0. \tag{4.13}$$

First we express (4.13) in parametric form,

$$x = t, \qquad y = t + 1, \qquad z = t^2; \qquad t > 0. \tag{4.14}$$

The tangent to C is $(1, 1, 2t)$ and $\mathbf{V} = (x, y, z) = (t, t + 1, t^2)$ on C. Hence $\mathbf{V}$ is nowhere tangent to C and we can try to solve the problem in a neighborhood of any point of C.

The functions $u_1 = y/x$ and $u_2 = z/x$ are two functionally independent first integrals of $\mathbf{V}$ defined in the domain $x > 0$ containing the curve C. Now,

$$U_1 = \frac{t + 1}{t}, \qquad U_2 = t$$

and eliminating t we obtain

$$U_2 - \frac{1}{U_1 - 1} = 0.$$

Substituting u_1 for U_1 and u_2 for U_2 in this equation we obtain the equation of the desired integral surface

$$\frac{z}{x} - \frac{1}{\frac{y}{x} - 1} = 0. \tag{4.15}$$

That equation (4.15) describes an integral surface of **V** is obvious, since the left hand side of (4.15) is a first integral of **V** (it is a function of the first integrals u_1 and u_2). To verify that the surface described by (4.15) contains the curve C, we substitute (4.14) into (4.15) and check that the result is an identity in t. Indeed, we get

$$\frac{t^2}{t} - \frac{1}{\frac{t+1}{t} - 1} \equiv 0, \ t > 0.$$

Note that (4.15) describes the desired integral surface of **V** in the domain defined by the inequalities $x > 0$, $y > x$.

Let us consider now the case in which the given curve C is an integral curve of the vector field **V**.

Theorem 4.3. Suppose that $\mathbf{V} = (P, Q, R)$ is tangent to C at every one of its points contained in a neighborhood Ω_0 of a point $(x_0, y_0, z_0) \in C$. Then in some neighborhood $\Omega_1 \subset \Omega_0$ of (x_0, y_0, z_0) there are infinitely many integral surfaces of **V** containing the part of C in Ω_1.

Proof. Let u_1 and u_2 be two functionally independent first integrals of **V** defined in a neighborhood $\Omega_1 \subset \Omega_0$ of (x_0, y_0, z_0). Since the part of C in Ω_1 is an integral curve of **V** the functions u_1 and u_2 must be constant on it. Hence on $C \cap \Omega_1$ we have

$$\begin{aligned} u_1(x, y, z) &= c_1 = u_1(x_0, y_0, z_0), \\ u_2(x, y, z) &= c_2 = u_2(x_0, y_0, z_0). \end{aligned} \tag{4.16}$$

Now let $F(u_1, u_2)$ be any C^1 function of two variables such that

$$F(c_1, c_2) = 0. \tag{4.17}$$

Then the equation

$$F(u_1(x, y, z), u_2(x, y, z)) = 0 \tag{4.18}$$

is the equation of an integral surface of **V** in Ω_1 containing $C \cap \Omega_1$ (why?). Since F can be chosen arbitrarily, subject only to the condition (4.17), the proof of the theorem is complete.

Example 4.3. Find the integral surfaces of $\mathbf{V} = (x, y, z)$ containing the half-line

$$x = y = z, \qquad x > 0. \tag{4.19}$$

Obviously this line is an integral curve of **V**. We use again the first integrals $u_1 = y/x$, $u_2 = z/x$. On the half line (4.19), $u_1 = 1$ and $u_2 = 1$. If $F(u_1, u_2)$ is any C^1 function of two variables such that $F(1, 1) = 0$, then

$$F\left(\frac{y}{x}, \frac{z}{x}\right) = 0$$

is the equation of an integral surface of **V** containing the half-line (4.19). For example, if we take $F(u_1, u_2) = u_1 - u_2$ or $F(u_1, u_2) = u_1 - u_2^2$ or $F(u_1, u_2) = \cos\left(\frac{\pi}{2} u_1 u_2\right)$ we obtain the integral surfaces

$$\frac{y}{x} - \frac{z}{x} = 0, \qquad \frac{y}{x} - \left(\frac{z}{x}\right)^2 = 0, \qquad \cos\left(\frac{\pi}{2}\frac{yz}{x^2}\right) = 0$$

all of which contain the half-line (4.19).

Problems

4.1. Find the integral surface of the vector field **V** containing the given curve C.

(a) $\mathbf{V} = (1, 1, z)$. $C: x = t,\ y = 0,\ z = \sin t;\ -\infty < t < \infty$

(b) $\mathbf{V} = (x, -y, 0)$. $C: x = t,\ y = t,\ z = t^2;\ 0 < t < \infty$

(c) $\mathbf{V} = (y - z, z - x, x - y)$. $C: x = t,\ y = 2t,\ z = 0;\ 0 < t < \infty$

(d) $\mathbf{V} = (y, -x, 2xyz)$. $C: x = t,\ y = t,\ z = t^2;\ 0 < t < \infty$.

4.2 In the following problems verify that the curve C is an integral curve of **V** and derive the formula for the infinitely many integral surfaces of **V** containing C.

(a) $\mathbf{V} = (1, 1, z)$. $C: x = t,\ y = 1 + t,\ z = e^t;\ 0 < t < \infty$

(b) $\mathbf{V} = (xz, yz, -xy)$. $C: x = t,\ y = -t,\ z = t;\ 0 < t < \infty$.

5. Applications to Plasma Physics and to Solenoidal Vector Fields

We present in this section two applications of the results of this chapter. The first is in plasma physics, the study of which has been stimulated in recent years by the possibility of thermonuclear reactors. The application consists of finding the general solution of a special case of the *Boltzmann equation* using the methods of this chapter. The second application is a proof of an important result concerning solenoidal vector fields, namely that such fields can be derived from vector potentials. The proof of this result provides a method for the determination of a vector potential for a given solenoidal vector field.

Application to Plasma Physics

The word "plasma" is used in physics to describe an ionized gas with a sufficiently high density so that the forces exerted by the gas particles on each other are not negligible in comparison with the forces exerted on the particles by external electromagnetic fields. The great radiation belts of the earth (the Van Allen belts) and streams of charged particles emitted by the sun are two examples of plasmas in nature. In the laboratory, plasmas occur when electricity is discharged through gases. In the last several years, the study of plasma physics has been stimulated by the possibility of controlled thermonuclear reactors. A thermonuclear reactor would make use of a gas at extremely high temperatures. At these temperatures the gas is fully ionized and is therefore a plasma. The main problem of thermonuclear reactors is how to contain the plasma. A material container cannot be used since its walls would be instantly vaporized. Instead it has been suggested that the plasma may be contained by a magnetic field. Hopefully, the study of plasma physics may lead to the successful design of such a "magnetic bottle."

The basic equation of plasma physics is known as the *Boltzmann equation*. It is a complicated equation, the understanding of which requires a considerable background in physics. The interested student may look at the book of Longmire.[2] From Longmire we take the following special case of the Boltzmann equation which is used in the study of a problem known as a static boundary-layer problem,

$$mv_1 \frac{\partial f}{\partial x} + e\left(\frac{v_2}{c}\frac{d\eta}{dx} - \frac{d\phi}{dx}\right)\frac{\partial f}{\partial v_1} - e\frac{v_1}{c}\frac{d\eta}{dx}\frac{\partial f}{\partial v_2} = 0. \tag{5.1}$$

In equation (5.1) f is the unknown function of the three independent variables x, v_1 and v_2. The functions ϕ and η are given functions of the variable x only, while m, e and c are constants. The partial differential equation (5.1) is an equation of the form (3.1). The associated system of o.d.e.'s is

$$\frac{dx}{mv_1} = \frac{dv_1}{e\left(\frac{v_2}{c}\frac{d\eta}{dx} - \frac{d\phi}{dx}\right)} = \frac{dv_2}{-e\frac{v_1}{c}\frac{d\eta}{dx}} \tag{5.2}$$

The equality of the first and third ratios (after canceling v_1) is an o.d.e. in x and v_2 which yields the first integral

$$f_1 = mv_2 + \frac{e}{c}\eta(x). \tag{5.3}$$

Multiplying the numerators and denominators of the second ratio in (5.2) by $2v_1$ and of the third ratio by $2v_2$ and adding the numerators and denominators of the resulting ratios yields the ratio

$$\frac{d(v_1^2 + v_2^2)}{-2ev_1\frac{d\phi}{dx}} \tag{5.4}$$

which is also equal to the ratios (5.2). The equality of the ratio (5.4) with the first ratio in (5.2) (after canceling v_1) is an o.d.e. in the variables x and $(v_1{}^2 + v_2{}^2)$ which yields the first integral

(5.5) $$f_2 = \frac{1}{2} m(v_1^2 + v_2^2) + e\phi(x).$$

Obviously, f_1 and f_2 are functionally independent and, according to Section 3, the general solution of (5.1) is given by

(5.6) $$f(x, v_1, v_2) = F\left(mv_2 + \frac{e}{c}\eta(x), \frac{1}{2} m(v_1^2 + v_2^2) + e\phi(x)\right),$$

where $F(f_1, f_2)$ is an arbitrary function of two variables. It turns out that the first integrals f_1 and f_2 have physical meaning; f_2 is the energy of a particle of mass m and f_1 is its canonical momentum. Moreover, the pair of equations

$$f_1 = c_1, \qquad f_2 = c_2$$

determine the trajectory of the particle.

Solenoidal Vector Fields

Let $\mathbf{V} = (P, Q, R)$ be a vector field defined in some domain Ω in R^3, with P, Q and R belonging to $C^1(\Omega)$. The *divergence* of $\mathbf{V}$, written div $\mathbf{V}$, is the function defined in Ω by

$$\operatorname{div} \mathbf{V} = \frac{\partial P}{\partial x} + \frac{\partial Q}{\partial y} + \frac{\partial R}{\partial z}.$$

$\mathbf{V}$ is said to be *solenoidal* in Ω if

$$\operatorname{div} \mathbf{V} = 0 \text{ in } \Omega.$$

The curl of $\mathbf{V}$, written curl $\mathbf{V}$, is the vector field defined in Ω by

$$\operatorname{curl} \mathbf{V} = \left(\frac{\partial R}{\partial y} - \frac{\partial Q}{\partial z}, \frac{\partial P}{\partial z} - \frac{\partial R}{\partial x}, \frac{\partial Q}{\partial x} - \frac{\partial P}{\partial y}\right).$$

The following theorem finds frequent application in many areas of engineering and physics.

Theorem 5.1. Let $\mathbf{V} = (P, Q, R)$ be a nonvanishing vector field defined in a domain Ω of R^3, with P, Q, R in $C^1(\Omega)$. If $\mathbf{V}$ is solenoidal in Ω, then given any point (x_0, y_0, z_0) in Ω, there is a neighborhood Ω_0 of (x_0, y_0, z_0) and a vector field $\mathbf{W}$ with C^1 components defined in Ω_0 such that

(5.7) $$\mathbf{V}(x, y, z) = \operatorname{curl} \mathbf{W}(x, y, z), \qquad (x, y, z) \in \Omega_0.$$

The vector field $\mathbf{W}$ is often called a *vector potential* for the given field $\mathbf{V}$. Although the theorem gives only a local result, it sometimes happens that one can find a vector potential $\mathbf{W}$ for $\mathbf{V}$ which is defined in the entire domain Ω. This occurs in Example 5.1.

Before giving the proof, we will list some identities from vector calculus which will be needed in the course of the proof. Let f be a C^1 function, and let $\mathbf{u}$, $\mathbf{v}$ be C^1 vector fields, all being defined in a common domain Ω. Then

(5.8) $$\operatorname{div}(f\mathbf{u}) = \operatorname{grad} f \cdot \mathbf{u} + f \operatorname{div} \mathbf{u},$$

(5.9) $$\operatorname{div}(\mathbf{u} \times \mathbf{v}) = (\operatorname{curl} \mathbf{u})\cdot\mathbf{v} - (\operatorname{curl} \mathbf{v})\cdot\mathbf{u},$$

(5.10) $$\operatorname{curl} (\operatorname{grad} f) = 0 \text{ (assume } f \in C^2 \text{ here)},$$

(5.11) $$\operatorname{curl} (f\mathbf{u}) = (\operatorname{grad} f) \times \mathbf{u} + f \operatorname{curl} \mathbf{u}.$$

Proof of Theorem 5.1. Let u_1, u_2 be two first integrals of $\mathbf{V}$ which are functionally independent in some neighborhood Ω_1 of (x_0, y_0, z_0). At each point of Ω_1, the vector field $\mathbf{V}$ is parallel to grad $u_1 \times$ grad u_2, so that we can write

(5.12) $$\mathbf{V}(x, y, z) = \lambda(x, y, z)(\operatorname{grad} u_1 \times \operatorname{grad} u_2)$$

for some function λ defined in Ω_1. The function λ is C^1 since

$$\lambda = \frac{\mathbf{V}\cdot(\operatorname{grad} u_1 \times \operatorname{grad} u_2)}{|\operatorname{grad} u_1 \times \operatorname{grad} u_2|^2},$$

and u_1, u_2 are actually C^2. This smoothness of u_1 and u_2, which we have not used before, follows from the manner in which u_1 and u_2 are obtained from the system of ordinary differential equations (1.10). Since $\mathbf{V}$ is solenoidal in Ω_1, it follows by applying identities (5.8), (5.9) and (5.10) that

(5.13) $$0 = \operatorname{div} \mathbf{V} = \operatorname{grad} \lambda\cdot(\operatorname{grad} u_1 \times \operatorname{grad} u_2) + \lambda[(\operatorname{curl} \operatorname{grad} u_1)\cdot\operatorname{grad} u_2 - (\operatorname{curl} \operatorname{grad} u_2)\cdot\operatorname{grad} u_1] = \operatorname{grad} \lambda\cdot(\operatorname{grad} u_1 \times \operatorname{grad} u_2).$$

Equation (5.13) shows that grad λ is perpendicular to grad $u_1 \times$ grad u_2 at each point of Ω_1, and so is perpendicular to $\mathbf{V}$ at each point of Ω_1, i.e.

(5.14) $$P\frac{\partial \lambda}{\partial x} + Q\frac{\partial \lambda}{\partial y} + R\frac{\partial \lambda}{\partial z} = 0 \quad \text{in} \quad \Omega_1.$$

Thus λ is a solution of the partial differential equation studied in Section 3, and we can apply the results of that section to express λ as a function of u_1 and u_2. Explicitly, Theorem 3.2 asserts that there is a neighborhood Ω_0 of (x_0, y_0, z_0) with $\Omega_0 \subset \Omega_1$, and a C^1 function $F(u_1, u_2)$ such that

(5.15) $$\lambda(x, y, z) = F(u_1(x, y, z), u_2(x, y, z)), \ (x, y, z) \in \Omega_0.$$

Now, let $G(u_1, u_2)$ be a function such that

(5.16) $$F(u_1, u_2) = \frac{\partial G}{\partial u_1}(u_1, u_2).$$

From (5.12), (5.15) and (5.16) we see that in Ω_0,

(5.17) $$\mathbf{V} = \left(\frac{\partial G}{\partial u_1} \operatorname{grad} u_1\right) \times \operatorname{grad} u_2 = \operatorname{grad} G \times \operatorname{grad} u_2.$$

In the last line of (5.17) we used the identities

$$\operatorname{grad} G(u_1, u_2) = \frac{\partial G}{\partial u_1} \operatorname{grad} u_1 + \frac{\partial G}{\partial u_2} \operatorname{grad} u_2, \qquad \operatorname{grad} u_2 \times \operatorname{grad} u_2 = 0.$$

To complete the proof we need only observe that (5.17) can be written in the form

$$\mathbf{V} = \text{curl }(G \text{ grad } u_2)$$

because of identities (5.10) and (5.11). If we set

$$\mathbf{W} = G \text{ grad } u_2 \tag{5.18}$$

then

$$\mathbf{V} = \text{curl }\mathbf{W} \quad \text{in} \quad \Omega_0.$$

Example 5.1. Let $\mathbf{V} = (y, -x, 0)$ be the vector field of Example 1.3, with Ω being R^3 minus the z-axis. $\mathbf{V}$ is clearly solenoidal in Ω. Two functionally independent first integrals of $\mathbf{V}$ were found in Example 1.3 to be

$$u_1 = x^2 + y^2, \qquad u_2 = z.$$

Easy calculations show that

$$\begin{aligned} \text{grad } u_1 \times \text{grad } u_2 &= 2(y, -x, 0) \\ &= 2\mathbf{V}, \end{aligned}$$

so that the proportionality factor λ in this case is simply

$$\lambda = \frac{1}{2}.$$

The function $F(u_1, u_2)$ in (5.15) is

$$F(u_1, u_2) \equiv \frac{1}{2},$$

and for $G(u_1, u_2)$ we can take the function $(1/2)u_1$. It follows from (5.18) that

$$\begin{aligned} \mathbf{W} &= \frac{1}{2} u_1 \text{ grad } u_2 \\ &= \frac{1}{2} (x^2 + y^2)\text{grad } z \\ &= (0, 0, \frac{1}{2} (x^2 + y^2)). \end{aligned}$$

Thus, for all points of Ω we have

$$\mathbf{V} = \text{curl }(0, 0, \frac{1}{2} (x^2 + y^2)).$$

Problems

5.1. Prove the vector identities (5.8)–(5.11).

5.2. Prove that a vector potential for $\mathbf{V}$ cannot be unique. [*Hint:* Consider the identity (5.10).]

5.3. Derive equation (5.13).

5.4. Derive equation (5.17).

5.5 Find vector potentials for

(a) $\mathbf{V} = (1, 1, 1)$ in $\Omega = R^3$,

(b) $\mathbf{V} = (x/(x^2 + y^2), y/(x^2 + y^2), 0)$ in $\Omega = R^3$ minus the z-axis.

References for Chapter II

1. Taylor, A. E.: *Advanced Calculus,* Boston: Ginn and Co., 1955.
2. Longmire, C. L.: *Elementary Plasma Physics,* New York: Interscience Publishers, 1963.

CHAPTER III

Theory and applications of quasi-linear and linear equations of first order

In this chapter we study quasi-linear (and linear) equations of the first order. The theory and method of solution of the initial-value problem for these equations is obtained as a direct application of the theory and method of construction of integral curves and surfaces of vector fields presented in Chapter II.

In Section 1 we define what is meant by a solution of a first order equation and we classify first order equations according to linearity. In Section 2 we define the general integral of a first order quasi-linear equation and describe a method for obtaining it. The general integral is a formula which yields most of the solutions of the equation. In Section 3 we describe the initial-value problem for a first order quasi-linear equation and obtain the condition under which there exists a unique solution to this problem. In Section 4 we show that if this condition is not satisfied then there is usually no solution to the problem, and in the special case in which there is a solution, there are actually infinitely many solutions. In Section 5 we apply the general theory to the study of conservation laws which are quasi-linear first order equations that arise in many areas of physics. The solutions of these equations frequently develop discontinuities known as shocks or shock waves, which are well-known phenomena in gas dynamics. Two examples, one in traffic flow and the other in gas dynamics, are discussed in detail in Section 6. Finally in Section 7 we present an important application of linear first order equations to probability, specifically to the study of certain stochastic processes. We discuss in detail two examples, one concerning a simple trunking problem in a telephone network and the other concerning the control of a tropical disease. Many other examples are described in the problems of the section.

1. First Order Partial Differential Equations

A *first order partial differential equation* in two independent variables x, y and one unknown z, is an equation of the form

$$F(x, y, z, z_x, z_y) = 0. \tag{1.1}$$

The function $F(x, y, z, p, q)$ is defined in some domain in R^5. We use (x, y, z, p, q) for the coordinates of a point in R^5. A *solution* of equation (1.1) in some domain Ω of R^2 is a function $z = f(x, y)$ defined and C^1 in Ω and such that the following two conditions are satisfied:

(i) For every $(x, y) \in \Omega$, the point (x, y, z, z_x, z_y) is in the domain of the function F.

(ii) When $z = f(x, y)$ is substituted in (1.1) the resulting equation is an identity in x, y for all $(x, y) \in \Omega$.

We classify first order p.d.e.'s according to the special form of the function F. An equation of the form

$$P(x, y, z)z_x + Q(x, y, z)z_y = R(x, y, z) \tag{1.2}$$

is called *quasi-linear*. Here the function F is a linear function in the derivatives z_x and z_y with the coefficients P, Q, R depending on the independent variables x and y as well as on the unknown z. An equation of the form

$$P(x, y)z_x + Q(x, y)z_y = R(x, y, z) \tag{1.3}$$

is called *almost linear*. Note that the coefficients of z_x and z_y are functions of the independent variables only. An equation of the form

$$a(x, y)z_x + b(x, y)z_y + c(x, y)z = d(x, y) \tag{1.4}$$

is called *linear*. Here the function F is linear in z_x, z_y and z with all coefficients depending on the independent variables x and y only. Finally, an equation which does not fit any of the above classifications is called *nonlinear*.

Example 1.1. The p.d.e.

$$z_x^2 + z_y^2 = 1 \tag{1.5}$$

is known as the equation of straight light rays. It is a nonlinear equation.

Example 1.2. The p.d.e.

$$a(z)z_x + z_y = 0 \tag{1.6}$$

where $a(z)$ is a given function of z, expresses a conservation law. It is a quasi-linear equation.

Example 1.3. The p.d.e.

$$xz_x + yz_y = nz \tag{1.7}$$

is satisfied by every function $z = f(x,y)$ which is homogeneous of degree n (see Problem 1.1). It is a linear equation. Equation (1.7) is called *Euler's relation*.

Example 1.4. Let $g(x, y, z)$ be a given C^1 function defined in some domain of R^3 and having non-vanishing gradient in this domain. Then g defines a one parameter family of surfaces

$$g(x, y, z) = c. \tag{1.8}$$

If the surface

(1.9) $$z = f(x, y)$$

is orthogonal to every member of the family of surfaces (1.8) (i.e., if at every point of (1.9) the normal to (1.9) is orthogonal to the normal to that surface of (1.8) passing through that point), then z satisfies the p.d.e.

(1.10) $$\frac{\partial g}{\partial x} z_x + \frac{\partial g}{\partial y} z_y = \frac{\partial g}{\partial z}.$$

(See Problem 1.2.) This is a quasi-linear equation.

Example 1.5. The p.d.e.

(1.11) $$xz_x + yz_y = z^2$$

is almost linear.

In this chapter we study quasi-linear first order p.d.e.'s. Note that linear and almost linear equations are special cases of quasi-linear equations.

Problems

1.1. Let f be a C^1 function in R^2 and suppose that, for some integer $n \geqq 1$, f satisfies the condition

(1.12) $$f(tx, ty) = t^n f(x, y)$$

for all $t \in R^1$ and all $(x, y) \in R^2$. Then the function f is called *homogeneous of degree n*.

(a) Give examples of functions which are homogeneous of degree 1, 2 and 3.

(b) Prove that if f is homogeneous of degree n then $z = f(x, y)$ satisfies the p.d.e. (1.7). [Hint: Differentiate (1.12) with respect to t and then substitute $t = 1$.]

1.2. Prove the assertion in Example 1.4.

2. The General Integral of $Pz_x + Qz_y = R$

In this section we describe a method for finding solutions of the quasi-linear equation

(2.1) $$P(x, y, z)z_x + Q(x, y, z)z_y = R(x, y, z).$$

We assume that the functions P, Q, R are defined and C^1 in some domain $\tilde{\Omega}$ of R^3 and do not vanish simultaneously at any point of this domain. A solution of equation (2.1) in a domain Ω of R^2 is a function $z = f(x, y)$ defined and C^1 in Ω and such that the following two conditions are satisfied:

(i) For every $(x, y) \in \Omega$, the point $(x, y, f(x, y))$ belongs to the domain $\tilde{\Omega}$ of the functions P, Q, R.

(ii) When $z = f(x, y)$ is substituted in (2.1), the resulting equation is an identity in x, y for all $(x, y) \in \Omega$.

A solution

$$z = f(x, y), \quad (x, y) \in \Omega \tag{2.2}$$

of equation (2.1) may be viewed as a surface in R^3, called a *solution surface* of equation (2.1). The normal to this surface is $(f_x, f_y, -1) = (z_x, z_y, -1)$. Equation (2.1) requires that this normal be orthogonal to $\mathbf{V} = (P, Q, R)$ at each point of (2.2). Thus, a surface S is a solution surface of equation (2.1) if S can be described by an equation of the form (2.2) and if at each point of S the vector $\mathbf{V} = (P, Q, R)$ is tangent to S. In other words, a solution surface of equation (2.1) is an integral surface of the vector field $\mathbf{V} = (P, Q, R)$ which can be described by an equation of the form (2.2). This suggests that in order to find solution surfaces (and hence solutions) of (2.1) we should look for integral surfaces of the vector field $\mathbf{V}$ or solution surfaces of the p.d.e.

$$Pu_x + Qu_y + Ru_z = 0 \tag{2.3}$$

which can be described by an equation of the form (2.2). Recall that a solution surface of (2.3) is a level surface, say

$$u(x, y, z) = 0, \quad (x, y, z) \in \tilde{\Omega}, \tag{2.4}$$

of a solution $u(x, y, z)$ of (2.3). If equation (2.4) can be solved for z in terms of x and y, then the resulting function is a solution of equation (2.1). We express this more precisely in the following lemma.

Lemma 2.1. Let u be in $C^1(\tilde{\Omega})$ and suppose that at every point of the level surface (2.4) the following two conditions are satisfied:

(i) $Pu_x + Qu_y + Ru_z = 0$

(ii) $u_z \neq 0$.

Then equation (2.4) implicitly defines z as a function of x and y and this function satisfies the p.d.e. (2.1).

Proof. By the implicit function theorem we have

$$z_x = -\frac{u_x}{u_z}, \quad z_y = -\frac{u_y}{u_z}$$

and hence

$$Pz_x + Qz_y = -\frac{Pu_x + Qu_y}{u_z} = -\frac{-Ru_z}{u_z} = R.$$

Lemma 2.1 indicates how to obtain solutions of equation (2.1) from solutions of equation (2.3). Since we already know the general solution of equation (2.3) (see Section 3, Chapter II), Lemma 2.1 yields a large class of solutions of equation (2.1).

Theorem 2.1. Let u_1 and u_2 be two functionally independent solutions of equation (2.3) in a domain $\tilde{\Omega}$ of R^3. Let $F(u_1, u_2)$ be an arbitrary C^1 function of two variables and consider the level surface

$$F(u_1(x, y, z), u_2(x, y, z)) = 0. \tag{2.5}$$

Then, on every part of this surface having normal with non-zero z-

component, equation (2.5) defines z implicitly as a function of x and y and this function is a solution of equation (2.1).

Definition 2.1. Equation (2.5) is called the *general integral* of equation (2.1) in $\tilde{\Omega}$.

It is known that not every solution of equation (2.1) can be obtained from the general integral (2.5) as described in Theorem 2.1. For this reason, equation (2.5) should not be called the general solution of (2.1).

In practice, the functions u_1 and u_2 used in the general integral (2.5) are obtained by solving the associated system of equations

$$\frac{dx}{P} = \frac{dy}{Q} = \frac{dz}{R} \tag{2.6}$$

as described in Chapter II, Section 2.

Example 2.1. Find the general integral of

$$xz_x + yz_y = z. \tag{2.7}$$

The associated system of equations is

$$\frac{dx}{x} = \frac{dy}{y} = \frac{dz}{z}$$

and we can take $u_1 = \frac{y}{x}$, $u_2 = \frac{z}{x}$. The general integral is

$$F\left(\frac{y}{x}, \frac{z}{x}\right) = 0 \tag{2.8}$$

where F is an arbitrary C^1 function of two variables. If we take $F(u_1, u_2) = u_1 - u_2$, (2.8) becomes

$$\frac{y}{x} - \frac{z}{x} = 0.$$

Solving for z we get $z = y$ which is obviously a solution of (2.7) in the whole of R^2. If we take $F(u_1, u_2) = u_1^2 - u_2$ we get the solution $z = y^2/x$ defined in either one of the domains $x > 0$ or $x < 0$. If we take $F(u_1, u_2) = u_1 - u_2^2$ then equation (2.8) becomes

$$\frac{y}{x} - \frac{z^2}{x^2} = 0.$$

The part of this surface with $z > 0$ defines z as a function of x and y,

$$z = \sqrt{xy}.$$

This is a solution of (2.7) in either one of the domains $x > 0, y > 0$ or $x < 0$, $y < 0$.

It should be remarked that if one of the functionally independent first integrals, say u_1, is independent of z, then, without loss of generality, the general integral (2.5) may be written in the form

$$u_2(x, y, z) = f(u_1(x, y)) \tag{2.9}$$

where f is an arbitrary C^1 function of a single variable.

Example 2.2. Consider the linear equation

$$a(x, y)z_x + b(x, y)z_y = 0 \tag{2.10}$$

where a and b are C^1 and do not vanish simultaneously. The general integral of (2.10) is given by

$$z = f(u(x, y)) \tag{2.11}$$

where f is an arbitrary C^1 function of a single variable and $u(x, y) = c$ is the general solution of the o.d.e.

$$\frac{dx}{a(x, y)} = \frac{dy}{b(x, y)}.$$

Indeed, the system of o.d.e.'s associated with (2.10) is

$$\frac{dx}{a(x, y)} = \frac{dy}{b(x, y)} = \frac{dz}{0}$$

and two functionally independent first integrals of this system are the functions $u(x, y)$ and z. It can be shown (see Problem 2.3) that (2.11) is actually the general solution of (2.10).

Problems

2.1. For each of the following equations find the general integral and compute three different solutions. Describe carefully the domain(s) of the (x, y)-plane in which each of these solutions is defined.
   (a) $x^2z_x + y^2z_y = 2xy$
   (b) $zz_x + yz_y = x$
   (c) $x^2z_x + y^2z_y = (x + y)z$
   (d) $z_y = 3y^2$
   (e) $(y + z)z_x + yz_y = x - y$
   (f) $xz_x + yz_y = xy(z^2 + 1)$
   (g) $x(y - z)z_x + y(z - x)z_y = z(x - y)$
   (h) $zz_y = -y$.

2.2. Show that the general integral of Euler's relation (1.7) leads to solutions of the form $z = x^n f(y/x)$ where f is a function of a single variable. Verify that these solutions are homogeneous functions of degree n.

2.3. Show that (2.11) is the general solution of (2.10). More precisely, prove the following assertion: Let $u(x, y) = c$ be the general solution of $dx/a = dy/b$ in a domain Ω of R^2, let $z(x, y)$ by any solution of (2.10) in Ω and let (x_0, y_0) be any point of Ω. Then there is a function $f(u)$ of a single variable such that $z(x, y) = f(u(x, y))$ for all (x, y) in some neighborhood of (x_0, y_0). [*Hint:* Use the fact that both u and z satisfy (2.10) and the fact that a and b do not vanish simultaneously to show that $\dfrac{\partial(z, u)}{\partial(x, y)} = 0$. Then apply Theorem V, Section 9.6 of Taylor.[1]]

2.4. The quasi-linear p.d.e.

$$z_x + y(1 - z)z_y = (y - 1)z$$

arises in the study of one-dimensional neutron transport theory (see Bellman, Kalaba and Wing[2]). Find one first integral. It is not a simple task to find a second first integral.

2.5. *Extension to n dimensions.* Consider the quasi-linear first order p.d.e. in the unknown z and n independent variables $x_1, \ldots, x_n$,

$$P_1(x, z)z_{x_1} + \cdots + P_n(x, z)z_{x_n} = R(x, z) \tag{2.12}$$

where $x = (x_1, \ldots, x_n)$. It is assumed that the functions $P_1, \ldots, P_n$, R are defined and C^1 in some domain $\tilde{\Omega}$ of R^{n+1} and do not vanish simultaneously at any point $(x, z) = (x_1, \ldots x_n, z)$ of $\tilde{\Omega}$.

(a) Define what is meant by a solution of (2.12) in some domain Ω of R^n.

In order to find solutions of (2.12) we look for solutions $u(x, z) = u(x_1, \ldots, x_n, z)$ of

$$P_1(x, z)u_{x_1} + \ldots + P_n(x, z)u_{x_n} + R(x, z)u_z = 0. \tag{2.13}$$

These solutions $u(x, z)$ are first integrals of the vector field $\mathbf{V}(x, z) = (P_1(x, z), \ldots, P_n(x, z), R(x, z))$ or of the associated system of o.d.e.'s

$$\frac{dx_1}{P_1(x, z)} = \ldots = \frac{dx_n}{P_n(x, z)} = \frac{dz}{R(x, z)}. \tag{2.14}$$

In practice, n functionally independent solutions of (2.13) are obtained by solving the system (2.14) using the methods described in Section 2 of Chapter II (see also Problem 3.3 of Chapter II). A level surface of a solution of (2.13), say

$$u(x_1, \ldots, x_n, z) = 0, \tag{2.15}$$

yields a solution of (2.12), if (2.15) can be solved for z.

(b) State and prove the extension of Lemma 2.1 to n dimensions.

(c) State and prove the extension of Theorem 2.1 which briefly says that the general integral of (2.12) is given by

$$F(u_1(x, z), \ldots, u_n(x, z)) = 0 \tag{2.16}$$

where $F(u_1, \ldots, u_n)$ is an arbitrary C^1 function of n variables and $u_1, \ldots, u_n$ are n functionally independent first integrals of (2.14). The general integral (2.16) implicitly defines most (but not necessarily all) solutions of (2.12).

2.6. For each of the following equations find the general integral and compute three different solutions.

(a) $$x_1z_{x_1} + x_1x_2z_{x_2} + x_1x_3z_{x_3} = z$$

(b) $$x_1z_{x_1} + x_1x_2z_{x_2} + zz_{x_3} = z.$$

3. The Initial Value Problem for Quasi-linear First Order Equations. Existence and Uniqueness of Solution

In this section we discuss the *initial value problem*, or *Cauchy problem*, for the first order quasi-linear p.d.e.

(3.1) $$P(x, y, z)z_x + Q(x, y, z)z_y = R(x, y, z).$$

Recall that the initial value problem for a first order ordinary differential equation asks for a solution of the equation which has a given value at a given point of R^1. The initial value problem for the partial differential equation (3.1) asks for a solution of (3.1) which has given values on a given curve in R^2. We first give a precise statement of the problem.

Initial Value Problem

Let C be a given curve in R^2 described parametrically by the equations

(3.2) $$x = x_0(t), \qquad y = y_0(t); \qquad t \in I$$

where $x_0(t)$, $y_0(t)$ are in $C^1(I)$. Let $z_0(t)$ be a given function in $C^1(I)$. The function $z_0(t)$ may be thought of as defining a function on the curve C. The initial value problem for equation (3.1) asks for a function $z = z(x, y)$ defined in a domain Ω of R^2 containing the curve C and such that:

(i) $z = z(x, y)$ is a solution of (3.1) in Ω.

(ii) On the curve C, z is equal to the given function z_0, i.e.

(3.3) $$z(x_0(t), y_0(t)) = z_0(t), \; t \in I.$$

The curve C is called the *initial curve* of the problem while the function z_0 is called the *initial data*. Equation (3.3) is called the *initial condition* of the problem.

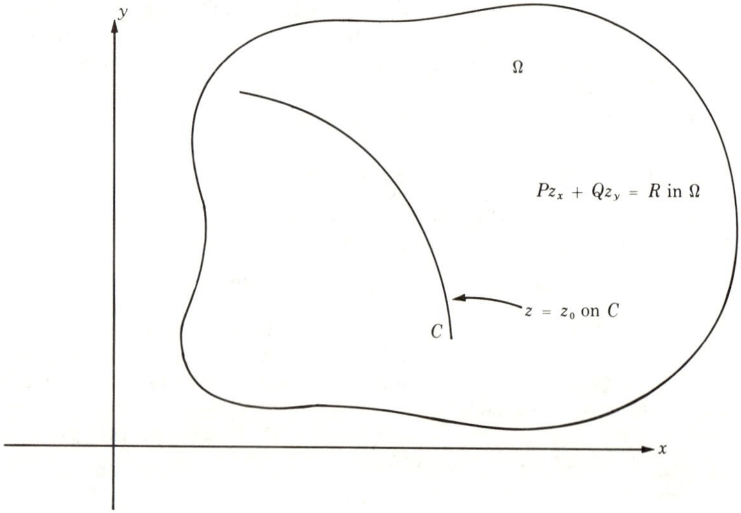

Fig. 3.1

If we view a solution $z = z(x, y)$ of (3.1) as a solution surface of (3.1), we can give a simple geometrical statement of the problem: Find a solution surface of (3.1) containing the given curve $\tilde{C}$ in R^3, described parametrically by the equations

$$x = x_0(t), \qquad y = y_0(t), \qquad z = z_0(t); \qquad t \in I. \tag{3.4}$$

The theorem that we state below asserts that under a certain condition we can solve this problem locally; i.e., we can find a unique solution of the problem in a neighborhood of any point of C at which a certain condition is satisfied. The solution can be found by using the method described in Section 4 of Chapter II for constructing an integral surface of the vector field $\mathbf{V} = (P, Q, R)$ containing a given curve.

Let (x_0, y_0, z_0) be a point of the curve $\tilde{C}$ corresponding to the parameter value $t = t_0 \in I$; i.e., $(x_0, y_0, z_0) = (x_0(t_0), y_0(t_0), z_0(t_0))$. Let $\tilde{\Omega}$ be a domain in R^3 containing (x_0, y_0, z_0) and let

$$u(x, y, z) = 0 \tag{3.5}$$

be an integral surface of the vector field $\mathbf{V} = (P, Q, R)$, or, equivalently, a solution surface of the equation

$$Pu_x + Qu_y + Ru_z = 0 \tag{3.6}$$

in $\tilde{\Omega}$ containing the part of $\tilde{C}$ in $\tilde{\Omega}$, i.e.

$$u(x_0(t), y_0(t), z_0(t)) = 0. \tag{3.7}$$

Suppose, furthermore, that

$$u_z(x_0, y_0, z_0) \neq 0. \tag{3.8}$$

Then, by Lemma 2.1, equation (3.5) implicitly defines a function $z = z(x, y)$ in a neighborhood U of (x_0, y_0), and this function is a solution of the initial value problem for (3.1) in U (see Fig. 3.2).

Combining the above observation with Theorem 4.2 of Chapter II we obtain the following basic theorem.

Theorem 3.1. Suppose that P, Q, R are of class C^1 in a domain $\tilde{\Omega}$ of R^3 containing the point (x_0, y_0, z_0) and suppose that

$$P(x_0, y_0, z_0)\frac{dy_0(t_0)}{dt} - Q(x_0, y_0, z_0)\frac{dx_0(t_0)}{dt} \neq 0. \tag{3.9}$$

Then in a neighborhood U of (x_0, y_0) there exists a unique solution of equation (3.1) satisfying the initial condition (3.3) at every point of C contained in U.

Proof. We note first that condition (3.9) implies that the vector $\mathbf{V} = (P, Q, R)$ is not tangent to the curve $\tilde{C}$ at the point (x_0, y_0, z_0) (why?). By Theorem 4.2 of Chapter II it follows that in a neighborhood of (x_0, y_0, z_0) there exits a unique integral surface of equation (3.6) containing the part of $\tilde{C}$ in this neighborhood. This integral surface can be written in the form (3.5). It remains to show that condition (3.8) is satisfied so that we can solve (3.5) for z. Condition (3.8) follows from condition (3.9). In fact, at the point (x_0, y_0, z_0), grad u is orthogonal to both $\mathbf{V}$ (by equation (3.6)) and

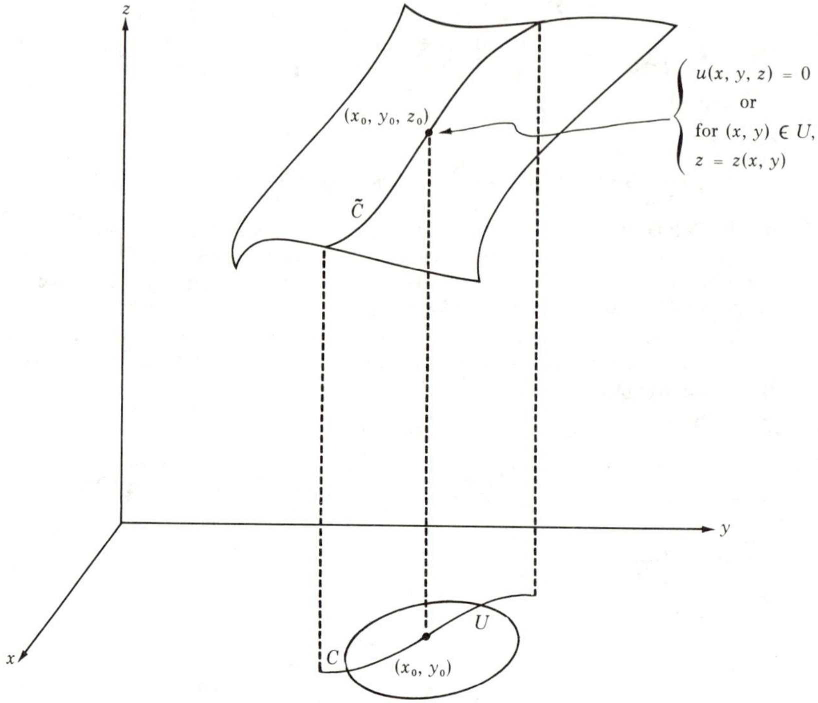

Fig. 3.2

to the tangent **T** to $\tilde{C}$ (by equation (3.7)). Hence, grad u is parallel to $\mathbf{V} \times \mathbf{T}$. Now, the left-hand side of (3.9) is precisely the z-component of $\mathbf{V} \times \mathbf{T}$ at (x_0, y_0, z_0). Hence, condition (3.9) implies that the z-component of grad u is different from zero at (x_0, y_0, z_0), which means that condition (3.8) is satisfied.

The uniqueness assertion of the theorem follows from the fact that every integral curve of **V** passing through any point of $\tilde{C}$ must lie on a solution surface of (3.1) containing $\tilde{C}$.

In geometrical language, condition (3.9) means that the projection of the vector $\mathbf{V}(x_0, y_0, z_0)$ on the (x, y)-plane is not tangent to the initial curve C at (x_0, y_0).

The method of construction of the solution to the initial value problem consists of viewing the initial condition as a given curve $\tilde{C}$ in R^3 and constructing, by the methods of Section 4, Chapter II, the integral surface of $\mathbf{V} = (P, Q, R)$ containing the curve $\tilde{C}$. Condition (3.9) of Theorem 3.1 guarantees that we can solve equation (3.5) of this integral surface for z in terms of x and y in a neighborhood U of the point (x_0, y_0). The size of the neighborhood U depends on the differential equation, on the initial curve C and on the initial data z_0. We illustrate the method of solution in the following example.

Example 3.1. Consider the quasi-linear equation

$$(y + z)z_x + yz_y = x - y. \tag{3.10}$$

Let the initial curve C be given by

$$y = 1, \qquad -\infty < x < \infty. \tag{3.11}$$

Find the solution $z = z(x, y)$ of (3.10) which on the initial curve C has the values

$$z = 1 + x. \tag{3.12}$$

First we express the initial condition (3.11)-(3.12) in parametric form. The curve C is given by

$$x = t, \qquad y = 1; \qquad -\infty < t < \infty \tag{3.13}$$

and on C the solution must take the values

$$z = 1 + t. \tag{3.14}$$

In geometrical terms the problem is to find the solution surface $z = z(x, y)$ of (3.10) containing the curve $\tilde{C}$ given by

$$x = t, \qquad y = 1, \qquad z = 1 + t; \qquad -\infty < t < \infty. \tag{3.15}$$

For equation (3.10) we have

$$\mathbf{V} = (P, Q, R) = (y + z, y, x - y),$$

and on the curve $\tilde{C}$,

$$P\frac{dy}{dt} - Q\frac{dx}{dt} = (1 + 1 + t)\,0 - 1 \times 1 = -1.$$

Thus, condition (3.9) is satisfied at every point of $\tilde{C}$ and by Theorem 3.1 we know that there exists a unique solution to the problem in a neighborhood of every point of C. We use the method described in Section 4, Chapter II, to find the solution. The system of equations associated with the vector field $\mathbf{V}$ is

$$\frac{dx}{y + z} = \frac{dy}{y} = \frac{dz}{x - y}.$$

This system was solved in Example 2.3 of Chapter II where we found the two first integrals

$$u_1 = \frac{x + z}{y}, \qquad u_2 = (x - y)^2 - z^2.$$

These first integrals are defined and are functionally independent in the domain $y > 0$ which contains the curve $\tilde{C}$. To find the integral surface of $\mathbf{V}$ containing $\tilde{C}$ we compute

$$U_1 = 1 + 2t, \qquad U_2 = -4t$$

and eliminating t we obtain

$$2U_1 - 2 + U_2 = 0.$$

The required integral surface is

$$2\frac{x+z}{y} - 2 + (x-y)^2 - z^2 = 0.$$

This equation has two solutions for z and in order to pick the one we want, we use the initial condition (3.11)-(3.12). Thus, we get

$$z = \frac{2}{y} + x - y. \tag{3.17}$$

It is left to the reader to verify that (3.17) satisfies the p.d.e (3.10) and the initial condition (3.11)-(3.12) and that it is therefore the required solution to the initial value problem. Note that the solution (3.17) is defined in the domain $y > 0$.

We close this section with an application of Theorem 3.1 to the following special initial value problem which arises frequently in applications,

$$P(x, y, z)z_x + z_y = R(x, y, z), \tag{3.18}$$

$$z(x, 0) = f(x), \tag{3.19}$$

where $f(x)$ is a given function defined for all $x \in R^1$. It is easy to verify that for this problem, condition (3.9) is always satisfied at every point of the initial curve, which in this case is the x-axis. Therefore, Theorem 3.1 yields the following existence and uniqueness result.

Corollary 3.1. Suppose that P and R are of class C^1 in R^3 and f is of class C^1 in R^1. Then in a neighborhood of every point of the x-axis there is a unique solution of the initial value problem (3.18), (3.19).

Problems

3.1. Solve the following initial value problems. Describe carefully the domain of the solutions.
  (a) $z_x + z_y = z$; $z = \cos t$ on the initial curve C: $x = t$, $y = 0$, $-\infty < t < \infty$.
  (b) $x^2 z_x + y^2 z_y = z^2$; $z = 1$ on the initial curve C: $y = 2x$.
  (c) $x(y - z)z_x + y(z - x)z_y = z(x - y)$; $z = t$ on the initial curve C: $x = t$, $y = 2t/(t^2 - 1)$, $0 < t < 1$.
  (d) $xz_x - yz_y = 0$; $z = x^2$ on the initial curve C: $y = x$, $x > 0$.
  (e) $yz_x - xz_y = 2xyz$; $z = t^2$ on the initial curve C: $x = t$, $y = t$; $t > 0$.
  (f) $xz_x + yz_y = z$; $z = 1$ on the initial curve C: $y = x^2$, $x > 0$.
  (g) $zz_x + yz_y = x$; $z = 2t$ on the initial curve C: $x = t$, $y = 1$; $-\infty < t < \infty$.

3.2. Answer the "why?" in the proof of Theorem 3.1.

3.3. Verify that for the problem (3.18), (3.19), condition (3.9) is always satisfied at every point of the initial line $y = 0$.

3.4. For each of the following two initial value problems

$$P(x, y, z)z_x + z_y = R(x, y, z), \qquad z(x, y_0) = f(x);$$
$$z_x + Q(x, y, z)z_y = R(x, y, z), \qquad z(x_0, y) = f(y);$$

formulate and prove existence and uniqueness results analogous to that stated in Corollary 3.1.

4. The Initial Value Problem for Quasi-linear First Order Equations. Nonexistence and Nonuniqueness of Solutions

In the previous section we proved the existence and uniqueness of solution of the initial-value problem for equation (3.1) in a neighborhood of any point (x_0, y_0) of the initial curve C at which condition (3.9) is satisfied. In geometrical language, condition (3.9) means that the projection of the vector $\mathbf{V}(x_0, y_0, z_0)$ on the (x, y)-plane is not tangent to the curve C at (x_0, y_0). In this section we will show that if condition (3.9) is violated, i.e., if

$$(4.1) \qquad P(x_0, y_0, z_0)\frac{dy_0(t_0)}{dt} - Q(x_0, y_0, z_0)\frac{dx_0(t_0)}{dt} = 0,$$

then usually there is no solution to the initial-value problem, and in a special case in which there is a solution, there are actually infinitely many solutions.

We assume here that P and Q do not vanish simultaneously. Note that condition (4.1) says that the components of the vectors $(P(x_0, y_0, z_0), Q(x_0, y_0, z_0))$ and $(dx_0(t_0)/dt, dy_0(t_0)/dt)$ are proportional, i.e.

$$(4.1)' \qquad \frac{\dfrac{dx_0(t_0)}{dt}}{P(x_0, y_0, z_0)} = \frac{\dfrac{dy_0(t_0)}{dt}}{Q(x_0; y_0, z_0)} = \mu$$

where μ is the constant of proportionality.

The reason why, under condition (4.1), there is usually no solution to the initial-value problem is that by using the p.d.e. (3.1) and condition (4.1) we can obtain information which can also be obtained from the initial condition (3.3). Thus the p.d.e. and the initial condition may contradict each other. Indeed, from the initial condition we know that at the point (x_0, y_0) the derivative of the solution along the initial curve C must be equal to $dz_0(t_0)/dt$. But this derivative must also be equal to

$$\begin{aligned} z_x(x_0, y_0)\frac{dx_0(t_0)}{dt} + z_y(x_0, y_0)\frac{dy_0(t_0)}{dt} &= \mu[P(x_0, y_0, z_0)z_x(x_0, y_0) + Q(x_0, y_0, z_0)z_y(x_0, y_0)] \\ &= \mu R(x_0, y_0, z_0), \end{aligned}$$

where in the first equality we used condition (4.1)′ and in the second equality we used the p.d.e. (3.1). Thus, if

$$(4.2) \qquad \frac{\dfrac{dz_0(t_0)}{dt}}{R(x_0, y_0, z_0)} \neq \mu$$

there cannot be a solution of the initial value problem. We have proved the following theorem.

Theorem 4.1. Under the conditions (4.1)′ and (4.2), there is no solution to the initial value problem (3.1)–(3.3) in any neighborhood of the point (x_0, y_0).

We want to emphasize here that in proving Theorem 4.1 we have shown that, under the condition (4.1), the left-hand side of the p.d.e. (3.1) evaluated at (x_0, y_0, z_0), is proportional to the derivative of z along the initial curve C (i.e., the directional derivative of z in the direction tangent to C). Since this derivative can also be computed from the initial data, there cannot be a solution to the initial value problem unless the two values agree.

In order to express the conditions of Theorem 4.1 geometrically, let $\mathbf{V}(t)$ denote the values of the vector field $\mathbf{V}$ on the curve $\tilde{C}$,

$$\mathbf{V}(t) = \mathbf{V}(x_0(t), y_0(t), z_0(t)), \qquad t \in I,$$

and let $\mathbf{T}(t)$ denote the tangent vector to $\tilde{C}$,

$$\mathbf{T}(t) = \left(\frac{dx_0(t)}{dt}, \frac{dy_0(t)}{dt}, \frac{dz_0(t)}{dt}\right).$$

Condition (4.1) means that the projection of $\mathbf{V}(t_0)$ on the (x, y)-plane is tangent to the initial curve C at (x_0, y_0). The conditions (4.1)′–(4.2) of Theorem 4.1 mean that the vectors $\mathbf{V}(t_0)$ and $\mathbf{T}(t_0)$ are not collinear whereas their projections on the (x, y)-plane are collinear (see Fig. 4.1). Alternatively, the conditions of Theorem 4.1 mean that $\mathbf{V}$ is not tangent to $\tilde{C}$ at (x_0, y_0, z_0) while its projection on the (x, y)-plane is tangent to C at (x_0, y_0).

If, under the conditions of Theorem 4.1, we try to find the solution to the initial-value problem by the method described in Section 3, we would

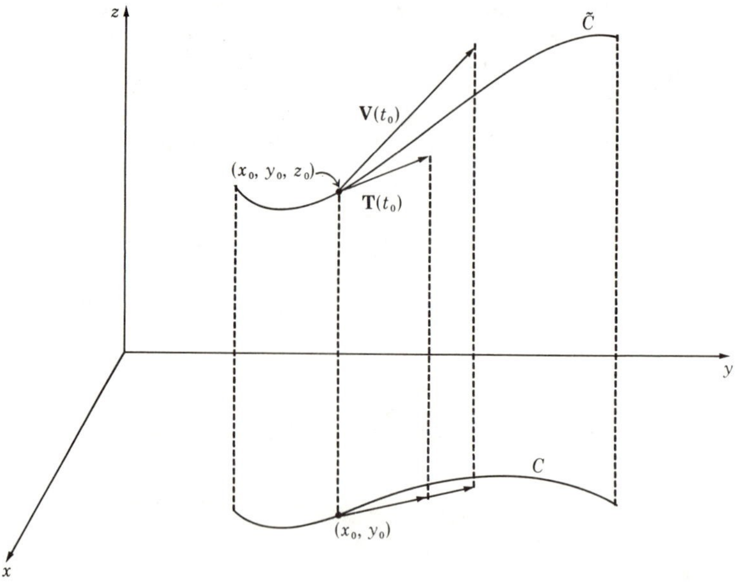

Fig. 4.1

find that the equation $u(x, y, z) = 0$ of the integral surface containing $\tilde{C}$ cannot be solved for z near (x_0, y_0, z_0) because $u_z(x_0, y_0, z_0) = 0$. This can be seen from Figure 4.1 since both vectors $\mathbf{V}(t_0)$ and $\mathbf{T}(t_0)$ must be tangent to $u(x, y, z) = 0$ at (x_0, y_0, z_0).

Let us suppose now that condition (4.1) holds and the p.d.e. and the initial conditions do not contradict each other, i.e.

$$\text{(4.3)} \qquad \frac{\dfrac{dx_0(t_0)}{dt}}{P(x_0, y_0, z_0)} = \frac{\dfrac{dy_0(t_0)}{dt}}{Q(x_0, y_0, z_0)} = \frac{\dfrac{dz_0(t_0)}{dt}}{R(x_0, y_0, z_0)} = \mu$$

or

$$\text{(4.3)}' \qquad \mathbf{T}(t_0) = \mu\mathbf{V}(t_0).$$

Here we want to consider only the special case in which condition (4.3) holds at every point of $\tilde{C}$ (or at least at every point of $\tilde{C}$ near (x_0, y_0, z_0)), i.e.

$$\text{(4.4)} \qquad \mathbf{T}(t) = \mu(t)\mathbf{V}(t), \qquad t \in I.$$

Condition (4.4) means that $\mathbf{V}$ is everywhere tangent to $\tilde{C}$, or that the curve $\tilde{C}$ is an integral curve of $\mathbf{V}$. The initial condition (3.3) requires that the desired solution surface of (3.1) passing through (x_0, y_0, z_0) must contain the integral curve of $\mathbf{V}$ passing through (x_0, y_0, z_0). But this requirement is always satisfied by every integral surface passing through (x_0, y_0, z_0) (recall Theorem 4.1 of Chapter II). Hence the initial condition is automatically satisfied and the initial value problem is reduced to the problem of finding a solution surface $z = f(x, y)$ of (3.1) passing through the point (x_0, y_0, z_0). But we can find infinitely many such solution surfaces by using the general integral (2.5). Thus we have proved the following theorem.

Theorem 4.2. Under the condition (4.4), the initial-value problem (3.1)–(3.3) has infinitely many solutions in a neighborhood of (x_0, y_0).

If, under the assumption (4.4), we try to find the solution of the initial-value problem by the method described in Section 3, we would find that there are infinitely many integral surfaces $u(x, y, z) = 0$ containing $\tilde{C}$ and since infinitely many of these are such that $u_z \neq 0$, we would obtain infinitely many solutions to the initial-value problem.

Problems

4.1. Consider the equation

$$zz_x + yz_y = x$$

and the initial curve

$$C\colon x = t, \qquad y = t; \qquad t > 0.$$

Decide whether there is a unique solution, there is no solution, or there are infinitely many solutions in a neighborhood of the point (1, 1), for each of the initial value problems with the following initial data:

(a) $z = 2t$ on C,
(b) $z = t$ on C,
(c) $z = \sin(\pi/2)t$ on C.

4.2. Consider the initial value problem

$$xz_x + yz_y = z;$$

$z = f(t)$ on the initial curve C: $x = t$, $y = t$; $t > 0$. Determine the class of functions $f(t)$ for which condition (4.4) is satisfied and hence there are infinitely many solutions to the problem. Find these solutions.

5. The Initial Value Problem for Conservation Laws. The Development of Shocks

Conservation laws are first order quasi-linear p.d.e.'s that arise in many physical applications (see Section 6 for examples). Let us consider the following initial value problem for a conservation law,

$$a(z)z_x + z_y = 0, \tag{5.1}$$

$$z(x, 0) = f(x), \tag{5.2}$$

where a and f are given C^1 functions. According to Corollary 3.1, this problem has a unique solution in a neighborhood of every point of the initial line $y = 0$. In order to find the solution we consider the system of o.d.e.'s associated with (5.1),

$$\frac{dx}{a(z)} = \frac{dy}{1} = \frac{dz}{0}.$$

Two functionally independent first integrals of this system are

$$u_1 = z, \qquad u_2 = x - a(z)y$$

and, therefore,

$$z = F(x - a(z)y)$$

is a general integral of (5.1). In order to satisfy the initial condition (5.2) we must take $F(x) = f(x)$. Thus, for sufficiently small $|y|$, the solution of (5.1), (5.2) is implicitly defined by the equation

$$z = f(x - a(z)y). \tag{5.3}$$

Using the implicit function theorem, it is easy to show (see Problem 5.1) that the solution of (5.1), (5.2) exists and is implicitly defined by (5.3) as long as the condition

$$1 + f'(x - a(z)y)a'(z)y > 0 \tag{5.4}$$

is satisfied. Note that (5.4) is always satisfied if $|y|$ is sufficiently small. By a solution of equation (5.1) we mean of course a differentiable function $z(x, y)$. From the formulas in Problem 5.1 we see that the derivatives z_x and z_y tend to infinity as the left side of (5.4) tends to zero. In fact when the left side of (5.4) becomes zero the solution develops a discontinuity known as a *shock*. The development of shocks is a well known phenome-

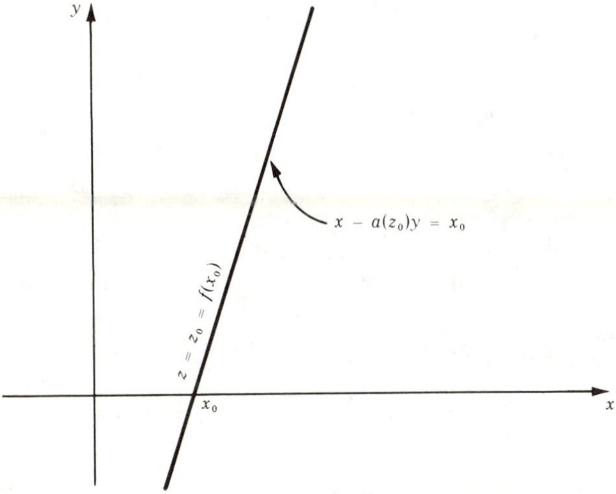

Fig. 5.1

non in gas dynamics. The mathematical analysis of shocks involves generalization of the concept of a solution of a p.d.e. to allow for discontinuities. It is also necessary to impose on the solution a certain condition to be satisfied across a discontinuity. (In gas dynamics this condition is known as the entropy condition, since it requires that the entropy of gas increase after crossing a discontinuity line.) In this book we do not go any further into the study of shocks. Instead we refer the interested student to the survey article by P. D. Lax.[3]

In order to visualize and compute the values of the solution defined implicitly by (5.3) and at the same time improve our understanding of the development of shocks, let us consider a fixed point x_0 of the x-axis and let $z_0 = f(x_0)$. Then the set of points (x, y, z) satisfying the pair of equations

$$x - a(z_0)y = x_0, \qquad z = z_0, \tag{5.5}$$

also satisfies equation (5.3). This means that the straight line in (x, y, z)-space defined by the pair of equations (5.5) lies on the surface defined by equation (5.3). It follows that along the line

$$x - a(z_0)y = x_0 \tag{5.6}$$

in the (x, y)-plane passing through the point $(x_0, 0)$, the solution z of the initial value problem (5.1), (5.2) is constant and equal to $z_0 = f(x_0)$ (see Fig. 5.1). In physical problems the variable y represents time and we are usually interested in the future behavior of the solution (after the initial instant $y = 0$). If no two lines of the form (5.6) intersect in the half-plane $y > 0$ we conclude that the solution exists as a differentiable function for all $y > 0$. If, however, two lines of the form (5.6) intersect when $y > 0$, then at the point of intersection we have an incompatibility since the solution cannot be equal to two distinct values. For example, let x_1 and x_2 be two points of the initial line $y = 0$, let $z_1 = f(x_1)$, $z_2 = f(x_2)$ and suppose that $a(z_1) > a(z_2)$. Then the lines

$$x - a(z_1)y = x_1, \qquad x - a(z_2)y = x_2,$$

intersect at the point (x_0, y_0) where

$$y_0 = \frac{x_2 - x_1}{a(z_1) - a(z_2)}$$

(see Fig. 5.2). At the point (x_0, y_0) we have an incompatibility since $z_1 \neq z_2$ and z cannot be equal to z_1 and z_2 at the same time. Hence the solution cannot exist as a differentiable function for $y \geq y_0$ and a shock has developed.

Lines of the form (5.6) are often called *characteristic lines* for the initial value problem (5.1), (5.2). (See Chapter V, Section 4.)

Example 5.1. The solution of the initial value problem

(5.7) $$zz_x + z_y = 0,$$

(5.8) $$z(x, 0) = -x,$$

exists and is implicitly defined by

(5.9) $$z = -(x - zy)$$

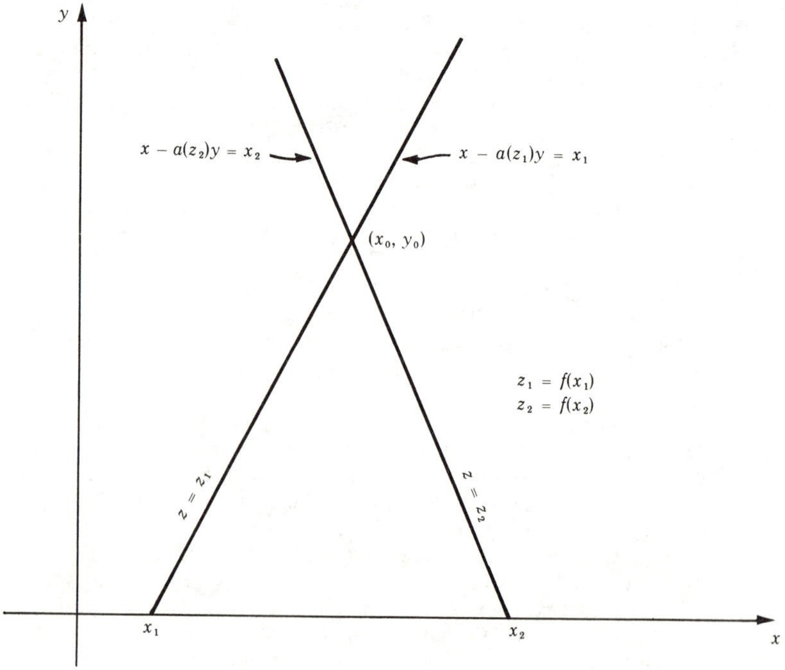

Fig. 5.2

as long as the condition

$$(5.10) \qquad 1 - y > 0,$$

is satisfied. In this case equation (5.9) can be easily solved for z,

$$(5.11) \qquad z = -\frac{x}{1 - y}, \qquad y < 1.$$

Clearly the solution breaks down and a shock is developed when $y = 1$. At a point x_0 of the x-axis, $z = z_0 = -x_0$, and the solution is constant and equal to $-x_0$ along the line

$$(5.12) \qquad x + x_0 y = x_0$$

passing through the point $(x_0, 0)$. Note that all lines (5.12) pass through the point $(0, 1)$.

Problems

5.1. Use the implicit function theorem to show that (5.3) implicitly defines z as a function of x and y provided that condition (5.4) is satisfied. Then, by implicit differentiation of (5.3) derive the formulas

$$z_x = \frac{f'(x - a(z)y)}{1 + f'(x - a(z)y)a'(z)y}, \qquad z_y = -\frac{a(z)f'(x - a(z)y)}{1 + f'(x - a(z)y)a'(z)y},$$

and verify that the function $z(x, y)$ implicitly defined by (5.3) satisfies the p.d.e. (5.1).

5.2. If $a(z) = a =$ constant, equation (5.1) is linear. Find the solution of (5.1), (5.2) in this case and show that shocks never develop. Also draw the lines in the (x, y)-plane along which the solution is constant.

5.3. Find the solution of (5.1), (5.2) if $f(x) = k =$ constant. Do shocks ever develop in this case?

5.4. Show that if the functions $a(z)$ and $f(x)$ are both nonincreasing or both nondecreasing, then the solution of (5.1), (5.2) exists and no shocks develop for $y \geqq 0$.

5.5. Find the solution of

$$zz_x + z_y = 0, \qquad z(x, 0) = x.$$

Draw the lines in the (x, y)-plane along which the solution is constant. Do shocks ever develop for $y \geq 0$?

5.6. For the problem

$$z^2 z_x + z_y = 0, \qquad z(x, 0) = x$$

derive the solution

$$z(x, y) = \begin{cases} x, & \text{when } y = 0 \\ \dfrac{\sqrt{1 + 4xy} - 1}{2y}, & \text{when } y \neq 0 \text{ and } 1 + 4xy > 0. \end{cases}$$

When do shocks develop? Use the Taylor series for $\sqrt{1 + \epsilon}$ about $\epsilon = 0$ to verify that $\lim_{y \to 0} z(x, y) = x$.

5.7. For the problem

$$zz_x + z_y = 0 \qquad z(x, 0) = x^2$$

derive the solution

$$z(x, y) = \begin{cases} x^2, & \text{when} \quad y = 0 \\ \dfrac{1 + 2xy - \sqrt{1 + 4xy}}{2y^2}, & \text{when} \quad y \neq 0 \quad \text{and} \quad 1 + 4xy > 0. \end{cases}$$

When do shocks develop? Use the Taylor series for $\sqrt{1 + \epsilon}$ about $\epsilon = 0$ to verify that $\lim_{y \to 0} z(x, y) = x^2$.

5.8. In the formulas for the solutions of Problems 5.6 and 5.7, "rationalize" the numerator to obtain formulas that show immediately that the solutions are continuous on the initial line $y = 0$.

5.9. Consider the initial value problem

$$a_1(z)z_{x_1} + \cdots + a_n(z)z_{x_n} + z_y = 0$$

$$z(x_1, \ldots, x_n, 0) = f(x_1, \ldots, x_n).$$

Derive the formula

$$z = f(x_1 - a_1(z)y, \ldots, x_n - a_n(z)y)$$

which implicitly defines the solution of the problem.

6. Applications to Problems in Traffic Flow and Gas Dynamics

We present here two applications of our analysis of the initial value problem for conservation laws. These laws arise in the study of many physical problems and, in particular, in the study of the flow of nonviscous compressible fluids. A conservation law asserts that the change in the total amount of a physical entity contained in any region of space is due to the flux of that entity across the boundary of that region. The first application deals with a conservation law which arises in the study of traffic flow along a highway. The second application deals with the one-dimensional, time-dependent flow of a compressible fluid under the assumption of constant pressure.

Traffic Flow Along a Highway

The model of traffic flow that we discuss here is based on the assumption that the motion of individual cars can be considered analogous to the flow of a continuous fluid. We take the x-axis along the highway and assume that the traffic flows in the positive direction. Let $\rho = \rho(x, t)$ denote the density (cars per unit length) of the traffic at the point x of the highway at time t, and let $q = q(x, t)$ denote the flow rate (cars per unit time) at which cars flow past the point x at time t. We derive a relation between ρ and q under the assumptions that cars do not enter or leave the highway at any one of its points and $\rho(x, t)$ and $q(x, t)$ are C^1 functions of x and t. Let $[x_1, x_2]$ with $x_2 > x_1$ be any segment of the highway. The total number of cars in this segment is given by

$$\int_{x_1}^{x_2} \rho\,(x, t)dx$$

and the time rate of change of the number of cars in the segment is

$$\frac{d}{dt}\int_{x_1}^{x_2} \rho(x, t)dx = \int_{x_1}^{x_2} \frac{\partial \rho}{\partial t}(x, t)dx.$$

This rate of change must also be equal to

$$q(x_1, t) - q(x_2, t)\,,$$

which measures the time rate of cars entering the segment at x_1 minus the time rate of cars leaving the segment at x_2. Therefore

$$\int_{x_1}^{x_2} \frac{\partial \rho}{\partial t}\,(x, t)dx = q(x_1, t) - q(x_2, t),$$

or

$$\int_{x_1}^{x_2} \frac{\partial \rho}{\partial t}(x, t)dx = -\int_{x_1}^{x_2} \frac{\partial q}{\partial x}(x, t)dx,$$

or

(6.1) $$\int_{x_1}^{x_2} \left[\frac{\partial \rho}{\partial t}\left(x, t\right) + \frac{\partial q}{\partial x}\left(x, t\right)\right]dx = 0.$$

Since the integrand in (6.1) is continuous and since (6.1) holds for every interval $[x_1, x_2]$ it follows that the integrand itself must vanish (see Problem 2.1 of Chapter VI),

(6.2) $$\frac{\partial \rho}{\partial t} + \frac{\partial q}{\partial x} = 0.$$

We now introduce in our analysis an additional assumption, the validity of which is supported by theoretical considerations as well as experimental data. According to this assumption, the flow rate q depends on x and t only through the traffic density ρ, i.e.

$$q(x, t) = G(\rho(x, t)),$$

or, simply,

(6.3) $$q = G(\rho),$$

for some function G. This assumption seems reasonable in view of the fact that the density of vehicles surrounding a given vehicle indeed controls the speed of that vehicle. The relationship between q and ρ depends on many factors such as road characteristics, weather conditions, speed limits, etc. In the book of Haight[4] various functional relations between q and ρ are discussed and compared. Some of these relations are derived from theoretical considerations while others are derived directly from experimental data. The relation

$$q = c\rho\left(1 - \frac{\rho}{\rho_1}\right) \tag{6.4}$$

is suggested by experimental data and is the relation that we will use here. ρ_1 is the maximum car density of the road (cars per unit length when traffic is bumper to bumper) and c is the mean free speed; i.e., the mean value of the free speeds of the cars on the highway (the free speed of a car is the speed at which it travels whenever it is free from interference from other cars). Normally, c is approximately equal to the speed limit of the highway. Note that according to (6.4) $q = 0$ when $\rho = 0$ and when $\rho = \rho_1$.

We now substitute (6.4) into (6.2) to obtain the equation

$$\frac{\partial\rho}{\partial t} + c\left(1 - 2\frac{\rho}{\rho_1}\right)\frac{\partial\rho}{\partial x} = 0. \tag{6.5}$$

Equation (6.5) can be simplified by dividing it by ρ_1 and introducing the normalized density $d = \rho/\rho_1$ to obtain

$$\frac{\partial d}{\partial t} + c(1 - 2d)\frac{\partial d}{\partial x} = 0. \tag{6.6}$$

Equation (6.6) is an example of a conservation law. If the initial normalized density is given by

$$d(x, 0) = f(x), \tag{6.7}$$

then, according to Section 5, the solution of the initial value problem (6.6), (6.7) is implicitly defined, for sufficiently small t, by the equation

$$d = f(x - ct(1 - 2d)). \tag{6.8}$$

In fact, if f is a C^1 function, the solution exists as a C^1 function and is implicitly defined by (6.8) as long as the condition

$$1 - 2ctf'(x - ct(1 - 2d)) > 0 \tag{6.9}$$

is satisfied. If this condition ever fails, shocks are developed in the sense that the derivatives of the car density become infinite and the density develops a jump discontinuity, as discussed in Section 5. If $f'(x) \leqq 0$ for all x, condition (6.9) is satisfied for all $t \geqq 0$. This leads to the (expected) conclusion that if the initial car density is constant or decreasing in the direction of traffic flow no shocks ever develop and the traffic continues to flow smoothly. However, if the initial car density is increasing over any length of the highway, a shock eventually develops as the following example illustrates.

Suppose that the initial car density is given by the function

$$f(x) = \begin{cases} 1/3, & \text{for } x \leq 0, \\ 1/3 + (5/12)x, & \text{for } 0 \leq x \leq 1, \\ 3/4, & \text{for } 1 \leq x, \end{cases} \tag{6.10}$$

the graph of which is shown in Figure 6.1. The derivative of this function has a jump at $x = 0$ and $x = 1$ and, strictly speaking, our theory does not apply since we required that $f(x)$ is C^1 for all x. We could of course smooth out $f(x)$ near $x = 0$ and $x = 1$ by replacing the corners in the graph of $f(x)$ by smoothly turning curves. However this smoothing would introduce a lot of difficulty in the computation of the solution of our problem. Fortunately, it turns out that the only effect of a jump in the derivative of the initial data is a jump in the derivative of the solution across a line in the (x, t)-plane. The solution is still implicitly defined, for sufficiently small t, by equation (6.8). In order to evaluate the solution we will use the fact that the solution is constant along certain lines on the (x, t)-plane (see Section 5). Since the time variable t is always multiplied by the free speed c, it is convenient to use ct in place of t in our computations. If $x_0 \leq 0$, then $d = d_0 = f(x_0) = 1/3$ along the line $x - ct(1 - 2d_0) = x_0$, or

(6.11) $$d = 1/3 \text{ along the lines } ct = 3(x - x_0),\ x_0 \leq 0.$$

Similarly,

(6.12) $$d = 3/4 \text{ along the lines } ct = -2(x - x_0),\ 1 \leq x_0.$$

In particular $d = 1/3$ along the line $ct = 3x$, while $d = 3/4$ along the line $ct = -2(x - 1)$. Since these two lines intersect at the point $(x, ct) = (2/5, 6/5)$, a shock appears at that point of the (x, ct)-plane as shown in Figure 6.2. If $0 \leq x_0 \leq 1$, then $d = d_0 = 1/3 + (5/12)x_0$ along the line $x - ct[1 - 2(1/3 + (5/12)x_0)] = x_0$, or

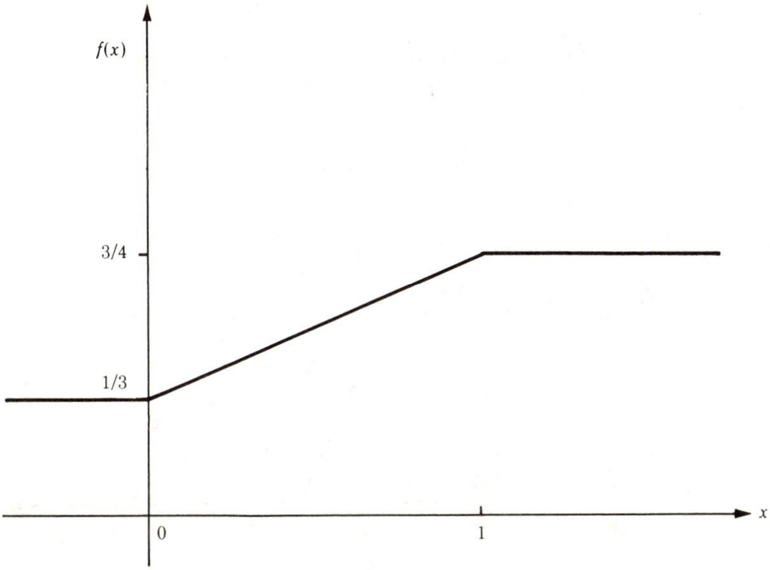

Fig. 6.1

(6.13) $d = {}^{1}/_{3} + ({}^{5}/_{12})x_0$ along the line $ct = \frac{6}{5}\frac{x - x_0}{\frac{2}{5} - x_0}$, $0 \le x_0 \le 1$.

Note that all the lines (6.13) pass through the point $(x, ct) = ({}^{2}/_{5}, {}^{6}/_{5})$ as shown in Figure 6.2. The lines $ct = 3x$ and $ct = -2(x - 1)$ divide the upper half of the (x, ct)-plane into four regions. In the "left" region, $d = {}^{1}/_{3}$, and in the "right" region $d = {}^{3}/_{4}$. In the "triangular" region with vertices (0, 0), (1, 0) and $({}^{2}/_{5}, {}^{6}/_{5})$, d is obtained from (6.13). In fact eliminating x_0 from (6.13) yields

$$(6.14)\quad d = \frac{1}{3} + \frac{5}{12}\frac{6x - 2ct}{6 - 5ct}, \qquad 0 \le \frac{6x - 2ct}{6 - 5ct} \le 1, \qquad 0 \le ct \le \frac{6}{5}.$$

Finally in the "shock" region the solution has a jump discontinuity and the values of the solution cannot be computed from our analysis. Figure 6.3 shows graphs of d versus x for four values of ct.

For other examples and further discussion of the development of shocks in traffic flow see the article by Richards.[5]

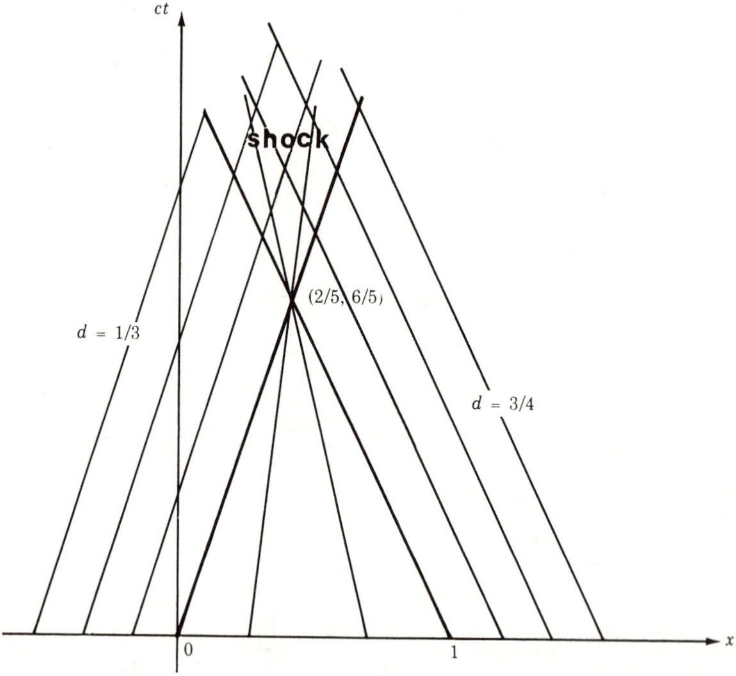

Fig. 6.2

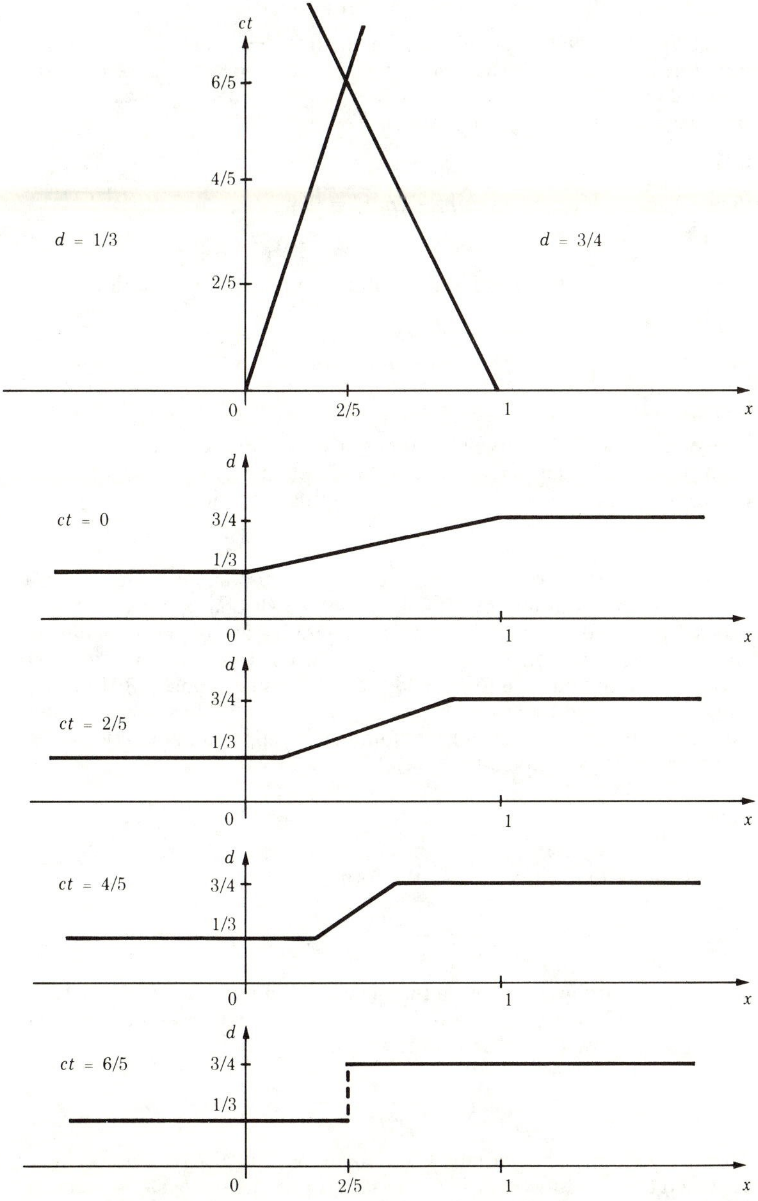
ct
6/5
4/5
2/5
d = 1/3
d = 3/4
0
2/5
1
x
d
ct = 0
3/4
1/3
0
1
x
d
3/4
ct = 2/5
1/3
0
1
x
d
ct = 4/5
3/4
1/3
0
1
x
d
ct = 6/5
3/4
1/3
0
2/5
1
x

Fig. 6.3

Compressible Fluid Flow under Constant Pressure

Let us consider the one-dimensional, time-dependent flow of a compressible fluid under the assumption of constant pressure p. If u denotes the fluid velocity, ρ the density and e the internal energy per unit volume, the basic equations of gas dynamics are

$$u_t + uu_x = 0, \tag{6.15}$$

$$\rho_t + (\rho u)_x = 0, \tag{6.16}$$

$$e_t + (eu)_x + pu_x = 0. \tag{6.17}$$

We want to solve these equations subject to the initial conditions

$$u(x, 0) = f(x) \tag{6.18}$$

$$\rho(x, 0) = g(x) \tag{6.19}$$

$$e(x, 0) = h(x) \tag{6.20}$$

where f, g and h are given C^1 functions. According to Section 5, the solution of the initial value problem (6.15), (6.18) always exists for sufficiently small t and is defined implicitly by the equation

$$u = f(x - ut). \tag{6.21}$$

If $f'(x) \geqq 0$ for all x, the solution exists as a C^1 function for all $t \geqq 0$. Otherwise the solution eventually develops discontinuities known as shocks, the study of which involves generalization of the concept of a solution (see Noh and Protter[6] for details). Once u is known, it can be substituted into equation (6.16) and the initial value problem (6.16), (6.19) can then be solved to obtain the density ρ. It is useful to obtain a formula for ρ in terms of u. To do this we note that u_x appears in equation (6.16), and from (6.21) we have,

$$u_x = \frac{f'(x - ut)}{1 + tf'(x - ut)}. \tag{6.22}$$

This suggests that a function of the form

$$\rho = \frac{G(x - ut)}{1 + tf'(x - ut)} \tag{6.23}$$

might be a solution of equation (6.16) (see also Problem 6.5). In order for (6.23) to satisfy the initial condition (6.19), the function G must be taken to be g. It is now left as an exercise (Problem 6.6) to show that

$$\rho = \frac{g(x - ut)}{1 + tf'(x - ut)} \tag{6.24}$$

satisfies not only the initial condition (6.19) but also the p.d.e. (6.16) provided that the function f is C^2. In view of our uniqueness theorem (see Section 5) concerning the solution of the initial value problem (6.16), (6.19), we conclude that the solution of this problem must be given by (6.24). Similarly, the solution of the initial value problem (6.17), (6.20) is given by

(6.25)
$$e = \frac{h(x - ut) + p}{1 + tf'(x - ut)} - p$$

(See Problem 6.7.)

Problems

6.1. Consider the traffic flow problem (6.6), (6.7) with

(6.26)
$$f(x) = \begin{cases} 0, & \text{for } x \le 0, \\ x/L, & \text{for } 0 \le x \le L, \\ 1, & \text{for } L \le x. \end{cases}$$

For this problem draw a figure like Figure 6.2 and obtain the solution,

$$d = \begin{cases} 0 & \text{in the ``left'' region} \\ \dfrac{x - ct}{L - 2ct} & \text{in the ``triangular'' region} \\ 1 & \text{in the ``right'' region.} \end{cases}$$

6.2. Consider the traffic problem (6.6), (6.7) and suppose that $f'(x)$ has a maximum positive value at some point x_1 of the x-axis,

$$f'(x_1) = \max_{-\infty<x<\infty} f'(x).$$

Then, as discussed in the text, a shock eventually develops at some positive time t. Let t_s be the positive time and x_s the location of the first appearance of the shock (t_s is sometimes known as the break-down or critical time). To compute t_s and x_s we use the fact that the shock first appears when the expression

$$1 - 2ctf'(x - ct(1 - 2d))$$

first becomes zero.

(a) Show that

$$f'(x - ct(1 - 2d)) = f'(x_1)$$

along the line

$$x - ct(1 - 2f(x_1)) = x_1$$

in the (x, ct)-plane, and that therefore (x_s, ct_s) must lie on this line.

(b) Show that

$$ct_s = \frac{1}{2f'(x_1)}$$

and

$$x_s = x_1 + \frac{1}{2f'(x_1)}(1 - 2f(x_1)).$$

(c) Refer to Figure 6.4 and show that

$$x_s = x_0 + ct_s.$$

6.3. Use the formulas of Problem 6.2 to compute x_s and ct_s for the traffic flow problem (6.6), (6.7) with initial car density given by (a) equation (6.10); (b) equation (6.26).

6.4. Consider the traffic flow problem (6.6), (6.7) with

$$f(x) = \begin{cases} 0, & \text{for } x \leq 0, \\ x^2, & \text{for } 0 \leq x \leq 1, \\ 1, & \text{for } 1 \leq x. \end{cases}$$

(a) Refer to Problem 6.2 and compute x_s and ct_s.
(b) Find d for $0 \leq ct < ct_s$ and $-\infty < x < \infty$.

6.5. Show that if u is a solution of (6.15) then $v = u_x$ is a solution of the equation $v_t + (uv)_x = 0$, which is equation (6.16) with ρ replaced by v. In view of (6.22), this also suggests (6.23) as a possible form of a solution of (6.16).

6.6. Show by direct substitution that (6.24) satisfies the p.d.e. (6.16) provided that f is a C^2 function.

6.7. Show that if $\epsilon = e + p$ (p is constant), then ϵ satisfies the equation $\epsilon_t + (\epsilon u)_x = 0$ which is of the same form as equation (6.16). Then use (6.23) to obtain the solution (6.25) of (6.17), (6.20).

6.8. Consider the initial value problem for gas dynamics (6.15), (6.18),

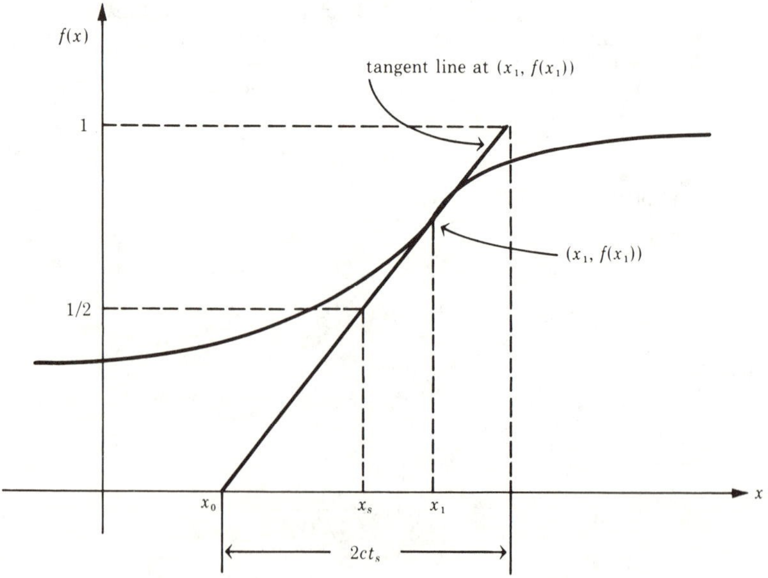

Fig. 6.4

with initial data given by

$$f(x) = \begin{cases} 1 & \text{for } x \leq 0\,, \\ 1 - x & \text{for } \ 0 \leq x \leq 1, \\ 0 & \text{for } 1 \leq x. \end{cases}$$

Follow the analysis of the traffic flow problem in the text to obtain the equation of the lines in the upper half (x, t)-plane along which the solution u is constant. Draw a figure like Figure 6.2 and obtain the solution

$$u = \begin{cases} 1 & \text{in the region } x < 1,\, 0 \leq t,\, x < t \\ \dfrac{1 - x}{1 - t} & \text{in the region } \ 0 \leq x \leq 1,\, 0 \leq t,\, t \leq x \\ 0 & \text{in the region } 1 < x,\, 0 \leq t,\, t < x. \end{cases}$$

Draw a figure like Figure 6.3 with graphs of u versus x at $t = 0$, $^1/_4$, $^1/_2$, $^3/_4$, 1.

6.9. Consider an infinitely long cylindrical pipe containing a fluid. Let the x-axis be along the axis of the pipe and suppose that the fluid is flowing in the positive x-direction. Let $\rho(x, t)$ and $q(x, t)$ be, respectively, the density (mass per unit pipe length) and rate of flow (mass per unit time) of the fluid at position x and time t. Suppose also that the walls of the pipe are composed of porous material allowing the fluid to leak. Let $H(x, t)$ denote the rate (mass per unit pipe length, per unit time) at which the fluid leaks out of the pipe.

(a) Follow the derivation of the traffic flow conservation law (6.2) to derive the equation

$$\frac{\partial \rho}{\partial t} + \frac{\partial q}{\partial x} = -H.$$

(b) Assume that q and H actually are functions of ρ given by

$$q = \frac{1}{2}\rho^2, \qquad H = \alpha\rho^2$$

where α is a positive constant. Obtain the p.d.e. for ρ,

$$\frac{\partial \rho}{\partial t} + \rho\frac{\partial \rho}{\partial x} = -\alpha\rho^2$$

and its first integrals,

$$u_1 = \rho e^{\alpha x}, \qquad u_2 = \frac{\rho}{1 - \alpha\rho t}.$$

If $\rho(x, 0) = 1$ show that

$$\rho(x, t) = \frac{1}{1 + \alpha t}.$$

6.10. For the initial value problem for gas dynamics (6.15), (6.18), show that if the initial velocity distribution $f(x)$ is nondecreasing, shocks never develop for $t \geq 0$. On the other hand, if $f'(x) < 0$ over some interval of the x-axis, a shock eventually develops at some positive t. Follow the analysis of Problem 6.2 to derive formulas for the time and location of the first appearance of a shock.

7. The Method of Probability Generating Functions. Applications to a Trunking Problem in a Telephone Network and to the Control of a Tropical Disease

We discuss in this section an important application of first order linear partial differential equations to some problems in probability, namely to problems that arise in the study of certain processes known as stochastic processes. The material can be understood by students without background in probability. In the text we discuss two applications of the method of probability generating functions and in the problems we describe applications to Poisson, Yule, Polya, birth and death, and other stochastic processes. The main references for the material of this section are Feller[7] Chapter XVII, Sections 5–7 and Chiang,[8] Chapters 2 and 3.

A Trunking Problem in a Telephone Network

We consider an idealized telephone network consisting of an infinite number of lines (trunklines), and assume that calls originate and terminate within the network during the time interval $[0, \infty)$ according to certain hypotheses which we describe below. The problem to be solved is the following. Given any non-negative integer n, find the probability $P_n(t)$ that exactly n lines are in use at time t, $0 < t < \infty$, assuming that the initial probabilities $P_n(0)$, $0 \leq n < \infty$, are known.

In stating the hypotheses concerning the initiation (birth) and termination (death) of phone calls within the network we will use the symbol $o(h)$ to denote any quantity which vanishes more rapidly than h as $h \to 0$; i.e., $\lim_{h \to 0} [o(h)/h] = 0$. The reasonableness and validity of the hypotheses are discussed in the book of Feller.[7] The hypotheses are: (i) if a line is occupied at time t, the probability of the conversation ending during the time interval $(t, t + h)$ is $\mu h + o(h)$, where μ is a constant; (ii) the probability of a call starting during the interval $(t, t + h)$ is $\lambda h + o(h)$, where λ is a constant; and (iii) the probability of two or more changes occurring (calls starting or ending) during the interval $(t, t + h)$ is $o(h)$.

The first step in the determination of the probabilities $P_n(t)$ consists of deriving a system of ordinary differential equations that are satisfied by $P_n(t)$. Let us suppose for a moment that t is fixed and that the probabilities $P_n(t)$ are known for all n, $0 \leq n < \infty$, and let us try to determine $P_n(t + h)$, the probability that n lines are in use at time $t + h$. Suppose first that $n \geq 1$. A moment's reflection should convince the reader that there will be n lines in use at time $t + h$ only if one of the following conditions is satisfied: (1) at time t, $n - 1$ lines are in use and one call originates during the time interval $(t, t + h)$; (2) at time t, $n + 1$ lines are in use and one call terminates during the interval $(t, t + h)$; (3) at time t, n lines are in use and no change occurs in the network during the interval $(t, t + h)$; and (4) two or more changes occur during the interval $(t, t + h)$. According to our hypotheses, the probability of the last event (4) is $o(h)$ while the probability of (1) is

$$[\lambda h + o(h)]P_{n-1}(t),$$

the probability of (2) is

$$(n + 1)[\mu h + o(h)]P_{n+1}(t),$$

and the probability of (3) is

$$[1 - \lambda h - n\mu h - o(h)]P_n(t).$$

Since the contingencies (1), (2) and (3) are mutually exclusive, their probabilities add. Therefore

$$\begin{aligned} P_n(t + h) &= \lambda h P_{n-1}(t) + (n + 1)\mu h P_{n+1}(t) \\ &\quad + (1 - \lambda h - n\mu h)P_n(t) + o(h). \end{aligned} \tag{7.1}$$

Using (7.1) to form the difference quotient $[P_n(t + h) - P_n(t)]/h$ and letting $h \to 0$, we obtain the ordinary differential equations

$$P_n'(t) = -(\lambda + n\mu)P_n(t) + \lambda P_{n-1}(t) + (n + 1)\mu P_{n+1}(t), \tag{7.2}$$

which must hold for all $n \geq 1$ and $0 < t < \infty$. For $n = 0$ a similar analysis leads to the equation

$$P_0'(t) = -\lambda P_0(t) + \mu P_1(t). \tag{7.3}$$

Since the initial probabilities $P_n(0)$, $0 \leq n < \infty$, are assumed to be known, the problem of finding the probabilities $P_n(t)$ for all $t > 0$ has been reduced to the initial value problem for the infinite system of ordinary differential equations (7.2), (7.3). The question of existence and uniqueness of solution of this initial value problem is not easy; the interested student can read the discussion and references cited in Feller.[7] Here, we describe a method for finding the solution of the problem by solving an initial value problem for a first order linear partial differential equation.

The function

$$G(t,s) = \sum_{n=0}^{\infty} P_n(t)s^n \tag{7.4}$$

is known as the *probability generating function* for the probabilities $P_n(t)$. As a consequence of the system of o.d.e.'s (7.2) it is easy to show that $G(t, s)$ must satisfy a linear first order partial differential equation. In fact, differentiating (7.4) we get

$$\frac{\partial G}{\partial s} = \sum_{n=1}^{\infty} nP_n(t)s^{n-1} = \sum_{n=0}^{\infty} (n + 1)P_{n+1}(t)s^n, \tag{7.5}$$

$$\frac{\partial G}{\partial t} = \sum_{n=0}^{\infty} P_n'(t)s^n. \tag{7.6}$$

Substitution of the expressions (7.2), (7.3) for $P_n'(t)$ into (7.6), followed by rearrangement and identification of the resulting series with the series in (7.4) and (7.5) yields the p.d.e. for G,

$$\frac{\partial G}{\partial t} + \mu(s - 1)\frac{\partial G}{\partial s} = \lambda(s - 1)G. \tag{7.7}$$

On the other hand, knowledge of the initial probabilities $P_n(0)$ leads to the initial condition for G along the line $t = 0$ of the (t, s)-plane,

$$G(0,s) = g(s), \tag{7.8}$$

where

$$g(s) = \sum_{n=0}^{\infty} P_n(0)s^n. \tag{7.9}$$

It is easy to obtain the solution of the initial value problem (7.7), (7.8). The associated system of o.d.e.'s of (7.7) is

$$\frac{dt}{1} = \frac{ds}{\mu(s - 1)} = \frac{dG}{\lambda(s - 1)G},$$

and two functionally independent first integrals are

$$u_1 = e^{-\mu t}(s - 1), \qquad u_2 = e^{-\frac{\lambda}{\mu}s}G. \tag{7.10}$$

Since u_1 does not depend on G, the general integral of (7.7) is

$$u_2 = f(u_1)$$

where f is an arbitrary C^1 function of a single variable. Substituting (7.10) in the general integral and solving for G we obtain the solutions of (7.7),

$$G(t, s) = e^{\frac{\lambda}{\mu}s} f(e^{-\mu t}(s - 1)). \tag{7.11}$$

The initial condition (7.8) determines the function f. In fact setting $t = 0$ in (7.11) and using (7.8) yields

$$g(s) = e^{\frac{\lambda}{\mu}s} f(s - 1),$$

and, consequently,

$$f(s) = g(s + 1)e^{-\frac{\lambda}{\mu}(s+1)}. \tag{7.12}$$

Finally, substituting (7.12) into (7.11) and simplifying we obtain the solution of the initial value problem (7.7), (7.8)

$$G(t, s) = g(1 + e^{-\mu t}(s - 1)) \exp\left[\frac{\lambda}{\mu}(s - 1)(1 - e^{-\mu t})\right]. \tag{7.13}$$

Once the probability generating function $G(t, s)$ has been found, the probabilities $P_n(t)$ can be found from the familiar formula for the coefficients of the Taylor series (7.4),

$$P_n(t) = \frac{1}{n!}\left[\frac{\partial^n}{\partial s^n} G(t, s)\right]_{s=0} \tag{7.14}$$

or by obtaining the series expansion of $G(t, s)$ in powers of s by some other means and then identifying the coefficients of the series with $P_n(t)$.

In order to illustrate the method of probability generating functions (p.g.f.), let us suppose that exactly one line is in use at time $t = 0$. This means that

$$P_1(0) = 1 \quad \text{and} \quad P_n(0) = 0 \quad \text{for} \quad n \neq 1, \tag{7.15}$$

and, therefore,

$$g(s) = \sum_{n=0}^{\infty} P_n(0)s^n = s. \tag{7.16}$$

Substitution of (7.16) into (7.13) yields the p.g.f.

$$G(t, s) = [1 + e^{-\mu t}(s - 1)] \exp\left[\frac{\lambda}{\mu}(s - 1)(1 - e^{-\mu t})\right]. \tag{7.17}$$

The probabilities $P_n(t)$ can be determined using formula (7.14). For $n = 0$ and $n = 1$ we have

$$P_0(t) = G(t, 0) = (1 - e^{-\mu t}) \exp\left[\frac{\lambda}{\mu}(e^{-\mu t} - 1)\right],$$

$$P_1(t) = \frac{\partial G}{\partial s}(t, 0) = \left[e^{-\mu t} + \frac{\lambda}{\mu}(1 - e^{-\mu t})^2\right] \exp\left[\frac{\lambda}{\mu}(e^{-\mu t} - 1)\right].$$

Clearly the computational labor to obtain $P_n(t)$ increases rapidly with n. An alternate method for obtaining $P_n(t)$ directly from (7.17) is outlined in Problem 7.4.

A more realistic model of a telephone network with a finite number of lines can be analyzed in a similar way. For details see Feller[7] Chapter XVII, Section 7.

A Problem in the Control of a Tropical Disease

Schistosomiasis is a parasitic infection that is estimated to affect more than two hundred million people in tropical and subtropical countries of the world. It is characterized by long term debility which is thought by many to be a significant obstacle to the advancement of many underdeveloped countries where large segments of the population are more or less permanently infected. The persistence of this infection in a locality depends on a complex cycle of events involving humans, certain parasitic flatworms (schistosomes) and particular species of snails. A probabilistic study of this cycle of events has been carried out in a paper by Nasell and Hirsch.[9] The results of this study make possible the comparison of the relative effectiveness of various procedures aimed at control or eradication of the disease. We present here a problem that appears in the paper[9] concerning the determination of a certain probability generating function.

The probability generating function $G(t, s)$ must satisfy the p.d.e.

$$\frac{\partial G}{\partial t} + \mu(s - 1)\frac{\partial G}{\partial s} = \frac{1}{2}\nu Y(t)(s - 1)G \tag{7.18}$$

and the initial condition

$$G(t_0, s) = s^m \tag{7.19}$$

along the line $t = t_0$ of the (t, s) plane. $Y(t)$ is a given continuous function, μ and ν are constants and m is a nonnegative integer. It is an easy exercise (see Problem 7.8) to obtain the first integrals of (7.18),

$$u_1 = e^{-\mu t}(s - 1), \qquad u_2 = Ge^{-\frac{1}{2}\beta(t)(s-1)}, \tag{7.20}$$

where

$$\beta(t) = e^{-\mu t}\int_0^t Y(\tau)e^{\mu\tau}\, d\tau.$$

Now, the general integral of (7.18) is

$$Ge^{-\frac{1}{2}\beta(t)(s-1)} = f(e^{-\mu t}(s - 1)), \tag{7.21}$$

where f is an arbitrary C^1 function. Solving (7.21) for G we obtain the solutions of (7.18),

$$G(t, s) = e^{\frac{1}{2}\beta(t)(s-1)} f(e^{-\mu t}(s - 1)). \tag{7.22}$$

The initial condition (7.19) determines the function f since it requires that

$$s^m = e^{\frac{1}{2}\beta(t_0)(s-1)} f(e^{-\mu t_0}(s - 1)). \tag{7.23}$$

Setting $z = e^{-\mu t_0}(s - 1)$ we have $s = 1 + ze^{\mu t_0}$ and (7.23) yields

$$f(z) = (1 + ze^{\mu t_0})^m \exp\left[-\frac{1}{2}\beta(t_0)ze^{\mu t_0}\right].$$

Therefore

$$f(e^{-\mu t}(s - 1)) = [1 + e^{-\mu(t-t_0)}(s - 1)]^m \cdot \exp\left[-\frac{1}{2}\beta(t_0)e^{-\mu(t-t_0)}(s - 1)\right],$$

and substituting in (7.22) we obtain the solution of the initial value problem (7.18), (7.19),

$$G(t,s) = [1 + e^{-\mu(t-t_0)}(s - 1)]^m \cdot \exp\left\{\frac{1}{2}[\beta(t) - \beta(t_0)e^{-\mu(t-t_0)}](s - 1)\right\}. \tag{7.24}$$

Problems

7.1. Derive equation (7.3).

7.2. Derive (7.7) from (7.4), (7.5) and (7.6). [Note that (7.5) expresses $\partial G/\partial s$ in two ways.]

7.3. Derive (7.13) from (7.11) and (7.12).

7.4. For fixed t, equation (7.17) has the form

$$G(t,s) = (a + bs)e^{c+ds}$$

where a, b, c, d are constants depending on t. Expand G in a Taylor series in s using the expansion for e^s and derive the formulas

$$P_n(t) = \frac{1}{n!}\frac{1}{\mu}\left(\frac{\lambda}{\mu}\right)^{n-1}(1 - e^{-\mu t})^{n-1}[\lambda(1 - e^{-\mu t})^2 + n\mu e^{-\mu t}]$$

$$\times \exp\left[\frac{\lambda}{\mu}(e^{-\mu t} - 1)\right],$$

for $0 \le t < \infty$ and $n = 0, 1, 2, \ldots$.

7.5. For the telephone network discussed in this section find the probability generating function and the probabilities $P_n(t)$ if at time $t = 0$, (a) two lines are in use, (b) m lines are in use, where m is a positive integer.

7.6. The *expectation* (mean value) $E(t)$ of the number of telephone lines in use at time t is defined by

$$E(t) = \sum_{n=0}^{\infty} nP_n(t).$$

It is a weighted mean of the number of lines that may be in use at time t, weighted by the corresponding probabilities. Show that

$$E(t) = \left.\frac{\partial G(t,s)}{\partial s}\right|_{s=1},$$

and calculate $E(t)$ if one line is in use at time $t = 0$.

7.7. In the formula of Problem 7.4 let $t \to \infty$ to show that

$$\lim_{t\to\infty} P_n(t) = \frac{e^{-\frac{\lambda}{\mu}}\left(\frac{\lambda}{\mu}\right)^n}{n!},$$

which is the Poisson distribution with parameter λ/μ.

7.8. Derive the first integrals (7.20) of equation (7.18).

7.9. *The Poisson process.* In many physical processes the occurrence of an event at a particular moment is independent of time and of the number of events that have already taken place. Examples are accidents occurring in a city, the splitting of atoms of a radioactive substance, breakage of chromosomes under harmful irradiation and phone calls arriving at a switchboard. Let $X(t)$ denote the total number of events occurring during the time interval $(0, t)$ and let $P_n(t)$ denote the probability that $X(t) = n$. A process is said to be a Poisson process if, for any $t \geq 0$, (i) the probability that an event occurs during the interval $(t, t+h)$ is $\lambda h + o(h)$ where λ is a constant, and (ii) the probability that more than one event occurs during $(t, t+h)$ is $o(h)$.

(a) Derive the system of o.d.e.'s

$$P_0'(t) = -\lambda P_0(t)$$

$$P_n'(t) = -\lambda P_n(t) + \lambda P_{n-1}(t), \qquad n = 1, 2, \ldots .$$

(b) Show that the p.g.f. $G(t,s) = \sum P_n(t)s^n$ satisfies the initial value problem,

$$\frac{\partial G}{\partial t} = -\lambda(1 - s)G$$

$$G(0, s) = 1.$$

(Note that $P_0(0) = 1$ and $P_n(0) = 0$ for $n = 1, 2, \ldots$.)

(c) Solve the initial value problem for G to show

$$G(t,s) = e^{-\lambda t(1-s)},$$

and obtain the probabilities

$$P_n(t) = \frac{e^{-\lambda t}(\lambda t)^n}{n!}, \qquad n = 0, 1, 2, \ldots$$

(This is the Poisson distribution with parameter λt.)

(d) Show that the expectation $E(t) = \lambda t$ (see Problem 7.6).

7.10. *The time-dependent Poisson process*. This is a Poisson process in which λ is not constant but is instead a function of t, $\lambda = \lambda(t)$. Follow the instructions of Problem 7.9 for this case and show that the formulas for $G(t, s)$, $P_n(t)$, and $E(t)$ are obtained from those of Problem 7.9 by replacing λt with $\int_0^t \lambda(\tau)d\tau$.

7.11. *The Yule process*. This process was first studied by Yule in connection with the mathematical theory of evolution. It is a simple example of what is known as a *pure birth process*. Consider a population of members (such as bacteria) which give birth to new members but do not die (bacteria may do this by splitting). Assume that during any short time interval $(t, t + h)$ each member has probability $\lambda h + o(h)$ to create a new member (λ is a constant), and that members give birth independently of each other. Then, if at time t the population size is n, the probability of increase of the population by exactly one during $(t, t + h)$ is $n\lambda h + o(h)$. Assume also that at time $t = 0$ the population size is n_0, so that if $P_n(t)$ is the probability that the population size is n at time t, then

$$P_{n_0}(0) = 1 \quad \text{and} \quad P_n(0) = 0 \quad \text{for} \quad n \neq n_0.$$

(a) Derive the system of o.d.e.'s,

$$P_{n_0}'(t) = -n_0\lambda P_{n_0}(t)$$

$$P_n'(t) = -n\lambda P_n(t) + (n-1)\lambda P_{n-1}(t), \qquad n > n_0.$$

(b) Introduce the p.g.f. $G(t,s) = \sum_{n=n_0}^{\infty} P_n(t)s^n$ and show that it must satisfy the initial value problem

$$\frac{\partial G}{\partial t} + \lambda s(1-s)\frac{\partial G}{\partial s} = 0$$

$$G(0, s) = s^{n_0}$$

(c) Show that

$$G(t, s) = s^{n_0}\left[\frac{e^{-\lambda t}}{1 - s(1 - e^{-\lambda t})}\right]^{n_0}$$

(d) Show that the expectation $E(t) = \sum_{n=n_0}^{\infty} nP_n(t)$ is

$$E(t) = n_0 e^{\lambda t}.$$

This is the familiar exponential population growth.

7.12. *The time-dependent Yule process.* This is a Yule process in which λ is a function of t, $\lambda = \lambda(t)$. Follow the instructions of Problem 7.11 for this case and show that the formulas for $G(t, s)$ and $E(t)$ are obtained from those of Problem 7.11 by replacing λt with $\int_0^t \lambda(\tau)d\tau$.

7.13. *The Polya process.* This is a pure birth process for which it is assumed that if at time t the population size is n, the probability of increase of the population by exactly one during the interval $(t, t + h)$ is $\lambda_n(t)h + o(h)$ where

$$\lambda_n(t) = \frac{\lambda + \lambda a n}{1 + \lambda a t},$$

with λ and a constants. Note that $a = 0$ corresponds to a Poisson process. Assume that initially the population size is n_0.

(a) Derive the system of o.d.e.'s

$$P_{n_0}'(t) = -\frac{\lambda + \lambda a n_0}{1 + \lambda a t} P_{n_0}(t)$$

$$P_n'(t) = -\frac{\lambda + \lambda a n}{1 + \lambda a t} P_n(t) + \frac{\lambda + \lambda a(n-1)}{1 + \lambda a t} P_{n-1}(t), \; n > n_0.$$

(b) Introduce the p.g.f. $G(t, s) = \sum_{n=n_0}^{\infty} \text{p}\, P_n(t)s^n$ and show that it must

satisfy the initial value problem

$$(1 + \lambda a t)\frac{\partial G}{\partial t} + \lambda a s(1-s)\frac{\partial G}{\partial s} = -\lambda(1-s)G$$

$$G(0, s) = s^{n_0}.$$

(c) Show that

$$G(t,s) = \left(\frac{s}{1 + \lambda at - \lambda ats}\right)^{n_0} (1 + \lambda at - \lambda ats)^{-1/a}$$

7.14. *The birth and death process.* This process allows for a population to decline as well as to grow, and therefore it provides a more realistic model for biological problems. Let $X(t)$ denote the size of population at time t and $P_n(t)$ denote the probability that $X(t) = n$. The basic hypotheses are the following: If $X(t) = n$, then during the interval $(t, t + h)$: (i) the probability of one birth occurring is $\lambda_n(t)h + o(h)$; (ii) the probability of one death occurring is $\mu_n(t)h + o(h)$; and (iii) the probability of more than one change (birth or death) is $o(h)$. Show that the $P_n(t)$ must satisfy the system of o.d.e.'s,

$$P_0'(t) = -[\lambda_0(t) + \mu_0(t)]P_0(t) + \mu_1(t)P_1(t)$$

$$P_n'(t) = -[\lambda_n(t) + \mu_n(t)]P_n(t)$$

$$+ \lambda_{n-1}(t)P_{n-1}(t) + \mu_{n+1}(t)P_{n+1}(t), \ n \geq 1.$$

7.15. In the telephone network discussed in this section the process of initiation and termination of calls may be considered as a birth and death process with the population size being the number of lines in use. Show that in this case $\lambda_n(t) = \lambda$ and $\mu_n(t) = n$ and verify that the system of o.d.e.'s of Problem 7.14 becomes system (7.2), (7.3).

7.16. *A birth and death process with linear growth.* Consider a population of living elements, such as bacteria, that can split or die. During any short time interval $(t, t + h)$ the probability of any living element splitting into two is $\lambda h + o(h)$ and the probability of it dying is $\mu h + o(h)$, where λ and μ are constants. Assume that at $t = 0$ the population size is n_0.

(a) Show that in the notation of Problem 7.14, $\lambda_n(t) = n\lambda$ and $\mu_n(t) = n\mu$ and write down the system of o.d.e.'s and the initial conditions for the $P_n(t)$.

(b) Show that the p.g.f. $G(t, s)$ must satisfy the initial value problem,

$$\frac{\partial G}{\partial t} + (1 - s)(\lambda s - \mu)\frac{\partial G}{\partial s} = 0$$

$$G(0, s) = s^{n_0}.$$

(c) If $\lambda \neq \mu$ obtain the solution

$$G(t, s) = \left[\frac{(\lambda s - \mu) + \mu(1 - s)e^{(\lambda-\mu)t}}{(\lambda s - \mu) + \lambda(1 - s)e^{(\lambda-\mu)t}}\right]^{n_0}$$

and show that the expectation is

$$E(t) = n_0 e^{(\lambda-\mu)t}.$$

(d) If $\lambda = \mu$ obtain the solution

$$G(t,s) = \left\{\frac{\alpha(t) + [1 - 2\alpha(t)]s}{1 - \alpha(t)s}\right\}^{n_0}$$

where $\alpha(t) = \lambda t/(1 + \lambda t)$, and show that the expectation is

$$E(t) = n_0.$$

References for Chapter III

1. Taylor, A. E.: *Advanced Calculus,* Boston: Ginn and Co., 1955.
2. Bellman, R., Kalaba, R., and Wing, G. M.: Invariant imbedding and neutron transport theory, I, *J. Math. Mech., 7:* 149–162, 1958.
3. Lax, P. D.: The formation and decay of shock waves, *Amer. Math. Monthly, 79:* 227–241, 1972.
4. Haight, F. A.: *Mathematical Theory of Traffic Flow,* New York: Academic Press, 1963.
5. Richards, P. I.: Shock waves on the highway, *Operations Res., 4:* 42–51, 1956.
6. Noh, W. F., and Protter, M. H.: Difference methods and the equations of hydrodynamics, *J. Math. Mech., 12:* 149–191, 1963.
7. Feller, W.: *An Introduction to Probability and Its Applications;* Vol. I, Ed. 3, 1968; Vol. II, Ed. 2, 1971; New York: John Wiley & Sons, Inc.
8. Chiang, C. L.: *Introduction to Stochastic Processes in Biostatistics,* New York: John Wiley & Sons Inc., 1968.
9. Nasell, I., and Hirsch, W. M.: The transmission dynamics of schistosomiasis, *Comm. Pure Appl. Math., 26:* 395–453, 1973.

CHAPTER IV

Series solutions
The Cauchy-Kovalevsky theorem

In this chapter we study one of the fundamental results in the theory of partial differential equations, the Cauchy-Kovalevsky theorem. This theorem asserts the existence of an analytic solution of the initial value problem for a p.d.e. when all functions involved in the problem are analytic. In Section 1 we review the Taylor series of a function of one or more variables and define analytic functions. In Section 2 we describe first how to compute the coefficients of the Taylor series of the solution of the initial value problem for a first order p.d.e. and then state the Cauchy-Kovalevsky theorem for this problem. In the problems of the second section we indicate the statement of the Cauchy-Kovalevsky theorem for equations of higher order and for systems of equations.

1. Taylor Series. Analytic Functions

Let f be a C^∞ function of a single variable x in an open interval I of R^1 and let x_0 be any point of I. The series

$$\sum_{n=0}^{\infty} \frac{f^{(n)}(x_0)}{n!}(x - x_0)^n \tag{1.1}$$

is called the *Taylor series* of the function f about the point x_0. In (1.1), $f^{(n)}$ denotes the nth derivative of f. For an arbitrary C^∞ function f, the Taylor series (1.1) may not converge at all, or, if it does converge, it may not converge to $f(x)$ (see Problem 1.1). The special C^∞ functions f which have Taylor series converging to $f(x)$ for all x near x_0, are called analytic at x_0.

Definition 1.1. Let $f \in C^\infty(I)$, where I is an open interval of R^1, and let x_0 be any point of I. If the Taylor series (1.1) of f about the point x_0 converges to $f(x)$ for every x in a neighborhood of x_0, then f is called *analytic at* x_0. If f is analytic at every point of I then f is called an *analytic function* in the interval I.

The Taylor series of the function $f(x) = e^x$ about the origin is

$$\sum_{n=0}^{\infty} \frac{1}{n!} x^n$$

and it is usually shown in courses on elementary calculus that this series converges to e^x for every $x \in R^1$. Hence the function e^x is analytic at the origin. In fact it is analytic on the whole real line R^1 and we can write

$$e^x = \sum_{n=0}^{\infty} \frac{1}{n!} x^n, \qquad x \in R^1.$$

Similarly, the functions sin x and cos x are analytic on R^1 and

$$\sin x = x - \frac{x^3}{3!} + \frac{x^5}{5!} - \ldots, \qquad x \in R^1$$

$$\cos x = 1 - \frac{x^2}{2!} + \frac{x^4}{4!} - \ldots, \qquad x \in R^1.$$

It is almost obvious that any polynomial is an analytic function on R^1 (see Problem 1.3). The function $f(x) = (1 - x)^{-1}$ is analytic for all $x \neq 1$ and its Taylor series about the origin converges to $f(x)$ in the interval $|x| < 1$,

$$\frac{1}{1-x} = 1 + x + x^2 + x^3 + \ldots, \quad |x| < 1.$$

We next consider functions of several variables. Let f be a C^∞ function defined in some domain Ω of R^n and let x^0 be any point of Ω. The series

$$(1.2) \qquad \sum_{(\alpha_1, \ldots, \alpha_n)} \frac{D_1^{\alpha_1} D_2^{\alpha_2} \ldots D_n^{\alpha_n} f(x^0)}{\alpha_1! \alpha_2! \ldots \alpha_n!} (x_1 - x_1^0)^{\alpha_1} (x_2 - x_2^0)^{\alpha_2} \ldots (x_n - x_n^0)^{\alpha_n}$$

is called the *Taylor series* of f about x^0. In (1.2), $D_j = \partial/\partial x_j$, and α_j is a non-negative integer, $j = 1, \ldots, n$. Thus

$$D_1^{\alpha_1} D_2^{\alpha_2} \ldots D_n^{\alpha_n} f = \frac{\partial^{\alpha_1 + \alpha_2 + \ldots + \alpha_n} f}{\partial x_1^{\alpha_1} \partial x_2^{\alpha_2} \ldots \partial x_n^{\alpha_n}}.$$

The summation in (1.2) is taken over all n-tuples of non-negative integers $(\alpha_1, \ldots, \alpha_n)$. The series (1.2) can be written in a shorter form if we introduce the notation

$$\alpha = (\alpha_1, \alpha_2, \ldots, \alpha_n)$$

$$x^\alpha = x_1^{\alpha_1} x_2^{\alpha_2} \ldots x_n^{\alpha_n}$$

$$D^\alpha = D_1^{\alpha_1} D_2^{\alpha_2} \ldots D_n^{\alpha_n}$$

$$\alpha! = \alpha_1! \alpha_2! \ldots \alpha_n!$$

$$|\alpha| = \alpha_1 + \alpha_2 + \ldots + \alpha_n.$$

Then the Taylor series of f about x^0 can be written in the form

$$\sum_{|\alpha| \geqq 0} \frac{D^\alpha f(x^0)}{\alpha!}(x - x^0)^\alpha. \tag{1.3}$$

As an example, if $n = 2$ the first few terms of the Taylor series (1.3) are

$$f(x_1^0, x_2^0) + D_1 f(x_1^0, x_2^0)(x_1 - x_1^0) + D_2 f(x_1^0, x_2^0)(x_2 - x_2^0)$$

$$+ \frac{1}{2!} D_1^2 f(x_1^0, x_2^0)(x_1 - x_1^0)^2 + \frac{1}{1!1!} D_1 D_2 f(x_1^0, x_2^0)(x_1 - x_1^0)(x_2 - x_2^0)$$

$$+ \frac{1}{2!} D_2^2 f(x_1^0, x_2^0)(x_2 - x_2^0)^2 + \ldots$$

We have written out only the terms of order $\leqq 2$ (i.e., $0 \leqq |\alpha| = \alpha_1 + \alpha_2 \leqq 2$) and the dots stand for terms of order $\geqq 3$ ($|\alpha| \geqq 3$).

Just as in the case of functions of a single variable, the Taylor series (1.3) of an arbitrary C^∞ function f may not converge at all, or, if it does converge, it may not converge to $f(x)$. The special C^∞ functions f which have Taylor series converging to $f(x)$ for all x near x^0, are called analytic at x^0.

Definition 1.2. Let $f \in C^\infty(\Omega)$ where Ω is a domain in R^n and let x^0 be any point of Ω. If the Taylor series (1.3) of f about x^0 converges to $f(x)$ for every x in a neighborhood of x^0, then f is called *analytic at* x^0. If f is analytic at every point of Ω then f is called an *analytic function* in Ω.

The Taylor series for the function

$$f(x_1, x_2) = e^{(x_1 + x_2)}$$

about the origin is

$$\sum_{\alpha \geqq 0} \frac{1}{\alpha!} x^\alpha = \sum_{\alpha_1, \alpha_2 = 0}^{\infty} \frac{1}{\alpha_1! \alpha_2!} x_1^{\alpha_1} x_2^{\alpha_2}$$

and this series converges for all $(x_1, x_2) \in R^2$. The function $e^{x_1 + x_2}$ is analytic in the whole of R^2 and we can write

$$e^{x_1 + x_2} = \sum_{\alpha_1, \alpha_2 = 0}^{\infty} \frac{1}{\alpha_1! \alpha_2!} x_1^{\alpha_1} x_2^{\alpha_2}, \qquad (x_1, x_2) \in R^2.$$

The function $\cos(2x_1 - x_2 + x_3^2)$ is analytic in the whole of R^3, while the function

$$f(x_1, \ldots, x_n) = \frac{1}{x_1^2 + \cdots + x_n^2 - 1}$$

is analytic in R^n except on the unit sphere. Finally, all polynomials in n variables $x_1, x_2, \ldots, x_n$ are analytic in R^n. Problem 1.8 indicates simple ways for recognizing analytic functions.

Problems

1.1. Let f be a function of a single variable defined by

$$f(x) = \begin{cases} 0 & \text{for} \quad x \leq 0 \\ e^{-1/x^2} & \text{for} \quad x > 0. \end{cases}$$

Prove that $f^{(n)}(0) = 0$ for all $n = 0, 1, \ldots$. Is this function analytic at the origin?

1.2. For each of the functions below write down the first five terms of their Taylor series about the indicated point x_0 and describe the largest interval containing x_0 in which they are analytic.

(a) $f(x) = \log x, \qquad x_0 = 1.$

(b) $f(x) = \dfrac{1}{1 - x^2}, \qquad x_0 = 0.$

(c) $f(x) = \cos x, \qquad x_0 = \dfrac{\pi}{4}.$

(d) $f(x) = \sin 2x, \qquad x_0 = 0.$

1.3. What is the Taylor series of the polynomial

$$a_0 + a_1 x + a_2 x^2 + \ldots + a_m x^m$$

about the origin?

1.4. It is often possible to obtain the Taylor series of a given function by using known Taylor series of simpler functions. For example, from

$$\sin x = x - \frac{x^3}{3!} + \frac{x^5}{5!} - \ldots$$

we obtain

$$\sin 2x = (2x) - \frac{(2x)^3}{3!} + \frac{(2x)^5}{5!} - \ldots$$

$$= 2x - \frac{2^3}{3!} x^3 + \frac{2^5}{5!} x^5 - \ldots$$

Find the Taylor series of the following functions about the origin.

(a) $\cos 3x^2$

(b) $\dfrac{1}{1 - 2x}$

(c) e^{x^2}.

1.5. For each of the functions below write down the terms up to order 2 (i.e. $0 \leq |\alpha| \leq 2$) of their Taylor series about the origin

(a) $f(x_1, x_2) = x_1^2 \sin(x_1 + x_2) - x_2^2$

(b) $f(x_1, x_2, x_3) = e^{x_1 x_2 x_3} + (x_1 - x_2)^2$

(c) $f(x, y) = x \log(1 + x + y) - y^2$.

1.6. Use the hint in problem 1.4 to obtain the Taylor series of the following functions about the origin.

(a) $f(x_1, x_2) = \dfrac{1}{1 - 4x_1^2 x_2}$

(b) $f(x_1, x_2, x_3) = e^{x_1 x_2 x_3}$

(c) $f(x_1, x_2) = \sin(x_1 x_2)$

1.7. Use the function in problem 1.1 and the hint in problem 1.4 to construct a function of n variables which is C^∞ in R^n but not analytic at the origin.

1.8. The following rules are useful for recognizing analytic functions: Finite sums and products of analytic functions are analytic; the quotient of two analytic functions is analytic except possibly at points where the denominator vanishes; compositions of analytic functions are analytic. Use these rules to determine the sets of points where the following functions are analytic.

(a) $f(x) = x^2 - 2x + \cos 4x$ (b) $f(x) = \tan x$

(c) $f(x) = \dfrac{1 + x^2}{(x - 2)(x + 1)}$ (d) $f(x, y) = x \sin(xy) - \cos y^2$

(e) $f(x, y) = \dfrac{1 + x^2 + y^2}{1 - xy}$ (f) $f(x, y) = e^{(x-y)^2}$

(g) $f(x, y, z) = \log(x + y + z)$. [*Hint:* $\log x$ is analytic for $x > 0$.]

2. The Cauchy-Kovalevsky Theorem

In this section we discuss a method of solution of the initial value problem for a partial differential equation by using Taylor series. The method consists of computing the coefficients of the Taylor series of the solution using the initial data and the partial differential equation. Of course this method can be useful only if the solution of the problem is an analytic function. The Cauchy-Kovalevsky theorem gives conditions under which the initial value problem has a solution which is an analytic function.

Let us consider first the following initial value problem for a first order ordinary differential equation in the unknown u and independent variable t,

$$\frac{du}{dt} = F(t, u) \tag{2.1}$$

$$u(0) = u_0. \tag{2.2}$$

Here u_0 is a given number and F is a function of two variables t and u. We are looking for the solution $u(t)$ of problem (2.1)-(2.2) defined in some interval of the t-axis containing the point $t = 0$. Let us assume that the function F is analytic in a neighborhood of the point $(t, u) = (0, u_0)$ of R^2, i.e., F has a Taylor series which converges to $F(t, u)$ for every point (t, u) in a neighborhood of the point $(0, u_0)$. Then the Cauchy-Kovalevsky theorem asserts that the initial value problem (2.1)-(2.2) has a solution $u(t)$

which is defined and analytic in an interval containing the point $t = 0$. In order to find the Taylor series for $u(t)$ about the point $t = 0$, we must compute the values of u and all its derivatives at $t = 0$. We can do this using equations (2.1) and (2.2). In fact from (2.2) we have

$$u(0) = u_0$$

and substituting $t = 0$ and $u = u_0$ into (2.1), we get

$$u^{(1)}(0) = F(0, u_0).$$

To find the second order derivative at $t = 0$ we differentiate (2.1) with respect to t,

$$u^{(2)} = F_t(t, u) + F_u(t, u)u^{(1)},$$

and substitute $t = 0$ and the previously obtained values of u and $u^{(1)}$ at $t = 0$,

$$u^{(2)}(0) = F_t(0, u_0) + F_u(0, u_0)F(0, u_0).$$

Next, we differentiate (2.1) twice with respect to t and then substitute $t = 0$ and the previously obtained values of u, $u^{(1)}$ and $u^{(2)}$ at $t = 0$ to get $u^{(3)}(0)$. It should be clear now that continuing in this way we can compute the values of u and all its derivatives at $t = 0$. The Cauchy-Kovalevsky theorem asserts that the series

$$\sum_{n=0}^{\infty} \frac{u^{(n)}(0)}{n!} t^n$$

converges for all t in an open interval I containing the point $t = 0$ and defines the solution

$$u(t) = \sum_{n=0}^{\infty} \frac{u^{(n)}(0)}{n!} t^n \tag{2.3}$$

of the initial value problem (2.1)-(2.2) in I.

Let us consider next the following initial value problem, or Cauchy problem, for a first order partial differential equation in the unknown u and two independent variables t and x,

$$\frac{\partial u}{\partial t} = F\left(t, x, u, \frac{\partial u}{\partial x}\right) \tag{2.4}$$

$$u(0, x) = \phi(x). \tag{2.5}$$

Note that the p.d.e. (2.4) is not the most general p.d.e. of the first order. The variable t plays a special role in two ways. First, the partial derivative $\partial u/\partial t$ appears in the equation and, second, the equation has been solved for this derivative. The function $F(t, x, u, p)$ is a function of four variables defined in some domain in R^4. In the initial condition (2.5) the given function ϕ is defined on some interval C of the x-axis containing the origin. Again, note that the initial curve C is not an arbitrary smooth curve in the (t, x)-plane (as was the case in the initial value problems considered in Chapter III), but it is special in that it lies on the x-axis ($t = 0$). We seek a solution $u(t, x)$ of the Cauchy problem (2.4)-(2.5) defined for (t, x) in some domain Ω of the (t, x)-plane containing the initial curve C (see Fig. 2.1).

Let us assume first that the given function $\phi(x)$ is analytic in a neighborhood of the origin of the x-axis. Then, from the initial condition (2.5) we can compute all the partial derivatives of u with respect to x at the origin,

$$\frac{\partial^n u}{\partial x^n}(0, 0) = \phi^{(n)}(0), \qquad n = 0, 1, 2, \ldots$$

Let us assume also that the function F is analytic in a neighborhood of the point $(0, 0, \phi(0), \phi^{(1)}(0))$ of R^4. Then the Cauchy-Kovalevsky theorem asserts that the problem (2.4)-(2.5) has a solution $u(t, x)$ which is defined and analytic in a neighborhood of the origin of the (t, x)-plane. In order to find the Taylor series for $u(t, x)$ about the origin, we must compute the values of all the partial derivatives of u at the origin. We can do this using (2.4) and (2.5). We have already seen that the derivatives of the form $\partial^n u/\partial x^n$ can be computed from the initial condition (2.5). Substituting in (2.4) $t = 0$, $x = 0$ and the previously obtained values of u and $\partial u/\partial x$ at (0, 0) we obtain the value of the derivative $\partial u/\partial t$ at the origin,

$$\frac{\partial u}{\partial t}(0, 0) = F(0, 0, \phi(0), \phi^{(1)}(0)).$$

To obtain the value of $\partial^2 u/\partial x \partial t$ we first differentiate (2.4) with respect to x,

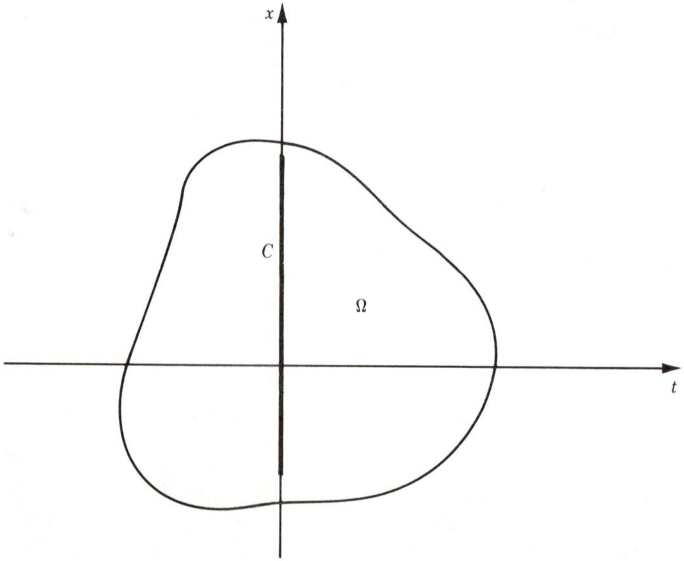

Fig. 2.1

$$\frac{\partial^2 u}{\partial x \partial t} = F_2(t, x, u, u_x)$$
$$+ F_3(t, x, u, u_x)u_x + F_4(t, x, u, u_x)u_{xx}$$

and then substitute $t = x = 0$ and the previously obtained values of u, u_x and u_{xx} at (0, 0). Here F_j denotes the partial derivative of F with respect to its jth variable, $j = 1, 2, 3, 4$. To obtain $\partial^3 u/\partial x^2 \partial t$ we differentiate (2.4) twice with respect to x and substitute $t = x = 0$ and the previously obtained values of u, u_x, u_{xx} and u_{xxx} at (0, 0). Continuing in this manner we can obtain the values of all partial derivatives $\partial^{n+1}u/\partial x^n \partial t$, $n = 0, 1, 2, \ldots$, at (0, 0). Next, to find $\partial^2 u/\partial t^2$ we differentiate (2.4) with respect to t,

$$\frac{\partial^2 u}{\partial t^2} = F_1(t, x, u, u_x)$$
$$+ F_3(t, x, u, u_x)u_t + F_4(t, x, u, u_x)u_{xt}$$

and substitute $t = x = 0$ and the previously obtained values of u, u_x, u_t and u_{xt} at the origin. It should be clear now that by successively differentiating (2.4) with respect to t and x and substituting the previously obtained values of u and its derivatives, we can obtain the values of all partial derivatives of u at the origin.

The Taylor series for $u(t, x)$ about the origin is

$$\sum_{(\alpha_t, \alpha_x)} \frac{D_t^{\alpha_t} D_x^{\alpha_x} u(0, 0)}{\alpha_t! \alpha_x!} t^{\alpha_t} x^{\alpha_x}$$

where the summation is taken over all pairs (α_t, α_x) of non-negative integers. The Cauchy-Kovalevsky theorem asserts that this series converges for all (t, x) in some neighborhood U of the origin and defines the solution

$$u(t, x) = \sum_{(\alpha_t, \alpha_x)} \frac{D_t^{\alpha_t} D_x^{\alpha_x} u(0, 0)}{\alpha_t! \alpha_x!} t^{\alpha_t} x^{\alpha_x} \tag{2.6}$$

of problem (2.4)-(2.5) in U. More precisely, the function defined by (2.6) satisfies the p.d.e. (2.4) for every (t, x) in U and the initial condition (2.5) for every point $(0, x)$ of C contained in U (see Fig. 2.2).

We will give the formal statement of the Cauchy-Kovalevsky theorem for the following initial value problem (or Cauchy problem) involving a first order p.d.e. in one unknown u and $n + 1$ independent variables $t, x_1, \ldots, x_n$,

$$\frac{\partial u}{\partial t} = F(t, x_1, \ldots, x_n, u, u_{x_1}, \ldots, u_{x_n}) \tag{2.7}$$

$$u(0, x_1, \ldots, x_n) = \phi(x_1, \ldots, x_n). \tag{2.8}$$

In the p.d.e. (2.7) the function $F(t, x_1, \ldots, x_n, u, p_1, \ldots, p_n)$ is a function of $2n + 2$ variables. Again note the special role of the variable t in (2.7); the derivative u_t appears in the equation and the equation is solved for this

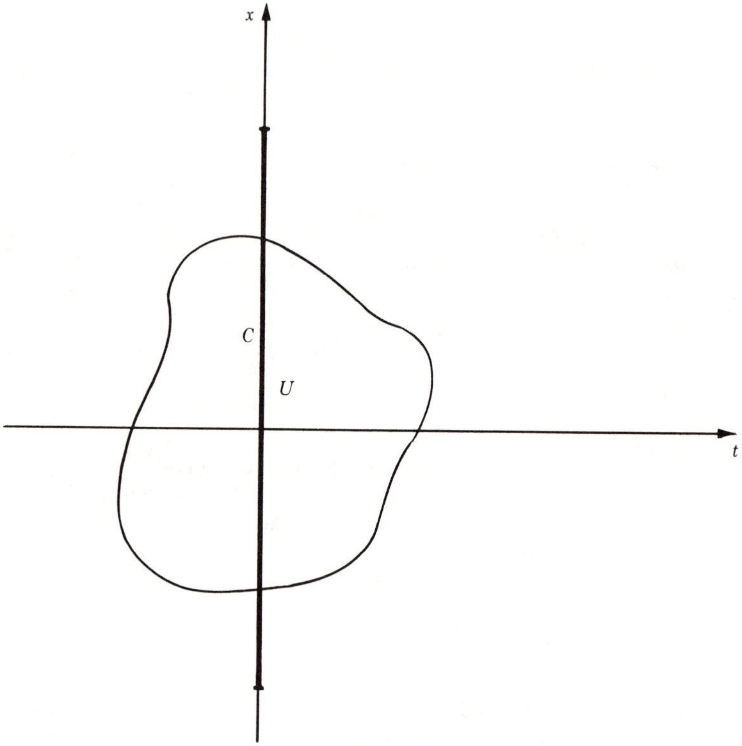

Fig. 2.2

derivative. In the initial condition (2.8) the given function ϕ is defined in some region S of the $(x_1, x_2, \ldots, x_n)$-space. We look for a solution $u(t, x_1, \ldots, x_n)$ of the Cauchy problem (2.7)-(2.8) defined for $(t, x_1, \ldots, x_n)$ in some domain Ω of R^{n+1} containing the region S on the hyperplane $t = 0$ on which the initial condition (2.8) is prescribed (see Fig. 2.3 where $n = 2$).

Theorem 2.1 (Cauchy-Kovalevsky). Suppose that the function ϕ is analytic in a neighborhood of the origin of R^n and suppose that the function F is analytic in a neighborhood of the point $(0, 0, \ldots, 0, \phi(0, \ldots, 0), \phi_{x_1}(0, \ldots, 0), \ldots, \phi_{x_n}(0, \ldots, 0))$ of R^{2n+2}. Then the Cauchy problem (2.7)-(2.8) has a solution $u(t, x_1, \ldots, x_n)$ which is defined and analytic in a neighborhood of the origin of R^{n+1} and this solution is unique in the class of analytic functions.

The theorem makes two assertions: (i) there exists an analytic solution in some neighborhood of the origin and (ii) this solution is unique in the class of analytic functions. In more precise language, the existence assertion states that there exists a function $u(t, x_1, \ldots, x_n)$ which is defined and analytic in a neighborhood U of the origin in R^{n+1}, and is such that at

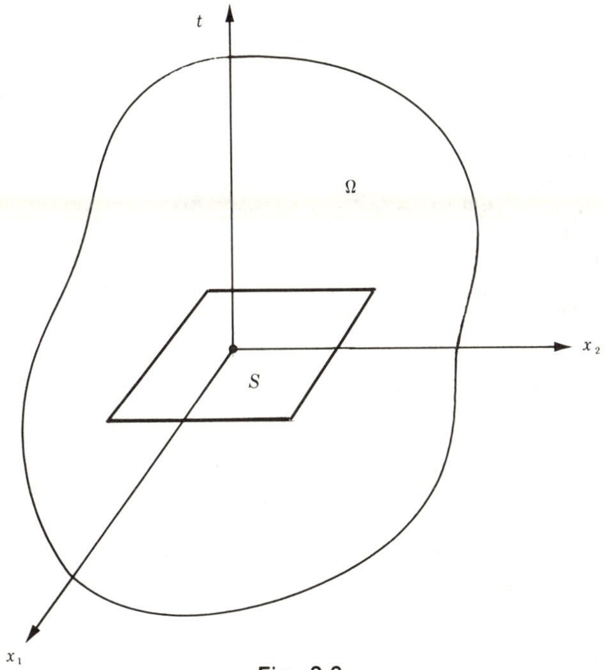

Fig. 2.3

every point $(t, x_1, \ldots, x_n)$ of U, $u(t, x_1, \ldots, x_n)$ satisfies the p.d.e. (2.7) and at every point $(0, x_1, \ldots, x_n)$ of the part of S contained in U, it satisfies the initial condition (2.8). The uniqueness assertion states that two analytic solutions of (2.7)-(2.8) must necessarily coincide in some neighborhood of the origin.

We do not give the details of the proof of the Cauchy-Kovalevsky theorem here but refer the interested reader to the book of Petrovskii.[1] The "existence" proof consists of showing that the coefficients of the Taylor series

$$(2.9) \qquad \sum_{(\alpha_t, \alpha_1, \ldots, \alpha_n)} \frac{D_t^{\alpha_t} D_1^{\alpha_1} \ldots D_n^{\alpha_n} u(0, \ldots, 0)}{\alpha_t! \alpha_1! \ldots \alpha_n!} t^{\alpha_t} x_1^{\alpha_1} \ldots x_n^{\alpha_n}$$

can be computed from (2.7) and (2.8) (as we have done for the case $n = 1$) and then proving that the series (2.9) converges in some neighborhood of the origin and that it satisfies (2.7) and (2.8) in this neighborhood. The proof of uniqueness of an analytic solution follows immediately by noting that any two analytic solutions must have Taylor series about the origin with coefficients exactly the coefficients in (2.9). Since the coefficients in (2.9) can be computed in a unique way from (2.7) and (2.8) it follows that any two analytic solutions must have the same Taylor series about the origin and hence they must coincide in some neighborhood of the origin.

Example 2.1. Find all terms of order less than or equal to three in the Taylor series about the origin of the solution of the initial value problem

$$u_t = uu_x \tag{2.10}$$

$$u(0, x) = 1 + x^2. \tag{2.11}$$

In this problem $\phi(x) = 1 + x^2$ and hence the function ϕ is analytic in a neighborhood of the origin of the x-axis (in fact it is analytic on the whole x-axis). Here $u_x(0, 0) = \phi'(0) = 0$. Moreover $F(t, x, u, p) = up$ and this function is analytic in a neighborhood of $(0, 0, 1, 0)$ of R^4 (in fact it is analytic in the whole of R^4). Hence, by the Cauchy-Kovalevsky theorem, the Cauchy problem (2.10)-(2.11) has an analytic solution in a neighborhood of the origin of the (t, x)-plane. We must compute all derivatives of u of order $\leqq 3$ at the origin. From (2.11) we have

$$u(0, x) = 1 + x^2, \qquad u_x(0, x) = 2x, \qquad u_{xx}(0, x) = 2, \qquad u_{xxx}(0, x) = 0$$

and hence

$$u(0, 0) = 1, \qquad u_x(0, 0) = 0, \qquad u_{xx}(0, 0) = 2, \qquad u_{xxx}(0, 0) = 0.$$

From (2.10) we have

$$u_t = uu_x, \qquad u_{tx} = uu_{xx} + u_x^2, \qquad u_{txx} = 3u_xu_{xx} + uu_{xxx}$$

and using the previously obtained values we get

$$u_t(0, 0) = 0, \qquad u_{tx}(0, 0) = 2, \qquad u_{txx}(0, 0) = 0.$$

Again from (2.10) we have

$$u_{tt} = u_tu_x + uu_{tx}, \qquad u_{ttx} = u_tu_{xx} + 2u_xu_{tx} + uu_{txx},$$

and using the previously obtained values we get

$$u_{tt}(0, 0) = 2, \qquad u_{ttx}(0, 0) = 0.$$

Finally from (2.10) we have

$$u_{ttt} = u_{tt}u_x + 2u_tu_{tx} + uu_{ttx}$$

and hence

$$u_{ttt}(0, 0) = 0.$$

The Taylor series for $u(t, x)$ about the origin is

$$u(t, x) = \sum_{(\alpha_t, \alpha_x)} \frac{D_t^{\alpha_t} D_x^{\alpha_x} u(0, 0)}{\alpha_t!\alpha_x!} t^{\alpha_t} x^{\alpha_x} \tag{2.12}$$

$$= 1 + t^2 + 2tx + x^2 + \ldots$$

where the dots stand for terms of order $\geqq 4$.

Since equation (2.10) is a quasi-linear first order equation, the theory developed in Chapter III is also applicable. According to Theorem 3.1 of Chapter III, there exists a unique solution of the initial value problem (2.10), (2.11) in a neighborhood of the point $(t, x) = (0, 0)$ (see Problem

2.4). Moreover, this solution can actually be found by the methods of Chapter III and is

$$u(t, x) = 2 \frac{1 + x^2}{1 - 2tx + \sqrt{1 - 4tx - 4t^2}}. \tag{2.13}$$

The Taylor series of $u(t, x)$ about $(0, 0)$ can be computed directly from (2.13), and it can be verified that the terms of order $\leqq 3$ are indeed given by (2.12).

The Cauchy-Kovalevsky theorem is a theorem of fundamental importance in the theory of partial differential equations. However its practical usefulness is often limited by the stringent requirement that the initial data and the right-hand side of the equation must be analytic and by the fact that it asserts the existence and uniqueness of the solution only in a (possibly very small) neighborhood of the origin.

We have stated the Cauchy-Kovalevsky theorem for initial value problems involving a single first order partial differential equation. In the problems we indicate the statement of the theorem for equations of higher order and for systems of equations in more than one unknown.

Problems

2.1. For each of the following initial value problems, verify first that the assumptions of the Cauchy-Kovalevsky theorem are satisfied and then find the terms of order $\leqq 2$ of the Taylor series of the solution about the origin. The answers are shown in brackets.

(a) $u_t = u^2 + u_x;\ u(0, x) = 1 + 2x$

$[u(t, x) = 1 + 3t + 2x + 5t^2 + 4tx + \ldots]$

(b) $u_t = u_x^2;\ u(0, x) = 1 + 2x - 3x^2$

$[u(t, x) = 1 + 4t + 2x - 48t^2 - 24tx - 3x^2 + \ldots]$

(c) $u_t = (\sin u)u_x;\ u(0, x) = \dfrac{\pi}{6} + x$

$\left[u(t, x) = \dfrac{\pi}{6} + \dfrac{1}{2}t + x + \dfrac{\sqrt{3}}{4}t^2 + \dfrac{\sqrt{3}}{2}tx + \ldots\right]$

(d) $u_t = e^{tx}u_x;\ u(0, x) = 1 - x + x^2$

$[u(t, x) = 1 - t - x + t^2 + 2tx + x^2 + \ldots]$

(e) $u_t = u_{x_1}u_{x_2};\ u(0, x_1, x_2) = x_1 + x_2 - 2x_1^2$

$[u(t, x_1, x_2) = x_1 + x_2 + t - 2t^2 - 4tx_1 - 2x_1^2 + \ldots]$

(f) $u_t = u^2 + u_{x_2}u_{x_3};\ u(0, x_1, x_2, x_3) = x_1 + x_2 + x_2^2 - x_3^2 + x_2x_3$

$[u(t, x_1, x_2, x_3) = x_1 + x_2 + x_2^2$
$- x_3^2 + x_2x_3 + tx_2 - 2tx_3 - t^2 + \ldots]$

2.2. Consider the initial value problem

$$u_t = \sin u_x; \qquad u(0, x) = \frac{\pi}{4} x.$$

Verify that the assumptions of the Cauchy-Kovalevsky theorem are satisfied and obtain the Taylor series of the solution about the origin. (Surprise!)

2.3. Consider the Cauchy problem

$$u_t = \cos u_x; \qquad u(0, x) = \frac{\pi}{4} x + \frac{\pi}{6} x^2.$$

Verify that the assumptions of the Cauchy-Kovalevsky theorem are satisfied and show that

$$u(t, x) = \frac{\pi}{4} x + \frac{\sqrt{2}}{2} t + \frac{\pi}{6} x^2 - \frac{\sqrt{2}}{2}\frac{\pi}{3} xt + \frac{\pi}{12} t^2$$

$$- \frac{\sqrt{2}}{4}\left(\frac{\pi}{3}\right)^2 x^2 t + \frac{1}{2}\left(\frac{\pi}{3}\right)^2 xt^2 - \frac{\sqrt{2}}{8}\left(\frac{\pi}{3}\right)^2 t^3 + \ldots$$

where the dots stand for terms of order $\geqq 4$.

2.4. (a) Restate the initial value problem (2.10), (2.11) in terms of the notation used in Chapter III and verify that the assumptions of Theorem 3.1 of Chapter III are satisfied.
(b) Derive the solution (2.13) by the methods of Chapter III.
(c) Obtain (2.12) directly from (2.13).

2.5. For the initial value problem (a) in Problem 2.1, use the methods of Chapter III to obtain the solution

$$u(t, x) = \frac{1 + 2(x + t)}{1 - t - 2t(x + t)}.$$

Obtain directly from this exact solution the terms of order $\leqq 2$ of its Taylor series.

2.6. For the initial value problem (c) in Problem 2.1, use the methods of Chapter III to obtain the relation

$$u = \frac{\pi}{6} + x + y \sin u,$$

which implicitly defines the solution u in a neighborhood of $(t, x) = (0, 0)$. Obtain directly from this relation the terms of order $\leqq 2$ of the Taylor series of the solution.

2.7. In this section we have considered only initial value problems in which the initial conditions are prescribed in a region of the hyperplane $t = 0$ containing the origin of the x-space. It is easy to show that a problem in which the initial condition is prescribed in a region of the hyperplane $t = t^0$ containing the point $x = x^0$ of the x-space, can be reduced to a problem of the above type by introducing a translation of coordinates.
(a) Show that the initial value problem

(2.14) $$u_t = F(t, x, u, u_x),$$

(2.15) $$u(t_0, x) = \phi(x),$$

where $\phi(x)$ is defined in an interval of the x-axis containing the point $x = x_0$, can be reduced to the problem (2.4), (2.5) by introducing the new independent variables

$$t' = t - t_0, \qquad x' = x - x_0.$$

(b) For the initial value problem (2.14), (2.15) state the Cauchy-Kovalevsky theorem asserting the existence of an analytic solution $u(t, x)$ in a neighborhood of the point (t_0, x_0).

2.8. Consider the following Cauchy problem for a second order p.d.e. in two independent variables,

(2.16) $$u_{tt} = F(t, x, u, u_t, u_x, u_{tx}, u_{xx})$$

(2.17) $$u(0, x) = \phi_0(x)$$

(2.18) $$u_t(0, x) = \phi_1(x).$$

Note that (2.16) is not the most general p.d.e. of the second order. The variable t is special in two ways. The derivative u_{tt}, which is the highest possible derivative with respect to t, appears in the equation, and the equation is solved for this derivative. The functions ϕ_0 and ϕ_1, known as the initial data, are defined on an interval C of the x-axis containing the origin (see Fig. 2.1). C is known as the initial curve. From the initial conditions (2.17) and (2.18) it follows that

$$\frac{\partial^n u}{\partial x^n}(0, 0) = \phi_0^{(n)}(0), \qquad \frac{\partial^{n+1} u}{\partial t \partial x^n}(0, 0) = \phi_1^{(n)}(0); \qquad n = 0, 1, 2, \ldots.$$

The statement of the *Cauchy-Kovalevsky theorem* for the Cauchy problem (2.16), (2.17), (2.18) is: Suppose that ϕ_0 and ϕ_1 are analytic in a neighborhood of the origin of R^1 and suppose that the function F is analytic in a neighborhood of the point $(0, 0, \phi_0(0), \phi_1(0), \phi_0^{(1)}(0), \phi_1^{(1)}(0), \phi_0^{(2)}(0))$ of R^7. Then the Cauchy problem (2.16), (2.17), (2.18) has a solution $u(t, x)$ which is defined and analytic in a neighborhood of the origin of R^2 and this solution is unique in the class of analytic functions.

Under the assumptions of this theorem show that it is possible to find the Taylor series for $u(t, x)$ about the origin, using the initial conditions (2.17), (2.18) and the p.d.e. (2.16).

2.9. Consider the Cauchy problem for the wave equation

$$u_{tt} = u_{xx}$$

$$u(0, x) = \phi_0(x)$$

$$u_t(0, x) = \phi_1(x).$$

Show that this problem is a special case of the general problem discussed in Problem 2.8. What does the Cauchy-Kovalevsky theorem say for this problem?

2.10. Consider the Cauchy problem for Laplace's equation

$$u_{xx} + u_{yy} = 0$$
$$u(x, 0) = \phi_0(x)$$
$$u_y(x, 0) = \phi_1(x).$$

Show that this problem is a special case of the general problem discussed in Problem 2.8. What does the Cauchy-Kovalevsky theorem say for this problem?

2.11. Consider the initial value problem for the heat equation,

$$u_t = u_{xx}$$
$$u(0, x) = \phi_0(x).$$

Show that this problem is *not* a special case of the general problem discussed in Problem 2.8 and that hence the Cauchy-Kovalevsky theorem is not applicable.

2.12. Consider the following Cauchy problem for a system of two first order partial differential equations in two unknowns u, v and two independent variables x, t,

$$u_t = F(t, x, u, v, u_x, v_x), \tag{2.19}$$

$$v_t = G(t, x, u, v, u_x, v_x), \tag{2.20}$$

$$u(0, x) = \phi(x),\ v(0, x) = \psi(x). \tag{2.21}$$

Note carefully the special role of the independent variable t: the derivatives u_t and v_t appear in the system and the system is solved for these derivatives. The statement of the *Cauchy-Kovalevsky theorem* for this problem is: Suppose that $\phi(x)$ and $\psi(x)$ are analytic in a neighborhood of the origin of R^1 and suppose that the functions F and G are analytic in a neighborhood of the point $(0, 0, \phi(0), \psi(0), \phi^{(1)}(0), \psi^{(1)}(0))$ of R^6. Then the Cauchy problem (2.19), (2.20), (2.21) has a solution $\{u(t, x), v(t, x)\}$, with each function $u(t, x)$ and $v(t, x)$ being defined and analytic in a neighborhood of the origin of R^2, and this solution is unique in the class of analytic functions.

Show that it is possible to find the Taylor series for u and v about the origin by using the equations (2.19), (2.20) and the initial conditions (2.21).

2.13. Consider the initial value problem for a conservation law,

$$a(u)u_x + u_y = 0,$$
$$u(x, 0) = f(x),$$

which was discussed in Section 5 of Chapter III. Assume that $a(u)$ and $f(x)$ are analytic functions of one variable for all values of that variable.

(a) Derive the series solution

$$u(x, y) = f(x) + \sum_{n=1}^{\infty} \frac{(-1)^n}{n!} \frac{d^{n-1}}{dx^{n-1}} \left\{[a(f(x))]^n \frac{df(x)}{dx}\right\} y^n \tag{2.22}$$

which is valid for sufficiently small values of $|y|$. [*Hint:* Start with the series expansion

$$u(x, y) = b_0(x) + \sum_{n=1}^{\infty} \frac{b_n(x)}{n!} y^n$$

where

$$b_n(x) = \left. \frac{\partial^n u(x, y)}{\partial y^n} \right|_{y=0} .$$

Then use induction to prove that for $n \geq 1$,

$$\frac{\partial^n u}{\partial y^n} = (-1)^n \frac{\partial^{n-1}}{\partial x^{n-1}} \left\{ [a(u)]^n \frac{\partial u}{\partial x} \right\}. \Big]$$

(b) If $a(u) = cu$ and $f(x) = ke^{-x^2/2}$ where c and k are constants, compute the leading terms of the series (2.22),

$$u(x, y) = ke^{-x^2/2}[1 + ckxe^{-x^2/2}y + \frac{c^2k^2}{2}(3x^2 - 1)e^{-x^2}y^2 + \ldots].$$

The series (2.22) is known as *Lagrange series*. It has been used by Banta[2] in studying sound waves of finite amplitude and by Ames and Jones[3] in studying a Monge-Ampére equation in connection with anisentropic flow of gas and longitudinal wave propagation in a moving threadline.

References for Chapter IV

1. Petrovskii, I. G.: *Partial Differential Equations*, Philadelphia: W. B. Saunders Co., 1967.
2. Banta, E. D.: Lossless propagation of one-dimensional finite amplitude sound waves, *J. Math. Anal. Appl., 10:* 166–173, 1965.
3. Ames, W. F., and Jones, S. E.: Integrated Lagrange expansions for a Monge-Ampére equation, *J. Math. Anal. Appl., 21:* 479–484, 1968.

CHAPTER V

Linear partial differential equations Characteristics, classification and canonical forms

In this chapter we define and discuss some general concepts associated with linear partial differential equations. In Section 1 we use a convenient notation to write down the general form of a linear p.d.e. and we define a characteristic surface for such an equation. In Section 2 we describe the characteristic surfaces of several important equations which will be studied in this book. In Section 3 we discuss the importance of characteristics by means of a very simple example. Specifically, we illustrate with this example how the characteristics are exceptional for the Cauchy problem, how they may be carriers of discontinuities of a solution or of its derivatives, how they play a crucial role in solving first order equations and how they can be used to introduce new coordinates in terms of which the equation has a particularly simple form called the canonical form of the equation. In Section 4 we discuss the theory and method of solution of the initial value problem for first order equations. In Section 5 we discuss the general Cauchy problem and state the Cauchy-Kovalevsky theorem and Holmgren's uniqueness theorem. In Section 6 we show how a first order equation can be reduced to its canonical form by a transformation of coordinates. Section 7 is devoted to second order equations in two independent variables. These equations are classified into three distinct types. Any equation of a particular type can be reduced by a transformation of coordinates to a canonical form associated with its type. Section 8 is devoted to the classification and reduction to canonical form of second order equations in two or more independent variables. Finally, in Section 9 we describe and illustrate the principle of superposition.

1. Linear Partial Differential Operators and Their Characteristic Curves and Surfaces

We recall the notation introduced in Chapter IV. By $x = (x_1, \ldots, x_n)$ we will denote a point in R^n and by D_j the partial differentiation operator

$\partial/\partial x_j$. Let $\alpha = (\alpha_1, \ldots, \alpha_n)$ denote an n-tuple of non-negative integers. Then we define

$$x^\alpha = x_1^{\alpha_1} x_2^{\alpha_2} \ldots x_n^{\alpha_n}$$

and

$$D^\alpha = D_1^{\alpha_1} D_2^{\alpha_2} \ldots D_n^{\alpha_n}$$

Let $|\alpha|$ denote the sum of the components of α, $|\alpha| = \alpha_1 + \alpha_2 + \cdots + \alpha_n$. Then x^α is a monomial of order $|\alpha|$ in the coordinates $x_1, \ldots, x_n$, and D^α is a partial differentiation operator of order $|\alpha|$. In the old notation

$$D^\alpha = \frac{\partial^{|\alpha|}}{\partial x_1^{\alpha_1} \partial x_2^{\alpha_2} \ldots \partial x_n^{\alpha_n}}.$$

For example, if $n = 3$ and $\alpha = (2, 1, 3)$, then $|\alpha| = 6$, $x^\alpha = x_1^2 x_2 x_3^3$ is a monomial of order 6 and

$$D^\alpha = D_1^2 D_2 D_3^3 = \frac{\partial^6}{\partial x_1^2 \partial x_2 \partial x_3^3}.$$

A linear partial differential equation of order m in R^n is an equation of the form

$$\sum_{|\alpha| \leq m} a^\alpha D^\alpha u = f \tag{1.1}$$

where the a^α and f are functions of $x \in R^n$. The function a^α is called the coefficient of the term $a^\alpha D^\alpha u$ and f is called the right hand side of the equation. The summation on the left is taken over all possible values of the "index vector" α with $|\alpha| \leqq m$. Thus m is the order of the derivatives of highest order appearing in the equation.

The linear partial differential operator on the left hand side of equation (1.1) will be denoted by $P(x, D)$,

$$P(x, D) = \sum_{|\alpha| \leqq m} a^\alpha(x) D^\alpha. \tag{1.2}$$

If the coefficients a^α are constant, we write $P(D)$ instead of $P(x, D)$.

Example 1.1. In R^2, the equation

$$D_1^2 u + \sin(x_1 x_2) D_2^2 u - x_2^2 D_1 D_2 u + x_1 D_2 u + e^{x_2} u = \cos(x_1 + x_2) \tag{1.3}$$

is a second order linear partial differential equation. The coefficients are

$$a^{(2,0)}(x) = 1, \qquad a^{(0,2)}(x) = \sin x_1 x_2, \qquad a^{(1,1)}(x) = -x_2^2,$$

$$a^{(1,0)}(x) = 0, \qquad a^{(0,1)}(x) = x_1, \qquad a^{(0,0)}(x) = e^{x_2}$$

and

$$f(x) = \cos(x_1 + x_2).$$

The operator in equation (1.3) is

(1.4) $$P(x, D) = D_1^2 + \sin(x_1x_2)D_2^2 - x_2^2D_1D_2 + x_1D_2 + e^{x_2}.$$

Example 1.2. The general linear first order partial differential operator in R^n has the form

$$P(x, D) = a^{(1,0,\ldots,0)}(x)D_1 + a^{(0,1,0,\ldots,0)}(x)D_2 + \cdots + a^{(0,\ldots,0,1)}(x)D_n + a^{(0,\ldots,0)}(x).$$

Of course we may prefer to use simpler notation. For example, the general linear first order operator in R^2 is

(1.5) $$P(x, D) = a_1(x)D_1 + a_2(x)D_2 + c(x).$$

Example 1.3. The general linear second order partial differential operator in R^2 has the form

$$P(x, D) = a^{(2,0)}(x)D_1^2 + a^{(1,1)}(x)D_1D_2 + a^{(0,2)}(x)D_2^2 + a^{(1,0)}(x)D_1 + a^{(0,1)}(x)D_2 + a^{(0,0)}(x).$$

Some important examples with constant coefficients are the *Laplacian operator* in two variables

(1.6) $$P(D) = D_1^2 + D_2^2,$$

the *wave operator* in one space variable

(1.7) $$P(D) = D_1^2 - D_2^2,$$

and the *heat operator* in one space variable

(1.8) $$P(D) = D_1^2 - D_2.$$

In (1.7) and (1.8), x_1 is a space variable and x_2 is a time variable. We plan to study these three operators in great detail. Another example is the *Tricomi operator* that appears in hydrodynamics,

(1.9) $$P(x, D) = x_2D_1^2 + D_2^2.$$

Example 1.4. The general linear second order partial differential operator in R^3 has the form

$$P(x, D) = a^{(2,0,0)}(x)D_1^2 + a^{(1,1,0)}(x)D_1D_2 + a^{(1,0,1)}(x)D_1D_3 + a^{(0,1,1)}(x)D_2D_3 + a^{(0,2,0)}(x)D_2^2 + a^{(0,0,2)}(x)D_3^2 + a^{(1,0,0)}(x)D_1 + a^{(0,1,0)}(x)D_2 + a^{(0,0,1)}(x)D_3 + a^{(0,0,0)}(x).$$

Important special cases with constant coefficients are the Laplacian in three space variables

(1.10) $$P(D) = D_1^2 + D_2^2 + D_3^2,$$

the wave operator in two space variables

(1.11) $$P(D) = D_1^2 + D_2^2 - D_3^2,$$

and the heat operator in two space variables

(1.12) $$P(D) = D_1^2 + D_2^2 - D_3.$$

In (1.11) and (1.12), x_1 and x_2 are space variables and x_3 is a time variable.

Example 1.5. In R^2, the *biharmonic operator*

(1.13) $$P(D) = D_1^4 + 2D_1^2D_2^2 + D_2^4$$

is a linear partial differential operator of order 4 which appears in the study of elasticity.

One of the major conclusions of the theory of partial differential equations is that most of the important properties of solutions of a linear partial differential equation depend only on the form of the highest order terms appearing in the equation. These terms form what is known as the principal part of the equation. The *principal part* of the general linear partial differential operator (1.2) is

(1.14) $$P_m(x, D) = \sum_{|\alpha|=m} a^\alpha(x)D^\alpha.$$

The principal part of the partial differential operator (1.4) is

$$P_2(x, D) = D_1^2 + \sin(x_1x_2)D_2^2 - x_2^2D_1D_2,$$

and the principal part of (1.5) is

(1.15) $$P_1(x, D) = a_1(x)D_1 + a_2(x)D_2.$$

The principal parts of the Laplacian and wave operators are equal to the operators themselves while the principal part of the heat operator (1.8) is

(1.16) $$P_2(D) = D_1^2.$$

A non-zero vector $\xi = (\xi_1, \ldots, \xi_n) \in R^n$ defines an (unsigned) direction in R^n. Note that for any real number $\lambda \neq 0$, the vectors ξ and $\lambda\xi$ define the same direction. A direction defined by the non-zero vector $\xi \in R^n$ is called *characteristic* at the point $x \in R^n$ with respect to the partial differential operator $P(x, D)$ given by (1.2), if

(1.17) $$P_m(x, \xi) = 0$$

where $P_m(x, D)$ given by (1.14) is the principal part of $P(x, D)$. Equation (1.17) is called the *characteristic equation* of $P(x, D)$ and its left hand side is obtained from (1.14) by replacing $D = (D_1, \ldots, D_n)$ with $\xi = (\xi_1, \ldots, \xi_n)$,

$$P_m(x, \xi) = \sum_{|\alpha|=m} a^\alpha(x)\xi^\alpha.$$

As an example, the characteristic equation of the operator (1.4) is

$$\xi_1^2 + \sin(x_1x_2)\xi_2^2 - x_2^2\xi_1\xi_2 = 0.$$

Thus the direction $(\xi_1, \xi_2) = (0, 1)$ is characteristic at the point $(x_1, x_2) = (2, \pi/2)$ with respect to this operator. The characteristic equation of the wave operator (1.11) is

$$\xi_1^2 + \xi_2^2 - \xi_3^2 = 0,$$

and the direction $(\xi_1, \xi_2, \xi_3) = (1, 1, \sqrt{2})$ is characteristic at every point (x_1, x_2, x_3) of R^3. Generally, if the coefficients of the principal part of an operator are constant then obviously the characteristic directions are also independent of the point x in R^n.

Let S be a smooth surface in R^n and let x^0 be a point of S. The surface S is said to be *characteristic at* x^0 *with respect to* $P(x, D)$ if a vector normal to S at x^0 defines a direction which is characteristic with respect to $P(x, D)$ at x^0. If the surface S is characteristic with respect to $P(x, D)$ at every one of its points then S is called a *characteristic surface*. Naturally, in R^2 a characteristic "surface" is a curve called a *characteristic curve*.

The line $x_2 = \pi/2$ in R^2 is characteristic at the point $(x_1, x_2) = (2, \pi/2)$ with respect to the operator (1.4), because the normal vector (0, 1) to the line defines a characteristic direction at the point $(2, \pi/2)$ with respect to the operator (1.4) (see Fig. 1.1).

The plane

$$x_1 + x_2 + \sqrt{2}\, x_3 = 0$$

in R^3 is a characteristic surface of the wave operator (1.11), because the normal $(1, 1, \sqrt{2})$ to the plane is everywhere characteristic with respect to (1.11) (see Fig. 1.2).

The partial differential operators that we study in this book and which appear in applications, either do not have any characteristic surfaces or else they have one-parameter families of surfaces each of which is characteristic. Characteristic surfaces play a fundamental role in the study of

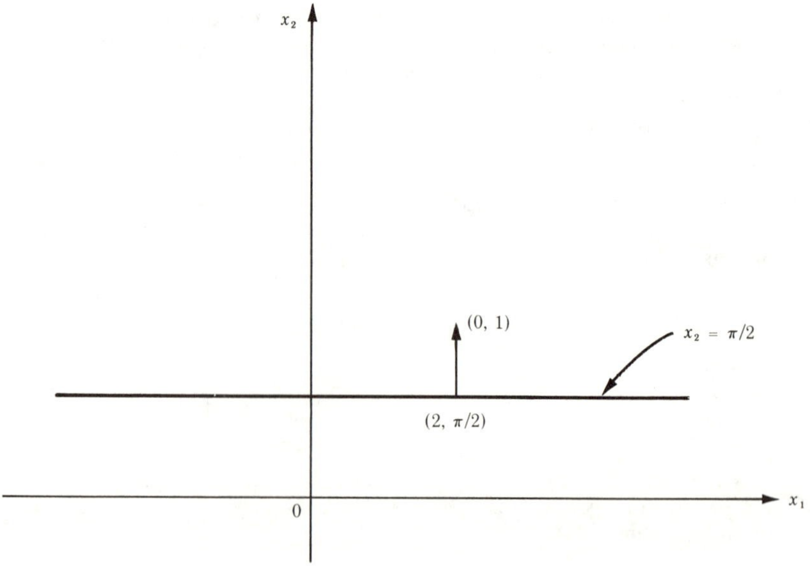

Fig. 1.1

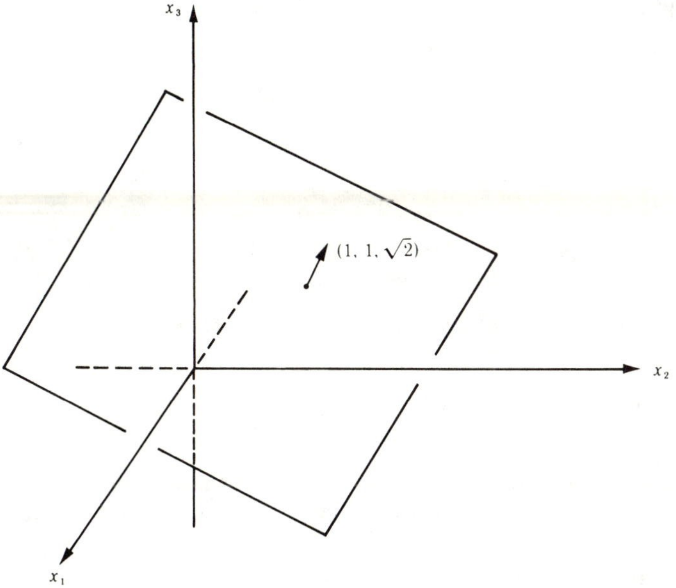

Fig. 1.2

partial differential equations. In Section 3 we will illustrate the importance of characteristics by means of a very simple example. In Section 2 we will discuss methods for finding characteristic surfaces and will find the characteristic surfaces of many important operators.

Problems

1.1. For each of the partial differential operators (1.6)–(1.13),
(a) write down the principal part $P_m(x, D)$,
(b) write down the characteristic equation (1.17).

1.2. Prove that if the vector $\xi = (\xi_1, \ldots, \xi_n)$ satisfies the characteristic equation (1.17) at some point $x \in R^n$, then, for any real number λ, the vector $\lambda\xi = (\lambda\xi_1, \ldots, \lambda\xi_n)$ also satisfies (1.17) at x.

2. Methods for Finding Characteristic Curves and Surfaces. Examples

The first step in trying to find the characteristic curves or surfaces of a linear partial differential operator is writing down its characteristic equation. If the coefficients of the principal part of the operator are constant then the characteristic equation is a homogeneous polynomial in $\xi_1, \ldots, \xi_n$ with constant coefficients. It may be possible to recognize the characteristic directions and determine the characteristic surfaces by simple geometric reasoning. The following five examples in R^2 illustrate this method.

Example 2.1. In R^2 let

$$P(x, D) = D_1 + c(x).$$

Here the order $m = 1$ and the principal part is

$$P_1(x, D) = D_1.$$

The characteristic equation is

$$\xi_1 = 0$$

so that the direction (0, 1) is the only characteristic direction at every point in R^2. The characteristic curves are the lines $x_2 =$ const.

Example 2.2. In R^2 consider the Laplace operator

$$P(D) = D_1^2 + D_2^2.$$

The characteristic equation is

$$\xi_1^2 + \xi_2^2 = 0$$

which can be satisfied only by $(\xi_1, \xi_2) = (0, 0)$. Consequently there are no characteristic directions and the Laplace operator has no characteristic curves.

Example 2.3. In R^2 consider the heat operator

$$P(D) = D_1^2 - D_2.$$

The principal part is

$$P_2(D) = D_1^2$$

and the characteristic equation is

$$\xi_1^2 = 0.$$

Just as in Example 2.1, the characteristic curves are the lines $x_2 =$ const.

Example 2.4. In R^2 consider the wave operator

$$P(D) = D_1^2 - D_2^2.$$

The characteristic equation is

$$\xi_1^2 - \xi_2^2 = 0$$

which is satisfied if $\xi_2 = \pm\xi_1$. The characteristic curves are straight lines making 45° angles with the axes; i.e., the lines $x_2 = x_1 + c_1$ and $x_2 = -x_1 + c_2$ (see Fig. 2.1). Note that through each point (x_1^0, x_2^0) pass exactly two characteristic curves.

Example 2.5. The equation

$$u_{xx} - \frac{1}{c^2} u_{tt} + au_t + bu = 0$$

where a, b, c are constants, is called the *telegraph equation*. Here we use

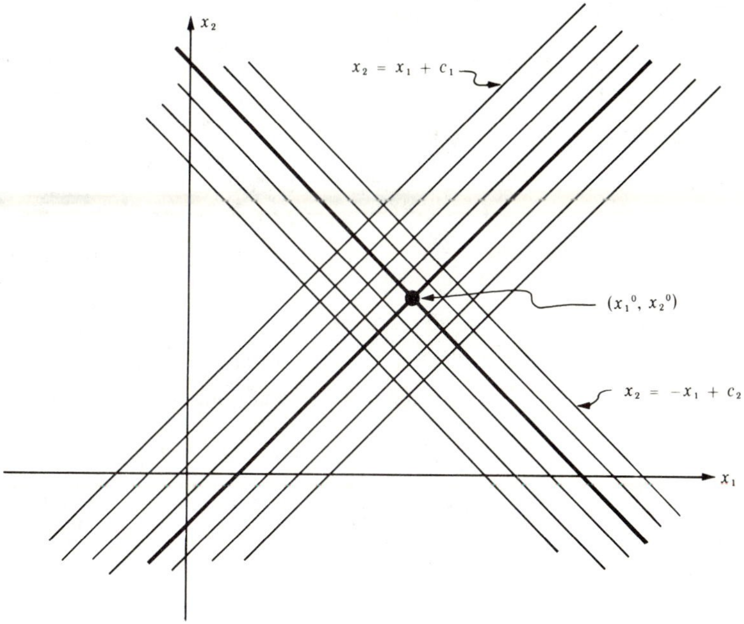

Fig. 2.1

(x, t) in place of (x_1, x_2). The principal part of the partial differential operator (p.d.o.) involved in the equation is

$$D_x^2 - \frac{1}{c^2} D_t^2.$$

The characteristic equation is

$$\xi_x^2 - \frac{1}{c^2} \xi_t^2 = 0$$

which is satisfied by the vectors $(\xi_x, \xi_t) = (1, \pm c)$. The characteristic curves are the straight lines $x + ct = c_1$ and $x - ct = c_2$. Through each point of the (x, t)-plane pass exactly two characteristic curves.

If the coefficients of the characteristic equation are not constant it may be necessary to use analytical methods for the determination of the characteristics. For example, in R^2, if the desired characteristic curves are expressed parametrically, then the characteristic equation leads to an ordinary differential equation which can be solved to yield the equations of the characteristics. This method is illustrated in the following two examples.

Example 2.6. In R^2 let

$$P(x, D) = a_1(x)D_1 + a_2(x)D_2 + c(x).$$

The order m is 1, the principal part is

$$P_1(x, D) = a_1(x)D_1 + a_2(x)D_2$$

and the characteristic equation is

$$a_1(x)\xi_1 + a_2(x)\xi_2 = 0.$$

Let C be a characteristic curve given parametrically by

$$x_1 = f_1(t), \qquad x_2 = f_2(t).$$

The tangent to this curve is given by $(dx_1/dt, dx_2/dt)$ and therefore $(dx_2/dt, -dx_1/dt)$ is normal to C. Hence

$$a_1(x_1, x_2)\frac{dx_2}{dt} - a_2(x_1, x_2)\frac{dx_1}{dt} = 0.$$

Thus, the characteristic curves can be obtained by solving the differential equation

$$a_1 dx_2 - a_2 dx_1 = 0.$$

For example the characteristic curves of $D_1 + D_2$ are solutions of the equation

$$dx_2 - dx_1 = 0$$

which are the lines $x_2 = x_1 + c$.

The characteristic curves of $D_1 + x_1D_2$ are solutions of

$$dx_2 - x_1dx_1 = 0$$

which are the parabolas $x_2 = x_1^2/2 + c$ (see Fig. 2.2).

Example 2.7. In R^2 the operator

$$P(x, D) = x_2D_1^2 + D_2^2$$

is called the Tricomi operator and appears in hydrodynamics. The characteristic equation is

$$x_2\xi_1^2 + \xi_2^2 = 0.$$

In the upper half plane, $x_2 > 0$, there are no characteristic directions and hence no characteristic curves. For $x_2 \leqq 0$, the characteristic directions at each point (x_1, x_2) are given by the vectors $(1, \pm\sqrt{-x_2})$. Just as in Example 2.6 we conclude that the characteristic curves are solutions of the equations

$$dx_1 = \pm\sqrt{-x_2}\, dx_2, \qquad x_2 \leqq 0.$$

The solutions of these equations are

$$x_1 - c = \pm\frac{2}{3}(-x_2)^{3/2}.$$

Thus, the characteristic curves are two one-parameter families of curves illustrated in Figure 2.3

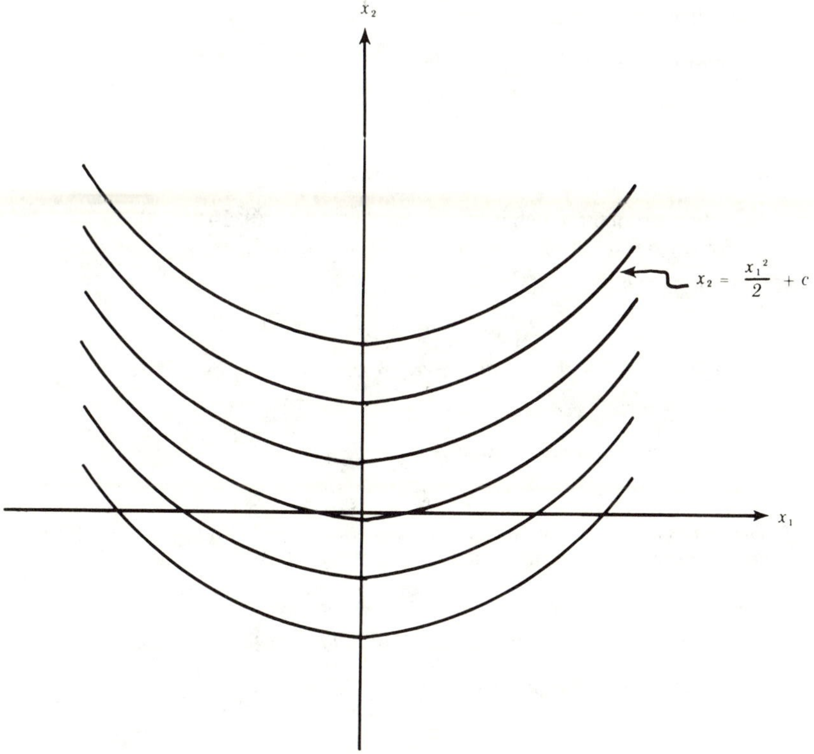

Fig. 2.2

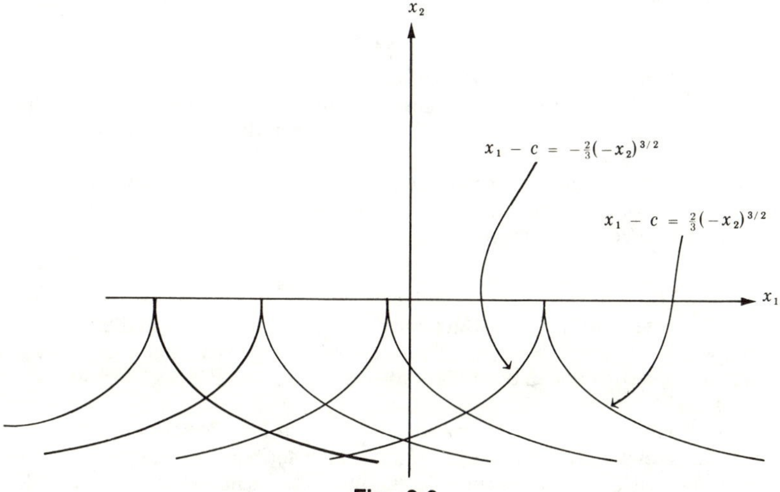

Fig. 2.3

We turn now to examples in higher dimensions.

Example 2.8. In R^n consider the Laplace operator

$$P(D) = D_1^2 + \cdots + D_n^2.$$

The characteristic equation is

$$\xi_1^2 + \cdots + \xi_n^2 = 0,$$

the only solution of which is $(\xi_1, \ldots, \xi_n) = (0, \ldots, 0)$. Hence, there are no characteristic directions and no characteristic surfaces.

Example 2.9. Consider the heat operator in R^{n+1},

$$P(D) = D_1^2 + \cdots + D_n^2 - D_t$$

where we use t for the $(n + 1)$st variable. The principal part is

$$P_2(D) = D_1^2 + \cdots + D_n^2$$

and the characteristic equation is

$$\xi_1^2 + \cdots + \xi_n^2 = 0.$$

The only characteristic direction is $(\xi_1, \ldots, \xi_n, \xi_t) = (0, \ldots, 0, 1)$ and the characteristic surfaces are the planes t = const.

Example 2.10. In R^{n+1} consider the wave operator

$$P(D) = D_1^2 + \cdots + D_n^2 - D_t^2$$

where we use t for the $(n + 1)$st variable. The characteristic equation is

$$\xi_1^2 + \cdots + \xi_n^2 - \xi_t^2 = 0.$$

If we look for vectors of unit length satisfying this equation, i.e., if we require that

$$\xi_1^2 + \cdots + \xi_n^2 + \xi_t^2 = 1,$$

then we must have $\xi_t = \pm 1/\sqrt{2}$. Since the components of a vector of unit length are the cosines of the angles that the vector makes with the corresponding coordinate axes, it follows that the characteristic directions make a 45° angle with the t-axis. Any n-dimensional surface with normal at each of its points making a 45° angle with the t-axis is characteristic. For example, the planes $t + x_1 = 0$ and $t - x_1 = 0$ are characteristic. The double conical surfaces

$$(t - t^0)^2 - (x_1 - x_1^0)^2 - \ldots - (x_n - x_n^0)^2 = 0$$

are characteristic surfaces which play a very important role in the study of the wave operator. They are called *characteristic cones*. Figure 2.4 shows a characteristic cone in three-dimensional space. Note that each point (x_1^0, x_2^0, t^0) is the apex of a characteristic cone.

In general, finding characteristic surfaces in three or more dimensions is a difficult matter. The analytical method described in Problem 2.5 leads to a first order partial differential equation which is nonlinear when $m \geq 2$.

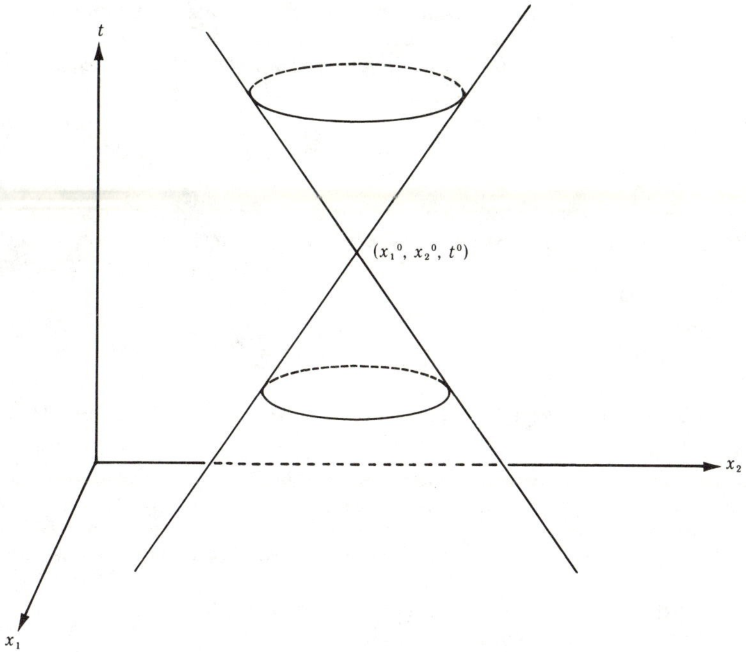

Fig. 2.4

Problems

2.1. Find the characteristic curves of each of the following operators in R^2:

(a) $P(x, D) = D_1^2 + x_2 D_2^2$

(b) $P(x, D) = D_1^2 + x_2^2 D_2^2$

(c) $P(D) = D_1^4 + 2D_1^2 D_2^2 + D_2^4$ (biharmonic operator)

(d) $P(x, D) = x_2^2 D_1^2 - 2x_1 x_2 D_1 D_2 + x_1^2 D_2^2 + x_2 D_1 + x_1 D_2$.

2.2. Find the characteristic curves of each of the equations:

(a) $y^3 u_{xx} + u_{yy} = 0$

(b) $u_x + 2xyu_y + e^x u = \cos(x + y)$.

2.3. Find the slopes of the characteristic curves in Example 2.5. Also draw the characteristic curves when (i) $c = \frac{1}{2}$, (ii) $c = 1$, and (iii) $c = 2$.

2.4. Describe the characteristic surfaces of the wave operator (with wave propagation speed c) in R^3

$$P(D) = D_1^2 + D_2^2 - \frac{1}{c^2} D_t^2.$$

2.5. Let F be a C^1 function of n variables with nonvanishing gradient and suppose that the level surfaces of F are characteristic surfaces of the general linear p.d.o. $P(x, D)$ given by (1.2). Show that F must satisfy the equation

$$\sum_{|\alpha|=m} a^{\alpha}(x)[\text{grad } F(x)]^{\alpha} = 0. \tag{2.1}$$

Thus, in order to find characteristic surfaces of $P(x, D)$ we must find solutions of the first order p.d.e. (2.1). Note that (2.1) is nonlinear if $m \geqq 2$. Also, verify that for the operator of Problem 2.4, equation (2.1) is

$$\left(\frac{\partial F}{\partial x_1}\right)^2 + \left(\frac{\partial F}{\partial x_2}\right)^2 - \frac{1}{c^2}\left(\frac{\partial F}{\partial t}\right)^2 = 0.$$

3. The Importance of Characteristics. A Very Simple Example

In this section we will illustrate the importance of characteristics by discussing the simplest possible partial differential operator, the operator $D_1 = \partial/\partial x$ in the (x, y)-plane. As we saw in the previous section, $(0, 1)$ is the only characteristic direction and the characteristics are the lines $y =$ const.

We first show that the characteristics are exceptional for the Cauchy (initial value) problem. The Cauchy problem for a first order partial differential equation in two independent variables asks for a solution u of the equation in a domain containing a curve C on which the values of u have been assigned. The curve C is called the initial curve (or initial manifold) of the problem and the assigned values of u on C are called the initial data. Suppose first that the initial curve C is nowhere characteristic with respect to D_1. Then the vector normal to C at every one of its points must have a non-zero component in the x direction and hence C must be given by an equation of the form (see Fig. 3.1)

$$x = \phi(y). \tag{3.1}$$

Consider now the initial value problem

$$D_1 u = 0, \tag{3.2}$$

$$u(\phi(y), y) = f(y), \tag{3.3}$$

where $f(y)$ is a given function. The differential equation (3.2) implies that along the lines $y =$ const., $u(x, y)$ is constant, independent of x. Hence $u(x, y) = u(\phi(y), y)$ and from the initial condition (3.3) we see that

$$u(x, y) = f(y)$$

is the unique solution to the problem (3.2), (3.3). Suppose now that the initial curve C is a characteristic curve, say the line $y = 0$, and consider the Cauchy problem

$$D_1 u = 0, \tag{3.2}$$

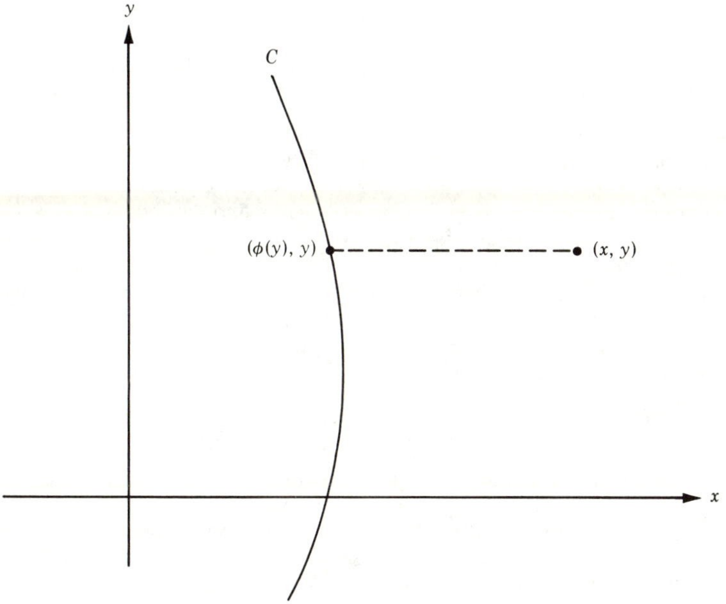

Fig. 3.1

(3.4) $$u(x, 0) = f(x),$$

where, again, f is a given function. If the function f is not identically constant then there cannot be any solution to the problem (3.2), (3.4) since the differential equation (3.2) contradicts the initial condition (3.4) on the initial line $y = 0$. On the other hand if $f(x) = c$ for all x, then for any function $g(y)$ satisfying the condition $g(0) = c$, the function

$$u(x, y) = g(y)$$

is a solution of the problem (3.2), (3.4). Thus, when the initial curve C is characteristic, either there is no solution to the Cauchy problem or there are infinitely many solutions; i.e., either there is no existence of solution or there is no uniqueness.

Another important feature of characteristics is that along a characteristic, a solution of the partial differential equation or its derivatives may admit discontinuities. We can illustrate this with the operator D_1. If f is a function of a single variable, then $u(x, y) = f(y)$ is a solution of the differential equation $D_1u = 0$. If f has a jump discontinuity at a point y_0, then the solution $u(x, y)$ has a jump discontinuity along the line $y = y_0$ which is a characteristic line. If $f'(y)$ has a jump discontinuity at y_0 then $\partial u/\partial y$ has a jump discontinuity along the characteristic $y = y_0$.

Next, characteristics play a crucial role in solving first order partial differential equations. For example a solution to the equation

(3.5) $$u_x = f(x, y)$$

is given by

$$u(x, y) = \int_{x_0}^{x} f(\xi, y)d\xi$$

where the integral is a line integral along a characteristic curve y = const. Note that along a characteristic curve y = const., the partial differential equation (3.5) is actually an ordinary differential equation. This fact is generally true for all linear first order p.d.e.'s and can be used to solve the initial value problem for these equations by solving initial value problems for o.d.e.'s.

Finally, we mention that characteristics can be used to introduce new coordinates in terms of which the differential equation has a particularly simple form, which is called the canonical form of the equation. This is done in detail for linear first order equations in Section 6 and second order equations in Section 7.

Problem

3.1. Consider the initial value problem for the equation $D_1 u = 0$ with initial curve the parabola $y = x^2$. Note that this curve is characteristic at (0, 0) but not characteristic at any other point. Show that unless the initial data satisfy a certain condition, the initial value problem has no global solution. However, if P is any point of the initial curve different from (0, 0), show that the initial value problem always has a solution in a (sufficiently small) neighborhood of P. Is this true for $P = (0, 0)$?

4. The Initial Value Problem for Linear First Order Equations in Two Independent Variables

In this section we consider the initial value problem (or Cauchy problem) for a general linear first order equation in two independent variables. Since a linear equation is a special case of a quasi-linear equation, all results concerning the existence and uniqueness of solution can be obtained as special cases of the corresponding results already proved in Chapter III. We reach the same conclusions as those obtained for the example discussed in the previous section. In brief, if the initial curve is not characteristic, there exists a unique solution. If the initial curve is characteristic, usually there is no solution and in the special case in which there is a solution there are actually infinitely many solutions. In Example 4.1 we recall from Chapter III the method of solution which is based on integrating the associated system of ordinary differential equations.

Initial Value Problem

Let the initial curve C be given parametrically by the equations

$$x = x_0(t), \qquad y = y_0(t); \qquad t \in I \tag{4.1}$$

where $x_0(t)$, $y_0(t)$ are in $C^1(I)$, and let the initial data be given by the function $\phi(t)$ which is also in $C^1(I)$. Find a function $u(x, y)$ defined in a domain Ω of R^2 containing C, such that

(i) $u = u(x, y)$ satisfies in Ω the p.d.e.

(4.2) $$a(x, y)u_x + b(x, y)u_y + c(x, y)u = f(x, y),$$

(ii) on the curve C, u is equal to the given function ϕ, i.e.

(4.3) $$u(x_0(t), y_0(t)) = \phi(t)$$

for every $t \in I$.

Concerning the p.d.e. (4.2) we assume throughout this section that the coefficients a, b, c and the right hand side f are of class C^1 in Ω and that the coefficients a and b of the principal part of (4.2) do not vanish simultaneously at any point of Ω; i.e., $a^2 + b^2 \neq 0$.

The following existence and uniqueness result follows directly from Theorem 3.1 of Chapter III.

Theorem 4.1. Let (x_0, y_0) be a point of the initial curve C and suppose that C is not characteristic at (x_0, y_0) with respect to the p.d.e. (4.2). Then in a neighborhood U of (x_0, y_0) there exists a unique solution of (4.2) satisfying the initial condition (4.3) at every point of C contained in U.

If t_0 is the value of the initial curve parameter t corresponding to the point (x_0, y_0), then the vector $\xi_0 = (dy_0(t_0)/dt, -dx_0(t_0)/dt)$ is normal to C at (x_0, y_0) (see Fig. 4.1), and the condition that C is not characteristic at (x_0, y_0) means that ξ_0 does not satisfy the characteristic equation of (4.2) at (x_0, y_0), i.e.

(4.4) $$a(x_0, y_0)\frac{dy_0(t_0)}{dt} - b(x_0, y_0)\frac{dx_0(t_0)}{dt} \neq 0.$$

This is precisely condition (3.9) of Theorem 3.1 of Chapter III specialized to the present linear case.

Briefly, Theorem 4.1 asserts the existence and uniqueness of solution of the initial value problem (4.2), (4.3) in a neighborhood of any point of the initial curve C at which C is not characteristic with respect to the equation.

The difference between the linear and quasi-linear case should be carefully noted. In the quasi-linear case, the basic condition (3.9) of Theorem 3.1 of Chapter III involves not only the differential equation and the initial curve but it involves also the initial data. In the linear case, the basic condition (4.4) involves the equation and the initial curve only and does not involve the initial data.

The word "characteristic" may be used (and it is often used) in the quasi-linear and nonlinear case as well as in the linear case. Thus the basic condition (3.9) in Theorem 3.1 of Chapter III may be expressed by saying that the initial curve C is not characteristic at (x_0, y_0) with respect to the differential equation and the given initial data. However in this book we have chosen to use the word characteristic only in the linear case.

The following special initial value problem arises frequently in applications:

(4.5) $$a(x, y)u_x + u_y + c(x, y)u = f(x, y)$$

(4.6) $$u(x, 0) = \phi(x).$$

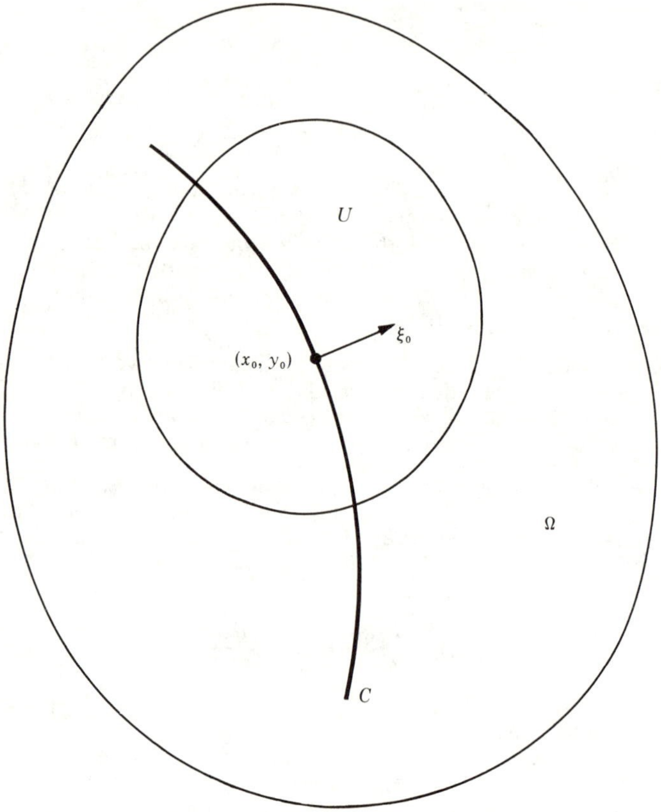

Fig. 4.1

Note that the initial curve of this problem is the x-axis. Since the vector (0, 1) is normal to the x-axis and since

$$a(x, 0)\cdot 0 + 1\cdot 1 \neq 0,$$

the x-axis is nowhere characteristic with respect to the equation (4.5). Therefore, Theorem 4.1 yields the following corollary.

Corollary 4.1. Let $(x_0, 0)$ be any point of the x-axis and suppose that a, c and f are of class C^1 in an open set containing $(x_0, 0)$ while ϕ is of class C^1 in an open interval containing x_0. Then in a neighborhood of $(x_0, 0)$ there is a unique solution of the initial value problem (4.5), (4.6).

In the following example we review the method developed in Chapter III for solving the initial value problem using the first integrals of the associated system of ordinary differential equations.

Example 4.1. Solve the initial value problem

(4.7) $$yu_x + u_y = x,$$

(4.8) $$u(x, 0) = x^2.$$

According to Corollary 4.1 there is a unique solution of this problem in a neighborhood of every point of the x-axis. Here we will be able to find a global solution valid in the whole (x, y)-plane. The system of o.d.e.'s associated with the p.d.e. (4.7) is

(4.9) $$\frac{dx}{y} = \frac{dy}{1} = \frac{du}{x}.$$

The first equality is an o.d.e., the general solution of which is

(4.10) $$x - \frac{y^2}{2} = c_1.$$

In the equality of the second and third ratios in (4.9) we eliminate x using (4.10) to obtain the o.d.e.,

$$\left(c_1 + \frac{y^2}{2}\right) dy = du,$$

the general solution of which is

$$c_1 y + \frac{y^3}{6} - u = c_2.$$

Eliminating c_1 using (4.10) we obtain

(4.11) $$xy - \frac{y^3}{3} - u = c_2.$$

Therefore, two functionally independent first integrals of (4.9) are

$$u_1 = x - \frac{y^2}{2}, \qquad u_2 = xy - \frac{y^3}{3} - u.$$

Since u_1 does not depend on u, the general integral of the p.d.e. (4.7) is given by $u_2 = F(u_1)$, or,

(4.12) $$xy - \frac{y^3}{3} - u = F\left(x - \frac{y^2}{2}\right),$$

where F is an arbitrary C^1 function of a single variable. The initial condition (4.8) determines F. In fact substituting $y = 0$ and $u = x^2$ in (4.12) we obtain

(4.13) $$F(x) = -x^2.$$

Therefore

$$xy - \frac{y^3}{3} - u = -\left(x - \frac{y^2}{2}\right)^2,$$

and solving for u we find the solution of (4.7), (4.8),

$$u = \left(x - \frac{y^2}{2}\right)^2 + xy - \frac{y^3}{3}. \tag{4.14}$$

We consider next the case in which the initial curve C, given by (4.1), is characteristic with respect to the p.d.e. (4.2) at the point $(x_0, y_0) = (x_0(t_0), y_0(t_0))$. Then the normal vector $\xi_0 = (dy_0(t_0)/dt, - dx_0(t_0)/dt)$ must satisfy the characteristic equation of (4.2) at (x_0, y_0); i.e.,

$$a(x_0, y_0)\frac{dy_0(t_0)}{dt} - b(x_0, y_0)\frac{dx_0(t_0)}{dt} = 0$$

or

$$\frac{\dfrac{dx_0(t_0)}{dt}}{a(x_0, y_0)} = \frac{\dfrac{dy_0(t_0)}{dt}}{b(x_0, y_0)}. \tag{4.15}$$

The following nonexistence result is a special case of Theorem 4.1 of Chapter III.

Theorem 4.2. Suppose that the initial curve C is characteristic with respect to (4.2) at (x_0, y_0) and that

$$\frac{\dfrac{d\phi(t_0)}{dt}}{f(x_0, y_0) - c(x_0, y_0)\phi(t_0)} \neq \mu \tag{4.16}$$

where μ is the common value of the ratios in (4.15). Then there is no solution to the initial value problem (4.2), (4.3) in any neighborhood of the point (x_0, y_0).

In Problem 4.5 we outline a direct proof of Theorem 4.2 based on the observation that the principal part $au_x + bu_y$ of a first order p.d.o. is equal to $\sqrt{a^2 + b^2}$ times the directional derivative of u in the direction of the vector field (a, b).

Finally, we consider the case in which the initial curve C is a characteristic curve with respect to equation (4.2). Then, if the initial data satisfy a certain condition, we have the following nonuniqueness result corresponding to Theorem 4.2 of Chapter III.

Theorem 4.3. Suppose that the condition

$$\frac{\dfrac{dx_0(t)}{dt}}{a(x_0(t), y_0(t))} = \frac{\dfrac{dy_0(t)}{dt}}{b(x_0(t), y_0(t))} = \frac{\dfrac{d\phi(t)}{dt}}{f(x_0(t), y_0(t)) - c(x_0(t), y_0(t))\phi(t)} \tag{4.17}$$

is satisfied for all $t \in I$ (or, at least, for all t in a neighborhood of t_0). Then in a neighborhood of $(x_0, y_0) = (x_0(t_0), y_0(t_0))$ the initial value problem (4.2), (4.3) has infinitely many solutions.

Problems

4.1. Verify by direct substitution that (4.14) is the solution of the initial value problem (4.7), (4.8) in the whole (x, y)-plane.

4.2. For each of the following two initial value problems

(i) $a(x, y)u_x + u_y + c(x, y)u = f(x, y),\ u(x, y_0) = \phi(x),$

(ii) $u_x + b(x, y)u_y + c(x, y)u = f(x, y),\ u(x_0, y) = \phi(y),$

formulate and prove existence and uniqueness results analogous to that stated in Corollary 4.1.

4.3. For each of the following initial value problems verify that there is a unique solution in a neighborhood of every point of the initial line. Then solve the problem.

(a) $2xyu_x + u_y - u = 0, \qquad u(x, 0) = x.$

(b) $u_x - yu_y - u = 1, \qquad u(0, y) = y.$

4.4. Show that the solution of the initial value problem

$$u_t + cu_x = 0, \qquad u(x, 0) = f(x),$$

where c is a positive constant, is given by

$$u(x, t) = f(x - ct).$$

If the graph of $f(x)$ is a "blip" with peak at $x = 0$ (see Fig. 1.3, Chapter VIII), sketch the solution for various values of the time variable t and interpret the solution as a wave traveling in the positive x-direction with speed c.

4.5. (i) If $\mathbf{V}$ is the vector field $\mathbf{V}(x, y) = (a(x, y), b(x, y))$ show that

$$au_x + bu_y = |\mathbf{V}|D_{\mathbf{V}}u.$$

where $D_{\mathbf{V}}u$ is the directional derivative of u in the direction of the vector $\mathbf{V}$.

(ii) The vector $\mathbf{T}(t) = (dx_0(t)/dt, dy_0(t)/dt)$ is tangent to the curve C given by (4.1). Show that

$$\frac{d}{dt}u(x_0(t), y_0(t)) = |\mathbf{T}|D_{\mathbf{T}}u.$$

(iii) Show that if C is characteristic with respect to (4.2) at (x_0, y_0), then $\mathbf{V}$ is tangent to C at (x_0, y_0) and in fact

$$\mathbf{T}_0 = \mu\mathbf{V}_0$$

where μ is the common value of the ratios in (4.15) and $\mathbf{T}_0 = \mathbf{T}(t_0)$, $\mathbf{V}_0 = \mathbf{V}(x_0, y_0)$. Consequently, at (x_0, y_0) we must have $D_{\mathbf{V}_0}u = D_{\mathbf{T}_0}u$.

(iv) Use (i), the p.d.e. (4.2) and the initial condition (4.3) to show that at (x_0, y_0)

$$D_{\mathbf{V}_0} u = \frac{1}{|\mathbf{V}_0|} [f(x_0, y_0) - c(x_0, y_0)\phi(t_0)].$$

(v) Use (ii) and the initial condition (4.3) to show that at (x_0, y_0)

$$D_{\mathbf{T}_0} u = \frac{1}{|\mathbf{T}_0|} \frac{d\phi(t_0)}{dt}$$

(vi) Prove Theorem 4.2.

5. The General Cauchy Problem. The Cauchy-Kovalevsky Theorem and Holmgren's Uniqueness Theorem

In this section we present a brief discussion of the general Cauchy problem and state the Cauchy-Kovalevsky theorem and Holmgren's uniqueness theorem for linear partial differential equations. Our purpose is not to give proofs but only to make the student aware of the content of these theorems.

General Cauchy Problem

Consider the linear p.d.e. of order m,

$$\sum_{|\alpha| \leq m} a^\alpha D^\alpha u = f \tag{5.1}$$

where the coefficients a^α and the right-hand-side f are functions of $x = (x_1, \ldots, x_n)$ in R^n. Let S be a given smooth surface in R^n and let $n = n(x)$ denote the unit vector normal to S at x. Suppose that on S the values of u and all of its directional derivatives in the direction n of order up to $m - 1$ are given, i.e.

$$u\,|_S = \phi_0, \qquad \left.\frac{\partial u}{\partial n}\right|_S = \phi_1, \ldots, \qquad \left.\frac{\partial^{m-1} u}{\partial n^{m-1}}\right|_S = \phi_{m-1}, \tag{5.2}$$

where $\phi_0, \phi_1, \ldots, \phi_{m-1}$ are given functions defined on S. Find a solution u of equation (5.1) defined in a domain Ω containing S and satisfying conditions (5.2) on S.

The surface S is called the *initial surface* of the problem and the conditions (5.2) are called the *initial conditions*. The given functions $\phi_0, \ldots, \phi_{m-1}$ which are defined on S are called the *initial data*.

The Cauchy-Kovalevsky theorem which we state below requires that all functions appearing in the statement of the problem as well as the initial surface S must be analytic. We have already defined analytic functions in Chapter IV. The surface S in R^n is said to be *analytic* if it is a level surface of an analytic function; i.e., if it is described by an equation of the form

$$F(x_1, \ldots, x_n) = 0$$

where F is an analytic function with nonvanishing gradient.

Theorem 5.1 (The Cauchy-Kovalevsky Theorem). Let x^0 be a point of the initial surface S. Suppose that the coefficients a^α, the right-hand side f, the initial data $\phi_0, \ldots, \phi_{m-1}$, and the initial surface S are all analytic in some neighborhood of x^0. Suppose furthermore that the initial surface S is not characteristic at x^0 with respect to equation (5.1), i.e.

$$\sum_{|\alpha|=m} a^\alpha(x^0)[n(x^0)]^\alpha \neq 0. \tag{5.3}$$

Then the Cauchy problem (5.1)-(5.2) has a solution $u(x)$ which is defined and analytic in a neighborhood of x^0, and this solution is unique in the class of analytic functions.

The theorem makes two assertions: (i) there exists an analytic solution in some neighborhood of x^0, and (ii) this solution is unique in the class of analytic functions. In more precise language the existence assertion states that there exists a function u which is defined and analytic in a neighborhood U of x^0 and is such that at every point $x \in U$, u satisfies the p.d.e. (5.1) and at every point x of the part of S contained in U, u satisfies the initial conditions (5.2). The uniqueness assertion states that two analytic solutions of (2.7)-(2.8) must necessarily coincide in some neighborhood of x^0. This uniqueness assertion still allows for the possibility that there may be more than one solution to the Cauchy problem if solutions which are not necessarily analytic are allowed. For example there may be two or more distinct solutions in the class of functions which are C^m in a neighborhood of x^0. It was proved by Holmgren that this cannot happen and that in fact any two C^m solutions of the Cauchy problem must necessarily coincide in a neighborhood of x^0.

Theorem 5.2. (Holmgren's Uniqueness Theorem). Suppose that all assumptions of the Cauchy-Kovalevsky theorem hold. Then any two solutions of the Cauchy problem (5.1)-(5.2) which are defined and are of class C^m in some neighborhood of x^0 must be equal in some neighborhood of x^0.

6. Canonical Form of First Order Equations

Consider the general linear first order p.d.e. in two independent variables,

$$a(x, y)u_x + b(x, y)u_y + c(x, y)u + d(x, y) = 0 \tag{6.1}$$

where the coefficients a, b, c, d are defined in some domain Ω of R^2. We assume that a and b are in $C^1(\Omega)$ and do not vanish simultaneously at any point of Ω. We will show that in a neighborhood U of any point (x_0, y_0) of Ω, we can introduce new coordinates ξ and η in terms of which the p.d.e. (6.1) takes the simple form

$$u_\xi + \gamma(\xi, \eta)u + \delta(\xi, \eta) = 0. \tag{6.2}$$

Thus, in the new coordinates the partial differential equation (6.1) becomes an ordinary differential equation with ξ as the independent variable and η as a parameter which may be treated as constant. Equation (6.2) is called the *canonical form* of equation (6.1). We also say that in the (ξ, η) coordinates the equation is in canonical form. Frequently the canonical form (6.2) can be easily integrated and, after returning to the original coordinates x and y, the general solution of the p.d.e. (6.1) can be obtained. Example 6.1 illustrates this procedure.

Let the new coordinates ξ, η be related to the old coordinates x, y by the equations

(6.3) $$\xi = \xi(x, y) \qquad \eta = \eta(x, y).$$

Since we are only interested in smooth nonsingular transformations of coordinates we must require that the functions $\xi(x, y)$, $\eta(x, y)$ be C^1 and that their Jacobian be different than zero, i.e.

(6.4) $$J \equiv \frac{\partial(\xi, \eta)}{\partial(x, y)} \equiv \xi_x\eta_y - \xi_y\eta_x \neq 0.$$

If condition (6.4) is satisfied at the point (x_0, y_0) of Ω, then we know that in a neighborhood of (x_0, y_0) we also have the inverse relations

(6.5) $$x = x(\xi, \eta) \qquad y = y(\xi, \eta).$$

Now from the chain rule, we have

(6.6) $$u_x = u_\xi\xi_x + u_\eta\eta_x, \qquad u_y = u_\xi\xi_y + u_\eta\eta_y$$

and substituting (6.5) and (6.6) into equation (6.1) we obtain the equation

(6.7) $$Au_\xi + Bu_\eta + cu + d = 0$$

where

(6.8) $$A = a\xi_x + b\xi_y, \qquad B = a\eta_x + b\eta_y.$$

From (6.8) we see that $B = 0$ if η is a solution of the first order partial differential equation

(6.9) $$a\eta_x + b\eta_y = 0.$$

Equation (6.9) has infinitely many solutions. We can find one of them by assigning initial data on a non-characteristic initial curve and solving the resulting initial value problem according to the method described in Chapter III or in Section 4 of this chapter. Supposing for example that $a(x_0, y_0) \neq 0$, we may assign

(6.10) $$\eta(x_0, y) = y.$$

Since the initial curve $x = x_0$ is not characteristic with respect to (6.9) at (x_0, y_0) (why?), there exists a unique solution of (6.9), (6.10) in a neighborhood U of (x_0, y_0). [If $b(x_0, y_0) \neq 0$ we simply reverse the roles of x and y.]

Let $\eta(x, y)$ be the solution of (6.9) and (6.10) in a neighborhood of (x_0, y_0). We are free to pick the function $\xi(x, y)$ subject only to the condition (6.4) that $J \neq 0$. From (6.10) we have

$$\eta_y(x_0, y_0) = 1$$

and if we pick

$$\xi(x, y) = x$$

condition (6.4) is satisfied at (x_0, y_0). Hence (by continuity) it is also satisfied in a neighborhood of (x_0, y_0). Let U be a neighborhood of (x_0, y_0) in which $\eta(x, y)$ is defined and at the same time $J \neq 0$. Then $A \neq 0$ in U. For if $A = 0$ at some point of U, then at that point (since $B = 0$ also) equations (6.8) would form a system of homogeneous linear equations in a and b with J being precisely the determinant of its coefficients. Since $J \neq 0$, both a and b must vanish at that point, contradicting our original assumption that a and b do not vanish simultaneously. Finally, since $B = 0$ and $A \neq 0$ in U we can divide equation (6.7) by A and obtain the desired canonical form (6.2).

It should be emphasized that the functions $\xi(x, y)$ and $\eta(x, y)$ describing the transformation of coordinates (6.3) which yields the canonical form (6.2) can be chosen in many (in fact infinitely many) ways. However, since $\eta(x, y)$ must satisfy equation (6.9), the level curves $\eta(x, y) =$ const. are always characteristic curves of equation (6.1). Thus, one set of the new coordinate curves are the characteristic curves of (6.1). The second set of coordinate curves, $\xi(x, y) =$ const., may be taken to be any one parameter family of smooth curves which are nowhere tangent to the characteristic curves (see Fig. 6.1). In the above discussion the second set of coordinate curves was chosen to be the set of lines parallel to the y-axis.

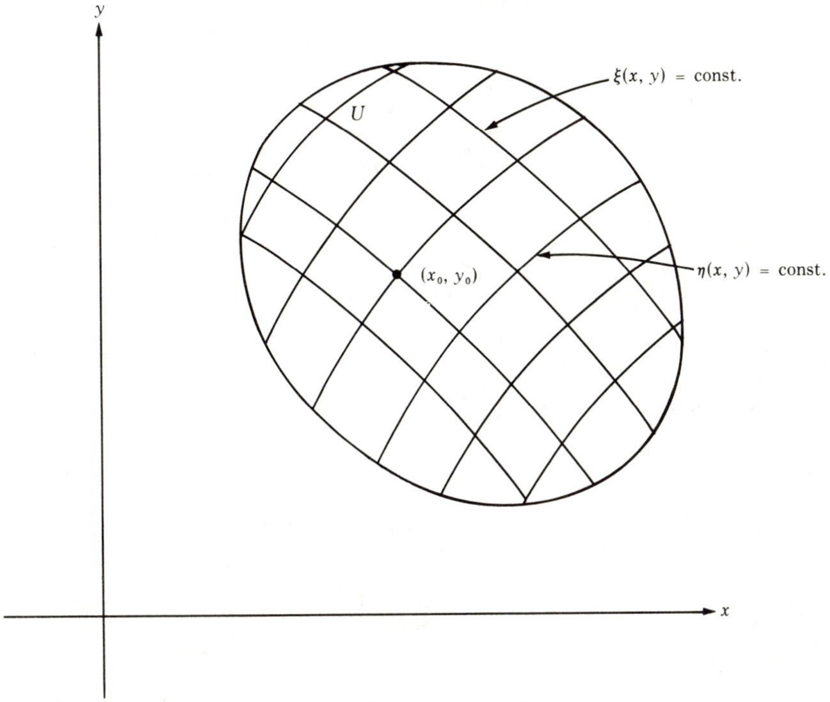

Fig. 6.1

Example 6.1. Let us consider the equation

(6.11) $$u_x + xu_y = y.$$

Here $a = 1$, $b = x$, $c = 0$, $d = -y$, and $\Omega = R^2$. We may take $(x_0, y_0) = (0, 0)$. The function η must satisfy

(6.12) $$\eta_x + x\eta_y = 0$$

and we may take as initial condition

(6.13) $$\eta(0, y) = y.$$

The general solution of $dx/1 = dy/x$ is $y - x^2/2 = c$, and, according to Example 2.2 of Chapter III, the general solution of (6.12) is $\eta = f(y - x^2/2)$. In order to satisfy (6.13) we must take $f(y) = y$ and thus we obtain the solution of (6.12), (6.13),

(6.14) $$\eta = y - \frac{x^2}{2},$$

in the whole of R^2. If we take

(6.15) $$\xi = x,$$

we see that the Jacobian is

$$J = \xi_x\eta_y - \xi_y\eta_x = 1.$$

Hence (6.14), (6.15) give a nonsingular transformation of coordinates in the whole of R^2 and the inverse relations are

$$x = \xi, \qquad y = \eta + \frac{\xi^2}{2}.$$

Now,

$$u_x = u_\xi + u_\eta(-x), \qquad u_y = u_\eta,$$

and in the new coordinates (ξ, η) the p.d.e. (6.11) becomes

(6.16) $$u_\xi = \eta + \frac{\xi^2}{2}.$$

The general solution of (6.16) is

(6.17) $$u = \eta\xi + \frac{\xi^3}{6} + f(\eta)$$

where $f(\eta)$ is an arbitrary function of η. Returning to the variables x and y we obtain the general solution of (6.11),

(6.18) $$u = xy - \frac{x^3}{3} + f\left(y - \frac{x^2}{2}\right)$$

Problems

6.1. Use the general solution (6.18) of (6.11) to find the solution of the initial value problems consisting of the p.d.e. (6.11) and each of the following initial conditions

(a) $u(0, y) = y^2$

(b) $u(0, y) = \sin y$

(c) $u(1, y) = 2y$.

6.2. For each of the following equations, first make a transformation of coordinates to obtain the canonical form. Then obtain the general solution of the equation by solving its canonical form.

(a) $u_x + u_y = u$

(b) $u_x + 2xyu_y = x$

(c) $2xyu_x + u_y - u = 0$

(d) $u_x - yu_y - u = 1$.

7. Classification and Canonical Forms of Second Order Equations in Two Independent Variables

The general linear second order partial differential equation in two independent variables is an equation of the form

$$au_{xx} + 2bu_{xy} + cu_{yy} + du_x + eu_y + fu + g = 0 \tag{7.1}$$

where a, b, c, d, e, f and g are functions of the variables x and y. In this section we assume that a, b and c are of class C^2 and do not vanish simultaneously. We will study equation (7.1) in domains Ω of R^2 in which the *discriminant*

$$\Delta = b^2 - ac \tag{7.2}$$

is either everywhere positive, everywhere negative, or everywhere zero in Ω. We will show that for every point $(x^0, y^0) \in \Omega$ we can find a neighborhood $U \subset \Omega$ of (x^0, y^0) and new coordinates ξ and η so that in U and in terms of the new coordinates the form of equation (7.1) is such that its principal part is particularly simple. We then say that the equation is in *canonical form* in U. It may be possible to find new coordinates for the whole domain Ω such that in terms of the new coordinates the equation is in canonical form in the whole of Ω. However this requires additional assumptions on a, b and c which will not be discussed here.

Let the new coordinates ξ, η be related to the old coordinates x, y by the equations

$$\xi = \xi(x, y) \qquad \eta = \eta(x, y). \tag{7.3}$$

Since we are only interested in smooth non-singular transformations of coordinates we require that the functions $\xi(x, y)$, $\eta(x, y)$ are C^2 and the Jacobian is not zero,

$$J \equiv \frac{\partial(\xi, \eta)}{\partial(x, y)} \equiv \xi_x\eta_y - \xi_y\eta_x \neq 0. \tag{7.4}$$

In a neighborhood of any point (x_0, y_0) of Ω where condition (7.4) is satisfied we also have the inverse relations

$$x = x(\xi, \eta), \qquad y = y(\xi, \eta). \tag{7.5}$$

Now, using the chain rule, we have

$$u_x = u_\xi \xi_x + u_\eta \eta_x, \qquad u_y = u_\xi \xi_y + u_\eta \eta_y \tag{7.6}$$

and

$$\begin{aligned} u_{xx} &= u_{\xi\xi}\xi_x^2 + 2u_{\xi\eta}\xi_x\eta_x + u_{\eta\eta}\eta_x^2 + \cdots \\ u_{xy} &= u_{\xi\xi}\xi_x\xi_y + u_{\xi\eta}\xi_x\eta_y + u_{\xi\eta}\xi_y\eta_x + u_{\eta\eta}\eta_x\eta_y + \cdots \\ u_{yy} &= u_{\xi\xi}\xi_y^2 + 2u_{\xi\eta}\xi_y\eta_y + u_{\eta\eta}\eta_y^2 + \cdots. \end{aligned} \tag{7.7}$$

In (7.7) we have written out only the terms involving second order derivatives of u. Here and throughout this section, dots stand for terms involving derivatives of u or order less than two. Substituting (7.5), (7.6) and (7.7) into equation (7.1) we obtain the form of the equation in the new ξ, η coordinates,

$$Au_{\xi\xi} + 2Bu_{\xi\eta} + Cu_{\eta\eta} + \cdots = 0 \tag{7.8}$$

where

$$\begin{aligned} A &= a\xi_x^2 + 2b\xi_x\,\xi_y + c\xi_y^2 \\ B &= a\xi_x\eta_x + b\xi_x\eta_y + b\xi_y\eta_x + c\xi_y\eta_y \\ C &= a\eta_x^2 + 2b\eta_x\eta_y + c\eta_y^2. \end{aligned} \tag{7.9}$$

We first note that using equations (7.9) we can prove the important relation,

$$B^2 - AC = (b^2 - ac)(\xi_x\eta_y - \xi_y\eta_x)^2 \tag{7.10}$$

or

$$\Delta' = \Delta J^2 \tag{7.11}$$

where Δ' is the discriminant of the p.d.e. in the new (ξ, η) coordinates,

$$\Delta' = B^2 - AC. \tag{7.12}$$

In view of (7.4), equation (7.11) shows that if we make a smooth nonsingular transformation of coordinates, the sign of the discriminant of equation (7.1) does not change. We restate this result in the following theorem.

Theorem 7.1. The sign of the discriminant of a second order p.d.e. in two independent variables is invariant under smooth nonsingular transformations of coordinates.

Theorem 7.1 shows that the fact that the discriminant is positive, zero, or negative is an intrinsic property of the equation which does not depend on the particular coordinate system used. This suggests that second order equations may be classified according to the sign of their discriminant.

Definition 7.1. Let Δ be the discriminant of a second order equation in two independent variables.

(a) If $\Delta > 0$ at the point (x_0, y_0), the equation is said to be *hyperbolic* at (x_0, y_0).

(b) If $\Delta = 0$ at the point (x_0, y_0), the equation is said to be *parabolic* at (x_0, y_0).

(c) If $\Delta < 0$ at the point (x_0, y_0), the equation is said to be *elliptic* at (x_0, y_0).

The equation is said to be hyperbolic, parabolic or elliptic in a domain Ω of R^2 if it is, respectively, hyperbolic, parabolic, or elliptic at every point of Ω.

Example 7.1. The wave equation

$$u_{xx} - u_{yy} = 0$$

is hyperbolic in R^2. The heat equation

$$u_{xx} - u_y = 0$$

is parabolic in R^2. Laplace's equation

$$u_{xx} + u_{yy} = 0$$

is elliptic in R^2. The Tricomi equation

$$yu_{xx} + u_{yy} = 0$$

is elliptic in the upper-half plane $y > 0$, parabolic on the line $y = 0$ and hyperbolic in the half-plane $y < 0$.

We will consider now each type of second order equation separately.

Theorem 7.2. Suppose that equation (7.1) is hyperbolic in a domain Ω. Then, in some neighborhood U of any point (x_0, y_0) of Ω, we can introduce new coordinates ξ and η in terms of which the equation has the canonical form

$$u_{\xi\eta} + \cdots = 0 \tag{7.13}$$

in U.

An alternative canonical form for hyperbolic equations can be obtained from (7.13) by performing a rotation of coordinates (see Problem 7.1). This form is

$$u_{\xi\xi} - u_{\eta\eta} + \cdots = 0. \tag{7.14}$$

Thus, in a neighborhood of any point, by introducing new coordinates, every hyperbolic equation in two independent variables can be brought to a canonical form having as principal part the same as that of the wave equation.

Proof of Theorem 7.2. In order to obtain the desired canonical form (7.13), we must choose the functions $\xi(x, y)$ and $\eta(x, y)$ so that the coefficients A and C in equation (7.8) vanish identically. Let (x_0, y_0) be any point of Ω. We may assume that not both a and c vanish at (x_0, y_0). Otherwise, we first introduce the new coordinates

$$x' = x + y, \qquad y' = x - y,$$

and verify that in the (x', y') coordinates, a and c are not both zero at $(x_0,$

y_0) (Problem 7.2). We may also assume that $a(x_0, y_0) \neq 0$. Otherwise, the roles of x and y should be interchanged in what follows. By continuity, $a(x, y) \neq 0$ in a neighborhood U of (x_0, y_0). Since $\Delta > 0$, the equation

$$a\lambda^2 + 2b\lambda + c = 0 \tag{7.15}$$

has two real and distinct roots $\lambda_1(x, y)$ and $\lambda_2(x, y)$ for all (x, y) in U. Now, let $\xi(x, y)$ and $\eta(x, y)$ be solutions of the first order p.d.e.'s

$$\xi_x = \lambda_1 \xi_y \tag{7.16}$$

$$\eta_x = \lambda_2 \eta_y \tag{7.17}$$

in U. Then, by substitution into the first and last of equations (7.9), it is easy to check that $A = C = 0$ in U. Hence, $\Delta' = B^2$ in U. Suppose now that $\xi(x, y)$ and $\eta(x, y)$, in addition to being solutions of (7.16) and (7.17), satisfy also condition (7.4) in U. Then, by (7.11), $\Delta' > 0$ in U. Hence, $B \neq 0$ in U, and dividing (7.8) by $2B$ we obtain the desired canonical form (7.13) in U.

In order to complete the proof of Theorem 7.2 it remains to show that in some neighborhood of (x_0, y_0), we can find solutions of (7.16) and (7.17) satisfying condition (7.4). Substituting (7.16) and (7.17) into (7.4) we obtain

$$J \equiv \xi_x \eta_y - \xi_y \eta_x = (\lambda_1 - \lambda_2)\xi_y \eta_y.$$

Since $\lambda_1 \neq \lambda_2$, it is only necessary to find solutions of (7.16) and (7.17) such that $\xi_y \neq 0$ and $\eta_y \neq 0$. Such solutions can be found by introducing appropriate initial conditions and solving the resulting initial value problems for ξ and η by the methods described in Chapter III or in Section 4 of this Chapter (see Problem 7.3).

It should be emphasized that the functions $\xi(x, y)$ and $\eta(x, y)$ describing the transformation of coordinates (7.3) can be chosen in many different ways. However, since these functions must satisfy equations (7.16) and (7.17), it is easy to verify (see Problem 7.4) that their level curves are characteristic curves of equation (7.1). We conclude that every hyperbolic equation has two distinct one-parameter families of characteristics and using these as new coordinate curves the equation takes the canonical form (7.13).

Example 7.2. Consider the wave equation

$$u_{xx} - u_{yy} = 0 \tag{7.18}$$

which is hyperbolic in R^2. In this equation $a = 1$, $b = 0$, $c = -1$ and equation (7.15) is

$$\lambda^2 - 1 = 0$$

which has the roots $\lambda_1 = 1$ and $\lambda_2 = -1$. The functions $\xi(x, y)$ and $\eta(x, y)$ must satisfy the equations

$$\xi_x = \xi_y, \qquad \eta_x = -\eta_y.$$

The functions

$$\xi = x + y, \qquad \eta = x - y$$

are solutions of these equations in R^2 and the Jacobian is

$$J \equiv \xi_x \eta_y - \xi_y \eta_x = -2 \neq 0.$$

In the ξ, η coordinates the wave equation has the canonical form

$$u_{\xi\eta} = 0. \tag{7.19}$$

From (7.19) we get (how?)

$$u_\xi = f(\xi) \tag{7.20}$$

where f is an arbitrary function of one variable. From (7.20) we get (how?)

$$u = F(\xi) + G(\eta) \tag{7.21}$$

where F and G are arbitrary functions of one variable. Equation (7.21) gives the general solution of (7.19). Returning to the (x, y)-variables, we obtain the general solution of the wave equation (7.18),

$$u = F(x + y) + G(x - y). \tag{7.22}$$

Theorem 7.3. Suppose that equation (7.1) is parabolic in a domain Ω. Then in some neighborhood U of any point (x_0, y_0) of Ω we can introduce new coordinates ξ and η in terms of which the equation has the canonical form

$$u_{\xi\xi} + \cdots = 0 \tag{7.23}$$

in U.

According to Theorem 7.3, in a neighborhood of any point, by introducing new coordinates, every parabolic equation in two independent variables can be brought to a canonical form having as principal part the same as that of the heat equation.

Proof of Theorem 7.3. Let (x_0, y_0) be any point of Ω. Since $\Delta = 0$, we may assume that not both a and c vanish at (x_0, y_0). Otherwise, b would also vanish at (x_0, y_0) contradicting our assumption that a, b and c do not vanish simultaneously. We may also assume that $a(x_0, y_0) \neq 0$. If $c(x_0, y_0) \neq 0$ instead, the remainder of the proof should be modified in an obvious way. By continuity, we have $a(x, y) \neq 0$ in a neighborhood U of (x_0, y_0) and since $\Delta = 0$, equation (7.15) has the single root $(-b/a)$. Let $\eta(x, y)$ be a solution of the equation

$$\eta_x = \left(-\frac{b}{a}\right)\eta_y \tag{7.24}$$

in U. By substitution into the last of equations (7.9), it is easy to check that $C = 0$ in U. As was pointed out in the proof of Theorem 7.2 it is possible to find a solution $\eta(x, y)$ of (7.24) such that $\eta_y \neq 0$ in U. For $\xi(x, y)$ we may use any function which is independent of $\eta(x, y)$ in U, for example, we may take $\xi(x, y) = x$. Then

$$J \equiv \xi_x \eta_y - \xi_y \eta_x = \eta_y \neq 0, \quad \text{in} \quad U.$$

Now, from (7.11) it follows that $\Delta' = 0$ in U. Since we already know that $C = 0$ in U, we have $\Delta' = B^2$ and hence $B = 0$ in U. Finally, from the first of equations (7.9) we have $A = a$ (since $\xi = x$) in U. Hence $A \neq 0$ and dividing (7.8) by A we obtain the desired canonical form (7.23) in U.

The functions $\xi(x, y)$ and $\eta(x, y)$ can be chosen in many different ways, but since $\eta(x, y)$ must satisfy equation (7.24), its level curves are characteristics of equation (7.1). Every parabolic equation has only one one-parameter family of characteristics. Using these characteristics as one set of new coordinate curves and any other one-parameter family of noncharacteristic curves as the second set of new coordinate curves, the equation takes the canonical form (7.23).

Theorem 7.4. Suppose that equation (7.1) is elliptic in a domain Ω. Then in some neighborhood U of any point (x_0, y_0) of Ω we can introduce new coordinates ξ and η in terms of which the equation has the canonical form

$$u_{\xi\xi} + u_{\eta\eta} + \cdots = 0 \tag{7.25}$$

in U.

According to Theorem 7.4, in a neighborhood of any point, by introducing new coordinates, every elliptic equation in two independent variables can be brought to a canonical form having as principal part the same as that of Laplace's equation.

The proof of Theorem 7.4 will not be given here since it is considerably more difficult than the proofs of the previous theorems. The interested student is referred to the book of Garabedian.[1] If the coefficients a, b and c are assumed to be analytic, the proof is simpler and involves analytic continuation into the domain of complex values of the variables x and y.

Elliptic equations have no characteristic curves.

Problems

7.1. Show that the alternative canonical form (7.14) of a hyperbolic equation can be obtained from the canonical form (7.13) by a 45° rotation of coordinates.

7.2. Suppose that both coefficients a and c of equation (7.1) vanish at the point (x_0, y_0). Show that in terms of the new coordinates $x' = x + y$, $y' = x - y$ the coefficients of $u_{x'x'}$ and $u_{y'y'}$ of the equation do not vanish at that point.

7.3. Show that in a neighborhood of (x_0, y_0) there exist solutions $\xi(x, y)$ and $\eta(x, y)$ of equations (7.16) and (7.17) satisfying the conditions $\xi_y \neq 0$ and $\eta_y \neq 0$.

7.4. Show that if λ_1 and λ_2 are roots of equation (7.15) and if $\xi(x, y)$ and $\eta(x, y)$ are solutions of (7.16) and (7.17), the level curves $\xi(x, y) =$ const. and $\eta(x, y) =$ const. are characteristic curves of equation (7.1).

7.5. Consider the second order equation

$$u_{xx} + cu_{yy} = 0$$

where c is constant. Find the type of the equation and sketch its characteristic curves for $c = -4, -1, -1/100, 0, 1$.

7.6. Prove the assertions made in this section concerning the existence or nonexistence of characteristic curves for second order hyperbolic, parabolic and elliptic equations in two independent variables. Specifically, prove that

(a) A hyperbolic equation has two one-parameter families of characteristics
(b) A parabolic equation has one one-parameter family of characteristics
(c) An elliptic equation has no characteristics.

7.7. For each of the following equations describe the regions in the (x, y)-plane where the equation is hyperbolic, parabolic or elliptic.
(a) $2u_{xx} + 4u_{xy} + 3u_{yy} - u = 0$
(b) $u_{xx} + 2xu_{xy} + u_{yy} + \sin(xy)u = 5$
(c) $yu_{xx} - 2u_{xy} + e^x u_{yy} + x^2 u_x - u = 0.$

8. Second Order Equations in Two or More Independent Variables

A second order linear p.d.e. in n independent variables is an equation of the form

(8.1)
$$\sum_{i,j=1}^{n} a_{ij} \frac{\partial^2 u}{\partial x_i \partial x_j} + \sum_{i=1}^{n} b_i \frac{\partial u}{\partial x_i} + cu = d$$

where the coefficients a_{ij}, b_i and c and the right hand side d are functions of the independent variables $x_1, \ldots, x_n$. Just as in the case of two independent variables, equation (8.1) can be classified according to a certain property of the "coefficient matrix" $[a_{ij}]$ of its principal part. Since $\partial^2 u/\partial x_i \partial x_j = \partial^2 u/\partial x_j \partial x_i$ we may assume without loss of generality that $[a_{ij}]$ is symmetric; i.e., that $a_{ij} = a_{ji}$, $i, j = 1, \ldots, n$. The property which may be used to classify equation (8.1) must be such that it remains invariant under smooth nonsingular transformations of coordinates.

For equation (7.1) in two independent variables the property used for its classification was the sign of the discriminant $\Delta = b^2 - ac$. We showed in Section 7 that the sign of Δ is invariant under smooth nonsingular transformations of coordinates. The coefficient matrix of the principal part of equation (7.1) is

(8.2)
$$\begin{bmatrix} a & b \\ b & c \end{bmatrix}.$$

The eigenvalues of this matrix are the roots of the equation

$$\begin{vmatrix} a - \lambda & b \\ b & c - \lambda \end{vmatrix} = 0$$

or

(8.3)
$$\lambda^2 - (a + c)\lambda - (b^2 - ac) = 0.$$

Let λ_1, λ_2 be the roots of (8.3). It is easy to check that λ_1 and λ_2 are real (in fact the eigenvalues of a symmetric matrix are always real) and that

$$\lambda_1 \lambda_2 = -(b^2 - ac)$$

or

(8.4)
$$\lambda_1 \lambda_2 = -\Delta.$$

(Do not confuse these λ_1 and λ_2 which are the eigenvalues of the matrix (8.2) with the roots of equation (7.15) of the previous section.) From equation (8.3) we conclude that:

(a) $\Delta > 0 \Leftrightarrow \lambda_1$ and λ_2 are non-zero and have opposite signs.

(b) $\Delta = 0 \Leftrightarrow$ at least one of λ_1, λ_2 is zero.

(c) $\Delta < 0 \Leftrightarrow \lambda_1$ and λ_2 are nonzero and have the same sign.

We can therefore classify second order equations in two independent variables according to the signs of the eigenvalues of the coefficient matrix of their principal part. In fact this is the classification scheme that we use for second order equations in two or more independent variables.

The eigenvalues of the coefficient matrix $[a_{ij}]$ are defined to be the roots of the equation

$$(8.5) \qquad \begin{vmatrix} a_{11}-\lambda & a_{12} & \cdots & a_{1n} \\ a_{21} & a_{22}-\lambda & \cdots & a_{2n} \\ \cdots & \cdots & \cdots & \cdots \\ a_{n1} & a_{n2} & \cdots & a_{nn}-\lambda \end{vmatrix} = 0.$$

It is known from linear algebra that since $[a_{ij}]$ is symmetric, its eigenvalues are all real. Moreover, using another theorem of linear algebra, it can be shown that, at a particular point x^0, the number of positive, zero and negative eigenvalues of the coefficient matrix of the principal part of equation (8.1) remains invariant under smooth non-singular transformations of coordinates. This shows that the following classification scheme for second order equations is independent of the particular coordinate system used.

Definition 8.1. Let $\lambda_1, \ldots, \lambda_n$ be the eigenvalues of the coefficient matrix $[a_{ij}]$ of the principal part of equation (8.1).

(a) If $\lambda_1, \ldots, \lambda_n$ are nonzero and have the same sign at the point x^0 the equation is said to be *elliptic* at x^0.

(b) If $\lambda_1, \ldots, \lambda_n$ are nonzero and all except one have the same sign at x^0, the equation is said to be *hyperbolic* at x^0.

(c) If $\lambda_1, \ldots, \lambda_n$ are nonzero and at least two of them are positive and two negative at x^0, the equation is said to be *ultrahyperbolic* at x^0.

(d) If any one of $\lambda_1, \ldots, \lambda_n$ is zero at x^0, the equation is called *parabolic* at x^0.

Equation (8.1) is said to be elliptic, hyperbolic, etc., in a domain Ω of R^n if it is respectively elliptic, hyperbolic, etc., at every point of Ω.

Example 8.1. Laplace's equation

$$u_{x_1x_1} + \cdots + u_{x_nx_n} = 0,$$

is elliptic in R^n. The wave equation

$$u_{x_1x_1} + \cdots + u_{x_{n-1}x_{n-1}} - u_{x_nx_n} = 0,$$

where x_n stands for the time variable t, is hyperbolic in R^n. The equation

$$u_{x_1x_1} + u_{x_2x_2} - u_{x_3x_3} - u_{x_4x_4} = 0,$$

is ultrahyperbolic in R^4. The heat equation

$$u_{x_1x_1} + \cdots + u_{x_{n-1}x_{n-1}} - u_{x_n} = 0,$$

where x_n stands for the time variable t, is parabolic in R^n.

We have seen in the previous section that by making a transformation of coordinates, every second order equation in two independent variables can be reduced to its canonical form in a whole neighborhood of any point. In general, this cannot be done for equations with variable coefficients in more than two independent variables. However, using a theorem from linear algebra it can be shown that it is always possible to make a linear transformation of coordinates and reduce equation (8.1) to a canonical form at any given single point. This canonical form is such that the coefficient matrix of its principal part is diagonal at that point. Explicitly, it can be shown that for any given point P in R^n, there is a linear transformation

$$\xi_i = \sum_{k=1}^{n} b_{ik}x_k, \qquad i = 1, \ldots, n, \tag{8.6}$$

such that in terms of the new coordinates $\xi_1, \ldots, \xi_n$, equation (8.1) has the form

$$\sum_{i,j=1}^{n} A_{ij}\frac{\partial^2 u}{\partial \xi_i \partial \xi_j} + \sum_{i=1}^{n} B_i \frac{\partial u}{\partial \xi_i} + Cu = D \tag{8.7}$$

where at P the values of the coefficients A_{ij} are

$$\begin{aligned} A_{ij}(P) &= 0 \qquad \text{if} \quad i \neq j \\ A_{ii}(P) &= +1, \ -1 \quad \text{or} \quad 0, \qquad i = 1, \ldots, n. \end{aligned} \tag{8.8}$$

Moreover the numbers $A_{11}, \ldots, A_{nn}$ differ from the eigenvalues of the coefficient matrix $[a_{ij}(P)]$ only by positive factors.

It should be emphasized that if the coefficients a_{ij} of equation (8.1) are variable, the coefficients A_{ij} of (8.7) are also variable and have the values given by (8.8) only at the given point P and not necessarily at any point other than P. If however the coefficients a_{ij} are constant, the A_{ij} are also constant and we have a canonical form which is valid everywhere.

Theorem 8.1. Suppose that the coefficients a_{ij} of equation (8.1) are constant in some domain Ω of R^n. Then there is a linear transformation of coordinates of the form (8.6) with nonsingular matrix $[b_{ij}]$, such that in terms of the new coordinates $\xi_1, \ldots, \xi_n$ equation (8.1) has the canonical form

$$\sum_{i=1}^{n} A_{ii}\frac{\partial^2 u}{\partial \xi_i^2} + \sum_{i=1}^{n} B_i \frac{\partial u}{\partial \xi_i} + Cu = D \tag{8.9}$$

in Ω, where $A_{11}, \ldots, A_{nn}$ are equal to one of the values $+1$, -1, or 0 in Ω. In particular if (8.1) is elliptic in Ω it can be reduced to the canonical form

$$\sum_{i=1}^{n} \frac{\partial^2 u}{\partial \xi_i^2} + \sum_{i=1}^{n} B_i \frac{\partial u}{\partial \xi_i} + Cu = D, \tag{8.10}$$

in Ω. If (8.1) is hyperbolic in Ω, it can be reduced to the canonical form

$$\sum_{i=1}^{n-1} \frac{\partial^2 u}{\partial \xi_i^2} - \frac{\partial^2 u}{\partial \xi_n^2} + \sum_{i=1}^{n} B_i \frac{\partial u}{\partial \xi_i} + Cu = D \tag{8.11}$$

in Ω.

We discuss now the forms of the most common second order elliptic, hyperbolic and parabolic equations that originate from physical problems. In order to do this we need the following definition.

Definition 8.2. Let $[a_{ij}]$, $i, j = 1, \ldots, n$ be a symmetric matrix. The following second order homogeneous polynomial in the variables $X_1, \ldots, X_n$,

$$Q(X) = \sum_{i,j=1}^{n} a_{ij} X_i X_j \tag{8.12}$$

is called the *quadratic form* associated with the symmetric matrix $[a_{ij}]$. As usual $X = (X_1, \ldots, X_n)$ is considered as a point in R^n. The quadratic form (8.12) is said to be *positive definite* if

$$\sum_{i,j=1}^{n} a_{ij} X_i X_j > 0, \qquad \text{for every} \quad X \neq 0 \quad \text{in} \quad R^n. \tag{8.13}$$

A well known theorem of linear algebra asserts that the eigenvalues of a symmetric matrix $[a_{ij}]$ are all positive if and only if the quadratic form associated with $[a_{ij}]$ is positive definite. In view of this theorem the following definition of ellipticity in a domain Ω is equivalent to the definition given previously. Assuming that the sign in front of equation (8.1) is chosen so that $a_{11} > 0$ in a domain Ω, equation (8.1) is said to be elliptic in Ω if the quadratic form

$$Q(x, X) = \sum_{i,j=1}^{n} a_{ij}(x) X_i X_j \tag{8.14}$$

is positive definite for every x in Ω.

Second order elliptic equations usually appear in the study of physical problems related to steady state phenomena. For example, if $u(x)$ is the steady state temperature at the point x of a nonhomogeneous, isotropic body, then at every point interior to the body, u must satisfy the second order elliptic equation

$$\sum_{i=1}^{n} \frac{\partial}{\partial x_i} \left[k(x) \frac{\partial u}{\partial x_i} \right] = 0. \tag{8.15}$$

The function $k(x)$ is always positive and is called the coefficient of thermal conductivity of the body at the point x. If the body is homogeneous, $k(x)$ is constant and equation (8.15) becomes Laplace's equation.

Wave propagation phenomena such as the propagation of sound or of electromagnetic waves are frequently described by second order hyperbolic equations of the general form

$$\frac{\partial^2 u}{\partial t^2} - \sum_{i,j=1}^{n} a_{ij} \frac{\partial^2 u}{\partial x_i \partial x_j} + \cdots = 0, \tag{8.16}$$

where the dots stand for terms of order less than two and the quadratic form associated with the matrix $[a_{ij}]$ is positive definite. In equation (8.16) there are $n + 1$ independent variables, n "space" variables $x_1, \ldots, x_n$ and one "time" variable t. It is left as an exercise to show that equation (8.16) is hyperbolic according to Definition 8.1 (see Problem 8.3).

Phenomena such as the flow of heat or the diffusion of a fluid through a porous medium are usually described by second order parabolic equations of the general form

$$\sum_{i,j=1}^{n} a_{ij} \frac{\partial^2 u}{\partial x_i \partial x_j} - \frac{\partial u}{\partial t} + \sum_{i=1}^{n} b_i \frac{\partial u}{\partial x_i} + cu = 0 \tag{8.17}$$

where the quadratic form associated with the matrix $[a_{ij}]$ is positive definite. In equation (8.17) there are $n + 1$ independent variables. Note carefully the special role of the time variable t. The principal part of the equation does not involve derivatives with respect to t and the coefficient of the first order derivative $\partial u/\partial t$ is -1. Equation (8.17) is obviously parabolic according to Definition 8.1, and because of its special character it is sometimes called *parabolic in the narrow sense*.

We close this section with a few remarks concerning characteristic surfaces of second order equations. We note first that if equation (8.1) is elliptic it has no characteristic surfaces. In fact, a nonzero vector $\xi = (\xi_1, \ldots, \xi_n) \in R^n$ defines a direction which is characteristic with respect to (8.1) if

$$\sum_{i,j=1}^{n} a_{ij}\xi_i\xi_j = 0. \tag{8.18}$$

Using the definition of ellipticity in terms of the positive definiteness of the quadratic form associated with $[a_{ij}]$ we see that (8.18) cannot be satisfied by a nonzero vector ξ. Hence a second order elliptic equation has no characteristic directions and hence no characteristic surfaces. This property of nonexistence of characteristics is used to define elliptic linear partial differential equations of any order (see Problem 8.4).

Consider next a parabolic equation of the form (8.17). A nonzero vector $\xi = (\xi_1, \ldots, \xi_n, \xi_t)$ in R^{n+1} defines a direction which is characteristic with respect to (8.17) if (8.18) is satisfied. Again, the positive definiteness of the quadratic form in (8.18) implies that $\xi_1 = \cdots = \xi_n = 0$ and, hence, $(\xi_1, \ldots, \xi_n, \xi_t) = (0, \ldots, 0, 1)$ is the only characteristic direction of (8.17). Hence, the hyperplanes t = const are the only characteristic surfaces of (8.17).

The characteristics of hyperbolic equations of the form (8.16) are more complicated. A non-zero vector $\xi = (\xi_1, \ldots, \xi_n, \xi_t)$ in R^{n+1} defines a direction which is characteristic with respect to (8.16) if

$$\xi_t^2 - \sum_{i,j=1}^{n} a_{ij}\xi_i\xi_j = 0.$$

There are infinitely many directions which satisfy this equation and the structure of the characteristic surfaces is even more complicated by the fact that the coefficients a_{ij} may be functions of x. Since the wave equation is a special case of (8.16), the student should recall the discussion of its characteristics in Example 10 of Section 2. Each point in R^{n+1} is the apex of a characteristic cone of the wave equation. This is a double cone with axis parallel to the t-axis and generators making a 45° angle with the t-axis. It divides the space R^{n+1} into three domains (except when $n = 1$). For the more general equation (8.16), each point in R^{n+1} is the apex of a characteristic "conoid." When the coefficients a_{ij} are variable, the characteristic conoid is not generated by straight lines, but it still divides R^{n+1} into three domains (except when $n = 1$).

Problems

8.1. Show that by changing the independent and dependent variables, every second order homogeneous elliptic equation with constant coefficients can be reduced to the canonical form

$$\sum_{i=1}^{n} \frac{\partial^2 v}{\partial \xi_i^2} + \lambda v = 0$$

where λ is a suitable constant. [*Hint:* In equation (8.10) with $D = 0$, set

$$v = u \exp\left(\sum_{i=1}^{n} \frac{B_i \xi_i}{2}\right).\Bigg]$$

8.2. Show that by changing the independent and dependent variables, every second order homogeneous hyperbolic equation with constant coefficients can be reduced to the canonical form

$$\sum_{i=1}^{n-1} \frac{\partial^2 v}{\partial \xi_i^2} - \frac{\partial^2 v}{\partial \xi_n^2} + \lambda v = 0$$

where λ is a suitable constant. [See Problem 8.1.]

8.3. Show that equation (8.16), with $[a_{ij}]$ positive definite, is hyperbolic.

8.4. Let $P_m(x, D)$ be the principal part of the linear partial differential operator of order m in n independent variables,

$$P(x, D) = \sum_{|\alpha| \leqq m} a^{\alpha}(x) D^{\alpha}.$$

$P(x, D)$ is said to be elliptic at the point $x^0 \in R^n$ if

$$P_m(x^0, \xi) = 0,\ \xi \in R^n \quad \Rightarrow \quad \xi = 0.$$

Note that this definition means that $P(x, D)$ is elliptic at x^0 if and only if $P(x, D)$ has no characteristic directions at x^0, and hence no characteristic surfaces passing through x^0. It can be shown that an elliptic operator must necessarily be of even order.

(a) Show that for $m = 2$ the above definition of ellipticity is equivalent to the one given in Definition 8.1.

(b) Show that the biharmonic operator

$$D_1^4 + 2D_1^2D_2^2 + D_2^4$$

is elliptic in R^2. This operator appears in the theory of elasticity.

9. The Principle of Superposition

Let $P = P(x, D)$ be a linear partial differential operator of order m in R^n,

$$P(x, D) = \sum_{|\alpha| \leq m} a^\alpha(x) D^\alpha. \tag{9.1}$$

Using the familiar fact that for any multi-index $\alpha = (\alpha_1, \ldots, \alpha_n)$ and any constants c_1 and c_2,

$$D^\alpha(c_1u_1 + c_2u_2) = c_1D^\alpha u_1 + c_2D^\alpha u_2, \tag{9.2}$$

it is easy to verify that

$$\sum_{|\alpha| \leq m} a^\alpha D^\alpha(c_1u_1 + c_2u_2) = c_1 \sum_{|\alpha| \leq m} a^\alpha D^\alpha u_1 + c_2 \sum_{|\alpha| \leq m} a^\alpha D^\alpha u_2 \tag{9.3}$$

or, more briefly,

$$P(c_1u_1 + c_2u_2) = c_1Pu_1 + c_2Pu_2. \tag{9.4}$$

In the relations (9.2)–(9.4) the functions u_1 and u_2 are any two functions which are sufficiently differentiable. (In any particular discussion, a function is said to be sufficiently differentiable if all the derivatives of the function appearing in that discussion exist.)

In the language of linear algebra, the relation (9.4) may be expressed by saying that P acts on functions u as a linear transformation. More precisely, if we consider only functions u in $C^m(\Omega)$, where Ω is a domain in R^n, then P is a linear transformation from the vector space $C^m(\Omega)$ to the vector space $C^0(\Omega)$.

As a consequence of the linearity property (9.4) of P, the solutions of the homogeneous equation

$$Pu = 0 \tag{9.5}$$

have the following *superposition property:* If u_1 and u_2 are any two solutions of the homogeneous equation (9.5) and c_1 and c_2 are arbitrary constants, then a linear combination $c_1u_1 + c_2u_2$ is also a solution of (9.5). In algebraic language this superposition property may be expressed by saying that the solutions of (9.5) form a vector space (which is usually called the null space, or kernel, of P).

The superposition property obviously holds for any finite number of solutions of the homogeneous equation (9.5). If $u_1, \ldots, u_k$ are solutions of (9.5) and $c_1, \ldots, c_k$ are arbitrary constants, then

$$c_1u_1 + \cdots + c_ku_k \tag{9.6}$$

is also a solution. The linear combination (9.6) is called a superposition of the solutions $u_1, \ldots, u_k$. Since the constants $c_1, \ldots, c_k$ may be chosen arbitrarily, by forming superpositions of a known collection of solutions of equation (9.5) we obtain a large collection of new solutions.

As an example, consider Laplace's equation in R^2,

$$(D_1^2 + D_2^2)u = 0$$

or

$$u_{xx} + u_{yy} = 0.$$

It is easy to verify that the functions

$$u_1(x, y) = 1, \qquad u_2(x, y) = x, \qquad u_3(x, y) = y$$

are solutions of the equation. By superposition, all polynomials

$$c_1 + c_2 x + c_3 y,$$

of degree $\leqq 1$ are also solutions of the equation.

It is natural to try to form superpositions of an infinite number of solutions of (9.5). Let $u_1, u_2, \ldots$ be solutions of equation (9.5) and suppose that the series

$$\sum_{k=1}^{\infty} c_k u_k$$

converges. Then the function

$$u = \sum_{k=1}^{\infty} c_k u_k \tag{9.7}$$

is also a solution of equation (9.5) provided that

$$P\left(\sum_{k=1}^{\infty} c_k u_k\right) = \sum_{k=1}^{\infty} c_k P u_k\,; \tag{9.8}$$

i.e., provided that P may be applied to the series term by term.

We may also form superpositions of a one-parameter family of solutions of (9.5). Suppose that for each value of a parameter λ in some interval I of R^1, the function $u(x, \lambda)$ is a solution of (9.5), i.e.

$$Pu(x, \lambda) = 0, \quad \text{for every} \quad \lambda \in I.$$

Suppose furthermore that for a real-valued function g defined on I, the integral

$$\int_I g(\lambda) u(x, \lambda) d\lambda \tag{9.9}$$

is convergent. Then the function

$$u(x) = \int_I g(\lambda) u(x, \lambda) d\lambda \tag{9.10}$$

is also a solution of (9.5) provided that

$$P\left[\int_I g(\lambda) u(x, \lambda) d\lambda\right] = \int_I g(\lambda) P u(x, \lambda) d\lambda,$$

i.e., provided that the order of application of P and of integration with respect to the parameter λ may be interchanged. We may also form superpositions of solutions of (9.5) depending on several parameters.

Finally, let $u(x, \lambda)$, $\lambda \in I$ be a one-parameter family of solutions of (9.5) and consider the superpositions

$$v(x, \lambda, h) = \frac{1}{h}[u(x, \lambda + h) - u(x, h)],\ h \neq 0$$

which are solutions of (9.5) depending also on the parameter h. Suppose that the limit

$$\lim_{h \to 0} v(x, \lambda, h) = \frac{\partial u}{\partial \lambda}(x, \lambda)$$

exists. Then the function

$$v(x, \lambda) = \frac{\partial}{\partial \lambda} u(x, \lambda)$$

is also a solution of (9.5) provided that

$$P\left[\frac{\partial}{\partial \lambda} u(x, \lambda)\right] = \frac{\partial}{\partial \lambda}\left[Pu(x, \lambda)\right].$$

All these methods of superposition, when they are valid, enable us to extend a known collection of solutions of a homogeneous equation to a much larger collection of solutions. We will see many examples of this in later chapters.

It should be pointed out that the principle of superposition is not valid for partial differential equations which are not linear. This can be easily seen by an example (see Problem 9.1). For this reason, it is much more difficult to obtain solutions to nonlinear partial differential equations.

Problems

9.1. Let P be the nonlinear partial differential operator in R^2 defined by

$$Pu = \left(\frac{\partial u}{\partial x} + \frac{\partial u}{\partial y}\right)^2 - u^2.$$

Show that the functions

$$u_1(x, y) = e^x, \qquad u_2(x, y) = e^{-y}$$

are solutions of the homogeneous equation

$$Pu = 0$$

while their sum

$$u(x, y) = e^x + e^{-y}$$

is not a solution.

9.2. Let P be a linear partial differential operator.

(a) Show that if u_1 and u_2 are solutions of the nonhomogeneous equations

$$Pu = f_1 \quad \text{and} \quad Pu = f_2,$$

respectively, then $u = c_1u_1 + c_2u_2$ is a solution of the equation

$$Pu = c_1f_1 + c_2f_2.$$

(b) Show that if, for each value of λ in an interval I of R^1, the function $u(x, \lambda)$ is a solution of the equation

$$Pu(x, \lambda) = f(x, \lambda),$$

and if, for some real-valued function g defined on I, the integrals

$$\int_I g(\lambda)u(x, \lambda)d\lambda, \qquad \int_I g(\lambda)f(x, \lambda)d\lambda$$

are convergent, then the function

$$u(x) = \int_I g(\lambda)u(x, \lambda)d\lambda$$

is a solution of the equation

$$Pu(x) = f(x)$$

where

$$f(x) = \int_I g(\lambda)f(x, \lambda)d\lambda.$$

Assume that the interchange of the order of application of P and of integration with respect to λ over I is valid.

9.3. (a) Verify that

$$u(x, y; \lambda) = e^{-\lambda y} \cos \lambda x, \quad -\infty < \lambda < \infty$$

is a one-parameter family of solutions of Laplace's equation in R^2.

(b) Calculate $v(x, y; \lambda) = \frac{\partial}{\partial \lambda} u(x, y; \lambda)$ and verify that $v(x, y; \lambda)$, $-\infty < \lambda < \infty$, is also a one-parameter family of solutions of Laplace's equation in R^2.

(c) For (x, y) in the upper half-plane $y > 0$, the improper integral

$$v(x, y) = \int_0^\infty u(x, y; \lambda)d\lambda = \int_0^\infty e^{-\lambda y} \cos \lambda x \, d\lambda$$

is convergent. Evaluate this integral and show by direct computation that $v(x, y)$ is a solution of Laplace's equation in the upper half-plane.

Reference for Chapter V

1. Garabedian, P. R.: *Partial Differential Equations,* New York: John Wiley & Sons, Inc., 1964.

CHAPTER VI

Equations of mathematical physics

In this chapter we discuss three of the most important partial differential equations of second order which arise in mathematical physics: the heat equation, the Laplace equation and the wave equation. In Section 1 we review the statement of the Divergence Theorem and we derive two useful integral identities known as the Green's identities. In Section 2 we derive the equation of heat conduction and describe various initial-boundary value problems associated with it. In Section 3 we describe physical phenomena, known as steady state phenomena, that are governed by Laplace's equation. In Section 4 we describe some physical phenomena that lead to the one-, two- and three-dimensional wave equation. Finally, in Section 5 we define what is a well-posed problem associated with a partial differential equation, and we give examples of some problems which are well-posed and others which are not.

1. The Divergence Theorem and the Green's Identities

The divergence theorem is one of the most useful theorems in the study of partial differential equations. This theorem is usually studied in a course on advanced calculus. In this section we review the statement of the theorem and present some immediate applications of it.

Let Ω be a bounded domain in R^3 satisfying the following conditions:
(a) The boundary $S = \partial\Omega$ of Ω consists of a finite number of smooth surfaces. (Recall that a smooth surface is a level surface of a C^1 function with nonvanishing gradient.)
(b) Any straight line parallel to any of the coordinate axes either intersects S at a finite number of points or has a whole interval common with S.

Let $\mathbf{n} = (n_x, n_y, n_z)$ be the unit normal vector to S directed in the direction exterior to Ω (see Fig. 1.1). Let

$$\mathbf{V}(x, y, z) = (P(x, y, z), Q(x, y, z), R(x, y, z)) \tag{1.1}$$

be a vector field defined in the closure $\bar{\Omega}$ of Ω such that each of the

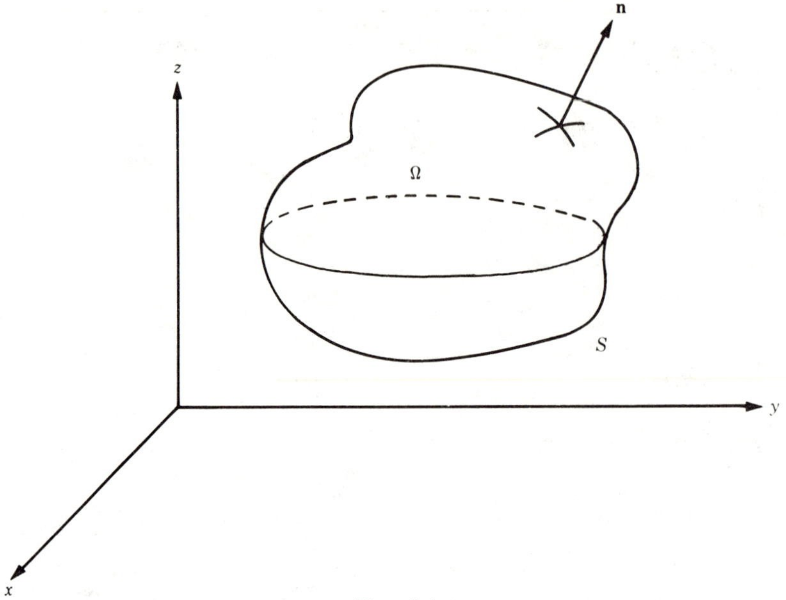

Fig. 1.1

component functions P, Q, R are in $C^1(\Omega)$ and in $C^0(\bar{\Omega})$, and suppose that the integral

$$\iiint_{\Omega} \left(\frac{\partial P}{\partial x} + \frac{\partial Q}{\partial y} + \frac{\partial R}{\partial z}\right) dxdydz$$

is convergent.

Under the above assumptions on Ω and $\mathbf{V}$ the divergence theorem asserts that

$$(1.2) \quad \iiint_{\Omega} \left(\frac{\partial P}{\partial x} + \frac{\partial Q}{\partial y} + \frac{\partial R}{\partial z}\right) dxdydz = \iint_{S} (Pn_x + Qn_y + Rn_z)d\sigma$$

where $d\sigma$ is the element of surface on S. The integrand on the left of equation (1.2) is known as the divergence of the vector field $\mathbf{V}$ and is denoted by

$$(1.3) \quad \operatorname{div} \mathbf{V} = \nabla \cdot \mathbf{V} = \frac{\partial P}{\partial x} + \frac{\partial Q}{\partial y} + \frac{\partial R}{\partial z},$$

where ∇ stands for the differentiation "vector" $\nabla = (\partial/\partial x, \partial/\partial y, \partial/\partial z) = (D_1, D_2, D_3)$. The integrand on the right of equation (1.2) is the component of $\mathbf{V}$ in the direction of the exterior normal to the boundary S. In vector notation, equation (1.2) takes the form

(1.4) $$\iiint\limits_{\Omega} \operatorname{div} \mathbf{V}\, dxdydz = \iint\limits_{S} \mathbf{V}\cdot\mathbf{n}\, d\sigma,$$

or, in even more compact notation,

(1.5) $$\int_{\Omega} \nabla\cdot\mathbf{V}\, dv = \int_{S} \mathbf{V}\cdot\mathbf{n}\, d\sigma.$$

In words, the divergence theorem asserts that if the domain Ω and the vector field $\mathbf{V}$ satisfy the conditions stated above, then the integral over Ω of the divergence of $\mathbf{V}$ is equal to the integral over the boundary S of Ω of the component of $\mathbf{V}$ in the direction of the exterior normal to S.

Conditions (a) and (b) are not the most general conditions on the domain Ω for the divergence theorem to hold. More general conditions can be found, for example, in the book of Kellogg.[1] Domains which satisfy these general conditions are called "normal." Certainly all the domains that we consider in this book are normal.

Two immediate applications of the divergence theorem are known as the Green's identities. We use here the usual notation of vector calculus. If u is a C^2 function of three variables,

(1.6) $$\nabla u = \operatorname{grad} u = \left(\frac{\partial u}{\partial x}, \frac{\partial u}{\partial y}, \frac{\partial u}{\partial z}\right)$$

and

(1.7) $$\nabla^2 u = \nabla\cdot\nabla u = \operatorname{div} \operatorname{grad} u = \frac{\partial^2 u}{\partial x^2} + \frac{\partial^2 u}{\partial y^2} + \frac{\partial^2 u}{\partial z^2}.$$

The partial differential operator ∇^2 is known as the *Laplace operator* and is also denoted by the symbol Δ,

(1.8) $$\Delta u = \nabla^2 u.$$

The following differential identity can be verified by direct computation (see Problem 1.1),

(1.9) $$u\nabla^2 w = \nabla\cdot(u\nabla w) - (\nabla u)\cdot(\nabla w).$$

Suppose now that u and w are in $C^2(\Omega)$ and in $C^1(\bar{\Omega})$ and that the integral

$$\int_{\Omega} u\nabla^2 w dv$$

is convergent. Then, integration of (1.9) over Ω yields

$$\int_{\Omega} u\nabla^2 w dv = \int_{\Omega} \nabla\cdot(u\nabla w)dv - \int_{\Omega} (\nabla u)\cdot(\nabla w)dv.$$

Applying the divergence theorem to the first integral on the right (with the vector field $\mathbf{V} = u\nabla w$) and using the fact that $\nabla w\cdot\mathbf{n}$ is the directional derivative $\partial w/\partial n$, we obtain the *first Green's identity*

(1.10) $$\int_{\Omega} u\nabla^2 w dv = \int_{S} u\frac{\partial w}{\partial n} d\sigma - \int_{\Omega} (\nabla u)\cdot(\nabla w)dv.$$

Interchanging u and w in (1.9) and subtracting the resulting identity from (1.9) yields

$$(1.11) \qquad u\nabla^2 w - w\nabla^2 u = \nabla\cdot(u\nabla w - w\nabla u).$$

If u and w are in $C^2(\Omega)$ and in $C^1(\bar{\Omega})$ and if the integral

$$\int_\Omega (u\nabla^2 w - w\nabla^2 u)dv$$

is convergent, then integration over Ω of identity (1.11) and application of the divergence theorem yields the *second Green's identity*

$$(1.12) \qquad \int_\Omega (u\nabla^2 w - w\nabla^2 u)dv = \int_S \left(u\frac{\partial w}{\partial n} - w\frac{\partial u}{\partial n}\right) d\sigma.$$

The Green's identities will be used in the study of Laplace's equation (Chapter VII).

The divergence theorem and the Green's identities are valid for vector fields and functions of any number of independent variables.

Problems

1.1. Verify the differential identity (1.9).

1.2. Let u be in $C^2(\Omega)$ and in $C^1(\bar{\Omega})$, where Ω is a normal bounded domain in R^n, and suppose that

$$\nabla^2 u = 0 \text{ in } \Omega,$$
$$u = 0 \text{ on } S,$$

where S is the boundary of Ω. Show that $u \equiv 0$ in Ω. [*Hint:* In the first Green's identity set $w = u$. Also use the fact that if the integral over Ω of a continuous nonnegative function is equal to zero, then the function is identically zero in Ω.]

1.3. Let u be in $C^2(\Omega)$ and in $C^1(\bar{\Omega})$, where Ω is a normal bounded domain in R^n, and suppose that

$$\nabla^2 u = 0 \text{ in } \Omega,$$
$$\frac{\partial u}{\partial n} = 0 \text{ on } S.$$

Show that $u \equiv$ constant in $\bar{\Omega}$.

1.4. Let $u \in C^2(\Omega) \cap C^1(\bar{\Omega})$ be a nontrivial solution of

$$\nabla^2 u + \lambda u = 0 \text{ in } \Omega,$$
$$u = 0 \text{ on } S,$$

where Ω is a bounded normal domain, and λ is a constant. Show that $\lambda \geq 0$.

2. The Equation of Heat Conduction

In this section we first derive the partial differential equation that must be satisfied by a function which describes the process of heat conduction

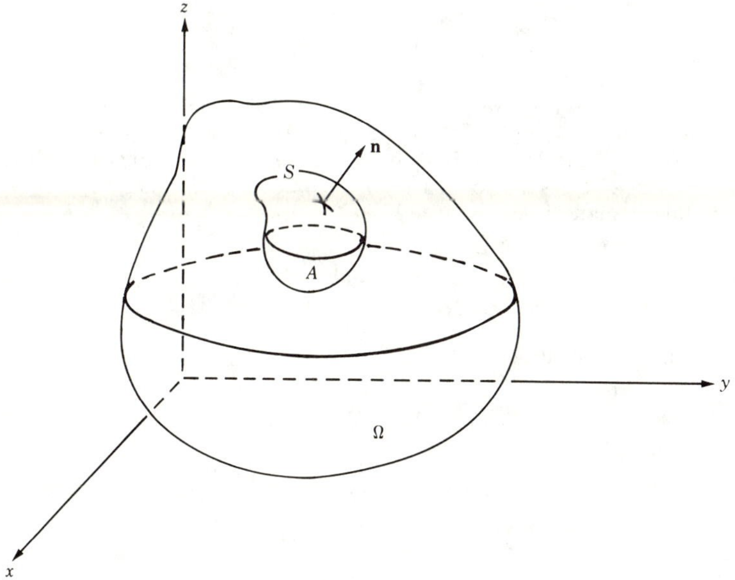

Fig. 2.1

in a body. We then discuss the supplementary conditions that must be specified in order to determine the temperature distribution in the body.

Let Ω denote the interior of the body and let the function $u(x, y, z, t)$ denote the temperature at the point (x, y, z) of the body at the time t. We assume that $u(x, y, z, t)$ is C^2 with respect to the space variables x, y, z and C^1 with respect to the time variable t.

The process of heat conduction is based on the following physical law. Let S be a smooth surface in Ω and let $\mathbf{n}$ denote a unit normal vector on S. The amount of heat (thermal energy) q that crosses S to the side of the normal $\mathbf{n}$ in the time interval from t_1 to t_2 is given by the formula

$$q = -\int_{t_1}^{t_2} \iint_S k(x, y, z) \frac{\partial u}{\partial n} \, d\sigma \, dt. \tag{2.1}$$

In (2.1) $\partial u/\partial n$ denotes the directional derivative of u in the direction of the normal $\mathbf{n}$ at the point (x, y, z) of S and at the instant t. The function $k(x, y, z)$ is positive and is called the *thermal conductivity* of the body at the point (x, y, z). We assume that the thermal conductivity $k(x, y, z)$ is a function of position (x, y, z) only and does not depend on the direction of the normal $\mathbf{n}$ to the surface S at the point (x, y, z). Bodies for which this assumption holds are said to be *isotropic* with respect to thermal conductivity.

Let us consider now within Ω a subregion A bounded by a smooth closed surface S with exterior unit normal $\mathbf{n}$ (see Fig. 2.1). The change in the amount of heat in the subregion A from $t = t_1$ to $t = t_2$ is given by

(2.2) $$\iiint_A c(x, y, z)\rho(x, y, z)[u(x, y, z, t_2) - u(x, y, z, t_1)]dxdydz.$$

In (2.2) $c(x, y, z)$ is the *specific heat* and $\rho(x, y, z)$ is the *density* of the body at the point (x, y, z). According to the law of conservation of thermal energy, this change of heat in A must be equal to the amount of heat that enters into A across the boundary S in the time interval from $t = t_1$ to $t = t_2$, and this amount of heat is given by

(2.3) $$\int_{t_1}^{t_2} \iint_S k(x, y, z) \frac{\partial u}{\partial n} d\sigma\, dt.$$

Equating the quantities (2.2) and (2.3) we obtain

(2.4) $$\iiint_A c(x, y, z)\rho(x, y, z)[u(x, y, z, t_2) - u(x, y, z, t_1)]dxdydz = \int_{t_1}^{t_2} \iint_S k(x, y, z) \frac{\partial u}{\partial n} d\sigma\, dt.$$

Now,

$$u(x, y, z, t_2) - u(x, y, z, t_1) = \int_{t_1}^{t_2} \frac{\partial u}{\partial t}(x, y, z, t)dt$$

and, since $\partial u/\partial n = \nabla u \cdot \mathbf{n}$, the divergence theorem applied to the vector field $\mathbf{V} = k\nabla u$ yields

$$\iint_S k \frac{\partial u}{\partial n} d\sigma = \iiint_A \nabla \cdot (k\nabla u)dxdydz.$$

Hence, equation (2.4) becomes

$$\int_{t_1}^{t_2} \iiint_A c\rho \frac{\partial u}{\partial t} dxdydzdt = \int_{t_1}^{t_2} \iiint_A \nabla \cdot (k\nabla u)dxdydzdt$$

or

(2.5) $$\int_{t_1}^{t_2} \iiint_A \left[c\rho \frac{\partial u}{\partial t} - \nabla \cdot (k\nabla u)\right]dxdydzdt = 0.$$

Since the integrand in (2.5) is continuous and since (2.5) is valid for all subregions A and all intervals $[t_1, t_2]$, it follows (see Problem 2.1) that the integrand must be zero for all (x, y, z) in Ω and for all t. Thus

$$c\rho \frac{\partial u}{\partial t} - \nabla \cdot (k\nabla u) = 0$$

or

$$c\rho \frac{\partial u}{\partial t} - \left[\frac{\partial}{\partial x}\left(k \frac{\partial u}{\partial x}\right) + \frac{\partial}{\partial y}\left(k \frac{\partial u}{\partial y}\right) + \frac{\partial}{\partial z}\left(k \frac{\partial u}{\partial z}\right)\right] = 0. \tag{2.6}$$

Equation (2.6) is known as the *equation of heat conduction* in an isotropic body. It is also known as the *heat equation* or the *equation of diffusion*. If the body, in addition to being isotropic, is also homogeneous, then k, ρ and c are constant and equation (2.6) takes the form

$$\frac{c\rho}{k}\frac{\partial u}{\partial t} - \left(\frac{\partial^2 u}{\partial x^2} + \frac{\partial^2 u}{\partial y^2} + \frac{\partial^2 u}{\partial z^2}\right) = 0. \tag{2.7}$$

Equation (2.7) can be simplified by changing the time scale: setting $t' = (k/c\rho)\, t$ and then dropping the prime, (2.7) becomes

$$\frac{\partial u}{\partial t} - \left(\frac{\partial^2 u}{\partial x^2} + \frac{\partial^2 u}{\partial y^2} + \frac{\partial^2 u}{\partial z^2}\right) = 0. \tag{2.8}$$

We conclude that if a function $u(x, y, z, t)$ is to describe the history of the temperature distribution in a homogeneous isotropic body during a certain interval of time, then $u(x, y, z, t)$ must satisfy equation (2.8) for all (x, y, z) in the interior Ω of the body and for all t in that time interval. However, equation (2.8) has infinitely many solutions. In order to select from this infinity of solutions the particular solution which describes the actual temperature distribution in the body, supplementary conditions must be specified.

From physical considerations it is reasonable to expect that the specification of the temperature distribution in the body at some instant of time t_0, together with the specification of the temperature distribution on the boundary $\partial\Omega$ of the body for all $t \geqq t_0$, completely determines the temperature distribution in the body for all time $t \geqq t_0$. The condition

$$u(x, y, z, t_0) = \phi(x, y, z), \; (x, y, z) \in \bar{\Omega} \tag{2.9}$$

which specifies the temperature distribution in the body at the instant t_0 is known as an *initial condition*. The function $\phi(x, y, z)$ is a given function defined in the closure $\bar{\Omega}$ of Ω. The condition

$$u(x, y, z, t) = f(x, y, z, t); \quad (x, y, z) \in \partial\Omega, t \geqq t_0 \tag{2.10}$$

which specifies the temperature distribution on the boundary $\partial\Omega$ of the body for all $t \geqq t_0$ is known as a *boundary condition*. The function $f(x, y, z, t)$ is a given function defined for (x, y, z) on the boundary $\partial\Omega$ and for all $t \geqq t_0$. The problem of finding the solution of the p.d.e. (2.8) which satisfies the initial condition (2.9) and the boundary condition (2.10) is known as an *initial-boundary value problem*. It can be shown, under some additional assumptions, that this problem has a unique solution $u(x, y, z, t)$ defined for all (x, y, z) in $\bar{\Omega}$ and all $t \geqq t_0$ (see Chapter IX). This function describes the history of the temperature distribution in the body for all $t \geqq t_0$.

Condition (2.10) is not the only boundary condition which, together with the initial condition (2.9), determines a unique solution of the heat equation. Instead of specifying the temperature on the boundary of the body, one might wish to specify the heat flux across the boundary. This leads to the boundary condition

$$\frac{\partial u}{\partial n}(x, y, z, t) = g(x, y, z, t); \quad (x, y, z) \in \partial\Omega, t \geqq t_0 \tag{2.11}$$

where $\partial u/\partial n$ denotes the directional derivative of u in the direction of the exterior normal $\mathbf{n}$ to $\partial\Omega$. The function $g(x, y, z, t)$ is a given function defined for (x, y, z) on $\partial\Omega$ and for $t \geqq t_0$. In the case of an insulated boundary, $g = 0$. Still another boundary condition can be specified. A knowledge of the temperature of the medium surrounding the body and of the heat flux across the boundary leads to the condition

$$\begin{aligned} \alpha(x, y, z)\frac{\partial u}{\partial n}(x, y, z, t) &+ \beta(x, y, z)u(x, y, z, t) \\ &= h(x, y, z, t); \quad (x, y, z) \in \partial\Omega, t \geqq t_0. \end{aligned} \tag{2.12}$$

The functions $\alpha(x, y, z)$ and $\beta(x, y, z)$ are given and defined for (x, y, z) on $\partial\Omega$ and $h(x, y, z, t)$ is given and defined for (x, y, z) on $\partial\Omega$ and $t \geqq t_0$.

Let us consider now a plate of constant thickness with its two plane surfaces insulated. If the initial temperature distribution does not vary across the thickness of the plate, then at any later time the temperature in the plate does not vary across its thickness, and if we choose the coordinate system with the z-axis perpendicular to the plate, the temperature in the plate is a function of x, y and t only. The heat equation (2.8) for the plate becomes

$$\frac{\partial u}{\partial t} - \left(\frac{\partial^2 u}{\partial x^2} + \frac{\partial^2 u}{\partial y^2}\right) = 0. \tag{2.13}$$

Finally let us consider a cylindrical rod with its cylindrical surface insulated and initial temperature constant in each cross section. If we choose the coordinate system with the center line of the rod along the x-axis, then the temperature does not vary over a cross section and so will be a function of x and t only. The heat equation for the rod is

$$\frac{\partial u}{\partial t} - \frac{\partial^2 u}{\partial x^2} = 0. \tag{2.14}$$

In closing this section it should be mentioned that equations (2.6) and (2.8) arise also in the study of the diffusion of a fluid through a porous medium and in the study of other diffusion processes involving liquids and gases.

Problems

2.1. Let $f(x_1, \cdots, x_n)$ be a continuous function in some domain Ω of R^n and suppose that for every subregion A in Ω,

$$\int \cdots \int_A f(x_1, \cdots, x_n)dx_1 \cdots dx_n = 0. \tag{2.15}$$

Show that f must be identically zero in Ω. [*Hint:* Suppose f is positive at some point P of Ω. Since f is continuous, f will be positive in some ball centered at P. Consider (2.15) when A is taken to be this ball.]

2.2. Derive equation (2.8) from (2.7).

2.3. Write down the initial-boundary value problem that must be solved to determine the history of temperature distribution in a cylindrical rod of length L with insulated cylindrical surface, given the initial temperature distribution in the rod at $t = t_0$ and the temperature at the two ends of the rod for all $t \geq t_0$.

3. Laplace's Equation

Laplace's equation

$$\frac{\partial^2 u}{\partial x^2} + \frac{\partial^2 u}{\partial y^2} + \frac{\partial^2 u}{\partial z^2} = 0 \tag{3.1}$$

arises in the study of a large class of physical phenomena known as *steady state phenomena*. These phenomena are characterized by the fact that they do not depend on the time variable t. Let us consider for example the steady state temperature distribution u in a homogeneous, isotropic body. Since the function u does not depend on the time variable t, $\partial u/\partial t = 0$ and the equation of heat conduction becomes Laplace's equation (3.1). If Ω denotes the interior of the body, the steady state temperature $u(x, y, z)$ must satisfy equation (3.1) at every point (x, y, z) in Ω.

Equation (3.1) has infinitely many solutions. In order to pick out the particular solution which describes the actual temperature distribution in the body, supplementary conditions must be specified. In contrast with the heat equation (2.8) which describes a time dependent phenomenon, no initial condition needs to be specified for equation (3.1). The time-independent forms of the boundary conditions (2.10), (2.11) and (2.12) are

$$u(x, y, z) = f(x, y, z), \qquad (x, y, z) \in \partial\Omega, \tag{3.2}$$

$$\frac{\partial u}{\partial n}(x, y, z) = g(x, y, z), \qquad (x, y, z) \in \partial\Omega, \tag{3.3}$$

$$\alpha(x, y, z)\frac{\partial u}{\partial n}(x, y, z) + \beta(x, y, z)u(x, y, z) = h(x, y, z), \qquad (x, y, z) \in \partial\Omega. \tag{3.4}$$

The problem of finding the solution of Laplace's equation (3.1) satisfying one of the boundary conditions (3.2), (3.3) or (3.4) is called a *boundary value problem*. More specifically, the problem of finding the solution of (3.1) satisfying the boundary condition (3.2) is known as the *Dirichlet problem*. The problem of solving (3.1) subject to the boundary condition (3.3) is known as the *Neumann problem*. Finally, the problem of solving (3.1) subject to the boundary condition (3.4) is known as the *mixed problem* or *third boundary value problem*. These problems will be studied in Chapter VII.

In the case of a plate of constant thickness with insulated plane surfaces, the steady state temperature u is a function of two variables only and so satisfies the two-dimensional Laplace's equation

$$\frac{\partial^2 u}{\partial x^2} + \frac{\partial^2 u}{\partial y^2} = 0, \tag{3.5}$$

(compare with equation (2.13)).

The two-dimensional Laplace equation also governs the shape of a stretched membrane such as the membrane of a drum. A membrane resists any further stretching (i.e., change in the area of any portion of the membrane) but does not resist any change in its shape. Let us suppose that the stretched membrane occupies a region of the (x, y)-plane bounded by a smooth closed curve C, and let Ω denote the interior of this region. The u-axis is taken to be orthogonal to the (x, y)-plane (see Fig. 3.1). Let the boundary C be given parametrically by the equations

$$x = x(s), \qquad y = y(s); \qquad s \in I.$$

Suppose now that each point of the boundary of the membrane is displaced along a line perpendicular to the (x, y)-plane and that the boundary is then fastened along a space curve $\tilde{C}$. The curve $\tilde{C}$ projects on the (x, y)-plane onto the curve C and is given by the equations

$$x = x(s), \qquad y = y(s), \qquad u = \phi(s); \qquad s \in I.$$

The membrane then takes the shape of a surface given by an equation of the form

$$u = u(x, y); \qquad (x, y) \in \bar{\Omega}.$$

We make now the following assumptions: (a) In displacing the membrane from the (x, y)-plane to its final shape $u = u(x, y)$, each point of the membrane moves only along a line parallel to the u-axis, (b) The membrane is bent only by a small amount so that the derivatives $\partial u/\partial x$ and $\partial u/\partial y$ are small. Under the assumptions (a) and (b) it can be shown that the function $u(x, y)$ must satisfy the two-dimensional Laplace's equation (3.5). Thus, in order to determine the final shape of the membrane we must solve the Dirichlet problem

$$\frac{\partial^2 u}{\partial x^2} + \frac{\partial^2 u}{\partial y^2} = 0; \qquad (x, y) \in \Omega$$

$$u(x, y) = \phi(x, y); \qquad (x, y) \in C.$$

Laplace's equation arises also in the study of force fields which are "derivable from a potential." For example let $\mathbf{F}$ be a force field due to a distribution of electric charges in space. $\mathbf{F}(x, y, z)$ is the force vector that would act on a unit charge placed at the point (x, y, z). It can be shown that $\mathbf{F}$ is derivable from a potential function u; i.e., there is a function u such that

$$\mathbf{F} = -\text{grad } u.$$

The potential u satisfies Laplace's equation at every point of space which is free from electric charge. A gravitational force field due to a mass distribution in space is also derivable from a potential and its potential

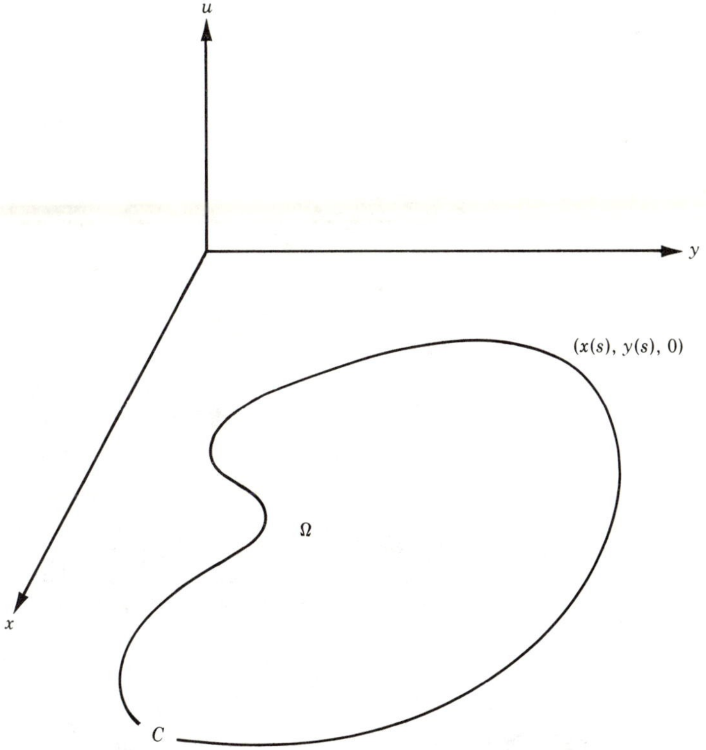

Fig. 3.1

function satisfies Laplace's equation at every point of space which is free from mass.

4. The Wave Equation

Vibration and wave propagation phenomena are governed by a partial differential equation known as the wave equation.

Let us consider first the vibrations of a stretched string such as the string of a guitar. Suppose that the length of the string is L and that, when the string is in equilibrium, it occupies the portion of the x-axis from $x = 0$ to $x = L$ (see Fig. 4.1). We assume that the string vibrates in a plane, the (x, u)-plane, and that each point of the string moves only along a line perpendicular to the x-axis (parallel to the u-axis). $u(x, t)$ denotes the displacement at the instant t of the point of the string located (when in equilibrium) at x. Under the additional assumption that $\partial u/\partial x$ is small (i.e., the vibrations of the string are small in amplitude) it can be shown that $u(x, t)$ must satisfy the p.d.e.

$$T\frac{\partial^2 u}{\partial x^2} - \rho\frac{\partial^2 u}{\partial t^2} = 0 \tag{4.1}$$

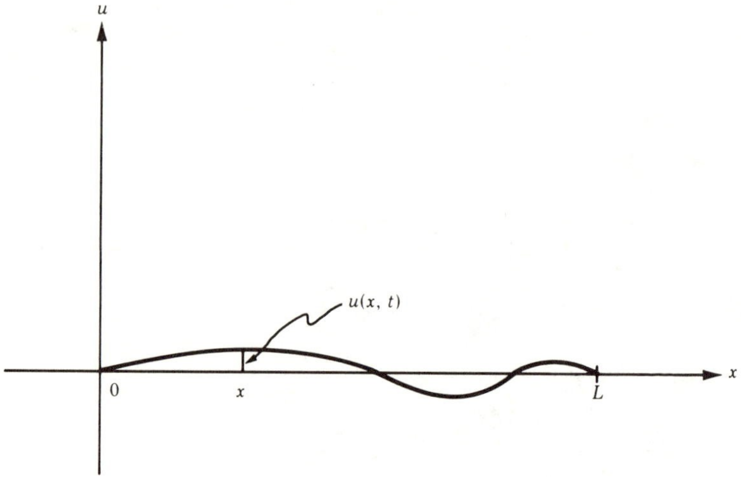

Fig. 4.1

where T is the tension of the string and ρ is its linear density. Equation (4.1) is known as the *equation of the vibrating string* or the *string equation*. It is also known as the *one-dimensional wave equation*. Setting $c = (T/\rho)^{1/2}$, equation (4.1) becomes

$$\frac{\partial^2 u}{\partial x^2} - \frac{1}{c^2}\frac{\partial^2 u}{\partial t^2} = 0. \tag{4.2}$$

As we will see in Chapter VIII, c is the speed of wave propagation in the string. Equation (4.2) can be simplified by changing the time scale: Setting $t' = ct$ and then dropping the prime, (4.2) becomes

$$\frac{\partial^2 u}{\partial x^2} - \frac{\partial^2 u}{\partial t^2} = 0. \tag{4.3}$$

The function $u(x, t)$ describing the history of motion of the string must satisfy equation (4.3) for every point x in the open interval $0 < x < L$ and for every t. Equation (4.3) has infinitely many solutions and in order to pick out the particular solution describing the actual vibration of the string supplementary conditions must be specified. Just as in the case of the heat equation, these conditions fall into two categories, initial conditions and boundary conditions. In contrast to the heat equation, two initial conditions need to be specified at the initial instant t_0,

$$u(x, t_0) = \phi(x), \qquad 0 \leqq x \leqq L \tag{4.4}$$

$$\frac{\partial u}{\partial t}(x, t_0) = \psi(x), \qquad 0 \leqq x \leqq L. \tag{4.5}$$

Condition (4.4) specifies the initial displacement of the string, while condition (4.5) specifies its initial velocity. Several types of boundary conditions at the ends $x = 0$ and $x = L$ of the string are possible, depending on the manner in which these ends are fastened or excited. These conditions specify the values of u or of the derivative $\partial u/\partial x$ at the ends of the string for all $t \geqq t_0$. For example, if both ends of the string are fixed, then,

$$u(0, t) = 0, \qquad u(L, t) = 0; \qquad t \geqq t_0. \tag{4.6}$$

The problem of finding the solution of the wave equation (4.3) subject to the initial conditions (4.4), (4.5) and to the boundary conditions (4.6) is an *initial-boundary value problem*.

If the string is "infinite" no boundary conditions need to be specified, and the problem of finding the solution of the wave equation (4.3) subject to the initial conditions

$$u(x, t_0) = \phi(x), \qquad -\infty < x < \infty \tag{4.7}$$

$$\frac{\partial u}{\partial t}(x, t_0) = \psi(x), \qquad -\infty < x < \infty \tag{4.8}$$

is an *initial value problem* or a *Cauchy problem* (compare with Chapter IV). The solution of this problem can be obtained using the general solution (7.22) of the wave equation which was derived in Chapter V.

Let us consider now the vibrations of a stretched membrane. Let $u(x, y, t)$ denote the displacement at the instant t of the point of the membrane located at (x, y) (see Fig. 3.1). Under the assumptions stated in Section 3, it can be shown that $u(x, y, t)$ must satisfy the equation

$$\frac{\partial^2 u}{\partial x^2} + \frac{\partial^2 u}{\partial y^2} - \frac{1}{c^2}\frac{\partial^2 u}{\partial t^2} = 0 \tag{4.9}$$

where $c = (T/\rho)^{1/2}$, T being the tension of the membrane and ρ its surface density. Equation (4.9) is known as the *equation of the vibrating membrane* or the *two-dimensional wave equation*. Just as in the case of the vibrating string, two initial conditions must be specified,

$$u(x, y, t_0) = \phi(x, y), \qquad (x, y) \in \bar{\Omega} \tag{4.10}$$

$$\frac{\partial u}{\partial t}(x, y, t_0) = \psi(x, y), \qquad (x, y) \in \bar{\Omega}. \tag{4.11}$$

Again, a variety of boundary conditions may be specified, depending on the manner in which the boundary of the membrane is fastened or excited. For example, if the boundary is fastened along a plane curve lying on the (x, y)-plane, the boundary condition that must be satisfied is

$$u(x, y, t) = 0; \qquad (x, y) \in \partial\Omega, \qquad t \geqq t_0. \tag{4.12}$$

Finally let us consider the propagation of acoustical (sound) waves. These are small vibrations of a gas, such as air, that occupies a region in three-dimensional space. Let Ω denote the interior of this region and let $u(x, y, z, t)$ denote the deviation from ambient (normal) pressure of the gas at the point (x, y, z) of Ω and at the instant t. Under some hypotheses, it can be shown that u must satisfy the p.d.e.

$$\frac{\partial^2 u}{\partial x^2} + \frac{\partial^2 u}{\partial y^2} + \frac{\partial^2 u}{\partial z^2} - \frac{1}{c^2}\frac{\partial^2 u}{\partial t^2} = 0, \tag{4.13}$$

where c is the speed of propagation of sound in the gas. Equation (4.13) is known as the *equation of acoustics* or the *three-dimensional wave equation*. The initial and boundary conditions associated with equation (4.13) are similar with those associated with the one and two-dimensional wave equations.

Many other vibration and wave propagation phenomena such as the propagation of electromagnetic waves are also described by the wave equation.

Problem

4.1. In Chapter V, Section 7, we have shown that the general solution of the one-dimensional wave equation (4.3) is given by

$$u(x, t) = F(x + t) + G(x - t)$$

where F and G are arbitrary functions of a single variable.

(a) Use this general solution to derive the following solution of the initial value problem (4.3), (4.7), (4.8), with $t_0 = 0$:

$$u(x, t) = \frac{1}{2}[\phi(x + t) + \phi(x - t)] + \frac{1}{2}\int_{x-t}^{x+t} \psi(\tau)d\tau. \tag{4.14}$$

(b) Show by direct substitution that (4.14) satisfies the wave equation (4.3) and the initial conditions (4.7) and (4.8), with $t_0 = 0$.

5. Well-Posed Problems

In the preceding three sections of this chapter we have seen how the study of many physical phenomena leads to problems involving partial differential equations. Let us recall two of the simplest of these problems. If $u(x, y)$ represents the steady state temperature distribution in a bounded homogeneous isotropic plate with insulated plane surfaces and if the temperature on the boundary of the plate is known, then u must be a solution of the boundary value problem

$$\frac{\partial^2 u}{\partial x^2} + \frac{\partial^2 u}{\partial y^2} = 0, \qquad (x, y) \in \Omega \tag{5.1}$$

$$u(x, y) = f(x, y), \qquad (x, y) \in \partial\Omega \tag{5.2}$$

where Ω is the interior of the plate and $\partial\Omega$ its boundary. If $u(x, t)$ represents the displacement of an "infinite" string and if the displacement and velocity of the string are known at the initial instant $t = t_0$, then u must be a solution of the initial value problem

$$\frac{\partial^2 u}{\partial x^2} - \frac{\partial^2 u}{\partial t^2} = 0; \qquad -\infty < x < \infty, \qquad t_0 < t \tag{5.3}$$

$$u(x, t_0) = \phi(x), \qquad -\infty < x < \infty \tag{5.4}$$

$$\frac{\partial u}{\partial t}(x, t_0) = \psi(x), \qquad \infty < x < \infty. \tag{5.5}$$

It is reasonable to expect that knowledge of the temperature at the boundary of the plate should completely determine the temperature at every point of the plate. It is the task of the mathematical analysis to show that this is actually the case. In order to do this we must show that the boundary value problem (5.1), (5.2) has a unique solution $u(x, y)$ defined in $\bar{\Omega}$. It should be noted carefully that this last statement consists of two parts: (i) the problem has a solution (existence of a solution) and (ii) the problem has at most one solution (uniqueness of the solution). Similarly, it is reasonable to expect that knowledge of the displacement and velocity of the string at the initial instant t_0 should completely determine the motion of the string for all subsequent time $t \geqq t_0$. Again, in order to prove this, the mathematical analysis should show that the initial value problem (5.3), (5.4), (5.5) has a unique solution $u(x, t)$ defined for all $-\infty < x < \infty$ and $t \geqq t_0$.

In subsequent chapters we will show that the problems (5.1), (5.2) and (5.3), (5.4), (5.5) have unique solutions. We will then describe methods for finding these solutions. However, in order to make sure that these solutions describe the physical phenomena under consideration in a satisfactory manner, we should also examine the dependence of the solutions on the data of the problems. Let us consider first the problem (5.1), (5.2). In practice, the function $f(x, y)$, which describes the temperature distribution on the boundary of the plate, is determined by experimental measurements which are subject to error. We must make sure that a small error in the data $f(x, y)$ does not produce a very large error in the solution $u(x, y)$ of the problem. In order to make this more precise let u be the solution of the problem with data f and let u' be the solution of the problem with different data f'. We say that the solution of problem (5.1), (5.2) *depends continuously on the data* of the problem if, given any number $\epsilon > 0$, there is a number $\delta > 0$ such that

$$\max_{(x,y)\in\bar{\Omega}} |u(x, y) - u'(x, y)| < \epsilon$$

provided that

$$\max_{(x,y)\in\partial\Omega} |f(x, y) - f'(x, y)| < \delta.$$

In other words, the solution of the problem is said to depend continuously on the data if the maximum of the change in the solution over $\bar{\Omega}$ can be made as small as we please by requiring that the maximum of the change in the data over $\partial\Omega$ is sufficiently small. The continuous dependence of the solution of problem (5.3), (5.4), (5.5) on the initial data ϕ and ψ is defined in a similar way (see Problem 5.1).

Definition 5.1. A problem involving a partial differential equation is said to be a *well-posed problem* if the following three requirements are satisfied:

(a) A solution of the problem exists.

(b) The solution is unique.
(c) The solution depends continuously on the data of the problem.

In studying a physical phenomenon by reducing it to a problem involving a p.d.e., it is not enough to establish that the problem has a unique solution. It is also necessary to know that the solution depends continuously on the data of the problem. Otherwise we cannot be sure that the solution of the problem describes the physical phenomenon with any required degree of accuracy.

The main goals of the study of partial differential equations are the following:

1. To determine the conditions under which a problem is well-posed.
2. To describe methods for finding the solution or an approximation to the solution of a well-posed problem.
3. To determine general properties of the solution.

We will show in Chapter VII that under certain assumptions the boundary value problem (5.1), (5.2) is well-posed. Under certain assumptions, the initial value problem (5.3), (5.4), (5.5) is also well-posed. In fact we have already established the existence of a solution in Problem 4.1 since (4.14) is a solution to the problem. In Chapter VIII we will show that the solution is unique and hence it is given by (4.14). Using this formula for the solution, we can also show that the solution depends continuously on the initial data (see Problem 5.1).

It should be emphasized that not every seemingly reasonable problem is well-posed. Fortunately, the study of the great majority of physical phenomena lead to initial, or boundary, or initial-boundary value problems which are well-posed.

It turns out that each partial differential equation has certain problems associated with it which are well-posed while other problems are not well-posed. In order to illustrate this let us consider again the boundary value problem (5.1), (5.2) and the initial value problem (5.3), (5.4), (5.5). These problems, which are well-posed, are strikingly different even though the Laplace equation and the wave equation differ only by a sign. It is natural to also examine the initial value problem (Cauchy problem) for the Laplace equation and the boundary value problem (Dirichlet problem) for the wave equation. It turns out that these problems are not well-posed. That the initial value problem for the Laplace equation is not well-posed was first shown by Hadamard (see Problem 5.2). We know from the Cauchy-Kovalevsky theorem that this problem has a unique solution if the initial data are assumed to be analytic. However, the problem fails to be well-posed because the solution does not depend continuously on the initial data. An example of a boundary value problem for the wave equation which is not well-posed is described in Problem 5.3. This problem is not well-posed because it has infinitely many solutions.

Problems

5.1. (a) Define what is meant by saying that the solution of the initial value problem (5.3), (5.4), (5.5) depends continuously on the initial data ϕ and ψ of the problem.

(b) Use formula (4.14), which gives the solution of the problem when $t_0 = 0$, to show that if ϕ and ψ vanish outside a finite interval, then the solution of the problem depends continuously on the initial data.

5.2. *Hadamard's example.*

(a) Consider the Cauchy problem for Laplace's equation in R^2,

$$\begin{cases} \dfrac{\partial^2 u}{\partial y^2} + \dfrac{\partial^2 u}{\partial x^2} = 0 \\ u(x, 0) = 0, \\ u_y(x, 0) = \dfrac{1}{n} \sin nx, \end{cases} \tag{5.6}$$

where n is a positive integer. Show that

$$u(x, y) = \frac{1}{n^2} \sinh ny \sin nx \tag{5.7}$$

is the solution to (5.6).

(b) Show that by taking n sufficiently large, the absolute value of the initial data in (5.6) can be made everywhere arbitrarily small, while the solution (5.7) takes arbitrarily large values even at points (x, y) with $|y|$ as small as we wish.

(c) Let f and g be analytic, and let u_1 be the solution to the Cauchy problem

$$\begin{aligned} \frac{\partial^2 u}{\partial y^2} + \frac{\partial^2 u}{\partial x^2} &= 0 \\ u(x, 0) &= f(x) \\ u_y(x, 0) &= g(x) \end{aligned} \tag{5.8}$$

and u_2 be the solution to the Cauchy problem

$$\begin{aligned} \frac{\partial^2 u}{\partial y^2} + \frac{\partial^2 u}{\partial x^2} &= 0 \\ u(x, 0) &= f(x) \\ u_y(x, 0) &= g(x) + \frac{1}{n} \sin nx. \end{aligned} \tag{5.9}$$

Show that

$$u_2(x, y) - u_1(x, y) = \frac{1}{n^2} \sinh ny \sin nx \tag{5.10}$$

(d) Study the difference in the initial data and the difference in the solutions of problems (5.8) and (5.9). Conclude that the solution to the Cauchy problem for Laplace's equation does not depend continuously on the initial data.

5.3. Consider the Dirichlet problem for the wave equation,

$$\frac{\partial^2 u}{\partial x^2} - \frac{\partial^2 u}{\partial t^2} = 0; \qquad 0 < x < L, \qquad 0 < t < T,$$

$$u(0, t) = u(L, t) = 0; \qquad 0 \leqq t \leqq T,$$

$$u(x, 0) = u(x, T) = 0; \qquad 0 \leqq x \leqq L$$

where the ratio T/L is a rational number, say $T/L = m/n$ where m and n are positive integers. Show that

$$u(x, t) = C \sin \frac{n\pi x}{L} \sin \frac{m\pi t}{T}$$

is a solution of the problem for every arbitrary constant C, and that the problem therefore has infinitely many solutions.

Reference for Chapter VI

1. Kellogg, O. D: *Foundations of Potential Theory,* New York: Dover Publications, Inc., 1953.

CHAPTER VII
Laplace's equation

This chapter is devoted to the study of Laplace's equation. This equation is of great interest to mathematicians and to engineers and scientists, because it arises in the study of many physical phenomena. In Section 1, harmonic functions are defined to be solutions of Laplace's equation which are twice continuously differentiable. In Sections 2 and 3 a large collection of harmonic functions is obtained using the methods of separation of variables, changes of variables and inversions with respect to circles and spheres. In Section 4, the boundary value problems associated with Laplace's equation are carefully described and are illustrated with physical examples. In Section 5 we use Green's second identity to prove a representation theorem which gives the value of a function at a point in terms of certain volume and surface integrals involving the function, its derivatives and the Laplacian of the function. This representation theorem is then used to prove certain fundamental properties of harmonic functions. Section 6 is devoted to the study of the well-posedness of the Dirichlet problem which asks for a function which is harmonic in a domain and has given values on the boundary of the domain. The solution of the Dirichlet problem for the unit disc is obtained in Section 7 using Fourier series. This solution is in series form. Summing this series yields the famous Poisson's integral formula for the solution. An elementary introduction to the subject of Fourier series is given in Section 8. In Section 9 the representation theorem of Section 5 is used to derive an integral formula for the solution of the Dirichlet problem for any domain, in terms of a function known as the Green's function for the domain. The Green's function and the solution of the Dirichlet problem for a ball in three-dimensional space are derived in Section 10. Section 11 is devoted to the study of further properties of harmonic functions. In particular the analyticity of harmonic functions is proved. In Section 12 it is shown how the Dirichlet problem for an unbounded domain can be transformed to the corresponding problem for a bounded domain using inversion with respect to circles and spheres. In Section 13 the Green's function for a region in three-dimensional space is interpreted as the potential due to a unit charge at a point of the region, if the boundary of the region is a grounded conducting surface. This interpretation leads to the method of electrostatic images for the construction of Green's functions. Section 14 describes the application of complex analysis to the study of Laplace's equation in two dimensions. In Section 15 the method of finite differences for obtaining approximate solutions of the Dirichlet problem is described. This numerical method is also applicable to problems associated with other p.d.e.'s. Finally, Section 16 is devoted to a brief study of the Neumann problem which asks for a function which is harmonic in a domain and has given normal derivative on the boundary of the domain.

1. Harmonic Functions

Laplace's equation

$$\frac{\partial^2 u}{\partial x_1^2} + \frac{\partial^2 u}{\partial x_2^2} + \cdots + \frac{\partial^2 u}{\partial x_n^2} = 0 \tag{1.1}$$

is the simplest and most important linear partial differential equation of elliptic type. As we saw in Chapter VI, this equation arises in the study of many physical phenomena. For this reason and also because the equation plays a fundamental role in the study of linear p.d.e.'s, equation (1.1) has been investigated by many mathematicians.

We are interested only in solutions of equation (1.1) which are twice continuously differentiable. Such solutions are called harmonic functions.

Definition 1.1. Let Ω be a domain in R^n. A function $u \in C^2(\Omega)$ which satisfies Laplace's equation in Ω is said to be *harmonic in* Ω.

Harmonic functions are not the only functions which satisfy Laplace's equation. In fact, there are functions which are not even continuous, but which nevertheless are solutions of Laplace's equation in the sense that their second order partial derivatives $\partial^2 u/\partial x_1{}^2, \cdots, \partial^2 u/\partial x_n{}^2$ exist, and the sum of these derivatives is equal to zero. (See Problem 1.4.) In more advanced books, a harmonic function is defined to be a continuous function which satisfies Laplace's equation. Our definition is only seemingly more restrictive as the following fundamental theorem shows.

Theorem 1.1. Let u be a continuous solution of Laplace's equation in a domain Ω. Then u is analytic in Ω.

According to this theorem, defining a harmonic function as a solution of Laplace's equation which is C^2 or merely continuous makes no real difference since in either case the function is actually analytic. All the theorems of this chapter concerning harmonic functions may be proved using only the assumption of continuity. We chose to assume that a harmonic function is C^2 in order to make the proofs of the theorems as simple as possible.

The proof of Theorem 1.1 is difficult and we will prove only a special case of it. The assertion of the theorem is valid for solutions of any elliptic equation with analytic coefficients.

The smoothness of solutions of elliptic equations is related to the fact that these equations have no characteristic surfaces. It is also related to the fact that elliptic equations describe steady state phenomena. Any initial uneveness is smoothed out by the time the steady state is attained.

Problems

1.1. Prove that all linear functions

$$u = a_1x_1 + a_2x_2 + \cdots + a_nx_n + a_0$$

are harmonic in R^n.

1.2. (a) Show that $u = xy$ and $u = x^2 - y^2$ are harmonic in R^2.
(b) Find all homogeneous polynomials of degree 2 which are harmonic in R^2.

1.3. (a) Show that $u = \log(x^2 + y^2)$ is harmonic in the domain consisting of R^2 excepting the origin.
(b) Show that $u = (x^2 + y^2 + z^2)^{-\frac{1}{2}}$ is harmonic in R^3 excepting the origin.

1.4. Let $z = x + iy$ be the complex variable. Consider the function u defined in R^2 by

$$u(x, y) = \begin{cases} \text{Re } e^{-1/z^4}, & \text{for } (x, y) \neq (0, 0) \\ 0, & \text{for } (x, y) = (0, 0). \end{cases}$$

Show that u satisfies Laplace's equation everywhere in R^2 and that u is not continuous at the origin.

2. Some Elementary Harmonic Functions. The Method of Separation of Variables

Our aim in this section is to obtain a collection of simple harmonic functions. Although this collection is rather small, it is possible to greatly enlarge its size by using the principle of superposition and other methods which will be described in the following sections. Recall that, according to the principle of superposition, if we have a collection of functions which are defined and harmonic in some domain Ω of R^n, then every linear combination of these functions is also harmonic in Ω.

Let us first recall (see Chapter VI) that the electrostatic potential at any point $(x, y, z) \neq (0, 0, 0)$, due to a unit charge located at the origin of R^3, is proportional to $1/r$ where r is the distance of (x, y, z) from the origin, $r = (x^2 + y^2 + z^2)^{\frac{1}{2}}$. It is well known in physics that the potential due to any distribution of charges satisfies Laplace's equation at any point of space free from charge. It takes a simple computation to show that the function

$$u = \frac{1}{r}, \qquad r \neq 0 \tag{2.1}$$

is a harmonic function in R^3 excepting the origin.

The function (2.1) is distinguished by its symmetry about the origin; it depends only on the radial distance r from the origin and does not depend on the angular variables θ and ϕ (see Fig. 2.1). This suggests that we try to find all harmonic functions in R^n which depend only on the radial variable

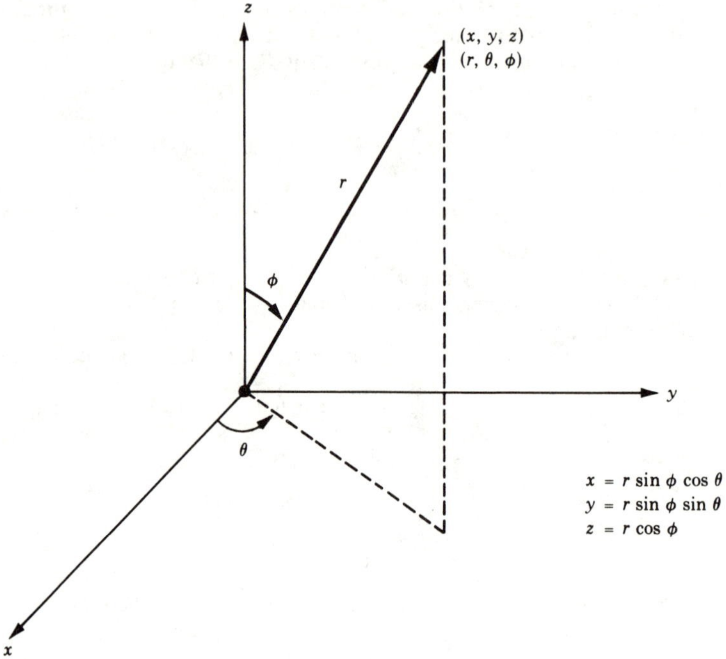

Fig. 2.1. Spherical coordinates in R^3

r. In order to do this we need the expression of Laplace's operator Δ in R^n in terms of spherical coordinates (polar coordinates in R^2). In R^2 we have

$$\Delta u = \frac{1}{r}\frac{\partial}{\partial r}\left(r\frac{\partial u}{\partial r}\right) + \frac{1}{r^2}\frac{\partial^2 u}{\partial \theta^2} \tag{2.2}$$

and in R^n with $n > 2$, we have

$$\Delta u = \frac{1}{r^{n-1}}\frac{\partial}{\partial r}\left(r^{n-1}\frac{\partial u}{\partial r}\right) + \frac{1}{r^2}\Lambda_n u \tag{2.3}$$

where Λ_n is a second order partial differential operator involving differentiations with respect to the angular variables only. In R^3,

$$\Lambda_3 u = \frac{1}{\sin\phi}\frac{\partial}{\partial\phi}\left(\sin\phi\frac{\partial u}{\partial\phi}\right) + \frac{1}{\sin^2\phi}\frac{\partial^2 u}{\partial\theta^2}\,. \tag{2.4}$$

We will not need the explicit expression of Λ_n for $n > 3$.

It is now easy to find the harmonic functions which depend only on r. In R^2, a harmonic function $u(r)$ must satisfy the equation

$$\frac{1}{r}\frac{\partial}{\partial r}\left(r\frac{\partial u}{\partial r}\right) = 0. \tag{2.5}$$

This is actually an ordinary differential equation of the second order which can be easily solved. The functions

$$1, \quad \log r \quad (n = 2) \tag{2.6}$$

are two linearly independent solutions of (2.5) and the general solution consists of all linear combinations of these functions. In R^n, with $n > 2$, a harmonic function $u(r)$ must satisfy the equation

$$\frac{1}{r^{n-1}} \frac{\partial}{\partial r}\left(r^{n-1} \frac{\partial u}{\partial r}\right) = 0, \tag{2.7}$$

and two linearly independent solutions of this equation are

$$1, \quad \frac{1}{r^{n-2}} \quad (n > 2). \tag{2.8}$$

Note that the first function in (2.6) and (2.8) is defined and harmonic in the whole of R^n, while the second function is defined and harmonic only in the complement of the origin of R^n. We will sometimes say that $\log r$ is a harmonic function in R^2 with pole at the origin and $r^{-(n-2)}$ is a harmonic function in R^n, $n > 2$, with pole at the origin.

We will now use the *method of separation of variables* or *Fourier method* to obtain other elementary harmonic functions. We do this first for R^2. The method consists of trying to find harmonic functions $u(r, \theta)$ which have the special form

$$u(r, \theta) = R(r)\Theta(\theta), \tag{2.9}$$

that is, we assume that $u(r, \theta)$ is the product of a function of r times a function of θ. Since we are only interested in real valued harmonic functions, the functions R and Θ are assumed to be real valued. Substituting (2.9) into Laplace's equation in polar coordinates, we obtain

$$R''\Theta + \frac{1}{r} R'\Theta + \frac{1}{r^2} R\Theta'' = 0,$$

where the primes denote ordinary differentiation. Dividing this equation by $R\,\Theta$, multiplying by r^2 and transferring the third term to the right side, we obtain

$$\frac{r^2R'' + rR'}{R} = -\frac{\Theta''}{\Theta}. \tag{2.10}$$

The left side of equation (2.10) is a function of r only, while the right side is a function of θ only. Now, it is not hard to see that, in order for a function of one variable to be equal to a function of another variable for all values of these variables, the two functions must be constant functions and in fact equal to the same constant. Hence (2.10) is equivalent to

$$\frac{r^2R'' + rR'}{R} = \mu = -\frac{\Theta''}{\Theta},$$

or to the pair of equations,

$$r^2R'' + rR' - \mu R = 0 \tag{2.11}$$

$$\Theta'' + \mu\Theta = 0 \tag{2.12}$$

where μ is some real constant. We conclude that for a function $u(r, \theta)$ of the form (2.9) to satisfy Laplace's equation, the functions R and Θ must satisfy the ordinary differential equations (2.11) and (2.12). We must therefore solve these equations.

Equation (2.11) is known as Euler's equation and has two linearly independent solutions,

$$R_\mu(r) = \begin{cases} 1, \log r & \text{if} \quad \mu = 0 \\ r^{\sqrt{\mu}}, r^{-\sqrt{\mu}} & \text{if} \quad \mu \neq 0. \end{cases} \tag{2.13}$$

Two linearly independent solutions of (2.12) are

$$\Theta_\mu(\theta) = \begin{cases} 1, \theta & \text{if} \quad \mu = 0 \\ \cos\sqrt{\mu}\theta, \quad \sin\sqrt{\mu}\theta & \text{if} \quad \mu \neq 0. \end{cases} \tag{2.14}$$

In (2.13) and (2.14) we use the subscript μ to indicate the dependence of the solutions on μ. When μ is negative, the functions in (2.13) and (2.14) are complex valued. The real and imaginary parts of these functions form pairs of real valued linearly independent solutions (see Problem 2.6).

It should not be assumed that, for every value of μ and for either choice of the functions in (2.13) and (2.14), the formula

$$u_\mu(r, \theta) = R_\mu(r)\Theta_\mu(\theta) \tag{2.15}$$

defines a harmonic function in any domain Ω of R^2. This is true only when (2.15) yields a "well defined" function which is C^2 in Ω. For example, we are frequently interested in finding harmonic functions in domains which contain curves encircling the origin. Examples of such domains are the whole R^2, an open disc given by $r < R$ and an open annulus given by $R_1 < r < R_2$. If Ω is such a domain and if Γ is any curve in Ω encircling the origin, it is clear that if we start at any point of Γ and travel once around Γ returning to the starting point, the angular variable θ changes by 2π. This means that in order for formula (2.15) to define a "single valued" function in Ω, the function $\Theta_\mu(\theta)$ must be periodic with period 2π; i.e., it must satisfy the condition

$$\Theta_\mu(\theta + 2\pi) = \Theta_\mu(\theta), \quad \text{for every } \theta. \tag{2.16}$$

It is left as an exercise to show that the functions $\Theta_\mu(\theta)$ given by (2.14), satisfy the periodicity conditions (2.16) only if

$$\sqrt{\mu} = n, \qquad n = 0, 1, 2, \dots,$$

and in fact, when $\mu = 0$, only the function 1 satisfies this condition. Thus, if Ω is a domain which contains curves encircling the origin, the only angular functions that can be used in (2.15) to define harmonic functions in Ω are

$$\Theta_n(\theta) = \cos n\theta, \sin n\theta; \qquad n = 1, 2, \dots. \tag{2.17}$$

The corresponding radial functions are

$$R_n(r) = \begin{cases} 1, \log r; & n = 0 \\ r^n, r^{-n}; & n = 1, 2, \dots, \end{cases} \tag{2.18}$$

and (2.15) yields the collection of functions,

$$u_n(r, \theta) = \begin{cases} 1, r^n \cos n\theta, r^n \sin n\theta; & n = 1, 2, \cdots \\ \log r, r^{-n} \cos n\theta, r^{-n} \sin n\theta; & n = 1, 2, \dots . \end{cases} \tag{2.19}$$

If Ω does not contain the origin of R^2 all functions in (2.19) are harmonic in Ω. If Ω contains the origin, only the functions on the first line are harmonic in Ω.

Let us suppose now that Ω is a domain of R^2 that does not contain curves encircling the origin. Then

$$u(r, \theta) = \theta, \tag{2.20}$$

with θ restricted to an appropriate range of values, defines a harmonic function in Ω. For example, if Ω is the right half plane $x > 0$, we may take $-\pi/2 < \theta < \pi/2$. In rectangular coordinates the harmonic function (2.20) is then

$$u(x, y) = \arctan\left(\frac{y}{x}\right), \qquad 0 < x < \infty, -\infty < y < \infty. \tag{2.21}$$

If Ω is the upper half plane $y > 0$, we may take $0 < \theta < \pi$ and in rectangular coordinates the corresponding harmonic function is

$$u(x, y) = \frac{\pi}{2} - \arctan\left(\frac{x}{y}\right), \qquad -\infty < x < \infty, 0 < y < \infty. \tag{2.22}$$

We now apply the method of separation of variables to obtain harmonic functions in domains of R^3. In this case we look for harmonic functions $u(r, \theta, \phi)$ of the form

$$u(r, \theta, \phi) = R(r)Y(\theta, \phi). \tag{2.23}$$

Substituting (2.23) into Laplace's equation in spherical coordinates and proceeding as above, we obtain the pair of equations

$$(r^2R')' - \mu R = 0 \tag{2.24}$$

$$\Lambda_3 Y + \mu Y = 0 \tag{2.25}$$

where μ is, again, a real constant. Two linearly independent solutions of equation (2.24) are

$$r^{\alpha_1}, r^{\alpha_2} \tag{2.26}$$

where α_1 and α_2 are roots of the equation

$$\alpha(\alpha + 1) - \mu = 0.$$

Equation (2.25) is a partial differential equation which is considerably more difficult to solve. It is useful to consider (θ, ϕ) as the coordinates of a point on the surface of the unit sphere $S(0, 1)$ centered at the origin of R^3. Instead of trying to find all solutions of equation (2.25) it is frequently sufficient to know only those solutions $Y(\theta, \phi)$ which are defined and C^2

on the whole of $S(0, 1)$. Such solutions must be periodic in θ, with period 2π, and at the poles of the sphere (i.e., at the points where $\phi = 0$ and $\phi = \pi$) the solutions must approach limits independent of θ. It can be shown (see Courant and Hilbert,[1] Vol. I, Chapter VII, §5) that equation (2.25) has nontrivial solutions satisfying these conditions only when μ is equal to one of the values

$$\mu_n = n(n + 1), \qquad n = 0, 1, 2, \dots .$$

For each such μ_n there are $2n + 1$ linearly independent solutions of (2.25) denoted by

$$Y_n^{(k)}(\theta, \phi), \quad k = 1, 2, \dots, 2n + 1.$$

These solutions are called the *Laplace spherical harmonics*. For $\mu = \mu_n$, the corresponding radial functions are

$$r^n, r^{-n-1}; \qquad n = 0, 1, 2, \dots,$$

and the corresponding harmonic functions (2.23) are

$$u_{n,k}(r, \theta, \phi) = \begin{cases} r^n Y_n^{(k)}(\theta, \phi); & k = 1, 2, \dots, 2n + 1; \quad n = 0, 1, 2, \dots \\ r^{-n-1} Y_n^{(k)}(\theta, \phi); & k = 1, 2, \dots, 2n + 1; \quad n = 0, 1, 2, \dots . \end{cases} \tag{2.27}$$

If Ω does not contain the origin of R^3 all functions in (2.27) are harmonic in Ω. Otherwise only the functions on the first line are harmonic in Ω.

The method of separation of variables may be used to find solutions of partial differential equations other than Laplace's equation. It may be used also with coordinate systems other than polar or spherical (see Problem 2.7).

Problems

2.1. Derive the linearly independent solutions (2.6) of equation (2.5) and the linearly independent solutions (2.8) of (2.7).

2.2. Derive the solutions (2.13) and (2.14) of equations (2.11) and (2.12).

2.3. Verify by direct substitution in Laplace's equation that the functions (2.19) are harmonic in appropriate domains of R^2.

2.4. Derive the separated equations (2.24) and (2.25).

2.5. Derive the solutions (2.26) of equation (2.24).

2.6. Find pairs of real valued linearly independent solutions of equations (2.11) and (2.12) when μ is negative. [You may need the following formulas: If $z = x + iy$ where x and y are real, then $e^z = e^x (\cos y + i \sin y)$. If $r > 0$, then

$$r^z = e^{z \log r}.]$$

2.7. (a) Use the method of separation of variables in rectangular coordinates to obtain the following harmonic functions which are bounded in the upper half plane $y > 0$ of R^2,

$$e^{-\lambda y} \cos \lambda x, \qquad e^{-\lambda y} \sin \lambda x; \qquad \lambda \geqq 0. \tag{2.28}$$

(b) By integration with respect to the parameter λ, obtain the functions

$$\frac{x}{x^2 + y^2}, \qquad \frac{y}{x^2 + y^2} \tag{2.29}$$

which are harmonic in the upper half plane $y > 0$ of R^2.

3. Changes of Variables Yielding New Harmonic Functions. Inversion with Respect to Circles and Spheres

In the previous section we obtained a collection of harmonic functions by the method of separation of variables. By the principle of superposition, all linear combinations of these functions are also harmonic. In this section we describe how to obtain new harmonic functions from known ones by changing variables.

We first consider harmonic functions in R^2. Let Ω and Ω' be two domains in R^2 and suppose that there is a one-to-one mapping from Ω to Ω' given by

$$x' = x'(x, y), \qquad y' = y'(x, y), \tag{3.1}$$

with the inverse mapping from Ω' to Ω given by

$$x = x(x', y'), \qquad y = y(x', y'). \tag{3.2}$$

We assume that the functions $x'(x, y)$ and $y'(x, y)$ are in $C^2(\Omega)$, while the functions $x(x', y')$ and $y(x', y')$ are in $C^2(\Omega')$. Let $u(x, y)$ be a given function defined in Ω and let $u(x', y')$ be defined in Ω' by the formula

$$u(x', y') = u(x(x', y'), y(x', y')). \tag{3.3}$$

The mapping (3.1), (3.2) may be thought of as a transformation of coordinates or change of variables. In Chapter V we used transformation of coordinates to reduce a second order equation to its canonical form. For example, if $u(x, y)$ satisfies an elliptic equation, then, in a neighborhood of any point, we can introduce new coordinates x', y' such that in terms of the new coordinates the function $u(x', y')$ defined by (3.3) satisfies an equation having the Laplacian as its principal part. In this section we are

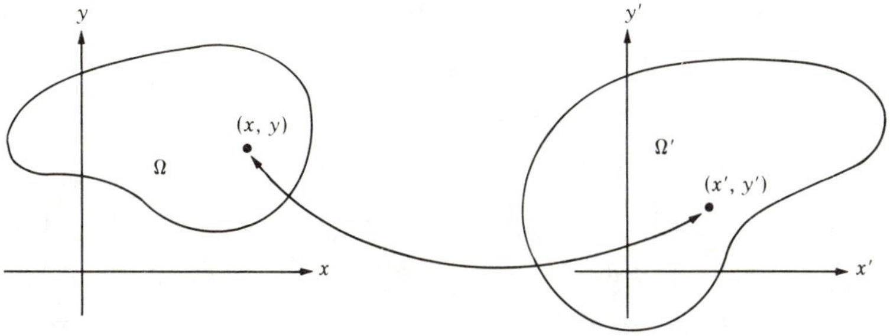

Fig. 3.1

interested in a different kind of question: Suppose that $u(x, y)$ satisfies Laplace's equation in Ω. For what coordinate transformations does $u(x', y')$ also satisfy Laplace's equation in Ω' with respect to the coordinates x', y'? It is easy to show, using the chain rule of differentiation, that this is the case for the following *elementary transformations:*

1. *Translations:*

$$x' = x + x_0, \qquad y' = y + y_0;$$
$$x = x' - x_0, \qquad y = y' - y_0,$$

where (x_0, y_0) is a fixed point of R^2.

2. *Rotations:*

$$x' = (\cos \alpha)x + (\sin \alpha)y, \qquad y' = -(\sin \alpha)x + (\cos \alpha)y;$$
$$x = (\cos \alpha)x' - (\sin \alpha)y', \qquad y = (\sin \alpha)x' + (\cos \alpha)y',$$

where α is a fixed angle.

3. *Reflections:* Here we mean reflections about any straight line in R^2. For example,

$$x' = x, \quad y' = -y; \qquad x = x', \quad y = -y',$$

represents a reflection about the x-axis, while

$$x' = -x, \quad y' = y; \qquad x = -x', \quad y = y',$$

represents a reflection about the y-axis and

$$x' = y, \quad y' = x; \qquad x = y', \quad y = x',$$

represents a reflection about the line $y = x$. A reflection about any straight line may be obtained as a combination of one of these reflections with a translation and a rotation.

4. *Similarity transformations:*

$$x' = \lambda x, \quad y' = \lambda y; \qquad x = \frac{1}{\lambda} x', \quad y = \frac{1}{\lambda} y',$$

where λ is a non-zero real constant.

Let us prove for example that if $u(x, y)$ is harmonic in Ω, then, under a rotation, $u(x', y')$ is harmonic in Ω'. We have

$$\frac{\partial u}{\partial x'} = \frac{\partial u}{\partial x} \cos \alpha + \frac{\partial u}{\partial y} \sin \alpha,$$

$$\frac{\partial^2 u}{\partial x'^2} = \frac{\partial^2 u}{\partial x^2} \cos^2 \alpha + 2 \frac{\partial^2 u}{\partial x \partial y} \cos \alpha \sin \alpha + \frac{\partial^2 u}{\partial y^2} \sin^2 \alpha,$$

$$\frac{\partial u}{\partial y'} = \frac{\partial u}{\partial x} (-\sin \alpha) + \frac{\partial u}{\partial y} \cos \alpha,$$

$$\frac{\partial^2 u}{\partial y'^2} = \frac{\partial^2 u}{\partial x^2} \sin^2 \alpha - 2 \frac{\partial^2 u}{\partial x \partial y} \sin \alpha \cos \alpha + \frac{\partial^2 u}{\partial y^2} \cos^2 \alpha,$$

and therefore

$$\frac{\partial^2 u}{\partial x'^2} + \frac{\partial^2 u}{\partial y'^2} = \frac{\partial^2 u}{\partial x^2} + \frac{\partial^2 u}{\partial y^2} = 0.$$

The proof for the other elementary transformations is even easier.

In order to obtain a new harmonic function from a known one by making one of these transformations, we follow this procedure. Let $u(x, y)$ be harmonic in Ω. First find the set Ω',

$$\Omega' = \{(x', y') \in R^2 : (x(x', y'), y(x', y')) \in \Omega\}.$$

Then find $u(x', y')$ by replacing x with $x(x', y')$ and y with $y(x', y')$ in $u(x, y)$. Finally, drop the primes since they are no longer necessary. Of course, if one knows what he is doing, he may not want to use primes at all.

Example 3.1. The function

$$\log r = \log [x^2 + y^2]^{1/2}$$

is harmonic in Ω consisting of R^2 except the origin. By a translation, the function

$$\log [(x' - x_0)^2 + (y' - y_0)^2]^{1/2}$$

is harmonic in Ω' consisting of R^2 except the point $(x', y') = (x_0, y_0)$. After dropping the primes, the function

$$\log [(x - x_0)^2 + (y - y_0)^2]^{1/2} \tag{3.4}$$

is harmonic in R^2 except the point (x_0, y_0). Sometimes we say that the function (3.4) is harmonic in R^2 with pole at (x_0, y_0). We may also want to use the vector notation $\mathbf{r} = (x, y)$, $\mathbf{r}_0 = (x_0, y_0)$ and write (3.4) as

$$\log |\mathbf{r} - \mathbf{r}_0|. \tag{3.5}$$

By a similarity transformation, the function

$$\log (\lambda r) = \log [(\lambda x)^2 + (\lambda y)^2]^{1/2}, \ \lambda > 0 \tag{3.6}$$

is also harmonic in R^2 except the origin.

Example 3.2. For rotations it is natural to use polar coordinates. By a rotation, the functions (2.19) become

$$\begin{cases} 1,\ r^n \cos n(\theta - \alpha),\ r^n \sin n(\theta - \alpha); & n = 1, 2, \ldots \\ \log r,\ r^{-n} \cos n(\theta - \alpha),\ r^{-n} \sin n(\theta - \alpha); & n = 1, 2, \ldots. \end{cases} \tag{3.7}$$

The functions on the first line are harmonic in R^2, while those on the second line are harmonic in R^2 except the origin.

The elementary transformations that we defined in R^2 have obvious analogues in R^3 and in higher dimensional spaces. For example, the function

$$(x^2 + y^2 + z^2)^{-1/2}$$

is harmonic in R^3 except the origin, and, by translation, the function

$$[(x - x_0)^2 + (y - y_0)^2 + (z - z_0)^2]^{-1/2}$$

is harmonic in R^3 except the point (x_0, y_0, z_0). In vector notation, $\mathbf{r} = (x, y, z)$, $\mathbf{r}_0 = (x_0, y_0, z_0)$, this function is written as

$$|\mathbf{r} - \mathbf{r}_0|^{-1} \tag{3.8}$$

and we say that this function is harmonic in R^3 with pole at $\mathbf{r}_0$.

Except for translations, all the other elementary transformations in R^n are given by equations of the form

$$x_i = \sum_{j=1}^{n} a_{ij}x_j', \qquad i = 1, \dots, n, \tag{3.9}$$

or, in matrix notation,

$$\mathbf{x} = A\mathbf{x}' \tag{3.10}$$

where

$$\mathbf{x} = \begin{bmatrix} x_1 \\ x_2 \\ \cdot \\ \cdot \\ \cdot \\ x_n \end{bmatrix}, \quad \mathbf{x}' = \begin{bmatrix} x_1' \\ x_2' \\ \cdot \\ \cdot \\ \cdot \\ x_n' \end{bmatrix}$$

and $A = [a_{ij}]$ is a $n \times n$ nonsingular matrix with inverse A^{-1} so that

$$\mathbf{x}' = A^{-1}\mathbf{x}. \tag{3.11}$$

A transformation of the form (3.10), (3.11) is called a *linear transformation of coordinates* in R^n and it is said to be given by the matrix A. We now ask the following question: Which linear transformations of coordinates preserve the harmonicity of a function? More precisely, which linear transformations of coordinates have the property that if $u(x_1, \dots, x_n)$ is any function which is harmonic with respect to $x_1, \dots, x_n$ then $u(x_1', \dots, x_n')$ is also harmonic with respect to $x_1', \dots, x_n'$? The answer to this question is given in the following theorem, the proof of which is left as an exercise (see Problem 3.1).

Theorem 3.1. A linear transformation of coordinates preserves the harmonicity of every harmonic function if and only if it is given by a matrix A of the form

$$A = \lambda B \tag{3.12}$$

where B is an orthogonal matrix and λ is a positive constant.

Recall that a matrix $B = [b_{ij}]$ is said to be orthogonal if

$$\sum_{k=1}^{n} b_{ik}b_{jk} = \begin{cases} 1 & \text{if } i = j \\ 0 & \text{if } i \neq j. \end{cases}$$

Recall also that orthogonal matrices define transformations which preserve distances and hence are compositions of rotations and reflections. Since (3.12) can be written as

$$A = (\lambda I)B$$

where I is the unit matrix, and since λI defines a similarity transformation, Theorem 3.1 asserts that the linear transformations which preserve harmonicity are compositions of similarity transformations, rotations and reflections.

We turn now to a discussion of another important and useful transformation which, for R^2, is known as *inversion with respect to a circle*. Let $S(0, a)$ denote the boundary of the circle in R^2 with center the origin and radius a. In polar coordinates, the points (r, θ) and (r^*, θ^*) are said to be *inverse with respect to* $S(0, a)$ if

$$rr^* = a^2, \qquad \theta^* = \theta. \tag{3.13}$$

Note that two points inverse with respect to $S(0, a)$ lie on the same radial line (see Fig. 3.2). Consider now the mapping which maps the point (r, θ) to (r^*, θ^*), given by

$$r^* = \frac{a^2}{r}, \qquad \theta^* = \theta \tag{3.14}$$

with the inverse mapping given by

$$r = \frac{a^2}{r^*}, \qquad \theta = \theta^*. \tag{3.15}$$

The mapping (3.14) is defined for all points (r, θ) in R^2 except the origin. It maps points outside the circle $S(0, a)$ to points inside $S(0, a)$ and vice-versa, while points that lie on the circle $S(0, a)$ remain fixed. A domain Ω lying outside of $S(0, a)$ is mapped to the domain Ω^* inside of $S(0, a)$. Now

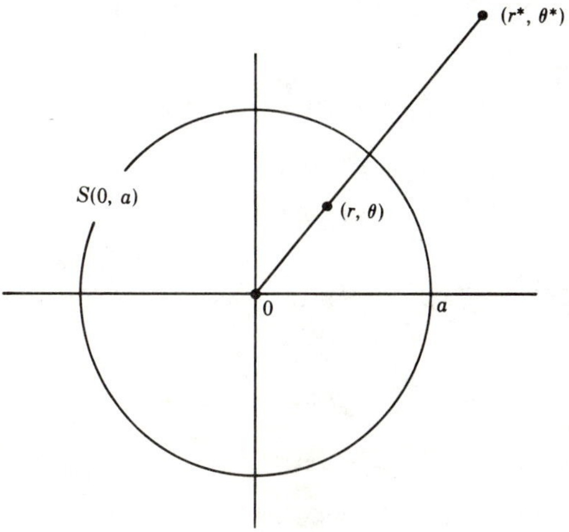

Fig. 3.2

let Ω be any domain in R^2 that does not contain the origin and let $u(r, \theta)$ be harmonic in Ω. Then the function $u(r^*, \theta^*)$ obtained from $u(r, \theta)$ by replacing r with a^2/r^* and θ with θ^*, is harmonic in Ω^*. The proof of this is left as an exercise (see Problem 3.2).

Inversion with respect to a sphere in R^3 is defined in a similar way. Let $S(0, a)$ be the surface of the sphere with center the origin and radius a. In spherical coordinates, the points (r, θ, ϕ) and (r^*, θ^*, ϕ^*) are said to be *inverse with respect to* $S(0, a)$ if

$$(3.16) \qquad rr^* = a^2, \qquad \theta^* = \theta, \qquad \phi^* = \phi.$$

Now let Ω be a domain in R^3 not containing the origin and suppose that the function $u(r, \theta, \phi)$ is harmonic in Ω. Let Ω^* be the image of Ω under the inversion (3.16) and define the function $u^*(r^*, \theta^*, \phi^*)$ in Ω^* by the formula

$$(3.17) \qquad u^*(r^*, \theta^*, \phi^*) = \frac{a}{r^*}\, u\left(\frac{a^2}{r^*}, \theta^*, \phi^*\right).$$

Then u^* is harmonic in Ω^* with respect to the variables r^*, θ^*, ϕ^*. The proof is again left as an exercise (see Problem 3.3).

In dealing with inversions, it is often necessary to use vector notation. If $\mathbf{r}$ and $\mathbf{r}^*$ are the position vectors of two points inverse with respect to $S(0, a)$ (in either R^2 or R^3) then

$$(3.18) \qquad \frac{\mathbf{r}^*}{r^*} = \frac{\mathbf{r}}{r}, \qquad |\mathbf{r}| = r, \qquad |\mathbf{r}^*| = r^*,$$

and hence

$$(3.19) \qquad \mathbf{r}^* = \frac{\mathbf{r}^*}{r^*}\, r^* = \frac{\mathbf{r}}{r}\, r^* = \frac{a^2}{r^2}\, \mathbf{r}.$$

Similarly,

$$(3.20) \qquad \mathbf{r} = \frac{\mathbf{r}}{r}\, r = \frac{\mathbf{r}^*}{r^*}\, r = \frac{a^2}{r^{*2}}\, \mathbf{r}^*.$$

In R^2, if $u(\mathbf{r})$ is harmonic in a domain Ω, then

$$(3.21) \qquad u\left(\frac{a^2}{r^{*2}}\, \mathbf{r}^*\right)$$

is harmonic in Ω^*. In R^3, if $u(\mathbf{r})$ is harmonic in a domain Ω, then

$$(3.22) \qquad u^*(\mathbf{r}^*) = \frac{a}{r^*}\, u\left(\frac{a^2}{r^{*2}}\, \mathbf{r}^*\right)$$

is harmonic in Ω^*.

Example 3.3. The function

$$(3.23) \qquad \log|\mathbf{r} - \mathbf{r}_0|$$

is harmonic in R^2 with pole at $\mathbf{r} = \mathbf{r}_0$. Inversion with respect to $S(0, a)$ yields

$$\log \left| \frac{a^2}{r^{*2}} \mathbf{r}^* - \mathbf{r}_0 \right| \tag{3.24}$$

or, after dropping the star,

$$\log \left| \frac{a^2}{r^2} \mathbf{r} - \mathbf{r}_0 \right| = \log \frac{a}{r} + \log \left| \frac{a}{r} \mathbf{r} - \frac{r}{a} \mathbf{r}_0 \right| .$$

This function is harmonic in R^2 with poles at the origin and at $\mathbf{r} = (a^2/r_0^2)\mathbf{r}_0$. (The pole at the origin corresponds to the "pole at infinity" of the function (3.23).) Therefore, the function

$$\log \left| \frac{a}{r} \mathbf{r} - \frac{r}{a} \mathbf{r}_0 \right| \tag{3.25}$$

is harmonic in R^2 with pole at $\mathbf{r} = (a^2/r_0^2)\mathbf{r}_0$. Note that the poles of (3.23) and (3.25) are inverse with respect to $S(0, a)$. If $r_0 < a$, then the pole of (3.25) is outside $S(0, a)$ and (3.25) is harmonic inside $S(0, a)$ (see Fig. 3.3).

Example 3.4. The function

$$\frac{1}{|\mathbf{r} - \mathbf{r}_0|} \tag{3.26}$$

is harmonic in R^3 with pole at $\mathbf{r} = \mathbf{r}_0$. Inversion with respect to $S(0, a)$ yields

$$\frac{a}{r^*} \frac{1}{\left| \frac{a^2}{r^{*2}} \mathbf{r}^* - \mathbf{r}_0 \right|},$$

or, after dropping the star and simplifying,

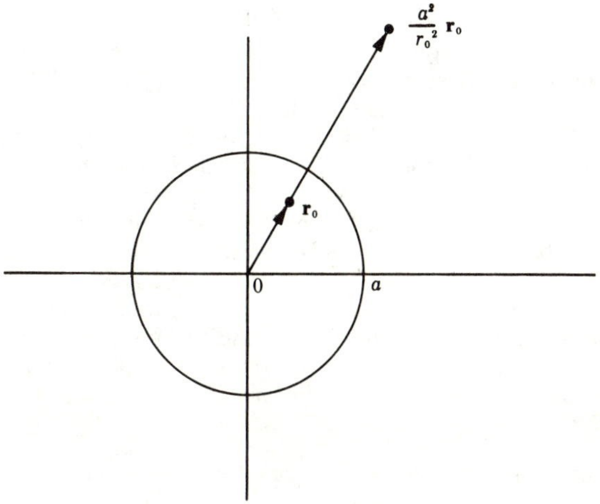

Fig. 3.3

$$\frac{1}{\left|\dfrac{a}{r}\mathbf{r} - \dfrac{r}{a}\mathbf{r}_0\right|}. \tag{3.27}$$

This function is harmonic in R^3 with pole at $\mathbf{r} = (a^2/r_0^2)\mathbf{r}_0$. The poles of (3.26) and (3.27) are inverse with respect to $S(0, a)$. If $r_0 < a$, then the pole of (3.27) is outside $S(0, a)$ and hence (3.27) is harmonic inside $S(0, a)$.

Problems

3.1. Prove Theorem 3.1. [First show that every linear transformation of the form (3.12) preserves harmonicity. Then show that a linear transformation which preserves the harmonicity of every harmonic function must necessarily be of the form (3.12). For this you may need the fact that the functions

$$x_i x_j, \quad x_i^2 - x_j^2; \qquad i \neq j;\ i, j = 1, \ldots, n$$

are harmonic.]

3.2. Use the chain rule and the form of the Laplacian in polar coordinates to prove that inversion with respect to a circle in R^2 preserves the harmonicity of a function.

3.3. Using the chain rule and the form of the Laplacian in spherical coordinates, prove that the function u^* defined by (3.17) is harmonic.

3.4. Show that applying inversion with respect to the unit circle on the collection of harmonic functions (2.19) does not yield any new harmonic functions.

3.5. Apply inversion with respect to the unit circle to the function $\log r$ in R^2. Apply inversion with respect to the unit sphere to the function $1/r$ in R^3.

4. Boundary Value Problems Associated with Laplace's Equation

We have seen in Chapter VI that Laplace's equation appears in the study of many physical phenomena. For example, if the function u represents the steady state temperature distribution in a homogeneous isotropic body, then, at every point interior to the body, u must satisfy Laplace's equation. Of course, this fact alone is not sufficient to determine u since there are infinitely many solutions of Laplace's equation. If we have additional information such as the temperature distribution on the boundary of the body or the heat flux across the boundary, then u must satisfy a condition on the boundary which is called a boundary condition. The problem of determining the function u satisfying Laplace's equation in the interior of the body and a boundary condition on its boundary is called a boundary value problem. In this section we state the three basic boundary value problems associated with Laplace's equation.

The Dirichlet Problem or First Boundary Value Problem

Let Ω be a bounded domain in R^n with smooth boundary $\partial\Omega$, and let f be a given function which is defined and continuous on $\partial\Omega$. Find a function u which is defined and continuous in the closure $\bar{\Omega}$ of Ω such that u is harmonic in Ω and u is equal to f on $\partial\Omega$. More explicitly, find a function u which is in $C^2(\Omega)$ and in $C^0(\bar{\Omega})$ and satisfies

$$\Delta u = 0 \qquad \text{in } \Omega, \tag{4.1}$$

$$u(x) = f(x), \qquad x \in \partial\Omega. \tag{4.2}$$

Equation (4.2) is called the *boundary condition* of the problem and the given function f is referred to as the *boundary data*.

In this definition of the Dirichlet problem, the conditions that we have imposed on Ω, $\partial\Omega$ and f are too stringent. We did this in order to make the discussion, at least initially, as simple as possible. Later on we will consider problems in which the domain Ω may be unbounded, the boundary $\partial\Omega$ may have corners and the boundary data function f may have discontinuities. When Ω is the exterior of a bounded region, the problem is called the *exterior Dirichlet problem*.

It is always useful to keep in mind a physical example. Let the function u describe the steady state temperature distribution in a homogeneous isotropic body the interior of which is the domain Ω, and let the given function f describe the temperature distribution on the surface of the body. In order to find the temperature distribution u we must solve the Dirichlet problem (4.1), (4.2).

Example 4.1. Solve the Dirichlet problem

$$\Delta u = 0 \quad \text{in} \quad \Omega$$

$$u(x) = c, \qquad x \in \partial\Omega$$

where Ω is a bounded domain in R^n, and c is a given constant.

In this problem $f(x) = c$. It is obvious that the constant function $u(x) = c$ is a solution to this problem. We will see later on in this chapter that this is

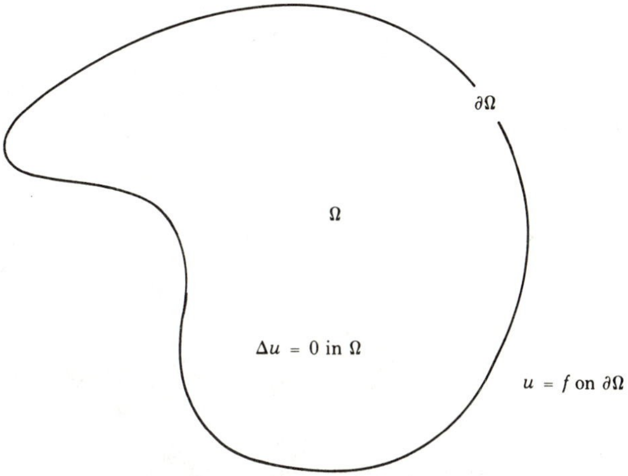

Fig. 4.1. The Dirichlet problem

the only solution to this problem. In terms of our physical example, this means that if the surface of a finite body is kept at a constant temperature c, the steady state temperature at every point inside the body is also equal to c.

Example 4.2. Let Ω be the unit disk $x^2 + y^2 < 1$ in R^2. Solve the Dirichlet problem

$$\Delta u = 0 \quad \text{in} \quad \Omega$$

$$u(1, \theta) = \cos\theta, \qquad 0 \leq \theta < 2\pi.$$

Here, it is convenient to use polar coordinates in the boundary condition. It is easy to check that the function given by

$$u(x, y) = x$$

or, in polar coordinates,

$$u(r, \theta) = r\cos\theta$$

is a solution to the problem. Again, as we will see, this is the only solution.

The Neumann Problem or Second Boundary Value Problem

Let Ω be a bounded domain in R^n with smooth boundary $\partial\Omega$, and let $\mathbf{n} = \mathbf{n}(x)$ be the outward unit normal vector to $\partial\Omega$ at the point x. Let f be a given function defined and continuous on $\partial\Omega$. Find a function u defined and continuous in $\bar{\Omega}$ such that u is harmonic in Ω and such that the outer normal derivative $\partial u/\partial n$ on $\partial\Omega$ is equal to f, i.e.

$$\Delta u = 0, \quad \text{in} \quad \Omega \tag{4.3}$$

$$\frac{\partial u(x)}{\partial n} = f(x), \qquad x \in \partial\Omega. \tag{4.4}$$

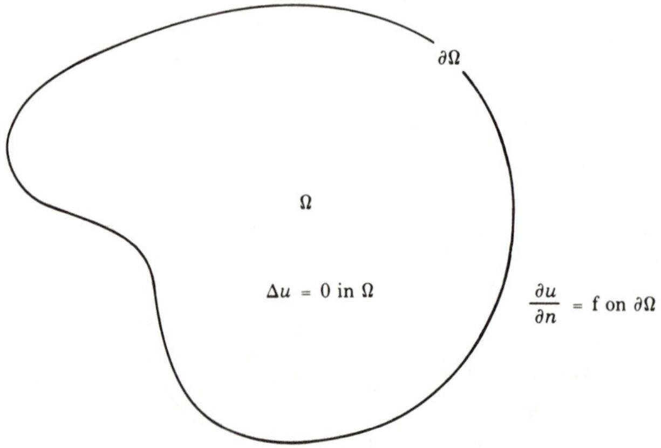

Fig. 4.2. The Neumann problem

A physical example associated with the Neumann problem is this: Find the steady state temperature distribution in a homogeneous isotropic body if the law of heat flux across its surface is known. If for example the surface of the body is insulated, the function f in the Neumann boundary condition (4.4) is zero.

Example 4.3. Solve the Neumann problem

$$\Delta u = 0, \quad \text{in} \quad \Omega,$$

$$\frac{\partial u(x)}{\partial n} = 0, \qquad x \in \partial\Omega,$$

where Ω is a bounded domain in R^n. It is obvious that all constant functions,

$$u(x) = c,$$

where c is any constant, are solutions of the problem. Thus, this problem has infinitely many solutions. We will show later on that the constant functions are the only solutions. In terms of our physical example, this means that the steady state temperature distribution inside a body with insulated surface is constant. This constant depends on the amount of heat trapped inside the body. In order to determine this constant temperature it is enough to know the temperature of the body at a single point.

A combination of the Dirichlet and Neumann boundary conditions also appears in problems of heat conduction and leads to the following boundary value problem.

The Mixed Problem or Third Boundary Value Problem

Let Ω be a bounded domain in R^n with smooth boundary $\partial\Omega$ and let $\mathbf{n} = \mathbf{n}(x)$ be the outward unit normal vector to $\partial\Omega$ at x. Let α, β and f be given functions defined and continuous on $\partial\Omega$. Find a function u defined and continuous in $\bar{\Omega}$ such that

$$\Delta u = 0, \quad \text{in} \quad \Omega, \tag{4.5}$$

$$\alpha(x)\frac{\partial u(x)}{\partial n} + \beta(x)u(x) = f(x), \qquad x \in \partial\Omega. \tag{4.6}$$

The three main goals of this chapter are the following:

1. To determine the conditions under which a boundary value problem is well-posed, i.e., the problem has a unique solution which depends continuously on the boundary data.

2. To describe methods for finding the solution of a well-posed problem.

3. To determine general properties of the solution.

It should be emphasized that not every seemingly reasonable problem is well-posed. We will see for example that the Neumann problem (4.3), (4.4) does not have any solution unless the function f is such that its integral over $\partial\Omega$ is equal to zero. Even when this necessary condition for the existence of a solution is satisfied, the problem may have infinitely

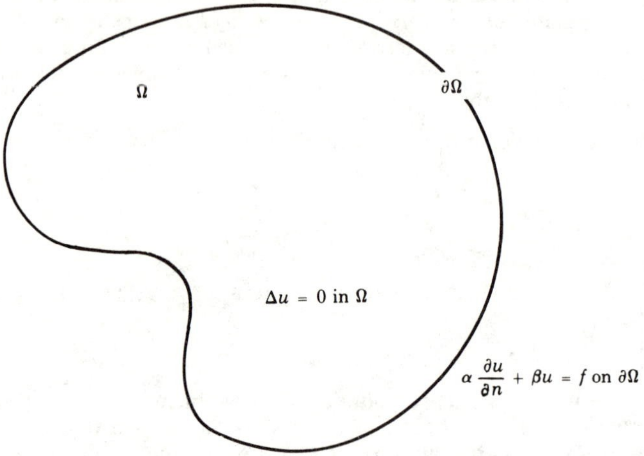

Fig. 4.3. The mixed problem

many solutions as is the case with the problem of Example 4.3. As another example, the exterior Dirichlet problem in two independent variables has infinitely many solutions unless we impose the condition that the solution must be bounded.

Once we know that a problem is well-posed we may try to find its solution. Except when the problem is particularly simple, we cannot expect to find a simple formula for the solution. However, we can always find numerical approximations to the solution, perhaps with the aid of a computer.

In the study of the boundary value problems associated with Laplace's equation the linearity of the Laplacian operator plays a very important role. Suppose for example that u_1 is a solution of the Dirichlet problem

$$\Delta u = 0 \quad \text{in} \quad \Omega; \qquad u = f_1 \quad \text{on} \quad \partial\Omega$$

and u_2 is a solution of the Dirichlet problem

$$\Delta u = 0 \quad \text{in} \quad \Omega; \qquad u = f_2 \quad \text{on} \quad \partial\Omega.$$

Then for any two constants c_1 and c_2, the linear combination $u = c_1 u_1 + c_2 u_2$ is a solution of the Dirichlet problem

$$\Delta u = 0 \quad \text{in} \quad \Omega; \qquad u = c_1 f_1 + c_2 f_2 \quad \text{on} \quad \partial\Omega.$$

In particular, if u_1 and u_2 are solutions of the same Dirichlet problem (4.1), (4.2) then the difference $u = u_1 - u_2$ is a solution of the Dirichlet problem with zero boundary data,

$$\Delta u = 0 \quad \text{in} \quad \Omega; \qquad u = 0 \quad \text{on} \quad \partial\Omega. \tag{4.7}$$

Thus, in order to prove uniqueness of solution of the Dirichlet problem

(4.1), (4.2) it is enough to show that the only solution to (4.7) is the function which is identically zero.

5. A Representation Theorem. The Mean Value Property and the Maximum Principle for Harmonic Functions

In this section we prove a representation theorem which gives the value of a C^2 function at a point in terms of certain volume and surface integrals involving the function, its derivatives and the Laplacian of the function. We then apply this theorem to show that a harmonic function has the mean value property: the value of a harmonic function at a point is equal to the average of its values over any sphere centered at that point. Finally, using the mean value property we show that a harmonic function satisfies the maximum principle: if u is harmonic in a bounded domain Ω and continuous in $\bar{\Omega}$, then u must attain its maximum and minimum values on the boundary $\partial\Omega$, and only on $\partial\Omega$, unless u is constant in $\bar{\Omega}$.

Theorem 5.1. (*Representation theorem, $n = 3$.*) Let Ω_0 be a bounded normal domain in R^3 and let $\mathbf{n}$ be the unit exterior normal to the boundary $\partial\Omega_0$ of Ω_0. Let u be any function in $C^2(\bar{\Omega}_0)$. Then the value of u at any point $\mathbf{r}_0 \in \Omega_0$ is given by the formula

$$u(\mathbf{r}_0) = \frac{1}{4\pi}\int_{\partial\Omega_0}\left[\frac{1}{|\mathbf{r}-\mathbf{r}_0|}\frac{\partial u(\mathbf{r})}{\partial n} - u(\mathbf{r})\frac{\partial}{\partial n}\frac{1}{|\mathbf{r}-\mathbf{r}_0|}\right]d\sigma - \frac{1}{4\pi}\int_{\Omega_0}\frac{\nabla^2 u(\mathbf{r})}{|\mathbf{r}-\mathbf{r}_0|}\,dv. \tag{5.1}$$

Proof. The proof is based on an application of the second Green's identity

$$\int_{\Omega_0}(u\nabla^2 w - w\nabla^2 u)\,dv = \int_{\partial\Omega_0}\left(u\frac{\partial w}{\partial n} - w\frac{\partial u}{\partial n}\right)d\sigma, \tag{5.2}$$

which was derived in Section 1, Chapter VI, using the divergence theorem. Identity (5.2) is certainly valid for any two functions u and w which are in $C^2(\bar{\Omega}_0)$. We want to apply (5.2) with u the function of the theorem, and w the harmonic function

$$w(\mathbf{r}) = \frac{1}{|\mathbf{r}-\mathbf{r}_0|} \tag{5.3}$$

with pole at $\mathbf{r}_0$. Since w has a singularity at $\mathbf{r}_0$ and $\mathbf{r}_0$ is a point of Ω_0, the identity (5.2) is not applicable. In order to avoid this difficulty we consider a new domain $\mathbf{\Omega}_\epsilon$ obtained from Ω_0 by removing the closed ball $\bar{B}(\mathbf{r}_0, \epsilon)$ of center $\mathbf{r}_0$ and radius ϵ,

$$\Omega_\epsilon = \Omega_0 - \bar{B}(\mathbf{r}_0, \epsilon). \tag{5.4}$$

Here ϵ is chosen to be any positive number such that $\bar{B}(\mathbf{r}_0, \epsilon) \subset \Omega_0$ (see Fig. 5.1).

Now, u and w are in $C_2(\bar{\Omega}_\epsilon)$ and (5.2), with Ω_0 replaced by Ω_ϵ, is applicable. Moreover, since w is harmonic in Ω_ϵ and since $\partial\Omega_\epsilon = \partial\Omega_0 \cup$

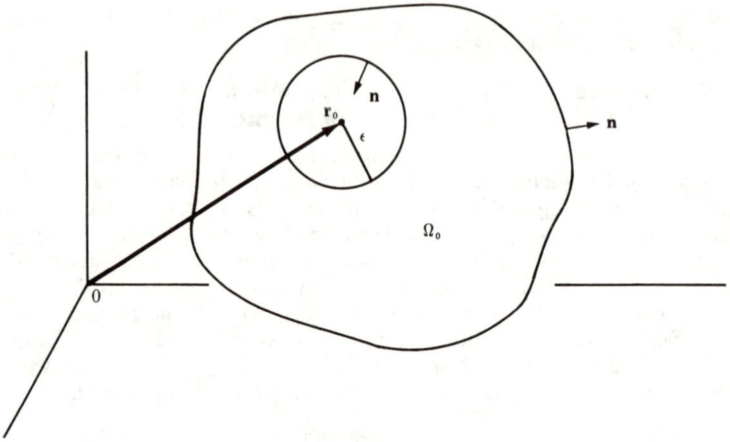

Fig. 5.1

$S(\mathbf{r}_0, \epsilon)$, where $S(\mathbf{r}_0, \epsilon)$ is the surface of the sphere of center $\mathbf{r}_0$ and radius ϵ, we have

$$(5.5)\quad \begin{aligned} -\int_{\Omega_\epsilon} \frac{\nabla^2 u(\mathbf{r})}{|\mathbf{r}-\mathbf{r}_0|}\,dv &= \int_{\partial\Omega_0}\left[u(\mathbf{r})\frac{\partial}{\partial n}\frac{1}{|\mathbf{r}-\mathbf{r}_0|} - \frac{1}{|\mathbf{r}-\mathbf{r}_0|}\frac{\partial u(\mathbf{r})}{\partial n}\right]d\sigma \\ &+ \int_{S(r_0,\epsilon)}\left[u(\mathbf{r})\frac{\partial}{\partial n}\frac{1}{|\mathbf{r}-\mathbf{r}_0|} - \frac{1}{|\mathbf{r}-\mathbf{r}_0|}\frac{\partial u(\mathbf{r})}{\partial n}\right]d\sigma. \end{aligned}$$

Equation (5.5) is valid for every sufficiently small $\epsilon > 0$. We will show that formula (5.1) is obtained from (5.5) by letting ϵ tend to 0. Since the integral

$$\int_{\Omega_0} \frac{\nabla^2 u(\mathbf{r})}{|\mathbf{r}-\mathbf{r}_0|}\,dv$$

is convergent (see Problem 5.1), it is clear that

$$\lim_{\epsilon\to 0}\left[-\int_{\Omega_\epsilon} \frac{\nabla^2 u(\mathbf{r})}{|\mathbf{r}-\mathbf{r}_0|}\,dv\right] = -\int_{\Omega_0} \frac{\nabla^2 u(\mathbf{r})}{|\mathbf{r}-\mathbf{r}_0|}\,dv.$$

The first integral on the right of (5.5) does not depend on ϵ. Hence, the limit as $\epsilon \to 0$ of the second integral on the right of (5.5) exists, and in order to obtain (5.1) it remains to show that

$$(5.6)\quad \lim_{\epsilon\to 0}\int_{S(r_0,\epsilon)}\left[u(\mathbf{r})\frac{\partial}{\partial n}\frac{1}{|\mathbf{r}-\mathbf{r}_0|} - \frac{1}{|\mathbf{r}-\mathbf{r}_0|}\frac{\partial u(\mathbf{r})}{\partial n}\right]d\sigma = 4\pi u(\mathbf{r}_0).$$

In order to prove (5.6) we note that for points $\mathbf{r}$ on $S(\mathbf{r}_0, \epsilon)$,

$$\frac{1}{|\mathbf{r}-\mathbf{r}_0|} = \frac{1}{\epsilon} \quad\text{and}\quad \frac{\partial}{\partial n}\frac{1}{|\mathbf{r}-\mathbf{r}_0|} = \frac{1}{\epsilon^2}$$

(why?). Hence

$$\int_{S(r_0,\epsilon)} \left[u(\mathbf{r}) \frac{\partial}{\partial n} \frac{1}{|\mathbf{r} - \mathbf{r}_0|} - \frac{1}{|\mathbf{r} - \mathbf{r}_0|} \frac{\partial u(\mathbf{r})}{\partial n} \right] d\sigma$$

$$= \int_{S(r_0,\epsilon)} \left[\frac{1}{\epsilon^2} u(\mathbf{r}) - \frac{1}{\epsilon} \frac{\partial u(\mathbf{r})}{\partial n} \right] d\sigma$$

$$= \int_{S(r_0,\epsilon)} \frac{1}{\epsilon^2} u(\mathbf{r}_0)\, d\sigma + \int_{S(r_0,\epsilon)} \left\{ \frac{1}{\epsilon^2} [u(\mathbf{r}) - u(\mathbf{r}_0)] - \frac{1}{\epsilon} \frac{\partial u(\mathbf{r})}{\partial n} \right\} d\sigma$$

$$= 4\pi u(\mathbf{r}_0) + \int_{S(r_0,\epsilon)} \left\{ \frac{1}{\epsilon^2} [u(\mathbf{r}) - u(\mathbf{r}_0)] - \frac{1}{\epsilon} \frac{\partial u(\mathbf{r})}{\partial n} \right\} d\sigma$$

and (5.6) will follow if we show that the last integral tends to 0 as $\epsilon \to 0$. We have

$$\left| \int_{S(r_0,\epsilon)} \left\{ \frac{1}{\epsilon^2} [u(\mathbf{r}) - u(\mathbf{r}_0)] - \frac{1}{\epsilon} \frac{\partial u(\mathbf{r})}{\partial n} \right\} d\sigma \right|$$

$$\leqq \frac{1}{\epsilon^2} \int_{S(r_0,\epsilon)} \left| u(\mathbf{r}) - u(\mathbf{r}_0) \right| d\sigma + \frac{1}{\epsilon} \int_{S(r_0,\epsilon)} \left| \frac{\partial u(\mathbf{r})}{\partial n} \right| d\sigma$$

$$\leqq 4\pi \max_{r \in S(r_0,\epsilon)} \left| u(\mathbf{r}) - u(\mathbf{r}_0) \right| + 4\pi\epsilon \max_{r \in \bar{\Omega}_0} \left| \text{grad } u(\mathbf{r}) \right|.$$

The first term on the right tends to 0 as $\epsilon \to 0$ since u is continuous in $\bar{\Omega}_0$. The second term also tends to 0 as $\epsilon \to 0$, since the maximum of $|\text{grad } u|$ in $\bar{\Omega}_0$ is finite. The proof of the theorem is now complete.

Theorem 5.2. (*Representation theorem, n = 2.*) Let Ω_0 be a bounded normal domain in R^2 and let $\mathbf{n}$ be the unit exterior normal to the boundary $\partial\Omega_0$ of Ω_0. Let u be any function in $C^2(\bar{\Omega}_0)$. Then the value of u at any point $\mathbf{r}_0 \in \Omega_0$ is given by the formula

(5.7)
$$u(\mathbf{r}_0) = \frac{1}{2\pi} \int_{\partial\Omega_0} \left[\log \frac{1}{|\mathbf{r} - \mathbf{r}_0|} \frac{\partial u(\mathbf{r})}{\partial n} - u(\mathbf{r}) \frac{\partial}{\partial n} \log \frac{1}{|\mathbf{r} - \mathbf{r}_0|} \right] ds - \frac{1}{2\pi} \int_{\Omega_0} \nabla^2 u(\mathbf{r}) \log \frac{1}{|\mathbf{r} - \mathbf{r}_0|}\, dx dy$$

where ds is the element of length on $\partial\Omega_0$.

The proof of Theorem 5.2 is very similar to the proof of Theorem 5.1 and is left for the problems (see Problem 5.2).

Representation theorems for $n > 3$ can be obtained in the same way as for $n = 2$ and $n = 3$. We use the second Green's identity with

$$w(\mathbf{r}) = \frac{1}{|\mathbf{r} - \mathbf{r}_0|^{n-2}}$$

which is a harmonic function in R^n with pole at $\mathbf{r}_0$.

An immediate application of the representation theorem shows that a harmonic function has the mean value property.

Definition 5.1. A function $u(\mathbf{r})$ defined in a domain Ω of R^n is said to have the *mean value property* in Ω, if the value of u at any point $\mathbf{r}_0$ of Ω is equal to the mean value (average) of the values of u over any sphere $S(\mathbf{r}_0, \delta)$ which, together with its interior $B(\mathbf{r}_0, \delta)$, belongs to Ω, i.e.

$$u(\mathbf{r}_0) = \frac{1}{|S(\mathbf{r}_0,\delta)|} \int_{S(r_0,\delta)} u(\mathbf{r})d\sigma \tag{5.8}$$

for every $\delta > 0$ such that $\bar{B}(\mathbf{r}_0, \delta) \subset \Omega$.

In (5.8), $|S(\mathbf{r}_0, \delta)|$ denotes the area of the sphere $S(\mathbf{r}_0, \delta)$. Thus, for $n = 2$, $|S(\mathbf{r}_0, \delta)| = 2\pi\delta$ and for $n = 3$, $|S(\mathbf{r}_0, \delta)| = 4\pi\delta^2$.

Theorem 5.3. (*Mean value theorem.*) Let u be a harmonic function in a domain Ω of R^n. Then u has the mean value property in Ω.

Proof. We give the proof only for $n = 3$. In the Representation Theorem 5.1, let $\Omega_0 = B(\mathbf{r}_0, \delta)$. Since $\bar{B}(\mathbf{r}_0, \delta) \subset \Omega$ and since u is harmonic in Ω, it follows that $u \in C^2(\bar{\Omega}_0)$ and the conditions of the theorem are satisfied. Formula (5.1) becomes

$$u(\mathbf{r}_0) = \frac{1}{4\pi} \int_{S(r_0,\delta)} \left[\frac{1}{|\mathbf{r} - \mathbf{r}_0|} \frac{\partial u(\mathbf{r})}{\partial n} - u(\mathbf{r}) \frac{\partial}{\partial n} \frac{1}{|\mathbf{r} - \mathbf{r}_0|}\right] d\sigma.$$

Now, on $S(\mathbf{r}_0, \delta)$, $\dfrac{1}{|\mathbf{r} - \mathbf{r}_0|} = \dfrac{1}{\delta}$ and $\dfrac{\partial}{\partial n} \dfrac{1}{|\mathbf{r} - \mathbf{r}_0|} = -\dfrac{1}{\delta^2}$

(why?). Hence

$$u(\mathbf{r}_0) = \frac{1}{4\pi\delta^2} \int_{S(r_0,\delta)} u(\mathbf{r})d\sigma + \frac{1}{4\pi\delta} \int_{S(r_0,\delta)} \frac{\partial u}{\partial n}(\mathbf{r})d\sigma.$$

By the Divergence Theorem, the second integral is zero since

$$\int_{S(r_0,\delta)} \frac{\partial u}{\partial n} d\sigma = \int_{S(r_0,\delta)} \nabla u \cdot \mathbf{n} d\sigma = \int_{B(r_0,\delta)} \nabla^2 u dv = 0.$$

The mean value property of harmonic functions yields a very important result known as the *maximum principle*.

Theorem 5.4. (*Maximum principle.*) Let Ω be a bounded domain in R^n and suppose that u is defined and continuous in $\bar{\Omega}$ and harmonic in Ω. Then u attains its maximum and minimum values on the boundary $\partial\Omega$ of Ω. Moreover, if u is not constant in $\bar{\Omega}$, then u attains its maximum and minimum values only on $\partial\Omega$.

We know that if u is continuous in the bounded closed set $\bar{\Omega}$, then u must attain its maximum and minimum values at points of $\bar{\Omega}$. The maximum principle asserts that if, in addition, u is harmonic in Ω, then u must attain its maximum and minimum values on $\partial\Omega$, and only on $\partial\Omega$ unless u is constant in $\bar{\Omega}$. Note carefully that this assertion is eqivalent to saying that if u attains its maximum or minimum values at a point of Ω (i.e., at an interior point of $\bar{\Omega}$) then u must be constant in $\bar{\Omega}$.

Proof of Theorem 5.4. We prove only the "maximum" part of the theorem since the proof for the "minimum" part is the same.

Let $\max_{x \in \bar{\Omega}} u(x) = M$ and suppose that $u(x^0) = M$ where x^0 is a point of Ω. We must show that $u(x) = M$ for every $x \in \Omega$. By continuity, this would imply that $u(x) = M$ for every $x \in \bar{\Omega}$.

We first prove the following *assertion A:* if $u(\bar{x}) = M$ at some point $\bar{x} \in \Omega$, then $u(x) = M$ for every x in the largest ball $B(\bar{x}, \bar{\delta})$ centered at $\bar{x}$ and contained in Ω (see Fig. 5.2). This assertion follows from the mean value property for harmonic functions: if $\delta < \bar{\delta}$, $u(\bar{x})$ must be equal to the average of the values of u over the sphere $S(\bar{x}, \delta)$. Since u is continuous and since the values of u on $S(\bar{x}, \delta)$ must be less than or equal to $M = u(\bar{x})$, it follows that the values of u on $S(\bar{x}, \delta)$ must be equal to M (see Problem 5.5). Since this is true for every δ between 0 and $\bar{\delta}$, the proof of assertion A is complete.

Now let y be any point of Ω. We must show that $u(y) = M$. Since Ω is connected there is a polygonal path C connecting x^0 with y and lying entirely in Ω (see Fig. 5.3). Since the minimum distance of C from $\partial\Omega$ is positive, it is possible to find a finite number of points

$$x^0, x^1, x^2, \ldots, x^n = y$$

on C which can be used as centers of balls $B(x^i, \delta_i)$ having the following two properties:

$$\text{(i) } B(x^i, \delta_i) \subset \Omega, \; i = 0, \ldots, n,$$

$$\text{(ii) } x^i \in B(x^{i-1}, \delta_{i-1}), \; i = 1, 2, \ldots, n.$$

Now, by assertion A, $u(x) = M$ for every $x \in B(x^0, \delta_0)$. Hence $u(x^1) = M$. Again by assertion A, $u(x) = M$ for every $x \in B(x^1, \delta_1)$. Hence $u(x^2) = M$.

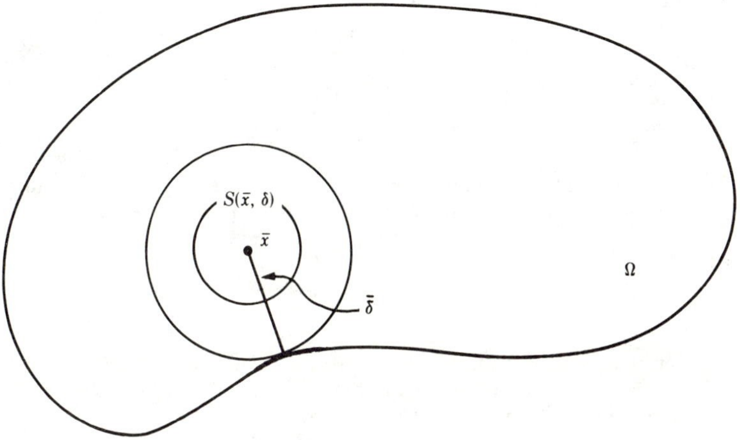

Fig. 5.2

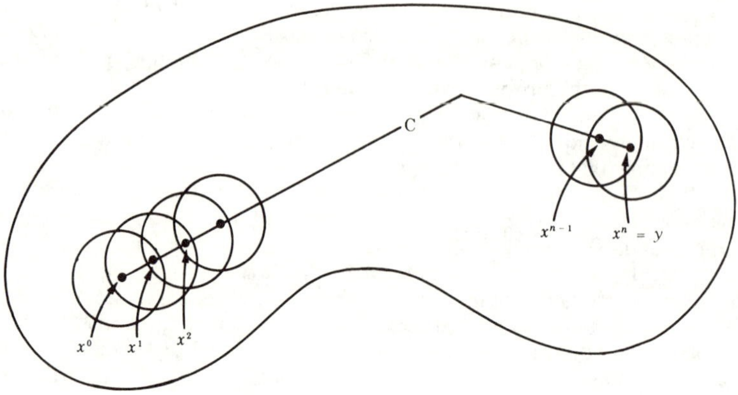

Fig. 5.3

Continuing in this way, we conclude after a finite number of steps that $u(y) = M$. The proof is complete.

Problems

5.1. Show that if $u \in C^2(\bar{\Omega}_0)$ with Ω_0 a bounded domain in R^3, then the integral

$$\int_{\Omega_0} \frac{\nabla^2 u(\mathbf{r})}{|\mathbf{r} - \mathbf{r}_0|} dv$$

is convergent.

5.2. Prove the two-dimensional representation theorem (Theorem 5.2).

5.3. Answer the "why?" in the proofs of Theorem 5.1 and 5.3.

5.4. Prove the mean value theorem for harmonic functions (Theorem 5.3) for $n = 2$.

5.5 Prove that if u has the mean value property in a domain Ω and if u assumes its maximum value M at a point $\mathbf{r}_0 \in \Omega$, then $u = M$ on every sphere $S(\mathbf{r}_0, \delta)$ for which $\bar{B}(\mathbf{r}_0, \delta) \subset \Omega$.
[*Hint:* Suppose that for some $\mathbf{r}_1 \in S_0 = S(\mathbf{r}_0, \delta)$, $u(\mathbf{r}_1) < M$. Then $u(\mathbf{r}) < M$ for every $\mathbf{r}$ in a subset S_1 of S_0 with positive area.]

5.6. Suppose that the function u has the mean value property in a domain Ω of R^3. Show that u also has the "volume" mean value property:

$$u(\mathbf{r}_0) = \frac{1}{\frac{4}{3}\pi\delta^3} \int_{B(r_0,\delta)} u(\mathbf{r}) dv$$

for every ball $B(\mathbf{r}_0, \delta)$ such that $\bar{B}(\mathbf{r}_0, \delta) \subset \Omega$.

5.7. Let Ω_0 be a bounded normal domain in R^3 and let u be in $C^1(\bar{\Omega}_0)$ and harmonic in Ω_0. Show that for any point $\mathbf{r}_0 \in \Omega_0$,

$$u(\mathbf{r}_0) = \frac{1}{4\pi} \int_{\partial\Omega} \left[\frac{1}{|\mathbf{r} - \mathbf{r}_0|} \frac{\partial u(\mathbf{r})}{\partial n} - u(\mathbf{r}) \frac{\partial}{\partial n} \frac{1}{|\mathbf{r} - \mathbf{r}_0|} \right] d\sigma.$$

[Make sure to check that, if u is harmonic in Ω_0, then in the representation theorem it is only necessary to assume that $u \in C^1(\bar{\Omega}_0)$.]

6. The Well-Posedness of the Dirichlet Problem

In this section we discuss the well-posedness of the Dirichlet problem: Find a function u in $C^2(\Omega) \cap C^0(\bar{\Omega})$ such that

$$\nabla^2 u = 0, \quad \text{in} \quad \Omega, \tag{6.1}$$

$$u = f, \quad \text{on} \quad \partial\Omega, \tag{6.2}$$

where Ω is a bounded domain in R^n and f is a given function which is defined and continuous on $\partial\Omega$. In order to prove that this problem is well posed we must show that (i) there exists a solution, (ii) the solution is unique, and (iii) the solution depends continuously on the boundary data f.

The uniqueness and the continuous dependence of the solution on the boundary data follow immediately from the maximum principle.

Theorem 6.1. (*Uniqueness.*) The Dirichlet problem (6.1), (6.2) has at most one solution.

Proof. We must show that any two solutions of the problem must be identical. Let u_1 and u_2 be any two solutions and consider their difference $\tilde{u} = u_1 - u_2$. Then $\tilde{u}$ is continuous in $\bar{\Omega}$, harmonic in Ω and zero on $\partial\Omega$. By the maximum principle $\tilde{u}$ must attain its maximum and minimum values on $\partial\Omega$. Hence $\tilde{u} \equiv 0$ in $\bar{\Omega}$ and consequently $u_1 \equiv u_2$ in $\bar{\Omega}$.

Theorem 6.2. (*Continuous dependence on data.*) Let f_1 and f_2 be two functions defined and continuous on $\partial\Omega$. Let u_1 be the solution of the Dirichlet problem (6.1), (6.2) with $f = f_1$ and let u_2 be the solution of the problem with $f = f_2$. For any $\epsilon > 0$, if

$$| f_1(x) - f_2(x) | < \epsilon, \quad \text{for all} \quad x \in \partial\Omega, \tag{6.3}$$

then

$$| u_1(x) - u_2(x) | < \epsilon, \quad \text{for all} \quad x \in \bar{\Omega}. \tag{6.4}$$

Proof. Let $\tilde{u} = u_1 - u_2$. Then $\tilde{u}$ is harmonic in Ω, continuous in $\bar{\Omega}$ and, by (6.3),

$$| \tilde{u}(x) | < \epsilon, \quad \text{for all} \quad x \in \partial\Omega.$$

By the maximum principle,

$$| \tilde{u}(x) | < \epsilon, \ , \text{for all} \quad x \in \bar{\Omega},$$

and (6.4) follows.

The question of existence of solution to the Dirichlet problem is a much more difficult one and the answer depends on the geometry of the domain Ω. Before stating conditions on the domain Ω which guarantee the existence of a solution, we will discuss briefly an example, due to H. Lebesgue, of a Dirichlet problem for a special domain which has no solution. The example illustrates how existence of a solution may fail.

Imagine a ball in R^3 with a deformable surface. At a point on the surface, push in a sharp spike and assume that near the tip of the spike the surface of the deformed ball takes the form of a "conical" surface obtained by rotating the curve

$$y = \begin{cases} e^{-\frac{1}{x}}, & \text{for } x > 0 \\ 0, & \text{for } x = 0 \end{cases}$$

about the x-axis. The interior of the resulting deformed ball is the domain Ω (see Fig. 6.1). Now, it is beyond the scope of this book to present a detailed proof of Lebesgue's result. Instead, we will try to make his result plausible by considering a heat conduction problem for a homogeneous isotropic body with interior the domain Ω. Suppose that the temperature distribution on $\partial\Omega$ is given by the continuous function f which is equal to zero at points of the spike, while f is equal to a large positive constant temperature T_0 at points away from the spike. Then the steady state temperature $u(x)$ should be close to T_0 for all x in Ω, but it is impossible for $u(x)$ to approach the temperature zero as x approaches the spike from within Ω. The spike does not have enough surface area to keep the temperature at surrounding points close to zero. For this Dirichlet problem, existence of a solution fails because the solution cannot possibly be continuous in the closure $\bar{\Omega}$ of Ω.

We describe now a geometrical condition which rules out the occurrence of very sharp spikes such as the one in Lebesgue's example. This condition guarantees the existence of a solution to the Dirichlet problem.

Consider the "conical" surfaces obtained by rotating the curves

$$y = x^k, \qquad x \geqq 0, \qquad k \geqq 1$$

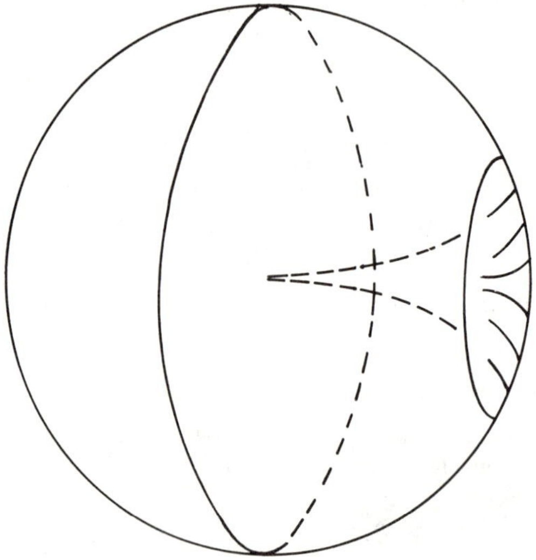

Fig. 6.1

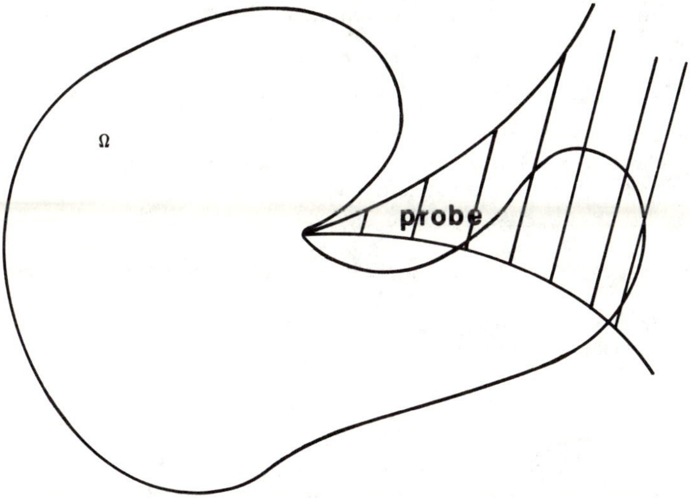

Fig. 6.2

about the x-axis. The surfaces bound spike-shaped volumes with the sharpness of the spikes increasing with k. These volumes will be called *probes*. The class of domains Ω in R^3 that we wish to consider satisfy the following condition:

Condition P. Each point $x \in \partial\Omega$ can be touched by the tip of some probe in such a manner that all points of the probe which lie at a distance not greater than some positive number ρ from the tip of the probe, lie outside of Ω (see Fig. 6.2).

Theorem 6.3. (*Existence, $n = 3$.*) If Ω is a bounded domain in R^3 satisfying condition P, then the Dirichlet problem always has a solution.

The ideas involved in the statement of this existence theorem can be extended to domains in R^n with $n > 3$. For $n = 2$, the probes in condition P may be replaced by straight line segments.

There are several proofs of existence theorems for the Dirichlet problem, all of which are difficult and cannot be presented here. A historical account of these proofs can be found in the book of Kellogg.[2] A proof due to Poincaré is presented in the book of Petrovskii.[3] The books of Kellogg and Petrovskii also describe a method using integral equations for constructing the solution. Lebesgue's example discussed above appears in the book of Kellogg.

7. Solution of the Dirichlet Problem for the Unit Disc. Fourier Series and Poisson's Integral

In this section we solve the Dirichlet problem for the unit ball in R^2, commonly known as the unit disc. It is convenient to use polar coordinates for the problem. Let

$$\Omega = B(0,1) = \{(r,\theta) \in R^2 \colon 0 \le r < 1, \qquad -\pi \le \theta \le \pi\}.$$

The problem is to find the function $u(r,\theta)$ in $C^2(\Omega) \cap C^0(\bar{\Omega})$ such that

$$(7.1) \qquad \nabla^2 u(r,\theta) = 0; \qquad 0 \le r < 1, \qquad -\pi \le \theta \le \pi,$$

$$(7.2) \qquad u(1,\theta) = f(\theta); \qquad -\pi \le \theta \le \pi,$$

where f is a given function in $C^0(\partial\Omega)$.

We will attempt to construct the solution of problem (7.1)-(7.2) by superposition of the harmonic functions

$$(7.3) \qquad r^n \cos n\theta, \qquad r^n \sin n\theta; \qquad n = 0,1,2, \ldots$$

which were obtained in Section 2 by the method of separation of variables. Each of the functions in (7.3) is harmonic in R^2 and, in particular, each satisfies (7.1). We proceed by assuming that the desired solution can be expressed in the form

$$(7.4) \qquad u(r, \theta) = \frac{A_0}{2} + \sum_{n=1}^{\infty} r^n(A_n \cos n\theta + B_n \sin n\theta).$$

(The factor 1/2 in the constant term $A_0/2$ is used for convenience later on.) The problem now is to determine the coefficients A_n, B_n; $n = 0, 1, 2, \ldots$, so that the series in (7.4) converges and defines a function $u(r, \theta)$ in $C^2(\Omega) \cap C^0(\bar{\Omega})$ which satisfies (7.1)-(7.2).

We first note that if the coefficients A_n, B_n; $n = 0, 1, 2, \ldots$ in (7.4) are bounded, i.e., if there is a constant $M > 0$ such that

$$(7.5) \qquad |A_n| \le M, \qquad |B_n| \le M \quad \text{for} \quad n = 0, 1, 2, \ldots,$$

then the series in (7.4) converges to a harmonic function in Ω (see Problem 7.1). The actual values of the coefficients are determined from the boundary condition (7.2). Indeed, in order for (7.4) to satisfy (7.2), the coefficients must be chosen so that

$$(7.6) \quad f(\theta) = \frac{A_0}{2} + \sum_{n=1}^{\infty} (A_n \cos n\theta + B_n \sin n\theta), \qquad -\pi \le \theta \le \pi.$$

Let us first consider the special case in which $f(\theta)$ is a trigonometric polynomial of the form

$$f(\theta) = \frac{a_0}{2} + \sum_{n=1}^{N} (a_n \cos n\theta + b_n \sin n\theta).$$

Then, clearly, the coefficients in (7.4) must be chosen to be

$$A_n = a_n,\ B_n = b_n \quad \text{for} \quad n = 0, 1, \ldots, N;$$
$$A_n = B_n = 0 \quad \text{for} \quad n = N+1, N+2, \ldots,$$

and the solution of (7.1), (7.2) is

$$u(r, \theta) = \frac{a_0}{2} + \sum_{n=1}^{N} r^n(a_n \cos n\theta + b_n \sin n\theta).$$

For example if

$$f(\theta) = -1 + 2\cos\theta + 5\sin 3\theta,$$

the solution of (7.1), (7.2) is

$$u(r, \theta) = -1 + 2r\cos\theta + 5r^3 \sin 3\theta.$$

This special case suggests that in the general case in which $f(\theta)$ is an arbitrary given function, we should try to express $f(\theta)$ as a series of the trigonometric functions $\cos n\theta$, $\sin n\theta$; $n = 0, 1, 2, \ldots$. More precisely, we should look for coefficients a_n, b_n; $n = 0, 1, 2, \ldots$, such that the trigonometric series

$$\frac{a_0}{2} + \sum_{n=1}^{\infty} (a_n \cos n\theta + b_n \sin n\theta) \tag{7.7}$$

converges to $f(\theta)$ for $-\pi \le \theta \le \pi$, allowing us to write

$$f(\theta) = \frac{a_0}{2} + \sum_{n=1}^{\infty} (a_n \cos n\theta + b_n \sin n\theta). \tag{7.8}$$

It turns out that if $f(\theta)$ satisfies certain conditions, then the representation (7.8) is possible. Suppose now that $f(\theta)$ can be expressed in the form (7.8). Then obviously the coefficients in (7.4) must be chosen to be

$$A_n = a_n, \qquad B_n = b_n; \qquad n = 0, 1, 2, \ldots,$$

and

$$u(r, \theta) = \frac{a_0}{2} + \sum_{n=1}^{\infty} r^n(a_n \cos n\theta + b_n \sin n\theta) \tag{7.9}$$

will be, hopefully, the desired solution of (7.1), (7.2).

The problem of representing a given function by a trigonometric series has a long history. It was first considered by Daniel Bernoulli (1700–1782) in his attempt to find solutions of the wave equation. In 1824, Fourier proved that the representation (7.8) is valid for functions f satisfying certain conditions. For this reason the trigonometric series (7.7) is known as a *Fourier series*.

Fourier series play an important role in the study of partial differential equations. In the next section we will digress from our study of Laplace's equation in order to present an introduction to some of the basic results of the theory of Fourier series and give the student a working knowledge of the subject. Meanwhile, we will use the theorems of the next section to complete our study of the solution of the Dirichlet problem for the unit disc. At this point the student may wish to study Section 8 before continuing with the present section.

We observe first that the assumption that f is in $C^0(\partial\Omega)$ means not only that $f(\theta)$ is continuous on the interval $[-\pi, \pi]$ but also that $f(-\pi) = f(\pi)$. (Remember that $\partial\Omega = \{(r, \theta) \in R^2 : r = 1, -\pi \le \theta \le \pi\}$ is the boundary of the unit circle, and if $f(-\pi) \neq f(\pi)$ then f would have a jump discontinuity at the point $(1, \pm\pi)$ of $\partial\Omega$, contrary to our assumption that f is continous on $\partial\Omega$.) Let us now make the additional assumption that the derivative f' of f is sectionally continuous on the interval $[-\pi, \pi]$. This means that $f'(\theta)$ is continuous in this interval except possibly at a finite number of points where $f'(\theta)$ may have finite jumps; see Definition 8.1.

According to Theorem 8.2, if a_n, b_n; $n = 0, 1, 2, \ldots$, are the Fourier coefficients of f given by

$$(7.10)\quad a_n = \frac{1}{\pi}\int_{-\pi}^{\pi} f(\theta)\cos n\theta\, d\theta, \qquad b_n = \frac{1}{\pi}\int_{-\pi}^{\pi} f(\theta)\sin n\theta\, d\theta,$$

the Fourier series (7.7) converges (uniformly) to $f(\theta)$ for $-\pi \le \theta \le \pi$ and the representation (7.8) is valid. We now claim that, in this case, (7.9) is actually the solution of the Dirichlet problem (7.1), (7.2). To check this we note first that the coefficients given by (7.10) are bounded by 2 max $|f|$ and therefore the series in (7.9) converges to a harmonic function in Ω. Hence $u(r, \theta)$ given by (7.9) is in $C^2(\Omega)$ and satisfies (7.1). Since the series in (7.8) converges uniformly to $f(\theta)$ for $-\pi \le \theta \le \pi$, the series in (7.9) converges uniformly in $\bar{\Omega}$. This follows from the fact that each term in (7.9) is obtained from the corresponding term in (7.8) by multiplying by r^n, and $r^n \le 1$ in $\bar{\Omega}$. Hence, $u(r, \theta)$ given by (7.9) must be continuous in $\bar{\Omega}$ since it is the sum of a uniformly convergent series of continuous functions in $\bar{\Omega}$. Finally, setting $r = 1$ in (7.9) and remembering (7.8), we see that the boundary condition (7.2) is satisfied. We restate our result in the form of a theorem.

Theorem 7.1. Suppose that $f \in C^0(\partial\Omega)$ and f' is sectionally continuous in the interval $[-\pi, \pi]$. Then the solution $u(r, \theta)$ of the Dirichlet problem (7.1), (7.2) is given by the series (7.9) with coefficients being the Fourier coefficients of f given by (7.10).

Example 7.1. Solve the Dirichlet problem

$$(7.11)\qquad \begin{aligned} \nabla^2 u(r, \theta) &= 0; \qquad 0 \le r < 1,\ -\pi \le \theta \le \pi, \\ u(1, \theta) &= |\theta|, \qquad -\pi \le \theta \le \pi. \end{aligned}$$

In this example, $f(\theta) = |\theta|$ is continuous on $\partial\Omega$ and

$$f'(\theta) = \begin{cases} -1 & \text{for } -\pi < \theta < 0 \\ 1 & \text{for } 0 < \theta < \pi \end{cases}$$

is sectionally continuous on $[-\pi, \pi]$. Therefore Theorem 7.1 is applicable. The Fourier coefficients of f are computed in Example 8.2 of the next section, where it is shown that $f(\theta)$ has the Fourier series representation (see equation (8.26))

$$|\theta| = \frac{\pi}{2} + \sum_{n=1}^{\infty} \frac{2}{\pi}\,\frac{(-1)^n - 1}{n^2}\cos n\theta, \qquad -\pi \le \theta \le \pi.$$

Therefore, the solution of problem (7.11) is

$$(7.12)\qquad u(r, \theta) = \frac{\pi}{2} + \sum_{n=1}^{\infty} \frac{2}{\pi}\,\frac{(-1)^n - 1}{n^2}\, r^n \cos n\theta.$$

Let us now drop the additional assumption that $f'(\theta)$ is sectionally continuous. Then the series in (7.9) still converges to a harmonic function in Ω, since its coefficients, which are the Fourier coefficients of f given by (7.10), are still bounded by 2 max $|f|$ (again by Problem 7.1). However, the series in (7.9) may not converge on $\partial\Omega$ because $r = 1$ on $\partial\Omega$ and the series is then the Fourier series of the function $f(\theta)$ which is merely assumed to be continuous. It is well known that there are continuous functions whose Fourier series do not converge. To circumvent this difficulty we define the function $u(r, \theta)$ to be given by the series (7.9) when $r < 1$ and to be equal to $f(\theta)$ when $r = 1$. Then it can be shown that $u(r, \theta)$ is continuous in $\bar{\Omega}$ and hence is the desired solution of the problem. We state this result in the following theorem.

Theorem 7.2. Suppose that f is in $C^0(\partial\Omega)$. Then the solution $u(r, \theta)$ of the Dirichlet problem (7.1), (7.2) is given by

$$(7.13) \qquad u(r, \theta) = \begin{cases} \dfrac{a_0}{2} + \sum_{n=1}^{\infty} r^n(a_n \cos n\theta + b_n \sin n\theta), & \text{for} \quad r < 1, \\ f(\theta), \quad \text{for} \quad r = 1, \end{cases}$$

with the coefficients being the Fourier coefficients of f given by (7.10).

We will now derive an alternate form of (7.9) which will express the solution of the Dirichlet problem (7.1), (7.2) as an integral rather than a series. Substitution of the formulas for the Fourier coefficients a_n, b_n (see Problem 8.5) into (7.9) yields, for $r < 1$,

$$\begin{aligned} u(r, \theta) &= \frac{1}{2\pi} \int_0^{2\pi} f(\phi) d\phi + \frac{1}{\pi} \sum_{n=1}^{\infty} r^n \left[\int_0^{2\pi} f(\phi) \cos n\phi \cos n\theta \, d\phi \right. \\ &\qquad \left. + \int_0^{2\pi} f(\phi) \sin n\phi \sin n\theta \, d\phi \right] \\ &= \frac{1}{2\pi} \int_0^{2\pi} f(\phi) d\phi + \frac{1}{\pi} \sum_{n=1}^{\infty} r^n \int_0^{2\pi} f(\phi) \cos n(\theta - \phi) d\phi \\ &= \frac{1}{2\pi} \int_0^{2\pi} f(\phi) \left[1 + 2 \sum_{n=1}^{\infty} r^n \cos n(\theta - \phi) \right] d\phi. \end{aligned}$$

The change in the order of integration and summation used in the last equality above is justified since the series in brackets converges uniformly for $r \le r_1$ with r_1 any number < 1. If we set

$$(7.14) \qquad P(r, \xi) = \frac{1}{2\pi} \left(1 + 2 \sum_{n=1}^{\infty} r^n \cos n\xi\right)$$

we have

$$(7.15) \qquad u(r, \theta) = \int_0^{2\pi} f(\phi) P(r, \theta - \phi) d\phi.$$

The function $P(r, \xi)$ is called the *Poisson kernel*. The sum of the series in

(7.14) defining $P(r, \xi)$ can be found in the following manner. Let z denote the complex variable in polar form,

$$z = re^{i\xi} = r(\cos \xi + i \sin \xi), \qquad |z| = r.$$

Then

$$z^n = r^n e^{in\xi} = r^n (\cos n\xi + i \sin n\xi)$$

and

$$1 + 2 \sum_{n=1}^{\infty} r^n \cos n\xi = \mathrm{Re}(1 + 2 \sum_{n=1}^{\infty} z^n); \qquad |z| < 1$$

where Re w denotes the real part of the complex number w. Now

$$\frac{1}{1 - z} = \sum_{n=0}^{\infty} z^n, \qquad |z| < 1$$

so that

$$\frac{1 + z}{1 - z} = 1 + 2 \sum_{n=1}^{\infty} z^n, \qquad |z| < 1.$$

Hence

$$P(r, \xi) = \frac{1}{2\pi} \mathrm{Re} \frac{1 + z}{1 - z} = \frac{1}{2\pi} \mathrm{Re} \frac{1 + re^{i\xi}}{1 - re^{i\xi}}, \qquad r < 1, \tag{7.16}$$

and a short calculation (see Problem 7.2) yields

$$P(r, \xi) = \frac{1}{2\pi} \frac{1 - r^2}{1 + r^2 - 2r \cos \xi}, \qquad r < 1. \tag{7.17}$$

Now, substitution of (7.17) into (7.15) yields the following integral formula for $u(r, \theta)$ when $r < 1$,

$$u(r, \theta) = \frac{1}{2\pi} \int_0^{2\pi} \frac{(1 - r^2)f(\phi)}{1 + r^2 - 2r \cos (\theta - \phi)} \, d\phi. \tag{7.18}$$

The integral in (7.18) is called *Poisson's integral.*

By direct computation, it can be shown that Poisson's integral defines a harmonic function in Ω $(r < 1)$ under the assumption that f is in $C^0(\partial\Omega)$. Moreover, as the point (r, θ) with $r < 1$ tends to any boundary point $(1, \theta_0)$, Poisson's integral tends to the value $f(\theta_0)$. (Detailed proofs of these assertions for the three-dimensional case will be given in Section 9.) Therefore, the function $u(r, \theta)$ defined to be given by Poisson's integral when $r < 1$, and equal to $f(\theta)$ when $r = 1$, is continuous in $\bar{\Omega}$ and hence is the desired solution of the problem.

Theorem 7.3. Suppose that f is in $C^0(\partial\Omega)$. Then the solution $u(r, \theta)$ of the Dirichlet problem (7.1), (7.2) is given by

$$u(r, \theta) = \begin{cases} \dfrac{1}{2\pi} \displaystyle\int_0^{2\pi} \frac{(1 - r^2)f(\phi)}{1 + r^2 - 2r \cos (\theta - \phi)} \, d\phi, & \text{for } r < 1 \\ f(\theta), & \text{for } r = 1. \end{cases} \tag{7.19}$$

The solution of the Dirichlet problem for a circle of radius a centered at the origin can be easily obtained from the solution of the problem for the unit circle using the fact, proved in Section 3, that under a similarity transformation of coordinates a harmonic function remains harmonic. Indeed, replacing r with r/a, the series solution for $B(0, 1)$ yields the series solution for $B(0, a)$,

$$u(r, \theta) = \frac{a_0}{2} + \sum_{n=1}^{\infty} \left(\frac{r}{a}\right)^n (a_n \cos n\theta + b_n \sin n\theta), \tag{7.20}$$

while the Poisson integral solution for $B(0, 1)$ yields the Poisson integral solution for $B(0, a)$,

$$u(r, \theta) = \frac{1}{2\pi} \int_0^{2\pi} \frac{(a^2 - r^2)f(\phi)}{a^2 + r^2 - 2ar \cos (\theta - \phi)} \, d\phi \tag{7.21}$$

Problems

7.1. Show that if the coefficients A_n, B_n in (7.4) are bounded, i.e., if they satisfy condition (7.5), then the series in (7.4) converges to a harmonic function in $\Omega = B(0, 1)$. [*Hint:* For $r \le r_1 < 1$, each term of the series in (7.4) is dominated by the corresponding term of the series

$$2M(1 + r_1 + r_1^2 + \dots). \tag{7.22}$$

The geometric series (7.22) converges to $2M(1 - r_1)^{-1}$ for $r_1 < 1$. Use the Weierstrass M-test to show that the series in (7.4) converges uniformly in each closed ball $\bar{B}(0, r_1)$ with $r_1 < 1$. Conclude that the series in (7.4) converges to a continuous function $u(r, \theta)$ in Ω. Then consider the series obtained by term-by-term differentiation of both of the series (7.4) and (7.22).]

7.2. Derive (7.17) from (7.16).

7.3. Derive (7.20) and (7.21).

7.4. Use (7.20) to show that the value at the origin of the solution of the Dirichlet problem for $B(0, a)$ is equal to the average of its values on the boundary $S(0, a)$ of $B(0, a)$. Do the same using (7.21).

7.5. (a) Consider the Dirichlet problem in an annulus Ω in R^2,

$$\Omega = \{(r, \theta) \in R^2; \quad 0 < a < r < 1, \quad -\pi \le \theta \le \pi\}$$

This problem requires one to find a function $u(r, \theta)$ in $C^2(\Omega) \cap C^0(\bar{\Omega})$ such that

$$\begin{aligned} \nabla^2 u(r, \theta) &= 0, \quad a < r < 1, \quad -\pi \le \theta \le \pi \\ u(1, \theta) &= f(\theta), \quad -\pi \le \theta \le \pi, \\ u(a, \theta) &= g(\theta), \quad -\pi \le \theta \le \pi. \end{aligned}$$

Assume the solution can be represented as a superposition of the functions 1, $\log r$, $r^n \cos n\theta$, $r^n \sin n\theta$, $r^{-n} \cos n\theta$, $r^{-n} \sin n\theta$, $n = 1, 2, \dots$, all of which are harmonic in Ω, and discuss the determination of the coefficients. Assume that f and g are C^1 functions.

(b) Solve the Dirichlet problem

$$\nabla^2 u(r,\theta) = 0, \qquad \frac{1}{2} < r < 1, \qquad -\pi \le \theta \le \pi,$$

$$u(1,\theta) = \sin\theta, \qquad -\pi \le \theta \le \pi,$$

$$u\left(\frac{1}{2}, \theta\right) = 1, \qquad -\pi \le \theta \le \pi.$$

7.6. Compute the values to two decimal places of the solution (7.12) of problem (7.11) at the following points (r,θ):

(a) $(0, \theta)$, (b) $(0.1, 0)$, (c) $(0.5, 0)$, (d) $(0.9, 0)$.

How many terms of the series do you need to use in each case? What is the value of the solution along the diameter $\theta = \pm \pi/2$?

7.7. (Study Section 8 first.) Find the series solution of the Dirichlet problem (7.1), (7.2) if

(a) $f(\theta) = \theta^2, \quad -\pi \le \theta \le \pi$

(b) $f(\theta) = |\sin\theta|, \quad -\pi \le \theta \le \pi$

(c) $f(\theta) = |\theta|(\pi - |\theta|), \quad -\pi < \theta < \pi$

(d) $f(\theta) = \begin{cases} \theta(\pi - \theta) & \text{for } 0 \le \theta \le \pi \\ -|\theta|(\pi - |\theta|) & \text{for } -\pi \le \theta \le 0. \end{cases}$

7.8. (Study Section 8 first.) Let $f(\theta)$ be sectionally continuous and have a sectionally continuous derivative on the interval $[-\pi, \pi]$. If necessary redefine f at points of discontinuity to be equal to the average of its limits from right and left. Also, let $f(\pm\pi) = \frac{1}{2}[f(\pi-0) + f(-\pi+0)]$. Show that (7.9), with coefficients being the Fourier coefficients of f, is a solution of the Dirichlet problem (7.1), (7.2) except that this solution is not necessarily continuous on the boundary ($r = 1$) of Ω. Find this solution of (7.1), (7.2) if

$$f(\theta) = \begin{cases} 0 & \text{for } -\pi < \theta < 0 \\ 1 & \text{for } 0 < \theta < \pi. \end{cases}$$

8. Introduction to Fourier Series

Let f be a function of one variable x defined on the interval $[-\pi,\pi]$. We are interested in the problem of representing f by means of a trigonometric series in the form

$$f(x) = \frac{a_0}{2} + \sum_{n=1}^{\infty} (a_n \cos nx + b_n \sin nx). \tag{8.1}$$

We would like to have answers to the following three questions: (a) For what functions f is the representation (8.1) possible? (b) In what sense is the representation valid; i.e., how does the series converge to $f(x)$? (c) How are the coefficients of the series determined from $f(x)$?

We can answer the last question first, assuming that the representation

(8.1) is valid in the sense that the series converges uniformly to $f(x)$ on the interval $[-\pi, \pi]$. We need the following formulas which can be easily proved by using elementary trigonometric identities (see Problem 8.1). For all positive integers n and k,

$$\int_{-\pi}^{\pi} \cos nx dx = \int_{-\pi}^{\pi} \sin nx dx = 0,$$

(8.2) $$\int_{-\pi}^{\pi} \cos nx \sin kx dx = 0,$$

$$\int_{-\pi}^{\pi} \cos nx \cos kx dx = \int_{-\pi}^{\pi} \sin nx \sin kx dx = 0, \text{ if } k \neq n,$$

and

(8.3) $$\int_{-\pi}^{\pi} \cos^2 nx dx = \int_{-\pi}^{\pi} \sin^2 nx dx = \pi, \qquad \int_{-\pi}^{\pi} 1\, dx = 2\pi.$$

In order to determine the coefficient a_k with $k \geq 1$, we multiply both sides of (8.1) by $\cos kx$ and integrate over $[-\pi, \pi]$. After interchanging the order of summation and integration (which is allowed under our assumption of uniform convergence of the series) we obtain

$$\int_{-\pi}^{\pi} f(x) \cos kx dx = \frac{a_0}{2} \int_{-\pi}^{\pi} \cos kx dx + \sum_{n=1}^{\infty} a_n \int_{-\pi}^{\pi} \cos nx \cos kx dx$$

$$+ \sum_{n=1}^{\infty} b_n \int_{-\pi}^{\pi} \sin nx \cos kx dx.$$

According to formulas (8.2), all except one of the terms on the right side of this equation are zero, so that

(8.4) $$\int_{-\pi}^{\pi} f(x) \cos kx dx = a_k \int_{-\pi}^{\pi} \cos^2 kx dx.$$

Using now (8.3) we obtain the formula for a_k, $k \geq 1$,

(8.5) $$a_k = \frac{1}{\pi} \int_{-\pi}^{\pi} f(x) \cos kx dx.$$

If $k = 0$, the above procedure leads to

(8.6) $$\int_{-\pi}^{\pi} f(x) dx = \frac{a_0}{2} \int_{-\pi}^{\pi} dx$$

in place of (8.4) and therefore

(8.7) $$a_0 = \frac{1}{\pi} \int_{-\pi}^{\pi} f(x) dx.$$

(The factor $1/2$ was used in the constant term a_0 of (8.1) in order to make formula (8.5) valid for all k including $k = 0$.) To obtain the formula for b_k, $k \geq 1$, we multiply both sides of (8.1) by sin kx and proceed as before to find

$$b_k = \frac{1}{\pi} \int_{-\pi}^{\pi} f(x) \sin kx dx. \tag{8.8}$$

We have shown so far that if the representation (8.1) is valid in the sense that the series converges uniformly to $f(x)$ on the interval $[-\pi, \pi]$, then the coefficients of the series must be given by formulas (8.5) and (8.8). But these formulas can be used to compute the coefficients a_n, b_n irrespective of whether or not the representation (8.1) is valid. It is only necessary for the function f to be such that the integrals in (8.5) and (8.8) exist. A large class of functions for which these integrals exist consists of the functions which are sectionally continuous on the interval $[-\pi, \pi]$. Roughly, a function is said to be sectionally continuous on an interval $[a, b]$ if it is continuous on $[a, b]$ except possibly at a finite number of points where it may have finite jumps. More precisely we have the following definition.

Definition 8.1. Suppose that the function f is defined and continuous at every point of an interval $[a, b]$ except possibly at the end points a, b and at a finite number of interior points $x_1, x_2, \ldots, x_{n-1}$ where

$$a = x_0 < x_1 < x_2 < \ldots < x_{n-1} < x_n = b. \tag{8.9}$$

Moreover, suppose that as x approaches the end points of each of the subintervals

$$(a, x_1), (x_1, x_2), \ldots, (x_{n-1}, b) \tag{8.10}$$

from the interior, the function f has finite limits. Then f is said to be *sectionally continuous* on the interval $[a, b]$.

We emphasize that in Definition 8.1 the function f is assumed to be continuous in each of the subintervals (8.10) but it may or may not be defined at the end points (8.9). However, each of the one-sided limits

$$\lim_{\substack{h \to 0 \\ h > 0}} f(x_i + h) \equiv f(x_i + 0), \qquad i = 0, 1, \ldots, n - 1; \tag{8.11}$$

$$\lim_{\substack{h \to 0 \\ h > 0}} f(x_i - h) \equiv f(x_i - 0), \qquad i = 1, 2, \ldots, n, \tag{8.12}$$

must exist (and be finite). The limits in (8.11) are called *limits from the right* and those in (8.12) *limits from the left*. The integral of f over $[a,b]$ exists and is equal to the sum of the integrals over each of the subintervals (8.10),

$$\int_a^b f(x)dx = \int_a^{x_1} f(x)dx + \int_{x_1}^{x_2} f(x)dx + \ldots + \int_{x_{n-1}}^{b} f(x)dx.$$

Note that continuous functions are special cases of sectionally continuous functions.

If the function f is sectionally continuous on the interval $[-\pi, \pi]$, then, for each $n = 0, 1, 2, \ldots$, the functions $f(x) \cos nx$ and $f(x) \sin nx$ are also sectionally continuous on $[-\pi, \pi]$ and the integrals in (8.5) and (8.8) exist. Therefore, for each function f which is sectionally continuous on $[-\pi, \pi]$, we can formally define a trigonometric series with coefficients given by (8.5) and (8.8).

Definition 8.2. Let f be a sectionally continuous function on the interval $[-\pi, \pi]$. The trigonometric series

$$\frac{a_0}{2} + \sum_{n=1}^{\infty} (a_n \cos nx + b_n \sin nx), \tag{8.13}$$

where

$$\begin{cases} a_n = \dfrac{1}{\pi}\displaystyle\int_{-\pi}^{\pi} f(x) \cos nx dx, \ n = 0, 1, 2, \dots, \\ b_n = \dfrac{1}{\pi}\displaystyle\int_{-\pi}^{\pi} f(x) \sin nx dx, \ n = 1, 2, \dots, \end{cases} \tag{8.14}$$

is called the *Fourier series associated with* f and the coefficients a_n, b_n are called the *Fourier coefficients of* f.

Note that since f is assumed to be sectionally continuous on $[-\pi, \pi]$, the Fourier coefficients of f are uniquely defined even if f is not defined at a finite number of points. In fact the values of f can be changed at a finite number of points without affecting the values of the Fourier coefficients of f.

Example 8.1. Let

$$f(x) = \begin{cases} 1 & \text{for} \quad -\pi < x < 0 \\ x & \text{for} \quad 0 < x < \pi. \end{cases} \tag{8.15}$$

This function is continuous on each of the intervals $(-\pi, 0)$ and $(0, \pi)$ and

$$f(-\pi + 0) = 1, \qquad f(0 - 0) = 1, \qquad f(0 + 0) = 0, \qquad f(\pi - 0) = \pi.$$

Therefore f is sectionally continuous on $[-\pi, \pi]$. The Fourier coefficients of f are

$$a_0 = \frac{1}{\pi}\int_{-\pi}^{\pi} f(x)dx = \frac{1}{\pi}\int_{-\pi}^{0} 1dx + \frac{1}{\pi}\int_{0}^{\pi} xdx = 1 + \frac{\pi}{2},$$

$$a_n = \frac{1}{\pi}\int_{-\pi}^{\pi} f(x) \cos nxdx = \frac{1}{\pi}\int_{-\pi}^{0} \cos nxdx + \frac{1}{\pi}\int_{0}^{\pi} x \cos nxdx$$

$$= \frac{1}{\pi}\left[x\frac{\sin nx}{n} + \frac{\cos nx}{n^2}\right]_0^{\pi} = \frac{-1 + (-1)^n}{\pi n^2}, \ n = 1, 2, \dots,$$

$$b_n = \frac{1}{\pi}\int_{-\pi}^{\pi} f(x) \sin nxdx = \frac{1}{\pi}\int_{-\pi}^{0} \sin nxdx + \frac{1}{\pi}\int_{0}^{\pi} x \sin nxdx$$

$$= -\frac{1 - (-1)^n}{\pi n} - \frac{(-1)^n}{n} = -\frac{1 + (\pi - 1)(-1)^n}{\pi n}, \qquad n = 1, 2, \dots .$$

The Fourier series associated with the function (8.15) is

$$\frac{2+\pi}{4} + \sum_{n=1}^{\infty}\left[\frac{-1+(-1)^n}{\pi n^2}\cos nx - \frac{1+(\pi-1)(-1)^n}{\pi n}\sin nx\right]. \tag{8.16}$$

The terms up to $n = 3$ of this series are

$$\frac{2+\pi}{4} + \frac{1}{2\pi}\cos 2x - \frac{2-\pi}{\pi}\sin x - \frac{1}{2}\sin 2x - \frac{2-\pi}{3\pi}\sin 3x.$$

Example 8.2. Let

$$f(x) = |x|, \qquad -\pi < x < \pi. \tag{8.17}$$

This function is continuous on $(-\pi, \pi)$ and

$$f(-\pi+0) = \pi, \qquad f(\pi-0) = \pi.$$

Therefore f is sectionally continuous on $[-\pi, \pi]$ and has Fourier coefficients

$$a_0 = \frac{1}{\pi}\int_{-\pi}^{0}(-x)dx + \frac{1}{\pi}\int_{0}^{\pi} xdx = \pi,$$

$$a_n = \frac{1}{\pi}\int_{-\pi}^{0}(-x)\cos nxdx + \frac{1}{\pi}\int_{0}^{\pi} x\cos nxdx$$

$$= \frac{1}{\pi}\left[-\frac{x\sin nx}{n} - \frac{\cos nx}{n^2}\right]_{-\pi}^{0} + \frac{1}{\pi}\left[\frac{x\sin nx}{n} + \frac{\cos nx}{n^2}\right]_{0}^{\pi}$$

$$= 2\,\frac{(-1)^n - 1}{\pi n^2}, \qquad n = 1, 2, \ldots,$$

$$b_n = \frac{1}{\pi}\int_{-\pi}^{0}(-x)\sin nxdx + \frac{1}{\pi}\int_{0}^{\pi} x\sin nxdx = 0, \qquad n = 1, 2, \ldots .$$

The Fourier series associated with (8.17) is

$$\frac{\pi}{2} + \sum_{n=1}^{\infty} 2\,\frac{(-1)^n - 1}{\pi n^2}\cos nx. \tag{8.18}$$

The terms up to $n = 3$ are

$$\frac{\pi}{2} - \frac{4}{\pi}\cos x - \frac{4}{9\pi}\cos 3x.$$

Example 8.3. Let

$$f(x) = x, \qquad -\pi < x < \pi. \tag{8.19}$$

This function is continuous on $(-\pi, \pi)$ and

$$f(-\pi+0) = -\pi, \qquad f(\pi-0) = \pi.$$

Therefore f is sectionally continuous on $[-\pi,\pi]$ and has Fourier coefficients

$$a_0 = \frac{1}{\pi}\int_{-\pi}^{\pi} x dx = 0,$$

$$a_n = \frac{1}{\pi}\int_{-\pi}^{\pi} x \cos nx dx = \frac{1}{\pi}\left[\frac{x \sin nx}{n} + \frac{\cos nx}{n^2}\right]_{-\pi}^{\pi} = 0, \qquad n = 1, 2, \ldots,$$

$$b_n = \frac{1}{\pi}\int_{-\pi}^{\pi} x \sin nx dx = \frac{1}{\pi}\left[-\frac{x \cos nx}{n} + \frac{\sin nx}{n^2}\right]_{-\pi}^{\pi}$$

$$= \frac{2}{n}(-1)^{n+1}, \qquad n = 1, 2, \ldots .$$

The Fourier series associated with (8.19) is

$$\sum_{n=1}^{\infty} \frac{2(-1)^{n+1}}{n} \sin nx. \tag{8.20}$$

The terms up to $n = 3$ are

$$2 \sin x - \sin 2x + \frac{2}{3} \sin 3x.$$

Examples 8.2 and 8.3 illustrate two general rules concerning the Fourier series of even and odd functions. The function f is said to be an *even function* if $f(-x) = f(x)$ for all x for which f is defined; f is an *odd function* if $f(-x) = -f(x)$. The function (8.17) is even on $(-\pi,\pi)$ and its Fourier series (8.18) has only cosine terms. The function (8.19) is odd on $(-\pi,\pi)$ and its Fourier series (8.20) has only sine terms. Generally, we have the following lemma, the proof of which is left to Problem 8.4.

Lemma 8.1. Let f be sectionally continuous on $[-\pi,\pi]$. If f is an even function, its Fourier coefficients are given by

$$a_n = \frac{2}{\pi}\int_0^{\pi} f(x) \cos nx dx,\ n = 0, 1, 2, \ldots ;\ b_n = 0,\ n = 1, 2, \ldots . \tag{8.21}$$

If f is an odd function, its Fourier coefficients are given by

$$a_n = 0,\ n = 0, 1, 2, \ldots ;\ b_n = \frac{2}{\pi}\int_0^{\pi} f(x) \sin nx dx,\ n = 1, 2, \ldots . \tag{8.22}$$

Thus, the Fourier series of an even function has only cosine (even) terms and for this reason it is called a *Fourier cosine series,* while the Fourier series of an odd function has only sine (odd) terms and it is called a *Fourier sine series*. Note that most functions are neither even nor odd and their Fourier series contain both cosine and sine terms, as illustrated by Example 8.1.

A theorem which gives conditions on the function f in order for f to be representable by its Fourier series, and which describes the sense in which the representation is valid is known as a *Fourier theorem*. Before stating such a theorem we observe that each term of the Fourier series of a function f is a periodic function of period 2π. Consequently, if the series converges to some value at a point x of the interval $(-\pi,\pi)$, it must also converge to the same value at every point of the x-axis of the form $x \pm 2n\pi$, $n = 1, 2, \ldots$, and if the series converges at either point $x = \pi$ or $x = -\pi$, it must converge to the same value at both points and at every point of the form $\pi \pm 2n\pi$, $n = 1, 2, \ldots$. Thus, convergence of the series at every point of the interval $[-\pi,\pi)$ guarantees the convergence of the series for all x, the sum of the series being a periodic function of period 2π. We conclude that if the Fourier series represents f on the basic interval $(-\pi,\pi)$ it must also represent the periodic extension of f on the whole real axis. On the other hand, if f were a periodic function of period 2π, then the Fourier series representation of f would be valid for all x if it were valid on the basic interval $(-\pi,\pi)$. It is convenient therefore to state the following basic Fourier theorem in terms of a periodic function.

Theorem 8.1. (*A Fourier theorem.*) Suppose that the function f and its derivative are sectionally continuous on $[-\pi,\pi]$ and that f is periodic of period 2π. Then f can be represented by its Fourier series (8.13) with coefficients given by (8.14), in the sense that at every point x where f is continuous,

$$f(x) = \frac{a_0}{2} + \sum_{n=1}^{\infty} (a_n \cos nx + b_n \sin nx), \tag{8.23}$$

and at every point x where f has a jump discontinuity,

$$\frac{1}{2}[f(x + 0) + f(x - 0)] = \frac{a_0}{2} + \sum_{n=1}^{\infty} (a_n \cos nx + b_n \sin nx). \tag{8.24}$$

If a function f and its derivative f' are sectionally continuous on an interval $[a,b]$ then f is said to be *sectionally smooth* on $[a,b]$. Roughly, the graph of such a function has at most a finite number of finite jumps and a finite number of corners. Theorem 8.1 requires that f be sectionally smooth on $[-\pi,\pi]$. Each of the functions in Examples 8.1, 8.2 and 8.3 is sectionally smooth on $[-\pi,\pi]$.

The conclusion of Theorem 8.1 is that the Fourier series of f converges to $f(x)$ at every point x where f is continuous, while at a point x where f has a jump discontinuity, the Fourier series converges to the average of the limits from the right and left. From the definition of continuity, at every point x where f is continuous, $f(x + 0) = f(x - 0) = f(x)$ and hence $(1/2)[f(x + 0) + f(x - 0)] = f(x)$, so that (8.24) holds true for every x, $-\infty < x < \infty$. An alternate statement of Theorem 8.1 is, therefore, the following: If f is periodic of period 2π and sectionally smooth on $[-\pi,\pi]$, then the Fourier series of f converges to $(1/2)[f(x + 0) + f(x - 0)]$ for every x, $-\infty < x < \infty$.

The proof of Theorem 8.1 is lengthy and technical. It can be found in any book on Fourier series such as Churchill[4] or Tolstov.[5]

If the function f is defined only on the interval $(-\pi,\pi)$, Theorem 8.1 may be applied to the periodic extension $\tilde{f}$ of f. For example, the periodic extension of the function (8.15) of Example 8.1 is defined by

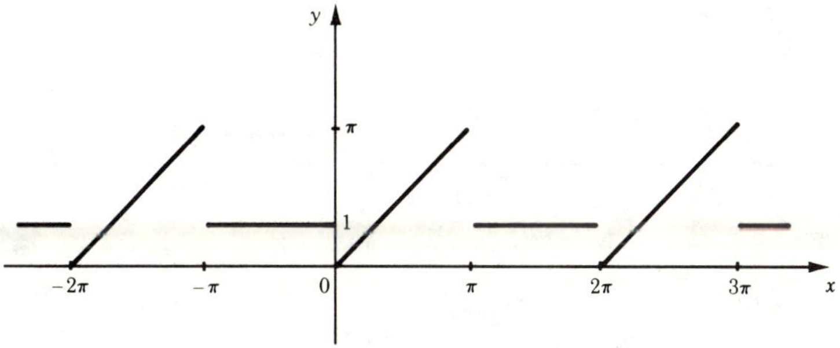

Fig. 8.1

$$\tilde{f}(x) = \begin{cases} 1 & \text{for} \quad -\pi < x < 0 \\ x & \text{for} \quad 0 < x < \pi \end{cases}; \qquad \tilde{f}(x + 2\pi) = \tilde{f}(x) \quad \text{for all} \quad x,$$

and its graph is shown in Figure 8.1. Since $\tilde{f}$ is sectionally smooth on $[-\pi,\pi]$, Theorem 8.1 is applicable and the Fourier series (8.17) converges to $\tilde{f}(x)$ at every point where $\tilde{f}$ is continuous, namely, at every $x \neq \pm n\pi$, $n = 0, 1, 2, \ldots$. At $x = 0$ the series converges to

$$\frac{1}{2}[\tilde{f}(0 + 0) + \tilde{f}(0 - 0)] = \frac{1}{2}[f(0 + 0) + f(0 - 0)] = \frac{1}{2}(0 + 1) = \frac{1}{2}$$

while at $x = \pi$ the series converges to

$$\frac{1}{2}[\tilde{f}(\pi + 0) + \tilde{f}(\pi - 0)] = \frac{1}{2}[f(-\pi + 0) + f(\pi - 0)] = \frac{1}{2}(1 + \pi).$$

Actually, it is easier to look at Figure 8.1 to determine the average of the limits from right and left at points of discontinuity. Figure 8.2(a) shows the graph of the function to which the Fourier series (8.16) converges for every x. Figures 8.2(b) and (c) show respectively the graphs of the functions to which the series (8.18) and (8.20) of Examples 8.2 and 8.3 converge.

The above discussion leads immediately to the following corollary concerning the Fourier series representation of a function on the finite interval $[-\pi,\pi]$.

Corollary 8.1. Suppose that f is sectionally smooth on the interval $[-\pi,\pi]$. Then f can be represented by its Fourier series (8.13) with coefficients given by (8.14) in the sense that the series converges to: (i) $f(x)$ at every interior point x where f is continuous, (ii) $\frac{1}{2}[f(x + 0) + f(x - 0)]$ at every interior point x where f has a jump discontinuity, and (iii) $\frac{1}{2}[f(\pi - 0) + f(-\pi + 0)]$ at each of the boundary points $x = -\pi$ and $x = \pi$.

Accordingly, from Examples 8.1, 8.2 and 8.3 we obtain the following Fourier series representations:

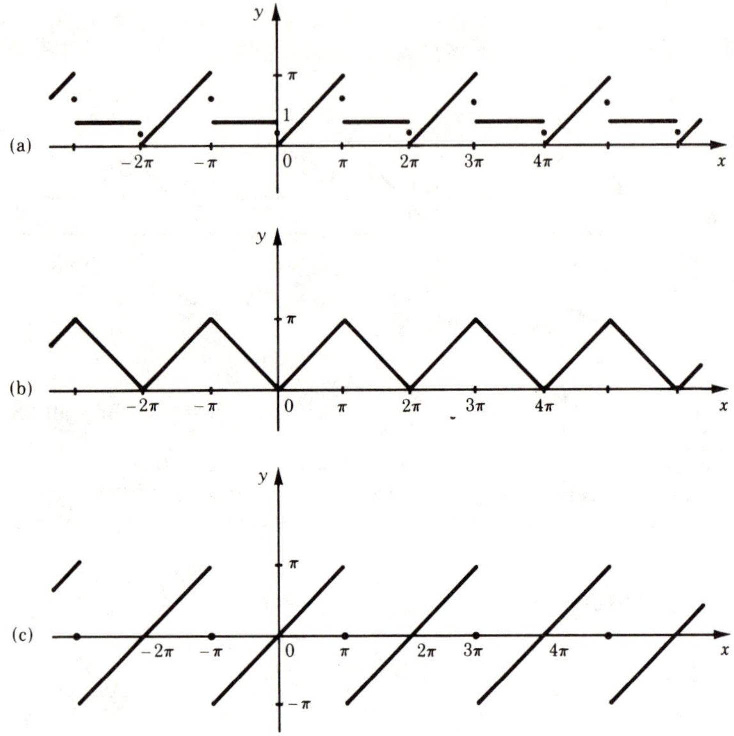

Fig. 8.2

(8.25) $$\frac{2+\pi}{4}+\sum_{n=1}^{\infty}\left[\frac{-1+(-1)^n}{\pi n^2}\cos nx-\frac{1+(\pi-1)(-1)^n}{\pi n}\sin nx\right]$$

$$=\begin{cases}1 & \text{for } -\pi<x<0\\ 1/2 & \text{for } x=0\\ x & \text{for } 0<x<\pi\\ (1+\pi)/2 & \text{for } x=\pm\pi.\end{cases}$$

(8.26) $$\frac{\pi}{2}+\frac{2}{\pi}\sum_{n=1}^{\infty}\frac{(-1)^n-1}{n^2}\cos nx=|x| \quad \text{for} \quad -\pi\le x\le\pi.$$

(8.27) $$2\sum_{n=1}^{\infty}\frac{(-1)^{n+1}}{n}\sin nx=\begin{cases}x & \text{for} \quad -\pi<x<\pi\\ 0 & \text{for} \quad x=\pm\pi.\end{cases}$$

Let us examine carefully the above three Fourier series representations. We observe that all three series converge to the function $f(x) = x$ on the interval $(0, \pi)$. In particular, the function $f(x) = x$ is represented on this interval by the Fourier cosine series in (8.26) and by the Fourier sine series in (8.27). This suggests that any function satisfying suitable conditions may be represented on the half-interval $(0, \pi)$ by a Fourier cosine series or by a Fourier sine series. That this is actually possible follows from the observation that a function f defined on $(0, \pi)$ can be extended to an even or an odd function on the interval $(-\pi, \pi)$. The even extension f_e of f is defined by

$$f_e(x) = \begin{cases} f(x) & \text{for} \quad 0 < x < \pi \\ f(-x) & \text{for} \quad -\pi < x < 0 \end{cases}$$

and the odd extension f_o by

$$f_o(x) = \begin{cases} f(x) & \text{for} \quad 0 < x < \pi \\ -f(-x) & \text{for} \quad -\pi < x < 0. \end{cases}$$

If f is sectionally smooth on $[0, \pi]$, f_e is sectionally smooth on $[-\pi, \pi]$ and since f_e is even, the Fourier cosine series representation of f_e on $[-\pi, \pi]$ is the Fourier cosine series representation of f on $[0, \pi]$. Similarly, the Fourier sine series representation of f_o on $[-\pi, \pi]$ is the Fourier sine series representation of f on $[0, \pi]$. We are thus lead to the following corollary.

Corollary 8.2 Suppose that f is sectionally smooth on $[0, \pi]$. Then f can be represented by the Fourier cosine series

$$(8.28) \qquad \frac{a_0}{2} + \sum_{n=1}^{\infty} a_n \cos nx; \qquad a_n = \frac{2}{\pi} \int_0^{\pi} f(x) \cos nx\, dx, \quad n = 0, 1, \ldots,$$

and by the Fourier sine series

$$(8.29) \qquad \sum_{n=1}^{\infty} b_n \sin nx; \quad b_n = \frac{2}{\pi} \int_0^{\pi} f(x) \sin nx\, dx, \qquad n = 1, 2, \ldots,$$

in the following sense: (i) At every interior point x where f is continuous both series converge to $f(x)$; (ii) at every interior point x where f has a jump discontinuity, both series converge to $\frac{1}{2}[f(x + 0) + f(x - 0)]$; (iii) at $x = 0$ the cosine series converges to $f(0 + 0)$ while the sine series converges to 0; and (iv) at $x = \pi$ the cosine series converges to $f(\pi - 0)$ while the sine series converges to 0.

It is actually easier to remember Corollary 8.2 in the following alternate form: If f is sectionally smooth on $[0, \pi]$, then at every point of the x-axis the Fourier cosine series (8.28) of f converges to the average of the limits from right and left of the even periodic extension of f, while the Fourier sine series (8.29) of f converges to the average of the limits from right and left of the odd periodic extension of f of period 2π.

Example 8.4. Let

$$(8.30) \qquad f(x) = 1, \qquad 0 < x < \pi.$$

Certainly this function is sectionally smooth on $[0, \pi]$. From (8.29)

$$b_n = \frac{2}{\pi}\int_0^{\pi} 1 \sin nx\, dx = \frac{2}{\pi}\left[-\frac{\cos nx}{n}\right]_0^{\pi} = \frac{2}{\pi}\frac{1-(-1)^n}{n},$$

and the Fourier sine series of (8.30) is

$$\frac{2}{\pi}\sum_{n=1}^{\infty}\frac{1-(-1)^n}{n}\sin nx. \tag{8.31}$$

This series converges to 1 for $0 < x < \pi$ and to 0 for $x = 0$ and $x = \pi$,

$$\frac{2}{\pi}\sum_{n=1}^{\infty}\frac{1-(-1)^n}{n}\sin nx = \begin{cases} 0 & \text{for} \quad x = 0 \quad \text{and} \quad x = \pi. \\ 1 & \text{for} \quad 0 < x < \pi \end{cases}$$

Figure 8.3 shows the graph of the function to which the series converges for all x. From (8.28),

$$a_n = \frac{2}{\pi}\int_0^{\pi} 1 \cos nx\, dx = \begin{cases} 2 & \text{for} \quad n = 0 \\ 0 & \text{for } n = 1, 2, \ldots \end{cases}$$

so that the Fourier cosine series of (8.30) consists of the single term 1 (a result that we could have guessed without any computations).

Example 8.5. The Fourier cosine and sine series representations of the function $f(x) = x(\pi - x)$, $0 \le x \le \pi$, are

$$x(\pi - x) = \frac{\pi^2}{6} - 2\sum_{n=1}^{\infty}\frac{1+(-1)^n}{n^2}\cos nx, \qquad 0 \le x \le \pi \tag{8.32}$$

$$x(\pi - x) = \frac{4}{\pi}\sum_{n=1}^{\infty}\frac{1-(-1)^n}{n^3}\sin nx, \qquad 0 \le x \le \pi. \tag{8.33}$$

The student should sketch the graphs of the functions to which the series in (8.32) and (8.33) converge for all x, $-\infty < x < \infty$.

We turn now to a few important results concerning the speed and manner of convergence of Fourier series.

The *Parseval relation*

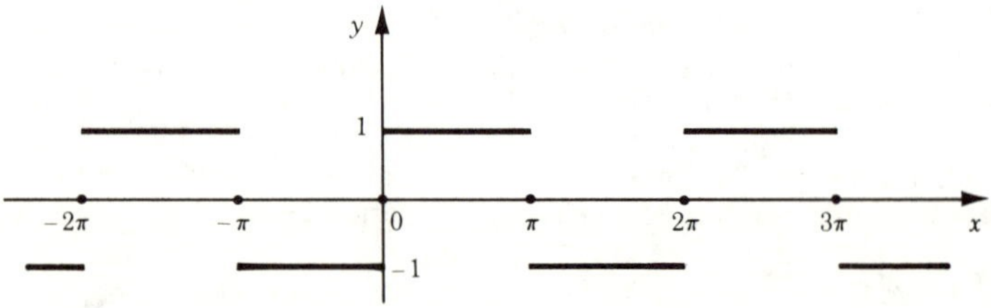

Fig. 8.3

$$\frac{a_0^2}{2} + \sum_{n=1}^{\infty} (a_n^2 + b_n^2) = \frac{1}{\pi} \int_{-\pi}^{\pi} [f(x)]^2 dx \tag{8.34}$$

is satisfied by every function f which is sectionally continuous on $[-\pi, \pi]$. An immediate consequence of this relation is that the Fourier coefficients a_n, b_n of f tend to 0 as $n \to \infty$,

$$\lim_{n\to\infty} a_n = 0, \qquad \lim_{n\to\infty} b_n = 0.$$

The speed of convergence of the coefficients to 0 depends on the function f. For example, the Fourier coefficients of the function (8.18) tend to 0 like $1/n^2$, while those of (8.20) tend to 0 like $1/n$. The following lemma describes a useful result in this respect.

Lemma 8.2. Suppose that $f \in C^k([-\pi, \pi])$, where k is some integer ≥ 1, and

$$f^{(p)}(-\pi) = f^{(p)}(\pi), \qquad p = 0, 1, \ldots, k-1.$$

Then, for $n = 1, 2, \ldots$,

$$|a_n| \leq 2Mn^{-k} \quad \text{and} \quad |b_n| \leq 2Mn^{-k},$$

where M is a constant that bounds $|f^{(k)}(x)|$,

$$|f^{(k)}(x)| \leqq M \quad \text{for} \quad -\pi \leq x \leq \pi.$$

Lemma 8.2 can be easily proved from the formulas for a_n and b_n using integration by parts. The lemma asserts that under the stated conditions on the function f, the Fourier coefficients of f tend to 0 like n^{-k}. Note that the conditions of the lemma are satisfied by a function f which is periodic of period 2π and belongs to $C^k(-\infty, \infty)$.

Since

$$|a_n \cos nx + b_n \sin nx| \leq |a_n| + |b_n|, \quad \text{for} \quad -\pi \leq x \leq \pi \quad \text{and} \quad n = 1, 2, \ldots,$$

the speed and manner of convergence of a Fourier series depends on the speed of convergence of its coefficients to 0 as $n \to \infty$. Suppose for example that the function f satisfies the conditions of Lemma 8.2 with $k = 2$. Then, according to Theorem 8.1,

$$f(x) = \frac{a_0}{2} + \sum_{n=1}^{\infty} (a_n \cos nx + b_n \sin nx),$$

for every x in $[-\pi, \pi]$. Moreover, according to Lemma 8.2,

$$|a_n| + |b_n| \leq \frac{4M}{n^2}, \qquad n = 1, 2, \ldots,$$

so that

$$|a_n \cos nx + b_n \sin nx| \leq \frac{4M}{n^2}, \quad \text{for} \quad -\pi \leq x \leq \pi \quad \text{and} \quad n = 1, 2, \ldots.$$

Since the series $\sum_{n=1}^{\infty} 4Mn^{-2}$ converges, it follows from the Weierstrass M-test (see Taylor)[6] that the Fourier series of f converges uniformly to $f(x)$ on the interval $[-\pi, \pi]$. Recall that this means that the rate of convergence to 0 of the remainder

$$R_N(x) = \sum_{N+1}^{\infty} (a_n \cos nx + b_n \sin nx)$$

as $N \to \infty$ is uniform over the interval $[-\pi, \pi]$. More precisely, given any positive number ϵ, there is an integer N_ϵ which depends on ϵ but *not* on x, such that

$$|R_N(x)| < \epsilon \quad \text{for} \quad N > N_\epsilon \quad \text{and} \quad -\pi \leq x \leq \pi.$$

A more careful analysis gives the following result on uniform convergence of Fourier series under weaker conditions on the function f.

Theorem 8.2. Suppose that the function f is continuous on the interval $[-\pi, \pi]$, that $f(-\pi) = f(\pi)$, and that the derivative f' is sectionally continuous on $[-\pi, \pi]$. Then the Fourier series representation

$$(8.35) \quad f(x) = \frac{a_0}{2} + \sum_{n=1}^{\infty} (a_n \cos nx + b_n \sin nx), \qquad -\pi \leq x \leq \pi,$$

is valid in the sense that the series converges absolutely and uniformly on the interval $[-\pi, \pi]$.

As an example, the Fourier series representation (8.26) is valid in the sense that the series converges to $|x|$ uniformly on $[-\pi, \pi]$. In fact the series converges uniformly on the whole x-axis to the function whose graph is shown in Figure 8.2(b).

Note that the conditions of the theorem are satisfied by a function f which is periodic of period 2π and which is continuous and has a sectionally continuous derivative on the interval $[-\pi, \pi]$. The corresponding results on uniform convergence of the Fourier cosine and sine series representation of a function f defined on $[0, \pi]$ can be easily obtained from Theorem 8.2 by considering the even and odd extensions of f.

Concerning the differentiability of Fourier series representations, it is useful to recall the following differentiability result which is valid for any function series: Suppose that $f(x) = \Sigma f_n(x)$ for x in some interval I and suppose that the differentiated series $\Sigma f_n'(x)$ converges uniformly on I. Then $f'(x) = \Sigma f_n'(x)$ for x in I. As an example, this result is applicable to (8.33) from which we obtain

$$(8.36) \qquad \pi - 2x = \frac{4}{\pi} \sum_{n=1}^{\infty} \frac{1 - (-1)^n}{n^2} \cos nx, \qquad 0 \leq x \leq \pi.$$

Observe that this result is not applicable to the series (8.26) or (8.32). However, the result described in Problem 8.20 is applicable to these series.

So far in our discussion of Fourier series representations we have dealt with functions which are either periodic of period 2π or can be extended to periodic functions of period 2π. Actually, the only thing special about the number 2π is that it makes the formulas simpler, due to the fact that all the trigonometric functions $\cos nx$ and $\sin nx$ are periodic of period 2π. For the Fourier series representation of a periodic function of period $2L$ the functions $\cos n\pi x/L$ and $\sin n\pi x/L$ should be used instead, since all of these functions are periodic of period $2L$. The corresponding formulas can be obtained by making the simple change of variable (change of scale),

$$t = \frac{\pi x}{L}, \qquad x = \frac{Lt}{\pi}.$$

If $f(x)$ is a periodic function of x of period $2L$ which is sectionally continuous on the interval $[-L, L]$, the function

$$F(t) = f\left(\frac{Lt}{\pi}\right)$$

is a periodic function of t of period 2π which is sectionally continuous on $[-\pi, \pi]$. According to Definition 8.2, the Fourier series associated with $F(t)$ is

$$\frac{a_0}{2} + \sum_{n=1}^{\infty} (a_n \cos nt + b_n \sin nt),$$

where

$$a_n = \frac{1}{\pi} \int_{-\pi}^{\pi} F(t) \cos nt \, dt, \quad b_n = \frac{1}{\pi} \int_{-\pi}^{\pi} F(t) \sin nt \, dt, \quad n = 0, 1, \ldots .$$

Returning to the variable x and remembering that $F(t) = f(Lt/\pi) = f(x)$, we obtain the Fourier series of $f(x)$

$$\frac{a_0}{2} + \sum_{n=1}^{\infty} \left(a_n \cos \frac{n\pi x}{L} + b_n \sin \frac{n\pi x}{L}\right) \tag{8.37}$$

where

$$\begin{aligned} a_n &= \frac{1}{L} \int_{-L}^{L} f(x) \cos \frac{n\pi x}{L} \, dx, \\ b_n &= \frac{1}{L} \int_{-L}^{L} f(x) \sin \frac{n\pi x}{L} \, dx; \qquad n=0,1,\ldots . \end{aligned} \tag{8.38}$$

It is easy to see that all the representation results of this section are valid for the Fourier series (8.37) with Fourier coefficients (8.38) provided that the words "period 2π" are replaced by "period $2L$" and the basic interval $[-\pi,\pi]$ is replaced by $[-L,L]$. In practice, the Fourier cosine and Fourier sine series representations of functions defined on the interval $[0,L]$,

$$(8.39)\quad \frac{a_0}{2}+\sum_{n=1}^{\infty} a_n \cos\frac{n\pi x}{L};\ a_n=\frac{2}{L}\int_0^L f(x)\cos\frac{n\pi x}{L}\,dx,\quad n=0,1,\ldots,$$

$$(8.40)\quad \sum_{n=1}^{\infty} b_n \sin\frac{n\pi x}{L};\quad b_n=\frac{2}{L}\int_0^L f(x)\sin\frac{n\pi x}{L}\,dx,\quad n=1,2,\ldots$$

are most frequently encountered.

Example 8.6. From (8.40) we obtain the Fourier sine series representation of the saw-tooth function,

$$(8.41)\quad \frac{4L}{\pi^2}\sum_{n=1}^{\infty}\frac{\sin(n\pi/2)}{n^2}\sin\frac{n\pi x}{L}=\begin{cases} x & \text{for } 0\le x\le \dfrac{L}{2} \\ L-x & \text{for } \dfrac{L}{2}\le x\le L, \end{cases}$$

and from (8.39) we obtain the Fourier cosine series representation of the rectangular pulse function,

$$(8.42)\quad \frac{a}{L}+\frac{4}{\pi}\sum_{n=1}^{\infty}\left(\frac{1}{n}\sin\frac{n\pi a}{2L}\cos\frac{n\pi}{2}\right)\cos\frac{n\pi x}{L} =\begin{cases} 0 & \text{for } 0\le x<\dfrac{L-a}{2} \\ 1 & \text{for } \dfrac{L-a}{2}<x<\dfrac{L+a}{2} \\ 0 & \text{for } \dfrac{L+a}{2}<x\le L, \end{cases}$$

where $0\le a\le L$. At the points $x=(L\pm a)/2$ the series in (8.42) converges to 1/2, which is the average of the limits from right and left. It is left to the student to draw the graphs of the sums of the series in (8.41) and (8.42) for all x.

We have described in this section only the more elementary aspects of Fourier series. We omitted the proofs of the basic Theorems 8.1, 8.2 and of the Parseval relation (8.34). These proofs and many other interesting and important results on Fourier series can be found in many existing books on the subject. We mention in particular the book of Tolstov[5] which contains advanced topics in addition to a leisurely introduction to Fourier series. We close with a short paragraph on a different kind of Fourier series representation of functions which are not necessarily sectionally continuous.

Consider the class of functions which are square integrable on the interval $[-\pi,\pi]$; i.e., funtions f such that the integral $\int_{-\pi}^{\pi}[f(x)]^2dx$ exists. This class of functions is larger than the class of sectionally continuous functions. For example, $|x|^{-1/4}$ is square integrable but not sectionally continuous on $[-\pi,\pi]$. Definition 8.2 can be used to define a Fourier series associated with every square integrable function. It can be shown that Parseval's relation is satisfied by every square integrable function. Moreover every square integrable function can be represented by its Fourier series in the sense of *convergence in the mean,*

$$\text{(8.43)}\quad \lim_{N\to\infty} \int_{-\pi}^{\pi} [f(x) - (\frac{a_0}{2} + \sum_{n=1}^{N} a_n\cos nx + b_n\sin nx)]^2 dx = 0.$$

Problems

8.1. Prove formulas (8.2) and (8.3).

8.2. Derive formula (8.8).

8.3. Prove that the constant term of the Fourier series associated with a function f is equal to the average of f on the interval $[-\pi,\pi]$.

8.4. Prove Lemma 8.1.

8.5. Prove that if f is sectionally continuous on $[-\pi,\pi]$ and periodic of period 2π, the Fourier coefficients of f defined by formulas (8.14) are also given by the formulas,

$$a_n = \frac{1}{\pi}\int_{-\pi+c}^{\pi+c} f(x)\cos nx\, dx, \qquad n = 0,1,2, \ldots ;$$

$$b_n = \frac{1}{\pi}\int_{-\pi+c}^{\pi+c} f(x)\sin nx\, dx, \qquad n = 1,2, \ldots ,$$

where c is any real number.

8.6. Rewrite the series (8.18) in the form

$$\frac{\pi}{2} - \frac{4}{\pi}\sum_{k=1}^{\infty} \frac{\cos(2k-1)x}{(2k-1)^2}$$

Do similar rewriting of the series (8.16), (8.31), (8.32) and (8.33).

8.7. Let f be defined and continuous on $[-\pi,\pi]$. State conditions under which the Fourier series of f converges to $f(x)$ for every x in $[-\pi,\pi]$.

8.8. Let f be defined and continuous on $[0,\pi]$. State conditions under which (a) the Fourier cosine series of f converges to $f(x)$ for every x in $[0,\pi]$; (b) the Fourier sine series of f converges to $f(x)$ for every x in $[0,\pi]$.

8.9. Derive (8.32) and (8.33).

8.10. Derive each of the following Fourier series representations. In each case sketch the graph of the function to which the series converges for all x.

(a) $$\frac{1}{2} + \frac{2}{\pi}\sum_{k=1}^{\infty} \frac{1}{2k-1}\sin(2k-1)x = \begin{cases} 0 & \text{for} \quad -\pi < x < 0 \\ 1 & \text{for} \quad 0 < x < \pi, \end{cases}$$

(b) $$\frac{2}{\pi} - \frac{4}{\pi}\sum_{k=1}^{\infty} \frac{1}{4k^2-1}\cos 2kx = \sin x, \quad \text{for} \quad 0 \le x \le \pi,$$

(c) $$\frac{\pi^2}{3} + 4\sum_{n=1}^{\infty} \frac{(-1)^n}{n^2}\cos nx = x^2, \quad \text{for} \quad -\pi \le x \le \pi,$$

(d) $$\frac{4\pi^2}{3} + 4\sum_{n=1}^{\infty} (\frac{1}{n^2}\cos nx - \frac{\pi}{n}\sin nx) = x^2, \text{ for } \ 0 < x < 2\pi.$$

8.11. Prove Lemma 8.2.

8.12. Show that if f is continuous and has a sectionally continuous derivative on $[0, \pi]$, then the Fourier cosine series representation

$$f(x) = \frac{a_0}{2} + \sum_{n=1}^{\infty} a_n \cos nx, \quad 0 \leq x \leq \pi;$$

$$a_n = \frac{2}{\pi} \int_0^{\pi} f(x) \cos nx \, dx, \quad n = 0, 1, \ldots ,$$

is valid in the sense of absolute and uniform convergence on $[0, \pi]$.

8.13. Show that if f is continuous and has a sectionally continuous derivative on $[0, \pi]$ and if $f(0) = f(\pi) = 0$, then the Fourier sine series representation

$$f(x) = \sum_{n=1}^{\infty} b_n \sin nx, \; 0 \leq x \leq \pi;$$

$$b_n = \frac{2}{\pi} \int_0^{\pi} f(x) \sin nx \, dx, \; n = 1, 2, \ldots ,$$

is valid in the sense of absolute and uniform convergence on $[0, \pi]$.

8.14. Are the representations (8.32) and (8.33) valid in the sense of uniform convergence on $[0, \pi]$?

8.15. Derive Parseval's relation (8.34) for functions which satisfy the conditions of Theorem 8.2.

8.16. Show that a Fourier series cannot converge uniformly on any interval in which the series converges to a discontinuous function.

8.17. If the coefficients of a Fourier series satisfy the conditions

$$|a_n| \leq Kn^{-k}, \qquad |b_n| \leq Kn^{-k}; \qquad n = 1, 2, \ldots ,$$

where $k > 1$, show that the series must converge uniformly on the whole x-axis.

8.18. Show that if a Fourier series converges to a discontinuous function, the series $\sum_{n=1}^{\infty} (|a_n| + |b_n|)$ must be divergent.

8.19. Decide whether or not the series in Problem 8.10 converge uniformly for all x.

8.20. Prove the following differentiability result: Suppose f is continuous on $[-\pi,\pi]$, $f(-\pi) = f(\pi)$, and f' and f'' are sectionally continuous on $[-\pi, \pi]$. Then

$$f(x) = \frac{a_0}{2} + \sum_{n=1}^{\infty} (a_n \cos nx + b_n \sin nx), \qquad -\pi \leq x \leq \pi$$

and

$$\frac{1}{2} [f'(x + 0) + f'(x - 0)] =$$

$$= \sum_{n=1}^{\infty} (-na_n \sin nx + nb_n \cos nx), \qquad -\pi < x < \pi.$$

[*Hint:* Apply Theorem 8.1 to f'. If α_n, β_n are the Fourier coefficients of f' show that $\alpha_n = nb_n$ and $\beta_n = -na_n$.]

8.21. Apply the result of Problem 8.20 to (8.26) to obtain

$$\frac{2}{\pi} \sum_{n=1}^{\infty} \frac{1 - (-1)^n}{n} \sin nx = \begin{cases} -1 & \text{for} \quad -\pi < x < 0 \\ 0 & \text{for} \quad x = 0 \\ +1 & \text{for} \quad 0 < x < \pi. \end{cases}$$

Do the same for (8.32) appropriately extended to $[-\pi, \pi]$, to obtain

$$2 \sum_{n=1}^{\infty} \frac{1 + (-1)^n}{n} \sin nx = \begin{cases} -\pi - 2x & \text{for} \quad -\pi < x < 0 \\ 0 & \text{for} \quad x = 0 \\ \pi - 2x & \text{for} \quad 0 < x < \pi. \end{cases}$$

What are the sums of these series for $x = \pm\pi$?

8.22. Derive (8.41) and (8.42). What do you obtain from (8.42) if $a = L$?

8.23. Find: (a) the Fourier cosine series representation of the saw-tooth function in (8.41); (b) the Fourier sine series representation of the rectangular pulse function in (8.42).

9. Solution of the Dirichlet Problem Using Green's Functions

In this section we use the representation theorem derived in Section 5 to obtain an integral formula for the solution of the Dirichlet problem. This formula involves a function known as the Green's function.

We give the details of the derivation of the formula only for the case of a domain Ω in R^3. For the derivation, we must assume that the solution u of the Dirichlet problem is in $C^2(\bar{\Omega})$ even though the problem only asks for u to be in $C^0(\bar{\Omega}) \cap C^2(\Omega)$.

Let Ω be a bounded normal domain in R^3 and suppose that the function u is in $C^2(\bar{\Omega})$ and harmonic in Ω. If $\mathbf{r}$ is any fixed point in Ω, the representation theorem yields the formula

$$u(\mathbf{r}) = \frac{1}{4\pi} \int_{\partial\Omega} \left[\frac{1}{|\mathbf{r}' - \mathbf{r}|} \frac{\partial u(\mathbf{r}')}{\partial n} - u(\mathbf{r}') \frac{\partial}{\partial n} \frac{1}{|\mathbf{r}' - \mathbf{r}|} \right] d\sigma \tag{9.1}$$

where $\mathbf{r}'$ is the variable point of integration on $\partial\Omega$. Formula (9.1) gives the value of u at any point $\mathbf{r} \in \Omega$ in terms of the values of u and of $\partial u/\partial n$ on $\partial\Omega$. However the Dirichlet problem specifies only the values of u on $\partial\Omega$. In order to get around this difficulty we consider a function h which is harmonic in Ω and is in $C^2(\bar{\Omega})$. Then, according to the second Green's identity applied to the functions h and u, we have

$$0 = \int_{\partial\Omega} \left[h(\mathbf{r}') \frac{\partial u(\mathbf{r}')}{\partial n} - u(\mathbf{r}') \frac{\partial h(\mathbf{r}')}{\partial n} \right] d\sigma. \tag{9.2}$$

Adding (9.1) and (9.2) we obtain the formula

$$u(\mathbf{r}) = \int_{\partial\Omega} \left\{ \left[\frac{1}{4\pi} \frac{1}{|\mathbf{r}' - \mathbf{r}|} + h(\mathbf{r}') \right] \frac{\partial u(\mathbf{r}')}{\partial n} - u(\mathbf{r}') \frac{\partial}{\partial n} \left[\frac{1}{4\pi} \frac{1}{|\mathbf{r}' - \mathbf{r}|} + h(\mathbf{r}') \right] \right\} d\sigma. \tag{9.3}$$

Now, in (9.3) the term involving $\partial u/\partial n$ will disappear if we require in addition that $h(\mathbf{r}')$ be equal to $-(1/4\pi)\,(1/|\mathbf{r}' - \mathbf{r}|)$ for $\mathbf{r}' \in \partial\Omega$. Formula (9.3) then becomes

$$u(\mathbf{r}) = -\int_{\partial\Omega} u(\mathbf{r}') \frac{\partial}{\partial n} \left[\frac{1}{4\pi} \frac{1}{|\mathbf{r}' - \mathbf{r}|} + h(\mathbf{r}') \right] d\sigma. \tag{9.4}$$

Thus, formula (9.4) gives the values of the solution u of the Dirichlet problem at every point $\mathbf{r} \in \Omega$, provided that we can find a function $h \in C^2(\bar{\Omega})$ which satisfies the special Dirichlet problem,

$$\Delta' h(\mathbf{r}') = 0, \quad \text{for} \quad \mathbf{r}' \in \Omega, \tag{9.5}$$

$$h(\mathbf{r}') = -\frac{1}{4\pi} \frac{1}{|\mathbf{r}' - \mathbf{r}|}, \quad \text{for} \quad \mathbf{r}' \in \partial\Omega, \tag{9.6}$$

for each $\mathbf{r} \in \Omega$, where Δ' denotes the Laplacian operator with respect to $\mathbf{r}'$. Since the solution of (9.5), (9.6) depends also on $\mathbf{r}$ we denote it by $h(\mathbf{r}', \mathbf{r})$. The function in brackets appearing in the integrand of formula (9.4) is known as the Green's function.

Definition 9.1. Let Ω be a domain in R^3. The function

$$G(\mathbf{r}', \mathbf{r}) = \frac{1}{4\pi} \frac{1}{|\mathbf{r}' - \mathbf{r}|} + h(\mathbf{r}', \mathbf{r}); \quad \mathbf{r}', \mathbf{r} \in \bar{\Omega},\ \mathbf{r}' \neq \mathbf{r}, \tag{9.7}$$

where $h(\mathbf{r}', \mathbf{r})$ satisfies (9.5) and (9.6) for every $\mathbf{r} \in \Omega$, is called the *Green's function for the Dirichlet problem for* Ω.

In terms of the Green's function, formula (9.4) becomes

$$u(\mathbf{r}) = -\int_{\partial\Omega} u(\mathbf{r}') \frac{\partial}{\partial n} G(\mathbf{r}', \mathbf{r})\, d\sigma, \quad \mathbf{r} \in \Omega. \tag{9.8}$$

Thus, the solution to the Dirichlet problem

$$\Delta u = 0 \quad \text{in} \quad \Omega, \tag{9.9}$$

$$u = f \quad \text{on} \quad \partial\Omega, \tag{9.10}$$

is given by the formula

$$u(\mathbf{r}) = -\int_{\partial\Omega} f(\mathbf{r}') \frac{\partial}{\partial n} G(\mathbf{r}', \mathbf{r})\, d\sigma, \quad \mathbf{r} \in \Omega. \tag{9.11}$$

We have derived formula (9.11) for the solution of the Dirichlet problem (9.9), (9.10) under the assumption that the solution u of the problem exists and is in $C^2(\bar{\Omega})$ and under the assumption that the Green's function $G(\mathbf{r}', \mathbf{r})$

exists. It can be shown by methods which are beyond the scope of this book (see Kellogg)[2] that, under suitable conditions on the domain Ω, the Green's function exists, and, for every $\mathbf{r} \in \Omega$, the solution $u \in C^0(\bar{\Omega})$ of the Dirichlet problem (9.9), (9.10) with $f \in C^0(\partial\Omega)$ is given by formula (9.11).

The existence of the Green's function was first hypothesized by G. Green on the basis of physical evidence in electrostatics.

At first glance, the student may rightfully doubt the usefulness of formula (9.11). After all, in order to use this formula to compute the solution of the Dirichlet problem (9.9), (9.10), we must first find the Green's function $G(\mathbf{r}', \mathbf{r})$, and this involves solving the Dirichlet problem (9.5), (9.6) for every $\mathbf{r} \in \Omega$. However this last problem involves boundary data of a special type. Thus, formula (9.11) reduces the Dirichlet problem with arbitrary data f to the Dirichlet problem with special data $-(1/4\pi)(1/|\mathbf{r}' - \mathbf{r}|)$. This in turn enables us to reduce the Dirichlet problem to the problem of solving an integral equation and therefore allows us to use the methods of the theory of integral equations. Finally, the Green's function is a useful tool in the theoretical study of the Dirichlet problem, especially in determining the properties of the solution of this problem.

For very simple domains it is actually possible to construct the Green's function explicitly. In the following section we will do this for a ball in R^3.

Now, let us examine briefly the properties of the Green's function $G(\mathbf{r}', \mathbf{r})$ for the Dirichlet problem for a domain Ω in R^3. By definition, as a function of $\mathbf{r}'$, $G(\mathbf{r}', \mathbf{r})$ vanishes on $\partial\Omega$ and is harmonic in Ω except at $\mathbf{r}' = \mathbf{r}$, where it has a pole. As $\mathbf{r}' \to \mathbf{r}$, $G(\mathbf{r}', \mathbf{r}) \to +\infty$ like the inverse of the distance of $\mathbf{r}'$ from $\mathbf{r}$. An important property of $G(\mathbf{r}', \mathbf{r})$ is its symmetry with respect to $\mathbf{r}'$ and $\mathbf{r}$,

$$G(\mathbf{r}', \mathbf{r}) = G(\mathbf{r}, \mathbf{r}'); \quad \mathbf{r}', \mathbf{r} \in \Omega, \ \mathbf{r}' \neq \mathbf{r}. \tag{9.12}$$

The proof of (9.12) is left as an exercise (see Problem 9.1). It follows from (9.12) that, as a function of $\mathbf{r}$, $G(\mathbf{r}', \mathbf{r})$ vanishes on $\partial\Omega$ and is harmonic in Ω except at $\mathbf{r} = \mathbf{r}'$. Consequently, the function $u(\mathbf{r})$ defined by (9.11) is harmonic in Ω (see Problem 9.2).

The Green's function for a domain in R^2 is defined in a similar way.

Definition 9.2. Let Ω be a domain in R^2. The function

$$G(\mathbf{r}', \mathbf{r}) = \frac{1}{2\pi} \log \frac{1}{|\mathbf{r}' - \mathbf{r}|} + h(\mathbf{r}', \mathbf{r}); \quad \mathbf{r}', \mathbf{r} \in \bar{\Omega}, \ \mathbf{r}' \neq \mathbf{r}, \tag{9.13}$$

where $h(\mathbf{r}', \mathbf{r})$ satisfies

$$\Delta' h(\mathbf{r}', \mathbf{r}) = 0, \quad \text{for} \quad \mathbf{r}' \in \Omega, \tag{9.14}$$

$$h(\mathbf{r}', \mathbf{r}) = -\frac{1}{2\pi} \log \frac{1}{|\mathbf{r}' - \mathbf{r}|}, \quad \text{for} \quad \mathbf{r}' \in \partial\Omega, \tag{9.15}$$

for every $\mathbf{r} \in \Omega$, is called the *Green's function for the Dirichlet problem for* Ω.

Using the representation theorem for $n = 2$, it can be shown, in exactly the same way as above, that the solution to the Dirichlet problem (9.9), (9.10) is given by the formula (9.11). Moreover, the properties of the two-dimensional Green's function are analogous to those of the three-dimensional one, the only difference being that, as $\mathbf{r}' \to \mathbf{r}$, $G(\mathbf{r}', \mathbf{r}) \to +\infty$, like the logarithm of the inverse of the distance of $\mathbf{r}'$ from $\mathbf{r}$.

For domains in R^n with $n > 3$ the Green's function is again defined in a similar way. As $\mathbf{r}' \to \mathbf{r}$, $G(\mathbf{r}', \mathbf{r}) \to +\infty$, like $1/|\mathbf{r}' - \mathbf{r}|^{n-2}$.

Problems

9.1. Prove the symmetry property (9.12) of the Green's function $G(\mathbf{r}', \mathbf{r})$. [*Hint:* Let $\mathbf{r}_1$, $\mathbf{r}_2$ be two distinct points of Ω, and consider the domain Ω_ϵ obtained from Ω by deleting two small balls $\bar{B}(\mathbf{r}_1, \epsilon)$ and $\bar{B}(\mathbf{r}_2, \epsilon)$ from Ω. Apply the second Green's identity to the harmonic functions $u_1(\mathbf{r}) = G(\mathbf{r}, \mathbf{r}_1)$, $u_2(\mathbf{r}) = G(\mathbf{r}, \mathbf{r}_2)$ in Ω_ϵ, and let ϵ tend to zero to obtain $G(\mathbf{r}_1, \mathbf{r}_2) = G(\mathbf{r}_2, \mathbf{r}_1)$, which is the desired symmetry property.]

9.2 Show that the function $u(\mathbf{r})$ defined by (9.11) is harmonic in Ω.

9.3. (a) Show that $G(\mathbf{r}', \mathbf{r}) \geqq 0$ for $\mathbf{r}' \in \Omega$, $\mathbf{r}' \neq \mathbf{r}$. [*Hint:* Apply the minimum principle to the harmonic function $G(\mathbf{r}', \mathbf{r})$ for $\mathbf{r}'$ in $\Omega_\epsilon = \Omega - B(\mathbf{r}, \epsilon)$ with ϵ arbitrarily small.]

(b) Show that $\left.\dfrac{\partial G}{\partial n}\right|_{\partial\Omega} \leqq 0$ for $\mathbf{r}' \in \partial\Omega$ where $\mathbf{n}$ is the exterior unit normal to $\partial\Omega$.

9.4. Consider the Dirichlet problem for the *Poisson equation* in a bounded normal domain Ω of R^3,

$$\nabla^2 u(\mathbf{r}) = -q(\mathbf{r}), \qquad \mathbf{r} \in \Omega \tag{9.16}$$

$$u(\mathbf{r}) = f(\mathbf{r}), \qquad \mathbf{r} \in \partial\Omega. \tag{9.17}$$

Equation (9.16) governs, for example, the steady state temperature distribution in Ω when heat sources, described by the function $q(\mathbf{r})$, are present in Ω. Assuming that $u \in C^2(\bar{\Omega})$, use the representation theorem of Section 5 to derive the following formula for the solution of (9.16), (9.17):

$$u(\mathbf{r}) = \int_\Omega q(\mathbf{r}')G(\mathbf{r}', \mathbf{r})dv - \int_{\partial\Omega} f(\mathbf{r}')\frac{\partial}{\partial n}G(\mathbf{r}', \mathbf{r})d\sigma \tag{9.18}$$

where $G(\mathbf{r}', \mathbf{r})$ is the Green's function for the Dirichlet problem for Ω, given by (9.7).

9.5. Show that the solution of the Dirichlet problem for the Poisson equation in a bounded normal domain Ω of R^2 is also given by formula (9.18).

10. The Green's Function and the Solution to the Dirichlet Problem for a Ball in R^3

Let Ω be the ball $B(0, a)$ in R^3 and let $\mathbf{r}$ be a fixed point in $B(0, a)$. In order to find the Green's function for the Dirichlet problem for $B(0, a)$ we must construct a function $h(\mathbf{r}', \mathbf{r})$ which, as a function of $\mathbf{r}'$, is harmonic in $B(0, a)$ and is equal to $-(1/4\pi)(1/|\mathbf{r}' - \mathbf{r}|)$ for $\mathbf{r}'$ on the boundary $S(0, a)$.

The function

$$\frac{1}{|\mathbf{r}' - \mathbf{r}|} \tag{10.1}$$

is harmonic in $B(0, a)$ with pole at $\mathbf{r}' = \mathbf{r}$. In Example 3.4, we saw that by inversion with respect to the sphere $S(0, a)$, the function (10.1) yields the function

$$\frac{1}{\left|\frac{a}{r'}\mathbf{r}' - \frac{r'}{a}\mathbf{r}\right|} \tag{10.2}$$

which is harmonic in $B(0, a)$ and is equal to $1/|\mathbf{r}' - \mathbf{r}|$ for $\mathbf{r}' \in S(0, a)$. Hence the desired function $h(\mathbf{r}', \mathbf{r})$ is

$$h(\mathbf{r}', \mathbf{r}) = -\frac{1}{4\pi}\frac{1}{\left|\frac{a}{r'}\mathbf{r}' - \frac{r'}{a}\mathbf{r}\right|} \tag{10.3}$$

and the Green's function for $B(0, a)$ is

$$G(\mathbf{r}', \mathbf{r}) = \frac{1}{4\pi}\left[\frac{1}{|\mathbf{r}' - \mathbf{r}|} - \frac{1}{\left|\frac{a}{r'}\mathbf{r}' - \frac{r'}{a}\mathbf{r}\right|}\right] \tag{10.4}$$

If Θ, $0 \leqq \Theta \leqq \pi$, denotes the angle between the vectors $\mathbf{r}'$ and $\mathbf{r}$, then using the cosine law we see that

$$G(\mathbf{r}', \mathbf{r}) = \frac{1}{4\pi}\left(\frac{1}{R} - \frac{1}{R'}\right) \tag{10.5}$$

where

$$R = |\mathbf{r}' - \mathbf{r}| = (r'^2 + r^2 - 2r'r\cos\Theta)^{1/2} \tag{10.6}$$

$$R' = \left|\frac{a}{r'}\mathbf{r}' - \frac{r'}{a}\mathbf{r}\right| = \left(a^2 + \frac{r'^2}{a^2}r^2 - 2r'r\cos\Theta\right)^{1/2} \tag{10.7}$$

Equations (10.5), (10.6), (10.7) clearly display the symmetry property of the Green's function, $G(\mathbf{r}', \mathbf{r}) = G(\mathbf{r}, \mathbf{r}')$.

Now, in order to use formula (9.11) for the solution of the Dirichlet problem for $B(0, a)$, we must compute the exterior normal derivative $\frac{\partial}{\partial n}G(\mathbf{r}', \mathbf{r})$ on the boundary $S(0, a)$. Since the exterior normal on $S(0, a)$ is in the radial direction, we have

$$\left.\frac{\partial G(\mathbf{r}', \mathbf{r})}{\partial n}\right|_{r' \in S(0,a)} = \left.\frac{\partial G(\mathbf{r}', \mathbf{r})}{\partial r'}\right|_{r'=a} \tag{10.8}$$

and a simple computation (see Problem 10.1) yields

$$\left.\frac{\partial G(\mathbf{r}', \mathbf{r})}{\partial r'}\right|_{r'=a} = -\frac{1}{4\pi a}\frac{a^2 - r^2}{(a^2 + r^2 - 2ar\cos\Theta)^{3/2}}. \tag{10.9}$$

Let us consider now the Dirichlet problem for $B(0, a)$ which asks for the function $u \in C^2(\Omega) \cap C^0(\bar{\Omega})$ satisfying

$$\nabla^2 u(\mathbf{r}) = 0 \quad \text{for} \quad \mathbf{r} \in B(0, a), \tag{10.10}$$

$$u(\mathbf{r}) = f(\mathbf{r}) \quad \text{for} \quad \mathbf{r} \in S(0, a), \tag{10.11}$$

where f is a given function in $C^0(S(0, a))$. According to formula (9.11), u is given by

$$u(\mathbf{r}) = \frac{1}{4\pi a} \int_{S(0,a)} \frac{(a^2 - r^2)f(\mathbf{r}')}{(a^2 + r^2 - 2ar \cos \Theta)^{3/2}}\, d\sigma, \quad \text{for} \quad \mathbf{r} \in B(0, a). \tag{10.12}$$

The integral in formula (10.12) is known as *Poisson's integral,* and the function

$$P(\mathbf{r}', \mathbf{r}) = \frac{1}{4\pi a} \frac{a^2 - r^2}{(a^2 + r^2 - 2\, ar \cos \Theta)^{3/2}} \tag{10.13}$$

is known as the *Poisson kernel.* In terms of the Poisson kernel, formula (10.12) becomes

$$u(\mathbf{r}) = \int_{S(0,a)} P(\mathbf{r}', \mathbf{r}) f(\mathbf{r}') d\sigma \quad \text{for} \quad \mathbf{r} \in B(0, a). \tag{10.14}$$

It is sometimes useful to write formula (10.12) in terms of spherical coordinates. If (r, θ, ϕ) are the spherical coordinates of $\mathbf{r} \in B(0, a)$ and if (a, θ', ϕ') are the spherical coordinates of the variable point of integration $\mathbf{r}'$ on $S(0, a)$, formula (10.12) becomes

$$u(r, \theta, \phi) = \frac{a}{4\pi} \int_0^{2\pi} \int_0^{\pi} \frac{(a^2 - r^2) f(\theta', \phi')}{(a^2 + r^2 - 2ar \cos \Theta)^{3/2}} \sin \phi' d\phi' d\theta', \quad \text{for} \quad r < a, \tag{10.15}$$

where (see Problem 10.2)

$$\cos \Theta = \cos \phi \cos \phi' + \sin \phi \sin \phi' \cos (\theta - \theta'). \tag{10.16}$$

In the previous section, we derived formula (9.11) under the assumption that the solution to the Dirichlet problem is in $C^2(\bar{\Omega})$. We then stated that, under suitable assumptions on the domain Ω, formula (9.11) gives the solution to the Dirichlet problem for $\mathbf{r} \in \Omega$ even when the solution is required to be only in $C^2(\Omega) \cap C^0(\bar{\Omega})$. We will now actually prove this last statement for the Dirichlet problem for $B(0, a)$.

Theorem 10.1. Let $u \in C^2(B(0, a)) \cap C^0(\bar{B}(0, a))$ be the solution to the Dirichlet problem (10.10), (10.11) with $f \in C^0(S(0, a))$. Then u is given by

$$u(\mathbf{r}) = \begin{cases} \displaystyle\int_{S(0,a)} P(\mathbf{r}', \mathbf{r}) f(\mathbf{r}') d\sigma, & \text{for} \quad r < a \\ f(\mathbf{r}), & \text{for} \quad r = a, \end{cases} \tag{10.17}$$

where $P(\mathbf{r}', \mathbf{r})$ is the Poisson kernel (10.13).

Before proving Theorem 10.1 we need to establish the following properties of the Poisson kernel:

(i) $P(\mathbf{r}', \mathbf{r}) \geqq 0$ for $\mathbf{r} \in B(0, a)$, $\mathbf{r}' \in S(0, a)$.

(ii) $P(\mathbf{r}', \mathbf{r})$ is harmonic as a function of $\mathbf{r} \in B(0, a)$ for $\mathbf{r}' \in S(0, a)$.

(iii) $\int_{S(0,a)} P(\mathbf{r}', \mathbf{r}) d\sigma = 1$ for every $\mathbf{r} \in \bar{B}(0, a)$.

Property (i) follows immediately from the definition (10.13) of $P(\mathbf{r}', \mathbf{r})$. Property (ii) follows from the fact that the Green's function $G(\mathbf{r}', \mathbf{r})$ is harmonic as a function of $\mathbf{r} \in B(0, a)$ for every $\mathbf{r}' \in S(0, a)$, and from the fact that $P(\mathbf{r}', \mathbf{r})$ is obtained from $G(\mathbf{r}', \mathbf{r})$ by differentiation involving the "parameter" $\mathbf{r}'$ (see Section 9 of Chapter V). Finally, property (iii) follows from the fact that the solution of the Dirichlet problem

$$\nabla^2 u(\mathbf{r}) = 0 \quad \text{for} \quad \mathbf{r} \in B(0, a), \qquad u(\mathbf{r}) = 1 \quad \text{for} \quad \mathbf{r} \in S(0, a)$$

is the function $u(\mathbf{r}) = 1$, $\mathbf{r} \in \bar{B}(0, a)$. Since this solution is in $C^2(\bar{B}(0, a))$ we know that formula (19.11) and hence formula (10.14) is valid. Formula (10.14) with $f = u = 1$ is precisely property (iii).

Proof of Theorem 10.1. Since the solution of the Dirichlet problem is unique, it is enough to show that the function $u(\mathbf{r})$, defined by (10.17), is harmonic in $B(0, a)$ and continuous in $\bar{B}(0, a)$. That $u(\mathbf{r})$ is harmonic in $B(0, a)$ follows immediately from the fact that in $B(0, a)$, $u(\mathbf{r})$ is a superposition of the harmonic functions $P(\mathbf{r}', \mathbf{r})$. It remains therefore to show that u is continuous in $\bar{B}(0, a)$ and since we already know that $u \in C^2(B(0, a))$ it is enough to show that for any point $\mathbf{r}_0 \in S(0, a)$,

$$\lim_{\substack{\mathbf{r} \to \mathbf{r}_0 \\ \mathbf{r} \in B(0,a)}} u(\mathbf{r}) = f(\mathbf{r}_0). \tag{10.18}$$

More explicitly we must show that

$$\lim_{\substack{\mathbf{r} \to \mathbf{r}_0 \\ \mathbf{r} \in B(0,a)}} \int_{S(0,a)} P(\mathbf{r}', \mathbf{r}) f(\mathbf{r}') d\sigma = f(\mathbf{r}_0). \tag{10.19}$$

Now, from property (iii) of $P(\mathbf{r}', \mathbf{r})$ we have

$$\int_{S(0,a)} P(\mathbf{r}', \mathbf{r}) f(\mathbf{r}_0) d\sigma = f(\mathbf{r}_0), \text{ for every } \mathbf{r} \in B(0, a),$$

and (10.19) is equivalent to

$$\lim_{\substack{\mathbf{r} \to \mathbf{r}_0 \\ \mathbf{r} \in B(0,a)}} \int_{S(0,a)} P(\mathbf{r}', \mathbf{r})[f(\mathbf{r}') - f(\mathbf{r}_0)] d\sigma = 0. \tag{10.20}$$

In order to prove (10.20) we note first that since f is continuous on $S(0, a)$, it is bounded on $S(0, a)$, i.e., there is a number $M > 0$ such that

$$|f(\mathbf{r}')| \leqq M \quad \text{for every} \quad \mathbf{r}' \in S(0, a). \tag{10.21}$$

Moreover, since f is continuous at $\mathbf{r}_0$, given any $\epsilon > 0$, there is a $\delta > 0$ such that

$$(10.22) \qquad |f(\mathbf{r}') - f(\mathbf{r}_0)| < \epsilon, \quad \text{for} \quad \mathbf{r}' \in S(0, a), |\mathbf{r}' - \mathbf{r}_0| < \delta.$$

Now let $C(\mathbf{r}_0, \delta)$ denote the part of the sphere $S(0, a)$ contained in the ball $B(\mathbf{r}_0, \delta)$,

$$C(\mathbf{r}_0, \delta) = S(0, a) \cap B(\mathbf{r}_0, \delta).$$

Then (10.22) can be written as

$$(10.23) \qquad |f(\mathbf{r}') - f(\mathbf{r}_0)| < \epsilon \quad \text{for} \quad \mathbf{r}' \in C(\mathbf{r}_0, \delta).$$

Now, we split the integral in (10.20) into two parts,

$$(10.24) \qquad \begin{aligned} &\int_{S(0,a)} P(\mathbf{r}', \mathbf{r})[f(\mathbf{r}') - f(\mathbf{r}_0)]d\sigma \\ &= \int_{C(\mathbf{r}_0,\delta)} P(\mathbf{r}', \mathbf{r})[f(\mathbf{r}') - f(\mathbf{r}_0)]d\sigma \\ &\qquad + \int_{S(0,a)-C(\mathbf{r}_0,\delta)} P(\mathbf{r}', \mathbf{r})[f(\mathbf{r}') - f(\mathbf{r}_0)]d\sigma. \end{aligned}$$

Using (10.23) and properties (i) and (iii) of $P(\mathbf{r}', \mathbf{r})$ we find that

$$(10.25) \qquad \begin{aligned} &\left| \int_{C(\mathbf{r}_0,\delta)} P(\mathbf{r}', \mathbf{r})[f(\mathbf{r}') - f(\mathbf{r}_0)]d\sigma \right| \\ &\leqq \epsilon \int_{C(\mathbf{r}_0,\delta)} P(\mathbf{r}', \mathbf{r})d\sigma \\ &\leqq \epsilon \int_{S(0,a)} P(\mathbf{r}', \mathbf{r})d\sigma = \epsilon, \quad \text{for every} \quad \mathbf{r} \in B(0, a). \end{aligned}$$

Since ϵ is an arbitrary positive number, (10.20) will follow if we can prove that

$$(10.26) \qquad \lim_{\substack{\mathbf{r}\to\mathbf{r}_0 \\ \mathbf{r}\in B(0,a)}} \int_{S(0,a)-C(\mathbf{r}_0,\delta)} P(\mathbf{r}', \mathbf{r})[f(\mathbf{r}') - f(\mathbf{r}_0)]d\sigma = 0.$$

From (10.21) we have

$$\left| \int_{S(0,a)-C(\mathbf{r}_0,\delta)} P(\mathbf{r}', \mathbf{r})[f(\mathbf{r}') - f(\mathbf{r}_0)]d\sigma \right| \leqq 2M \int_{S(0,a)-C(\mathbf{r}_0,\delta)} P(\mathbf{r}', \mathbf{r})d\sigma$$

and hence it is enough to show that

$$(10.27) \qquad \lim_{\substack{\mathbf{r}\to\mathbf{r}_0 \\ \mathbf{r}\in B(0,a)}} \int_{S(0,a)-C(\mathbf{r}_0,\delta)} P(\mathbf{r}', \mathbf{r})d\sigma = 0.$$

Since we are only interested in what happens when $\mathbf{r}$ is near $\mathbf{r}_0$, let us only consider $\mathbf{r} \in B(0, a) \cap B(\mathbf{r}_0, \delta/2)$. Then for $\mathbf{r}' \in [S(0, a) - C(\mathbf{r}_0, \delta)]$

$$|\mathbf{r}' - \mathbf{r}| \geqq |\mathbf{r}' - \mathbf{r}_0| - |\mathbf{r} - \mathbf{r}_0| \geqq \frac{\delta}{2}$$

and

$$P(\mathbf{r}', \mathbf{r}) = \frac{1}{4\pi a}\frac{a^2 - r^2}{|\mathbf{r}' - \mathbf{r}|^3} \leqq \left(\frac{2}{\delta}\right)^3 \frac{1}{4\pi a}(a^2 - r^2).$$

Consequently,

$$\int_{S(0,a)-C(\mathbf{r}_0,\delta)} P(\mathbf{r}', \mathbf{r})d\sigma \leqq \left(\frac{2}{\delta}\right)^3 \frac{1}{4\pi a}(a^2 - r^2)\int_{S(0,a)} 1\, d\sigma = \left(\frac{2}{\delta}\right)^3 a(a^2 - r^2)$$

for $\mathbf{r} \in B(0, a) \cap B(\mathbf{r}_0, \delta/2)$, and because of the factor $(a^2 - r^2)$, (10.27) follows. The proof of the theorem is now complete.

A careful review of the proof of Theorem 10.1 shows that we did not fully use the assumption that f is continuous everywhere on $S(0, a)$. We only used the fact that f is bounded on $S(0, a)$ and that f is continuous at the point $\mathbf{r}_0 \in S(0, a)$. It follows that under the assumption that f is piecewise continuous (continuous everywhere except along a finite number of curves on $S(0, a)$ where f may have finite jumps), the Poisson integral (10.14) defines a function $u(\mathbf{r})$ which is harmonic for $\mathbf{r} \in B(0, a)$ and which is such that at every point $\mathbf{r}_0 \in S(0, a)$ where f is continuous,

$$\lim_{\substack{\mathbf{r}\to\mathbf{r}_0 \\ \mathbf{r}\in B(0,a)}} u(\mathbf{r}) = f(\mathbf{r}_0).$$

This observation enables us to solve the Dirichlet problem with piecewise continuous boundary data using the Poisson integral (10.14), with the understanding that the desired solution $u(\mathbf{r})$ is not in $C^0(\bar{B}(0, a))$, but is harmonic in $B(0, a)$ and approaches the value $f(\mathbf{r}_0)$ as $\mathbf{r} \to \mathbf{r}_0 \in S(0, a)$, at every point $\mathbf{r}_0 \in S(0, a)$ where f is continuous. A simple example is the problem where f is equal to 1 on the upper hemisphere of $S(0, a)$ and 0 on the lower hemisphere.

The Poisson integral solution of the Dirichlet problem for the ball $B(0, a)$ in R^2 was obtained in Section 7 using separation of variables, Fourier series and summing the resulting series solution. This solution can be obtained also by determining the Green's function for $B(0, a)$ in R^2 and using formula (9.11) (see Problem 10.4). The same can be done for a ball in R^n with $n > 3$.

Problems

10.1. Derive formula (10.9).

10.2. Derive formula (10.16). [*Hint:* $\cos \Theta = \mathbf{r}' \cdot \mathbf{r}/(r'r)$.]

10.3. Use Theorem 10.1 to show that the value at the origin of the solution of the Dirichlet problem for $B(0, a)$ in R^3 is equal to the average of its values on the boundary $S(0, a)$.

10.4. Use Example 3.3 to derive the Green's function for the Dirichlet problem for $B(0, a)$ in R^2,

$$G(\mathbf{r}', \mathbf{r}) = \frac{1}{2\pi}\left[\log\frac{1}{|\mathbf{r}' - \mathbf{r}|} - \log\frac{1}{\left|\dfrac{a}{r'}\mathbf{r}' - \dfrac{r'}{a}\mathbf{r}\right|}\right].$$

Then use formula (9.11) to derive the Poisson integral solution (7.21).

11. Further Properties of Harmonic Functions

The Poisson integral solution of the Dirichlet problem for a ball in R^n is not very useful for the actual computation of the values of the solution in the ball, since the integrations involved are rather complicated. However, as we will see in this section, the Poisson integral is very useful in deriving important properties of harmonic functions.

Let u be harmonic in a domain Ω of R^n and suppose that B is any ball whose closure $\bar{B}$ is contained in Ω. Then the value of u at any point in B is given by the Poisson integral involving the values of u on the boundary S of B. Since the harmonicity of a function is invariant under translation of coordinates, we can always take the center of B to be the origin. Suppose then that the ball $B(0, a)$ is such that its closure $\bar{B}(0, a)$ is contained in the domain Ω where u is harmonic. Then, for every $\mathbf{r} \in B(0, a)$,

(11.1) $$u(\mathbf{r}) = \int_{S(0,a)} P(\mathbf{r}', \mathbf{r})u(\mathbf{r}')d\sigma$$

where $\mathbf{r}'$ is the variable point of integration on $S(0, a)$ and $P(\mathbf{r}', \mathbf{r})$ is the Poisson kernel. In previous sections we derived the expressions for $P(\mathbf{r}', \mathbf{r})$ for $n = 2$ and $n = 3$. For $n = 2$

(11.2) $$P(\mathbf{r}', \mathbf{r}) = \frac{1}{2\pi a}\frac{a^2 - r^2}{a^2 + r^2 - 2ar\cos(\theta - \phi)}, \qquad (n = 2),$$

where (r, θ) and (a, ϕ) are the polar coordinates of $\mathbf{r}$ and $\mathbf{r}'$, respectively. For $n = 3$

(11.3) $$P(\mathbf{r}', \mathbf{r}) = \frac{1}{4\pi a}\frac{a^2 - r^2}{(a^2 + r^2 - 2ar\cos\Theta)^{3/2}}, \quad (n = 3),$$

where $r = |\mathbf{r}|$, $a = |\mathbf{r}'|$ and Θ is the angle between $\mathbf{r}$ and $\mathbf{r}'$. For $n > 3$ the expressions for $P(\mathbf{r}', \mathbf{r})$ are similar.

The two important theorems that we present in this section are valid for all n. Their proofs, which are based on the Poisson integral (11.1), vary only in minor details for different values of n. We will only give the proofs either for $n = 2$ or for $n = 3$.

Theorem 11.1. (*Liouville's theorem.*) A function $u(\mathbf{r})$ which is harmonic for every $\mathbf{r} \in R^n$ cannot have an upper bound or a lower bound unless $u(\mathbf{r})$ is constant.

Proof. We give the proof for $n = 2$. Let us assume that $u(\mathbf{r})$ is harmonic in R^2 and has a lower bound, i.e., there is a number M such that $u(\mathbf{r}) \geqq M$ for every $\mathbf{r} \in R^2$. We must show that $u(\mathbf{r}) =$ const. in R^2. The function $u'(\mathbf{r}) = u(\mathbf{r}) - M$ is also harmonic in R^2 and is such that $u'(\mathbf{r}) \geqq 0$ for every $\mathbf{r} \in R^2$. If we show that $u'(\mathbf{r}) = \text{const.}$ in R^2, it would follow that $u(\mathbf{r}) = \text{const.}$ in R^2. Let us then drop the prime and assume that $u(\mathbf{r})$ is harmonic and $u(\mathbf{r}) \geqq 0$ for every $\mathbf{r} \in R^2$. We will show that for any point $\mathbf{r} \in R^2$, $u(\mathbf{r}) = u(0)$ and hence $u(\mathbf{r})$ is constant.

Let $\mathbf{r}$ be a fixed point in R^2 and let a be a number greater than $r = |\mathbf{r}|$ so that the circle $B(0, a)$ contains $\mathbf{r}$. Then, by (11.1) and (11.2) we have

$$(11.4) \qquad u(\mathbf{r}) = \frac{1}{2\pi}\int_0^{2\pi} \frac{a^2 - r^2}{a^2 + r^2 - 2ar\cos(\theta - \phi)}\, u(a, \phi)d\phi.$$

Since $-1 \leqq \cos(\theta - \phi) \leqq 1$,

$$\frac{a - r}{a + r} \leqq \frac{a^2 - r^2}{a^2 + r^2 - 2ar\cos(\theta - \phi)} \leqq \frac{a + r}{a - r},$$

and since $u(a, \phi) \geqq 0$,

$$\frac{a - r}{a + r}\, u(a, \phi) \leqq \frac{a^2 - r^2}{a^2 + r^2 - 2ar\cos(\theta - \phi)}\, u(a, \phi) \leqq \frac{a + r}{a - r}\, u(a, \phi).$$

Integrating with respect to ϕ, in view of (11.4), we obtain

$$\frac{a - r}{a + r}\frac{1}{2\pi}\int_0^{2\pi} u(a, \phi)d\phi \leqq u(\mathbf{r}) \leqq \frac{a + r}{a - r}\frac{1}{2\pi}\int_0^{2\pi} u(a, \phi)d\phi,$$

and using the mean value theorem for harmonic functions,

$$\frac{a - r}{a + r}\, u(0) \leqq u(\mathbf{r}) \leqq \frac{a + r}{a - r}\, u(0).$$

Now letting $a \to +\infty$ we obtain

$$u(0) \leqq u(\mathbf{r}) \leqq u(0)$$

and hence $u(\mathbf{r}) = u(0)$.

We have shown that if $u(\mathbf{r})$ has a lower bound then it is a constant. If instead, $u(\mathbf{r})$ had an upper bound, then $-u(\mathbf{r})$ would have a lower bound; consequently $-u(\mathbf{r})$, and hence $u(\mathbf{r})$, would be a constant. The proof of the theorem is complete.

An immediate corollary of Liouville's theorem is the following.

Corollary 11.1. The only functions which are bounded and harmonic in all of R^n are the constant functions.

Another important result which follows from the Poisson integral solution of the Dirichlet problem is the fact that a harmonic function is necessarily analytic in its domain of definition. Recall that a function u is analytic in a domain Ω of R^n, if, at every point of Ω, u has a Taylor series which converges to u in a neighborhood of that point.

Theorem 11.2. A function u which is harmonic in a domain Ω of R^n is analytic in Ω.

Proof. We give the proof for $n = 3$. Let Q be any point of Ω. We must show that u is analytic at Q. Since harmonicity and analyticity are invariant under translation of coordinates, we may assume that the point Q is the origin. Hence, we must show that u has a Taylor series expansion,

$$u(x, y, z) = \sum_{i,j,k=0}^{\infty} a_{ijk} x^i y^j z^k \tag{11.5}$$

which is valid for (x, y, z) in some neighborhood of the origin. To do this we use the Poisson integral representation of u near the origin. Let $a > 0$ be sufficiently small so that $\bar{B}(0, a) \subset \Omega$. Then, for every $\mathbf{r} \in B(0, a)$, $u(\mathbf{r})$ is given by (11.1), where $P(\mathbf{r}', \mathbf{r})$ is given by (11.3). Let (x, y, z) and (ξ, η, ζ) be the rectangular coordinates of $\mathbf{r}$ and $\mathbf{r}'$, respectively. Suppose that we can show that the Poisson kernel $P(\mathbf{r}', \mathbf{r})$ has a Taylor series expansion

$$P(\mathbf{r}', \mathbf{r}) = \sum_{i,j,k=0}^{\infty} b_{ijk}(\xi, \eta, \zeta) x^i y^j z^k \tag{11.6}$$

which converges uniformly for $(x, y, z) \in B(0, \delta)$, for some $\delta > 0$, and $(\xi, \eta, \zeta) \in S(0, a)$. Then substitution of (11.6) into (11.1) and interchange of the order of integration and summation (which is allowed by the uniformity of the convergence of (11.6)) would yield (11.5) valid for $(x, y, z) \in B(0, \delta)$. Now since

$$P(\mathbf{r}', \mathbf{r}) = \frac{a^2 - r^2}{4\pi a^4} \left[1 - \frac{2ar\cos\Theta - r^2}{a^2}\right]^{-3/2}$$

and since $a^2 - r^2 = a^2 - (x^2 + y^2 + z^2)$ is a polynomial, it is enough to show that

$$\left[1 - \frac{2ar\cos\Theta - r^2}{a^2}\right]^{-3/2} = \sum_{i,j,k=0}^{\infty} c_{ijk}(\xi, \eta, \zeta) x^i y^j z^k \tag{11.7}$$

with the convergence being uniform for $(x, y, z) \in B(0, \delta)$ and $(\xi, \eta, \zeta) \in S(0, a)$. In order to show this we use the binomial expansion

$$(1 - w)^{-3/2} = \sum_{m=0}^{\infty} \alpha_m w^m \tag{11.8}$$

where $\alpha_0 = 1$ and $\alpha_m > 0$ for all m. The series in (11.8) converges absolutely and uniformly for $|w| \leqq \rho$ where ρ is any number <1. If we set

$$w = \frac{2ar\cos\Theta - r^2}{a^2} \tag{11.9}$$

and if $\mathbf{r} \in B(0, a)$, we have

$$|w| = \frac{|2ar\cos\Theta - r^2|}{a^2} \leqq \frac{2ar + r^2}{a^2} = \frac{2a + r}{a^2} r \leqq \frac{3}{a} r.$$

If we now restrict $\mathbf{r}$ to be in $B(0, a/4)$, we have

$$|w| \leqq \frac{3}{4} \quad \text{for} \quad (x, y, z) \in B(0, a/4) \quad \text{and} \quad (\xi, \eta, \zeta) \in S(0, a).$$

Hence

(11.10) $$\left[1 - \frac{2ar \cos \Theta - r^2}{a^2}\right]^{-3/2} = \sum_{m=0}^{\infty} \alpha_m w^m$$

with the series convergent uniformly for $(x, y, z) \in B(0, a/4)$ and $(\xi, \eta, \zeta) \in S(0, a)$. From (11.9),

$$w = \frac{1}{a^2}[2(\xi x + \eta y + \zeta z) - (x^2 + y^2 + z^2)],$$

so that the series in (11.10) is a series of polynomials with positive coefficients. We now use a result from the theory of analytic functions of a complex variable, according to which the series in (11.10) can be rearranged by expanding $[2(\xi x + \eta y + \zeta z) - (x^2 + y^2 + z^2)]^m$ and grouping together terms of the form $x^i y^j z^k$. This rearrangement would yield (11.7) with the convergence being uniform for $(x, y, z) \in B(0, a/4)$ and $(\xi, \eta, \zeta) \in S(0, a)$. The proof of the theorem is now complete.

Problems

11.1. The functions

$$1, r \cos \theta, r \sin \theta, r^2 \cos 2\theta, r^2 \sin 2\theta, \ldots$$

are harmonic in all of R^2 (see Section 2). Show directly that, except for the function 1, all the others are not bounded either from above or from below.

11.2. Prove Liouville's theorem for $n = 3$.

11.3. Prove Theorem 11.2 for $n = 2$.

12. The Dirichlet Problem in Unbounded Domains

Up to this point of the chapter we have studied the Dirichlet problem under the assumption that Ω, the domain in which Laplace's equation is to hold, is a bounded domain. In this section we turn our attention to unbounded domains. We will see that, in general, it is possible to use inversion with respect to a sphere in order to transform a Dirichlet problem for an unbounded domain into a Dirichlet problem for a bounded domain, provided that the solution in the unbounded domain satisfies a certain condition at infinity.

We begin with a simple example which shows that uniqueness for the Dirichlet problem in an unbounded domain may fail to hold if the solution is not required to satisfy any additional condition. Let Ω be the complement of the closed unit ball in R^n, $\Omega = R^n - \bar{B}(0, 1)$, and consider the problem of finding a function u in $C^2(\Omega) \cap C^0(\bar{\Omega})$ such that

(12.1) $$\nabla^2 u(\mathbf{r}) = 0, \qquad \mathbf{r} \in \Omega,$$

(12.2) $$u(\mathbf{r}) = 1, \qquad \mathbf{r} \in \partial\Omega = S(0, 1).$$

It is easy to verify that the functions

$$u_1(\mathbf{r}) = 1, \qquad u_2(\mathbf{r}) = 1 + \log r, \quad \text{if} \quad n = 2,$$

and

$$u_1(\mathbf{r}) = 1, \qquad u_2(\mathbf{r}) = r^{2-n}, \quad \text{if} \quad n \geqq 3,$$

are solutions of the problem. In fact superposition of these two solutions yields infinitely many solutions of the problem. We are faced then with the following question: What additional condition must be imposed on the solution u in order to guarantee uniqueness for the Dirichlet problem in unbounded domains? As we will see, this additional condition arises naturally when the method of inversion with respect to a sphere is used to transform the Dirichlet problem for an unbounded domain to a problem for a bounded domain. The condition limits the behavior of $u(\mathbf{r})$ for large values of r and thus it may be viewed as a (boundary) condition at infinity.

The domain of the above example is known as an exterior domain. A domain Ω in R^n is called an *exterior domain* if it is the complement of the closure of a bounded domain. More specifically, if G is a bounded domain such that

$$\Omega = R^n - \bar{G}$$

then Ω is called the *domain exterior to G*. Note that $\partial\Omega = \partial G$ and $\Omega \cup G \cup \partial G = R^n$. For simplicity, our discussion of the method of inversion with respect to a sphere will be limited to exterior domains.

Let $\Omega \subset R^n$ be a domain exterior to a bounded domain G, let f be a given function in $C^0(\partial\Omega)$ and suppose that $u(\mathbf{r})$ is a function in $C^2(\Omega) \cap C^0(\bar{\Omega})$ satisfying

(12.3) $$\nabla^2 u(\mathbf{r}) = 0, \qquad \mathbf{r} \in \Omega,$$

(12.4) $$u(\mathbf{r}) = f(\mathbf{r}), \quad \mathbf{r} \in \partial\Omega.$$

Since translations and similarity transformations preserve harmonicity, we may assume without loss of generality that the closed unit ball $\bar{B}(0, 1)$ is contained in $\bar{G}$. Let Ω^* be the domain consisting of the origin together with the set of points obtained by inversion of Ω with respect to $S(0, 1)$ (see Fig. 12.1). Ω^* is a bounded domain contained in $\bar{B}(0, 1)$. Each point $\mathbf{r}^* \in \Omega^*$, other than the origin, corresponds to a unique point $\mathbf{r} \in \Omega$ according to the inversion formulas

(12.5) $$r\mathbf{r}^* = r^*\mathbf{r}, \qquad rr^* = 1.$$

The origin, $\mathbf{r}^* = 0$, corresponds to infinity. At this point we restrict the discussion to the case $n = 3$ and let

(12.6) $$u^*(\mathbf{r}^*) = \frac{1}{r^*}\, u(\mathbf{r}), \qquad \mathbf{r}^* \in \Omega^*, \qquad \mathbf{r}^* \neq 0,$$

and

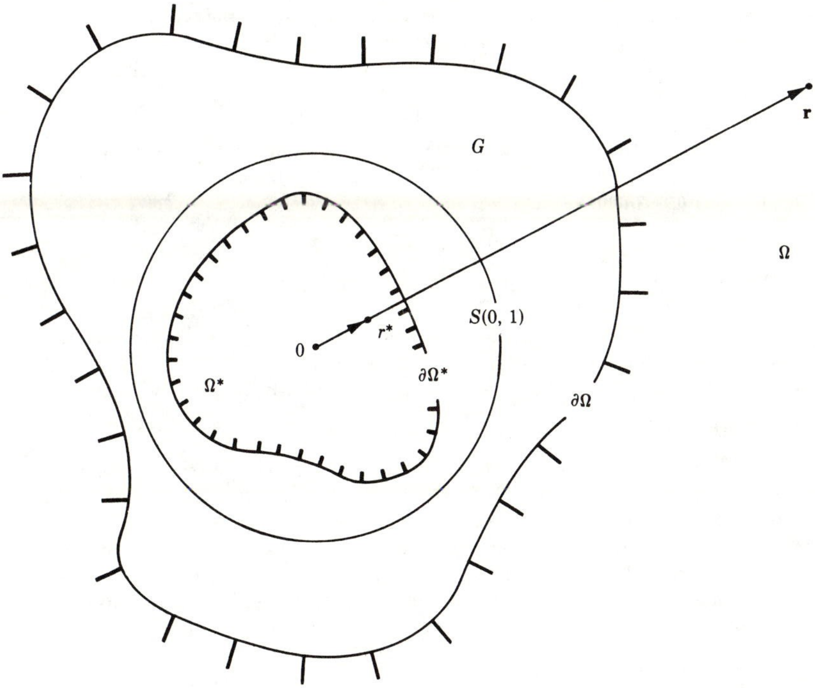

Fig. 12.1

$$f^*(\mathbf{r}^*) = \frac{1}{r^*} f(\mathbf{r}) \qquad \mathbf{r}^* \in \partial\Omega^*. \tag{12.7}$$

The function u^* is defined and harmonic in Ω^* excepting the origin, f^* is defined and continuous on $\partial\Omega^*$, and u^* assumes the values of f^* on $\partial\Omega^*$. If u^* were defined and harmonic in all of Ω^* including the origin, then u^* could be regarded as the solution of a Dirichlet problem in Ω^*. The following theorem will enable us to define u^* at the origin in such a way that the resulting function is harmonic in the whole of Ω^*, provided that u satisfies a certain condition at infinity.

Theorem 12.1. Let $v(\mathbf{r})$ be defined and harmonic in a neighborhood Ω_0 of the origin of R^n except at the origin itself. Then $v(\mathbf{r})$ can be defined at the origin in such a way that $v(\mathbf{r})$ will be harmonic throughout Ω_0 (including the origin) provided that the following condition is satisfied:

$$\begin{cases} v(\mathbf{r}) \quad \text{is bounded for} \quad \mathbf{r} \in \Omega_0 - \{0\}, \quad \text{if} \quad n = 2, \\ r^{n-2}\,|\,v(\mathbf{r})\,| \leqq \mu(\mathbf{r}) \quad \text{for} \quad \vec{\mathbf{r}} \in \Omega_0 - \{0\}, \quad \text{if} \quad n \geqq 3, \end{cases} \tag{12.8}$$

where $\mu(\mathbf{r}) \to 0$ as $\vec{\mathbf{r}} \to 0$.

Theorem 12.1 is known as a theorem on removable singularities of harmonic functions; a proof may be found in the book of Petrovskii.[3]

For the case $n = 3$ under discussion, it follows from (12.6) that

$$r^* \mid u^*(\mathbf{r}^*) \mid = \mid u(\mathbf{r}) \mid.$$

Since $r \to \infty$ as $r^* \to 0$, Theorem 12.1 asserts that $u^*(\mathbf{r}^*)$ can be defined at $\mathbf{r}^* = 0$ in such a way that u^* will be harmonic throughout Ω^*, provided that $u(\mathbf{r})$ approaches 0 as $\mathbf{r} \to \infty$,

$$u(\mathbf{r}) \to 0 \quad \text{as} \quad \mathbf{r} \to \infty. \tag{12.9}$$

(Note that (12.9) means: given $\epsilon > 0$, there is $R > 0$ such that $|u(\mathbf{r})| < \epsilon$ for all $\mathbf{r} \in \Omega$ such that $|\mathbf{r}| > R$.)

In exactly the same way, it is easy to see that for $n > 3$ the required condition at ∞ is again condition (12.9), while for $n = 2$ the condition at ∞ is that u remains bounded; i.e., there is a constant $M > 0$ such that

$$\mid u(\mathbf{r}) \mid \leqq M \quad \text{for all} \quad \mathbf{r} \in \Omega. \tag{12.10}$$

We have shown that by inversion with respect to a sphere the exterior Dirichlet problem (12.3), (12.4), subject to the condition at infinity (12.9) when $n \geqq 3$ and (12.10) when $n = 2$, can be transformed to a Dirichlet problem for a bounded domain. It follows that all the results that we already know for bounded domains, such as theorems on uniqueness, existence and continuous dependence on data, lead to corresponding results for exterior domains. For example, we have the following uniqueness theorem.

Theorem 12.2. Let Ω be an exterior domain in R^n and let f be a given function in $C^0(\partial\Omega)$. There is at most one function u in $C^2(\Omega) \cap C^0(\bar{\Omega})$ which satisfies (12.3), (12.4) and the condition at infinity (12.9) if $n \geqq 3$ or (12.10) if $n = 2$.

Example 12.1. If $n = 2$, the only solution of (12.1), (12.2) which remains bounded is the function $u(\mathbf{r}) = 1$. If $n = 3$, the only solution of (12.1), (12.2) which approaches zero as $\mathbf{r} \to \infty$ is the function $u(\mathbf{r}) = 1/r$.

Example 12.2. Let Ω be the domain exterior to the unit disc $B(0, 1)$ in R^2 and consider the exterior Dirichlet problem

$$\nabla^2 u(r, \theta) = 0, \qquad r > 1, \qquad -\pi \leqq \theta \leqq \pi, \tag{12.11}$$

$$u(1, \theta) = f(\theta), \qquad -\pi \leqq \theta \leqq \pi, \tag{12.12}$$

$$u(r, \theta) \quad \text{is bounded in} \quad \Omega, \tag{12.13}$$

where f is in $C^0(\partial\Omega)$, and f' is sectionally continuous. By inversion with respect to $S(0, 1)$, $u^*(r^*, \theta)$ satisfies the interior Dirichlet problem for the unit disc which was solved in Section 7. $u^*(r^*, \theta)$ is given by either the series (7.9) or the Poisson integral (7.18) with r replaced by r^*. Since $r^* = 1/r$ and $u^*(r^*, \theta) = u(r, \theta)$, the solution of the exterior problem (12.11), (12.12), (12.13) is given by the series

$$(12.14)\quad u(r,\theta) = \frac{a_0}{2} + \sum_{n=1}^{\infty} r^{-n}(a_n \cos n\theta + b_n \sin n\theta), \qquad r \geq 1,$$

where a_n, b_n; $n = 0, 1, \ldots$, are the Fourier coefficients of f, or by the (Poisson) integral,

$$(12.15)\quad u(r,\theta) = -\frac{1}{2\pi}\int_0^{2\pi} \frac{(1-r^2)f(\phi)d\phi}{1+r^2-2r\cos(\theta-\phi)}, \qquad r \geq 1.$$

The method of inversion with respect to a sphere can be applied to unbounded domains other than exterior domains, provided that the complement of the considered domain contains a ball. Since the boundaries of such domains may extend to infinity, the behavior of the boundaries and of the Dirichlet data as infinity is approached must be carefully studied and taken into account in formulating well-posed problems.

Problems

12.1. Derive condition (12.10) for the case $n = 2$.

12.2. For $n \geqq 3$, use the maximum principle for bounded domains to prove uniqueness and continuous dependence on data for the exterior Dirichlet problem (12.3), (12.4), (12.9). [*Hint:* Consider the domain $\Omega_R = \Omega \cap B(0,R)$.]

12.3. Find the steady state temperature distribution in an infinite uniform atmosphere exterior to a spherical solid of radius R, if the surface of the solid is kept at the constant temperature T_0, assuming that the temperature approaches 0 at infinity.

12.4. Find the steady state temperature distribution in an infinite uniform atmosphere exterior to an infinite circular cylinder of radius R, if the surface of the cylinder is kept at constant temperature T_0, assuming that the temperature remains bounded.

12.5. Find the series solution of the exterior Dirichlet problem of Example 12.2 if $f(\theta)$ is given as in Problem 7.7.

12.6. Use inversion with respect to $S(0, 1)$ to obtain the solution $u(r,\theta,\phi)$ of the exterior Dirichlet problem,

$$\begin{aligned} &\nabla^2 u(r,\theta,\phi) = 0, \quad r > 1, && 0 \leqq \theta \leqq 2\pi, && 0 \leqq \phi \leqq \pi \\ &u(1,\theta,\phi) = \sin\phi\cos\theta, && 0 \leqq \theta \leqq 2\pi, && 0 \leqq \phi \leqq \pi \\ &u(\mathbf{r}) \to 0 \quad \text{as} \quad \mathbf{r} \to \infty \end{aligned}$$

where (r,θ,ϕ) are the spherical coordinates of $\mathbf{r} \in R^3$.

12.7. Show that without the condition that u remains bounded, the Dirichlet problem for the upper half-plane $y > 0$,

$$\begin{aligned} \nabla^2 u(x,y) &= 0, \quad -\infty < x < \infty, \quad y > 0, \\ u(x,0) &= 1, \quad -\infty < x < \infty, \end{aligned}$$

has infinitely many solutions. What is the unique bounded solution of this problem?

12.8. (a) Show that inversion with respect to the circle $S(0, 1)$ maps the half plane $\Omega = \{(x,y) \in R^2, y > 1\}$ onto the disc with center $(0, 1/2)$ and radius $1/2$. Polar coordinates are useful here.

(b) Show that inversion with respect to $S(0, 1)$ maps the quarter plane $\Omega = \{(x,y) \in R^2, x > 1, y > 1\}$ onto the lens-shaped domain bounded by the two circles

$$\left(x - \frac{1}{2}\right)^2 + y^2 = \frac{1}{4}; \quad x^2 + \left(y - \frac{1}{2}\right)^2 = \frac{1}{4}.$$

12.9. Consider the lens-shaped domain of problem 12.8(b). Obtain the solution

$$u(x, y) = \frac{2}{\pi} \arctan \frac{x^2 + y^2 - y}{x^2 + y^2 - x}$$

of the Dirichlet problem which requires $u = 1$ on the top boundary of the lens, and $u = 0$ on the bottom boundary of the lens. [*Hint:* The solution of the corresponding Dirichlet problem in the quarter space $x > 1, y > 1$ is easily found to be

$$u(x, y) = \frac{2}{\pi} \phi = \frac{2}{\pi} \arctan \frac{y - 1}{x - 1},$$

where ϕ is the polar angle with $(1, 1)$ as the center of the polar coordinate system. See Section 2, in particular, the discussion following (2.20).]

13. Determination of the Green's Function by the Method of Electrostatic Images

Let us recall the definition of the Green's function for the Dirichlet problem for a domain Ω of R^3,

$$G(\mathbf{r}', \mathbf{r}) = \frac{1}{4\pi} \frac{1}{|\mathbf{r}' - \mathbf{r}|} + h(\mathbf{r}', \mathbf{r}); \quad \mathbf{r}', \mathbf{r} \in \bar{\Omega}, \mathbf{r}' \neq \mathbf{r}. \tag{13.1}$$

For each fixed $\mathbf{r} \in \Omega$, the function $h(\mathbf{r}', \mathbf{r})$ is harmonic in Ω as a function of $\mathbf{r}'$ and satisfies the boundary condition

$$h(\mathbf{r}', \mathbf{r}) = -\frac{1}{4\pi} \frac{1}{|\mathbf{r}' - \mathbf{r}|}, \qquad \mathbf{r}' \in \partial\Omega. \tag{13.2}$$

Thus, as a function of $\mathbf{r}'$, $G(\mathbf{r}', \mathbf{r})$ is harmonic in Ω excepting the point $\mathbf{r}' = \mathbf{r}$, and vanishes on $\partial\Omega$. Moreover as $\mathbf{r}' \to \mathbf{r}$, $G(\mathbf{r}', \mathbf{r}) \to \infty$ like the inverse of the distance between $\mathbf{r}'$ and $\mathbf{r}$. We have seen in Section 9 that the solution of the Dirichlet problem

$$\nabla^2 u = 0 \quad \text{in} \quad \Omega, \qquad u = f \quad \text{on} \quad \partial\Omega \tag{13.3}$$

can be represented in terms of $G(\mathbf{r}', \mathbf{r})$ by the formula

$$u(\mathbf{r}) = -\int_{\partial\Omega} f(\mathbf{r}') \frac{\partial}{\partial n} G(\mathbf{r}', \mathbf{r}) d\sigma, \qquad \mathbf{r} \in \Omega, \tag{13.4}$$

where $\partial/\partial n$ represents differentiation in the direction of the exterior normal to $\partial\Omega$. The problem of solving (13.3) is thus reduced to the problem of determining the Green's function $G(\mathbf{r}', \mathbf{r})$. It can be shown that, under appropriate conditions, formula (13.4) gives the solution of the Dirichlet problem even when Ω is an unbounded domain.

A useful method for the determination of the Green's function $G(\mathbf{r}', \mathbf{r})$ is based on interpreting $G(\mathbf{r}', \mathbf{r})$ as the electrostatic potential due to a unit charge located at the point $\mathbf{r}$ of the region Ω if the boundary $\partial\Omega$ of the region is a grounded conducting surface. (The potential is zero on a grounded surface). The first term in the formula (13.1) for $G(\mathbf{r}', \mathbf{r})$ represents the potential due to a unit charge at the point $\mathbf{r}$. This unit charge induces a distribution of charges on the conducting surface $\partial\Omega$, and the second term $h(\mathbf{r}', \mathbf{r})$ in (13.1) represents the potential due to this induced charge distribution on $\partial\Omega$. The determination of $h(\mathbf{r}', \mathbf{r})$ thus depends on first finding the induced charge distribution on $\partial\Omega$, which in itself is a difficult problem. The *method of electrostatic images* enables us to circumvent this problem. Instead of viewing $h(\mathbf{r}', \mathbf{r})$ as the potential due to the induced charge distribution on $\partial\Omega$, we consider $h(\mathbf{r}', \mathbf{r})$ as being the potential due to imaginary charges located in the complement of Ω. These charges, which are called the *electrostatic images* of the unit charge at the point $\mathbf{r}$ of Ω, must be introduced in the complement of Ω in such a manner that the potential $h(\mathbf{r}', \mathbf{r})$ due to these charges satisfies condition (13.2). In other words, at each point of the boundary $\partial\Omega$ of Ω, the potential due to the electrostatic images must be equal to the negative of the potential due to the unit charge at $\mathbf{r}$. The total potential $G(\mathbf{r}', \mathbf{r})$ would then be equal to zero on $\partial\Omega$, and $\partial\Omega$ could be regarded as a grounded conducting surface. In many cases, the geometry of $\partial\Omega$ is simple enough that the choice of electrostatic images is obvious.

Example 13.1. Let Ω be the upper-half space in R^3,

$$\Omega = \{(x,y,z)\colon z > 0\}$$

and consider a unit charge at the point $\mathbf{r} = (x, y, z)$ of Ω as shown in Figure 13.1. If we introduce a negative unit charge at the point $\mathbf{r}^* = (x, y, -z)$, the resultant potential due to the two charges will be zero on the boundary $z = 0$ of Ω. Thus, the necessary electrostatic image of the unit charge at $\mathbf{r}$ is a negative unit charge at the point $\mathbf{r}^*$ which is the mirror image of $\mathbf{r}$ with respect to the boundary of Ω. The resulting Green's function is then

$$G(\mathbf{r}', \mathbf{r}) = \frac{1}{4\pi}\frac{1}{|\mathbf{r}' - \mathbf{r}|} - \frac{1}{4\pi}\frac{1}{|\mathbf{r}' - \mathbf{r}^*|}. \tag{13.5}$$

Indeed, when $\mathbf{r}'$ is on $\partial\Omega$, $|\mathbf{r}' - \mathbf{r}| = |\mathbf{r}' - \mathbf{r}^*|$ and $G(\mathbf{r}', \mathbf{r}) = 0$. In terms of coordinates

$$G(\mathbf{r}', \mathbf{r}) = \frac{1}{4\pi}\frac{1}{[(x' - x)^2 + (y' - y)^2 + (z' - z)^2]^{1/2}} - \frac{1}{4\pi}\frac{1}{[(x' - x)^2 + (y' - y)^2 + (z' + z)^2]^{1/2}}.$$

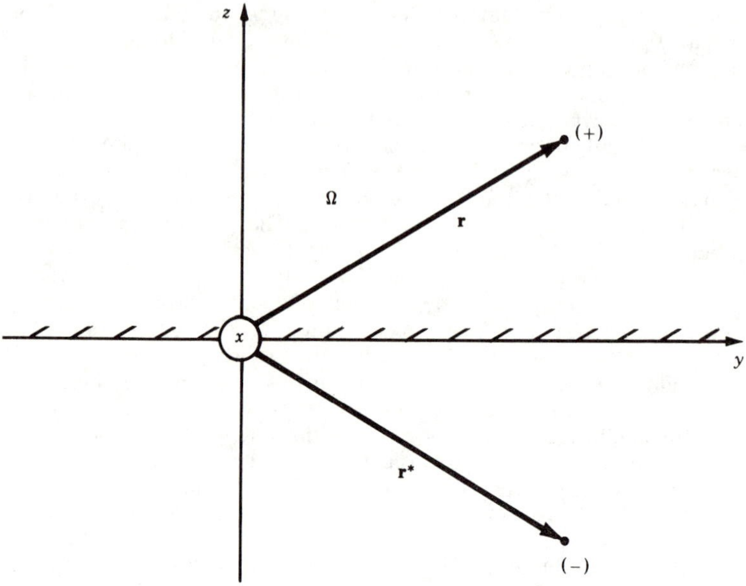

Fig. 13.1

It is left as an exercise to show that formula (13.4) for the solution of the Dirichlet problem (13.3) becomes in this case

$$u(x, y, z) = \frac{z}{2\pi}\int_{-\infty}^{\infty}\int_{-\infty}^{\infty} \frac{f(x', y')dx'dy'}{[(x' - x)^2 + (y' - y)^2 + z^2]^{3/2}} \tag{13.6}$$

Example 13.2. Let Ω be the quarter space in R^3,

$$\Omega = \{(x, y, z): y > 0, z > 0\},$$

and consider a unit charge at the point $\mathbf{r} = (x, y, z)$ of Ω as shown in Figure 13.2. The necessary electrostatic images in this case are: a negative unit charge at $\mathbf{r}_1^* = (x, -y, z)$, a positive unit charge at $\mathbf{r}_2^* = (x, -y, -z)$, and a negative unit charge at $\mathbf{r}_3^* = (x, y, -z)$. It is easy to see from the geometry of Figure 13.2 that the resulting potential due to all four charges vanishes on the boundary of Ω and the desired Green's function for the Dirichlet problem for Ω is

$$G(\mathbf{r}', \mathbf{r}) = \frac{1}{4\pi}\left[\frac{1}{|\mathbf{r}' - \mathbf{r}|} - \frac{1}{|\mathbf{r}' - \mathbf{r}_1^*|} + \frac{1}{|\mathbf{r}' - \mathbf{r}_2^*|} - \frac{1}{|\mathbf{r}' - \mathbf{r}_3^*|}\right]. \tag{13.7}$$

Example 13.3. Let Ω be the ball $B(0, a)$ and consider a unit charge at the point $\mathbf{r}$ of Ω as shown in Figure 13.3. Using some geometrical arguments it can be seen that the necessary electrostatic image must be located at the

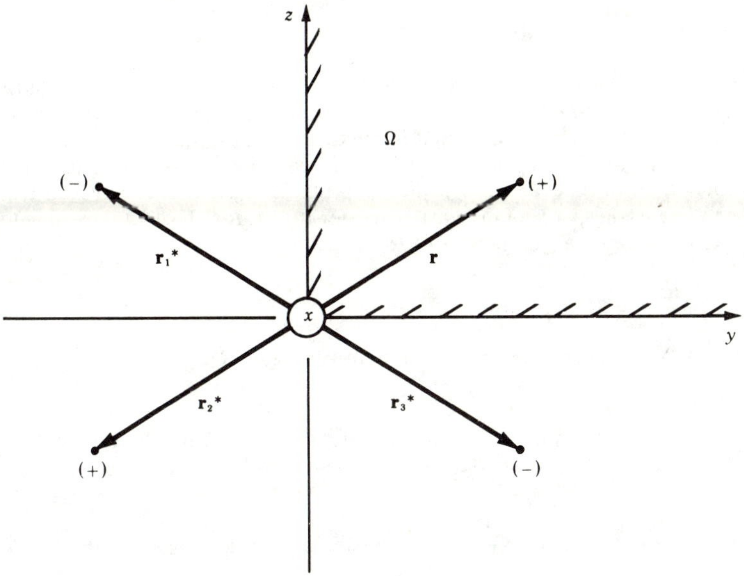

Fig. 13.2

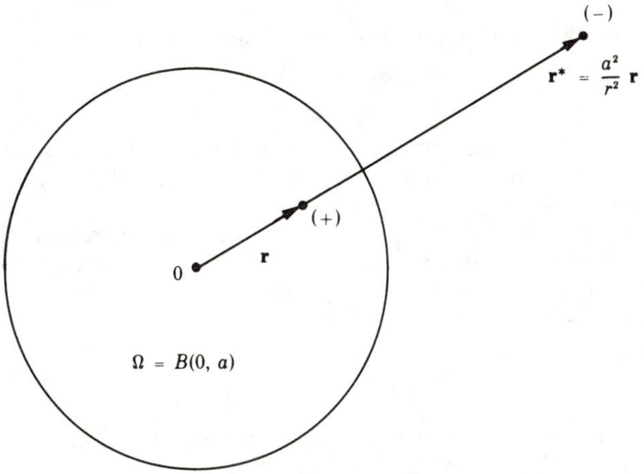

Fig. 13.3

point $\mathbf{r}^* = \dfrac{a^2}{r^2}\,\mathbf{r}$ and must be a negative charge of magnitude a/r. (The point $\mathbf{r}^*$ is inverse to $\mathbf{r}$ with respect to the sphere $S(0, a)$.) The resulting Green's function is then

(13.8) $$G(\mathbf{r}', \mathbf{r}) = \frac{1}{4\pi}\frac{1}{|\mathbf{r}' - \mathbf{r}|} - \frac{1}{4\pi}\frac{a/r}{|\mathbf{r}' - \mathbf{r}^*|}.$$

It is left as an exercise to show that when $\mathbf{r}'$ is on the boundary $S(0, a)$ of Ω, $G(\mathbf{r}', \mathbf{r}) = 0$. After some manipulation, formula (13.8) can be written in the form (10.4) which was obtained in Section 10.

The method of images can be used with domains in spaces of dimension other than three even though the concepts of electrostatics have no real significance in spaces of dimension greater than three. The potential due to a unit "charge" at the point $\mathbf{r}$ of R^n is

$$\frac{1}{2\pi}\log\frac{1}{|\mathbf{r}' - \mathbf{r}|} \quad \text{for} \quad n = 2 \quad \text{and} \quad \frac{1}{S_n}\frac{1}{|\mathbf{r}' - \mathbf{r}|^{n-2}} \text{ for } \quad n \geqq 3,$$

where S_n is the surface area of the unit sphere in R^n.

Problems

13.1. Derive formula (13.6) for the solution of the Dirichlet problem

$$\nabla^2 u = 0; \qquad -\infty < x, y < \infty, \qquad 0 < z < \infty,$$
$$u(x, y, 0) = f(x, y), \qquad -\infty < x, y < \infty.$$

Note that if $f(x, y)$ vanishes outside a bounded region of the (x, y)-plane, then $u(\mathbf{r}) \to 0$ as $\mathbf{r} \to \infty$.

13.2. Prove that the Green's function (13.8) vanishes when $\mathbf{r}'$ is on the sphere $S(0, a)$. You will need to make some careful geometric arguments.

13.3. Find the Green's function for a domain Ω between two parallel planes

$$\Omega = \{(x, y, z): 0 < z < 1\}.$$

You will need an infinite sequence of electrostatic images. [*Hint:* introduce charges so that each plane $z = k, k = 0, \pm 1, \pm 2, \ldots$ is at zero potential.]

13.4. Find the Green's function for the Dirichlet problem for a "wedge" domain Ω bounded by two planes intersecting at an angle $\pi/4$. Use cylindrical coordinates with the z-axis coinciding with the line of intersection of the planes so that

$$\Omega = \{(\rho, \theta, z): \rho > 0, 0 < \theta < \frac{\pi}{4}, -\infty < z < \infty\}.$$

Will the method of images work for any wedge of angle π/k, $k = 1, 2, \ldots$?

13.5. Find the Green's function for the upper hemisphere

$$\Omega = \{(x, y, z): (x, y, z) \in B(0, a), z > 0\}.$$

13.6. Use the method of images to find the Green's function for the upper-half plane in R^2. Then use (13.4) to derive the formula

$$u(x, y) = \frac{y}{\pi}\int_{-\infty}^{\infty}\frac{f(x')}{(x' - x)^2 + y^2}\,dx',$$

for the solution of the Dirichlet problem

$$\nabla^2 u = 0; \qquad -\infty < x < \infty, \qquad 0 < y < \infty,$$
$$u(x, 0) = f(x), \qquad -\infty < x < \infty.$$

What does the formula give when $f(x) \equiv 1$?

13.7. Find the Green's function for the square in R^2

$$\Omega = \{(x, y)\colon 0 < x < 1, 0 < y < 1\}.$$

14. Analytic Functions of a Complex Variable and Laplace's Equation in Two Dimensions

There are important relations between analytic functions of a complex variable and harmonic functions of two real variables which can be successfully exploited in solving boundary value problems for Laplace's equation in two dimensions. Some knowledge of the subject of complex analysis is necessary for the understanding of this section. We will only sketch the basic results and give a simple illustration of the method of conformal mappings. For a more detailed study of this topic the student should refer to any applications-oriented book on complex analysis such as Churchill[7] (see in particular Chapters 4 and 8–11).

Let $z = x + iy$ and suppose that the function

$$f(z) = u(x, y) + iv(x, y) \tag{14.1}$$

is an analytic function of z in some domain Ω of R^2. The functions

$$u(x, y) = \operatorname{Re} f(z), \qquad v(x, y) = \operatorname{Im} f(z) \tag{14.2}$$

are real valued analytic functions of the two real variables x and y; they are known, respectively, as the real and imaginary parts of $f(z)$. A necessary condition for analyticity of $f(z)$ is that u and v satisfy the *Cauchy-Riemann equations*

$$u_x = v_y, \qquad u_y = -v_x, \qquad (x, y) \in \Omega. \tag{14.3}$$

An immediate consequence of these equations is that each of the functions u and v satisfies Laplace's equation in Ω,

$$u_{xx} + u_{yy} = 0, \qquad v_{xx} + v_{yy} = 0; \qquad (x, y) \in \Omega. \tag{14.4}$$

Thus, the real and imaginary parts of an analytic function are harmonic functions. For example, the function

$$f(z) = e^z = e^{x+iy} = e^x \cos y + ie^x \sin y,$$

is an analytic function of z in the whole z-plane, and the real and imaginary parts

$$u(x, y) = e^x \cos y, \qquad v(x, y) = e^x \sin y$$

are harmonic functions of x and y in R^2.

Two harmonic functions $u(x, y)$ and $v(x, y)$, which are the real and imaginary parts of an analytic function of a complex variable, are said to be *conjugate harmonic*. It can be shown that given a harmonic function $u(x, y)$

in a simply connected (no holes) domain Ω, its conjugate harmonic in Ω can be determined up to a constant of integration. (See Problem 14.2.)

Besides being rich sources of harmonic functions, analytic functions of a complex variable play an important role in the study of the two-dimensional Laplace's equation because of the following important property: Harmonic functions remain harmonic under changes of variables defined by analytic functions of a complex variable. Indeed, let $w = f(z)$ and $z = F(w)$ be analytic functions which are inverses of each other. Then the derivatives $dw/dz = f'(z)$ and $dz/dw = F'(w)$ are non-zero and $dw/dz = 1/(dz/dw)$. Moreover, if $z = x + iy$, $w = u + iv$, and if the functions $u(x, y)$, $v(x, y)$ and $x(u, v)$, $y(u, v)$ are the real and imaginary parts of $f(z)$ and $F(z)$, respectively, then the relations

$$u = u(x, y), \qquad v = v(x, y)$$
$$x = x(u, v), \qquad y = y(u, v) \tag{14.5}$$

define a (nonsingular) transformation of coordinates (see Problems 14.4 and 14.5). Suppose now that $U(x, y)$ is a C^2 function of the variables x and y. Then, using the chain rule and the Cauchy-Riemann equations, it can be shown that

$$U_{xx} + U_{yy} = (U_{uu} + U_{vv})|f'(z)|^2. \tag{14.6}$$

Thus, if $U(x, y)$ is harmonic as a function of x and y, the transformed function $U(u, v) = U(x(u, v), y(u, v))$ will also be harmonic as a function of u and v.

In practice, the above transformation property of harmonic functions can be very useful. Consider for example the Dirichlet problem for a domain Ω of the (x, y)-plane

$$\nabla^2 U = 0 \quad \text{in} \quad \Omega; \qquad U = \phi \quad \text{on} \quad \partial\Omega.$$

Suppose that the function $w = f(z)$ is defined and analytic in the domain Ω

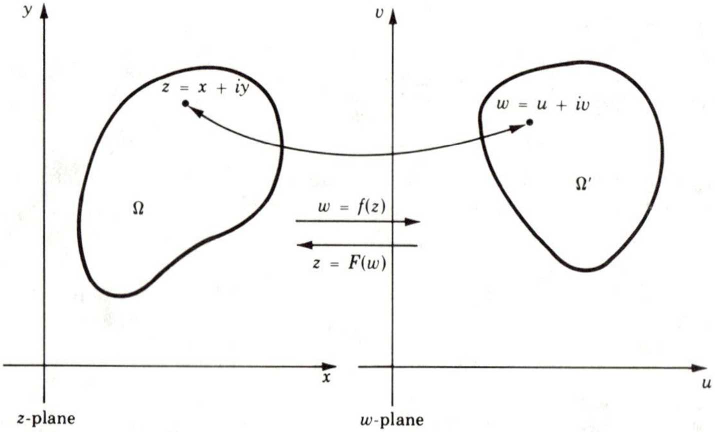

Fig. 14.1

of the z-plane, maps Ω to the domain Ω' of the w-plane and has an inverse $z = F(w)$ which is defined and analytic in Ω' and maps Ω' back to Ω. A mapping of Ω to Ω' defined by such a function $w = f(z)$ is said to be *conformal* because it preserves the angle between any pair of intersecting curves in Ω; i.e., the image curves in Ω' intersect at the same angle. Under such a conformal mapping, the Dirichlet problem for $U(x, y)$ in Ω is transformed to a Dirichlet problem for the transformed function $U(u, v)$ in Ω'. If this latter problem can be solved, then the solution of the original Dirichlet problem can be obtained by returning to the z-plane. We illustrate this method with a very simple example.

Example 14.1. Find the solution of the Dirichlet problem for the unit disc

$$\nabla^2 U(r, \theta) = 0 \qquad 0 \leqq r < 1, \qquad -\pi \leqq \theta \leqq \pi$$

$$U(1, \theta) = \begin{cases} 1 & 0 < \theta < \pi \\ -1 & -\pi < \theta < 0. \end{cases}$$

The problem is indicated in the z-plane of Figure 14.2. The function

$$w = \frac{1 + z}{1 - z} \tag{14.7}$$

is analytic in the circular domain $\Omega = \{z \;:\; |z| < 1\}$ and maps Ω to the half-plane $\Omega' = \{w : u = \text{Re}\, w > 0\}$ of the w-plane, with the boundary of Ω being mapped to the boundary of Ω' as indicated in the figure. The transformed problem in the w-plane is

$$\nabla^2 U(u, v) = 0 \qquad 0 < u < \infty, \qquad -\infty < v < \infty$$

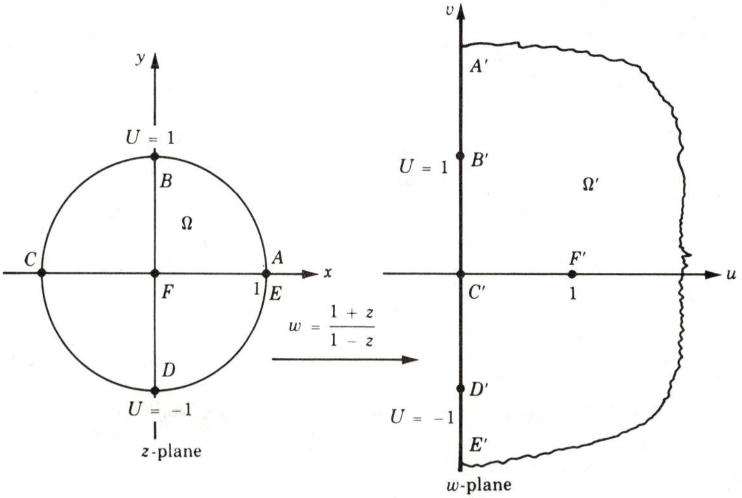

Fig. 14.2

$$U(0, v) = \begin{cases} 1 & \text{for} \quad 0 < v < \infty \\ -1 & \text{for} \quad -\infty < v < 0. \end{cases}$$

The bounded solution of this problem is (see 2.20)

$$U(u, v) = \frac{2}{\pi} \text{arc tan} \left(\frac{v}{u}\right). \tag{14.8}$$

To find the solution of the original problem we must return to the z-plane i.e., we must change variables from u, v to x, y. From (14.7) we easily find that

$$u = \frac{1 - x^2 - y^2}{(1 - x)^2 + y^2}, \qquad v = \frac{2y}{(1 - x)^2 + y^2}.$$

Therefore, the solution of the original problem is

$$U(x, y) = \frac{2}{\pi} \text{arc tan} \left(\frac{2y}{1 - x^2 - y^2}\right). \tag{14.9}$$

The extent of the practical applicability of this method depends of course on the information that we have at our disposal concerning the transformation of domains by conformal mappings. Such information can be found for example in Appendix 2 of Churchill,[7] or in the book of Kober,[8] which is an extensive catalogue of conformal transformations.

In theory, at least, the method of conformal mappings can be used to solve the Dirichlet problem for any simply connected two-dimensional domain Ω with piecewise smooth boundary ($\Omega \neq R^2$). The celebrated *Riemann mapping theorem* (see Ahlfors,[9] Section 4.2) asserts that such a domain can be mapped conformally onto the open unit disc with the boundary of the domain being mapped to the circumference of the disc. The Dirichlet problem for Ω can thus be transformed to the Dirichlet problem for the unit disc. As we know, the solution of this last problem is given by Poisson's integral.

Problems

14.1. Derive (14.4) from (14.3).

14.2. Let $u(x, y)$ be harmonic in a simply connected domain Ω. Use the Cauchy-Riemann equations to obtain the formula for the conjugate harmonic

$$v(x, y) = \int_{(x_0, y_0)}^{(x, y)} (u_x dy - u_y dx)$$

where (x_0, y_0) is any fixed point of Ω and the integration is along any path in Ω joining (x_0, y_0) and (x, y). Note that if Ω is not simply connected (i.e., Ω has a hole) then $v(x, y)$ may be multiple-valued.

14.3. Find a conjugate harmonic of $u(x, y) = x^2 - y^2$ in R^2.

14.4. Show that if $u(x, y)$ and $v(x, y)$ are conjugate harmonic in Ω, then grad $u(x, y)$ and grad $v(x, y)$ are orthogonal at every point of Ω. In particular, if one of these gradients is $\neq 0$ in Ω, then the other gradient is also $\neq 0$ in Ω and the level curves

$$u(x, y) = c_1, \qquad v(x, y) = c_2$$

are orthogonal.

14.5. Show that under the assumptions stated in the text, the Jacobian of the transformation (14.5) is nonvanishing.

14.6. Derive the relation (14.6).

14.7. Show that as a consequence of the Cauchy-Riemann equations (14.3), the directional derivative of u in any given direction ($\cos\alpha$, $\sin\alpha$) is equal to the directional derivative of v in the direction ($\cos(\alpha + \pi/2)$, $\sin(\alpha + \pi/2)$) which is obtained by rotating the given direction 90° counterclockwise.

15. The Method of Finite Differences

The method of finite differences is a widely used numerical method for finding approximate values of solutions of problems involving partial differential equations. The basic idea of the method consists of approximating the partial derivatives of a function by finite difference quotients. Suppose for example that the function $u(x, y)$ is of class C^2 in a domain Ω of R^2. Then, from Taylor's formula (see for example Taylor,[6] pp. 227–228), we have

$$u(x + h, y) = u(x, y) + hu_x(x, y) + \frac{h^2}{2} u_{xx}(\bar{x}, y),$$

provided that the line segment joining the points (x, y) and $(x + h, y)$ lies in Ω. The point $(\bar{x}, y)$ is some point on this segment. It follows that

$$\left| u_x(x, y) - \frac{1}{h}[u(x + h, y) - u(x, y)] \right| = \frac{h}{2} \left| u_{xx}(\bar{x}, y) \right|,$$

and if u_{xx} is bounded in Ω, i.e., if there is a constant $M > 0$ such that

$$|u_{xx}(x, y)| \leqq M \quad \text{for} \quad (x, y) \in \Omega,$$

then the derivative $u_x(x, y)$ at $(x, y) \in \Omega$ can be approximated by the finite difference quotient

$$\frac{1}{h}[u(x + h, y) - u(x, y)]$$

with an error which is not larger than $(M/2)h$. (It is common practice to say in this case that the error is $0(h)$.) Thus the smaller the h, the smaller the error will be. The process of replacing partial derivatives by finite difference quotients is known as a *discretization process* and the associated error the *discretization error*. Note that several discretizations are possible. For example, u_x can be also approximated by

$$\frac{1}{h}[u(x, y) - u(x - h, y)] \quad \text{or} \quad \frac{1}{2h}[u(x + h, y) - u(x - h, y)]$$

with the error still being $0(h)$.

We will illustrate the method of finite differences by applying it to the Dirichlet problem for a bounded domain in R^2. The discretization process approximates Laplace's operator Δ by a finite difference operator Δ_h, in effect replacing Laplace's equation by a system of linear algebraic equations involving the values of the desired solution at a finite set of points of the domain. As we will see, the solution of this system always exists and can be computed with the aid of various numerical schemes and digital computers.

The method of finite differences is applicable to domains of very general shape and in any number of dimensions. It can be used to obtain numerical approximations of solutions of well-posed problems associated with any partial differential equation. However, our treatment of the method is only introductory in nature. The student who is interested in the highly developed area of numerical methods should consult specialized books such as Forsythe and Wasow[10] or Smith.[11]

Our problem is to find approximate values of the solution $u(x, y)$ of the Dirichlet problem

$$(15.1) \qquad u_{xx}(x, y) + u_{yy}(x, y) = 0, \qquad (x, y) \in \Omega$$

$$(15.2) \qquad u(x, y) = f(x, y), \qquad (x, y) \in \partial\Omega$$

where Ω is a bounded domain in R^2 with boundary $\partial\Omega$ consisting of a finite number of smooth curves, and f is a given function which is defined and continuous on $\partial\Omega$. The method of finite differences will produce, for each $h > 0$, a collection of approximate values $u_h(P)$ at certain points P of Ω, such that

$$\lim_{h \to 0} u_h(P) = u(P).$$

The procedure is as follows.

For simplicity we pick the coordinate axes so that the domain Ω lies in the first quadrant. Then, with h being a fixed small positive number, we draw two families of lines parallel to the x- and y-axes,

$$x = mh, \qquad y = nh; \qquad m, n = 0, 1, 2, \ldots .$$

The domain Ω is thus covered by a *mesh* of squares of side h as shown in Figure 15.1. The number h is called the *mesh size*. The points having coordinates (mh, nh); $m, n = 0, 1, 2, \ldots$, are called *nodes* of the mesh. The four nodes

$$E = ((m+1)h, nh), \quad N = (mh, (n+1)h),$$

$$W = ((m-1)h, nh), \quad S = (mh, (n-1)h)$$

are called the neighbors of the node $P = (mh, nh)$. The notation suggests the location of the neighbor nodes E, N, W, S relative to the central node P. A node P is said to be an *interior node* if P and its four neighbor nodes all lie in $\bar{\Omega}$. P is called a *boundary node* if P is in $\bar{\Omega}$ while at least one of its neighbors is not in $\bar{\Omega}$. The dots in Figure 15.1 indicate boundary nodes.

At each point P which is either an interior or boundary node, we will

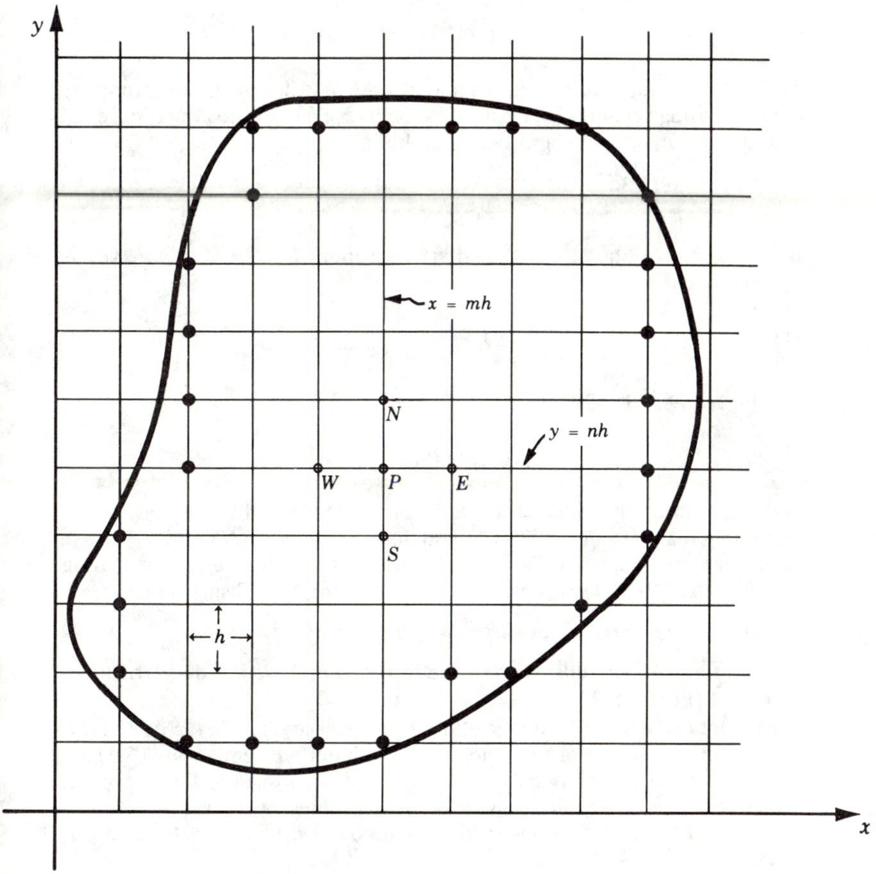

Fig. 15.1

compute an approximate value $u_h(P)$ of the solution $u(P)$ of problem (15.1), (15.2). We do this as follows. First, if P is a boundary node, we let $u_h(P)$ be equal to the value of the data f at the point of $\partial\Omega$ nearest to P. If there is more than one point on $\partial\Omega$ which is closest to P, the values of f at these points will not differ very much since f is continuous, and $u_h(P)$ can be set equal to any one of these values. Having assigned the values of u_h at the boundary nodes, the values of u_h at interior nodes will be determined by requiring that at each interior node P, u_h should satisfy the *approximate* Laplace equation

(15.3) $$\Delta_h u_h(P) = 0.$$

In (15.3) Δ_h is the *five-point Laplace difference operator* defined by

$$(15.4) \qquad \Delta_h v(P) = \frac{1}{h^2}[v(E) + v(N) + v(W) + v(S) - 4v(P)].$$

The difference operator Δ_h is obtained from the differential operator Δ by approximating second order partial derivatives by the following difference expressions: $v_{xx}(x, y)$ is approximated by

$$\frac{1}{h}\left[\frac{v(x+h, y) - v(x, y)}{h} - \frac{v(x, y) - v(x-h, y)}{h}\right]$$

or, in terms of the node P and its neighbors E, N, W, S, $v_{xx}(P)$ is approximated by

$$(15.5) \qquad \frac{1}{h^2}[v(E) - 2v(P) + v(W)].$$

Similarly, $v_{yy}(P)$ is approximated by

$$(15.6) \qquad \frac{1}{h^2}[v(N) - 2v(P) + v(S)],$$

and the sum of the difference expressions (15.5) and (15.6) is $\Delta_h v(P)$. Assuming that v is of class C^4 in $\bar{\Omega}$ and that the segments NS and EW are in Ω, it can be shown, using Taylor's formula, that the error in approximating $\Delta v(P)$ by $\Delta_h v(P)$ is $0(h^2)$, i.e., there is a constant $M > 0$ such that

$$(15.7) \qquad |\Delta_h v(P) - \Delta v(P)| \leqq Mh^2.$$

The constant M depends on the bounds of certain fourth order derivatives of v in $\bar{\Omega}$ (Problem 15.1).

Now let I denote the number of interior nodes and let P_i, $i = 1, 2, \ldots, I$, denote some indexing of these nodes. At each node P_i, equation (15.3) is a linear equation involving $u_h(P_i)$ and the values of u_h at the four neighbor nodes of P_i, some of which may be boundary nodes. We have, therefore, a system of I linear equations in the I unknowns $u_h(P_i)$, $i = 1, 2, \ldots, I$, namely

$$(15.8) \qquad \Delta_h u_h(P_i) = 0 \qquad i = 1, 2, \ldots, I.$$

After canceling the factor h^{-2} and transferring to the right side values of u_h at boundary nodes, wherever such values appear (remember that these values of u_h have already been determined), the system (15.8) takes the form

$$(15.9) \qquad A\mathbf{u}_h = \mathbf{b}.$$

A is an $I \times I$ matrix with entries the numbers 0, 1 or -4; $\mathbf{u}_h$ is a column vector with entries the unknowns $u_h(P_i)$, $i = 1, 2, \ldots, I$; $\mathbf{b}$ is a column vector with entries either 0 or linear combinations (with coefficient -1) of previously determined values of u_h at boundary nodes. The precise form of the matrix A is complicated; it depends on h, the geometry of Ω and the ordering chosen for interior nodes. In general, A is a large matrix since $I \sim$ Area $(\Omega)/h^2$. In spite of the complicated form of system (15.9), it can be proved rather easily that this system always has a unique solution. The proof is based on the *discrete mean value property*

$$(15.10) \qquad u_h(P) = \frac{1}{4}\,[u_h(E) + u_h(N) + u_h(W) + u_h(S)]$$

which is possessed by every solution of the approximate Laplace equation (15.3) at every interior node P (Problem 15.2). In order to prove that the system (15.9) has a unique solution $\mathbf{u}_h$, we need only show that $\mathbf{b} = \mathbf{0}$ (i.e., $A\mathbf{u}_h = \mathbf{0}$) implies that $\mathbf{u}_h = 0$. This follows from (15.10) and the way in which the values of u_h at boundary nodes appear in $\mathbf{b}$ (see Problem 15.4). Thus, in theory at least, we can always solve the system (15.9) to obtain the set of approximate values $u_h(P_i)$ at the I interior nodes.

A more involved use of the discrete mean value property (15.10) shows that

$$(15.11) \qquad \lim_{h \to 0} u_h(P) = u(P),$$

i.e., the approximate values $u_h(P)$ obtained by the above procedure converge to the exact values $u(P)$ of the solution of (15.1), (15.2) as the mesh size h shrinks to zero (see Petrovskii).[3] It is assertion (15.11), of course, that justifies the use of this approximation method. In practice the limiting process (15.11) cannot be carried out. One must be satisfied with carrying out the approximation procedure for a small enough value of h, taking into account the cost of computing time and the limitation imposed on the size of the system (15.9) by the computer storage capacity.

From a practical point of view, the above approximation method leaves much to be desired. Even writing out the system (15.9) is a rather formidable task. Fortunately, approximate solutions of (15.9) can be obtained by a *successive approximation scheme* which requires neither the determination of the matrix A nor the direct solution of a large system of linear equations. This scheme proceeds as follows.

First, it is necessary to select an appropriate indexing of the interior nodes. The first interior node P_1 must be chosen so that at least one of its neighbor nodes is a boundary node. The second interior node P_2 must have as one of its neighbors either the node P_1 or a boundary node, and so forth. After the indexing has been completed in this manner, we make an initial approximation by assigning values of u_h at the points $P_1, \ldots, P_I$. We denote the values of this initial approximation by

$$(15.12) \qquad u_h^{(0)}(P_i), \qquad i = 1, 2, \ldots, I.$$

In choosing the initial values (15.12) one usually employs some sort of interpolation scheme which involves the previously determined values of u_h at the boundary nodes and takes into account the maximum principle. The successive approximations will converge, however, regardless of the particular choice of the initial approximation (15.12).

Suppose now that the values of the kth successive approximation

$$(15.13) \qquad u_h^{(k)}(P_i), \qquad i = 1, 2, \ldots, I,$$

have been computed and are stored in I storage locations, with the ith location containing the value $u_h{}^{(k)}(P_i)$, $i = 1, 2, \ldots, I$. The values of the $(k + 1)$st successive approximation

$$u_h^{(k+1)}(P_i), \qquad i = 1, 2, \dots, I, \tag{15.14}$$

are then computed by sweeping through the storage locations from $i = 1$ to $i = I$, replacing the current value $u_h^{(k)}(P_i)$ in the ith location by the mean value of the four current values at the neighbor nodes of P_i (some of which may have already been recomputed during the current sweep). This procedure thus utilizes newly computed values as soon as they become available and are needed. Each sweep of the I storage locations will produce the values of the next successive approximation. The values of each new successive approximation more nearly satisfy the discrete mean value property (15.10) and the system (15.9). It can be shown that

$$\lim_{k \to \infty} u_h^{(k)}(P_i) = u_h(P_i) \qquad i = 1, 2, \dots, I \tag{15.16}$$

(see Petrovskii).[3] In practice, of course, one stops the procedure after a finite number of sweeps, this number being determined, for example, by comparing the values of successive approximations.

We illustrate the above described methods in the following example.

Example 15.1. Let Ω be the square $0 < x < \pi$, $0 < y < \pi$ and consider the Dirichlet problem

$$u_{xx} + u_{yy} = 0, \qquad (x, y) \in \Omega \tag{15.17}$$

$$\begin{aligned} &u(0, y) = 0, \qquad u(\pi, y) = 0; \qquad 0 \leqq y \leqq \pi \\ &u(x, 0) = x(\pi - x), \qquad u(x, \pi) = 0; \qquad 0 \leqq x \leqq \pi. \end{aligned} \tag{15.18}$$

We have deliberately chosen this simple problem because it can be easily solved by the method of separation of variables and Fourier series. The solution is given by the series

$$u(x, y) = \frac{8}{\pi} \sum_{n=1}^{\infty} \frac{1}{(2n-1)^3} \frac{\sinh (2n-1)(\pi - y)}{\sinh (2n-1)\pi} \sin (2n-1)x. \tag{15.19}$$

Using rather simple estimates on the error resulting from truncation of the series, the values of the solution can be computed within any specified degree of accuracy. The results of the method of finite differences can then be compared with these values.

We will solve the problem (15.17), (15.18) by the method of finite differences with $h = 2^{-4}\pi$. With this value of h there are 64 boundary nodes, all of which lie on the boundary of Ω, and 225 interior nodes ($I = 225$). The interior nodes form an array of 15 rows with each row containing 15 nodes. With the indexing requirements of the successive approximation scheme in mind, we index the interior nodes starting with the lower-left node $P_1 = (2^{-4}\pi, 2^{-4}\pi)$, moving from left to right along each row. Upon completion of a row we move up to the left-hand end of the next higher row.

Since the boundary nodes lie on the boundary of Ω, the values of u_h at these nodes are uniquely determined from the boundary condition (15.18). The linear system (15.9) for the determination of the values of u_h at interior nodes involves in this case a 15 × 15 coefficient matrix A. For obvious reasons we will not display the explicit form of this system. The reader, however, should write out equations (15.8) for the nodes P_1 and P_2 (Problem 15.7).

We will use the successive approximation scheme for finding approximate solutions of the system (15.9). The values of the initial approximation (15.12) are obtained by interpolating between boundary values on each of the segments

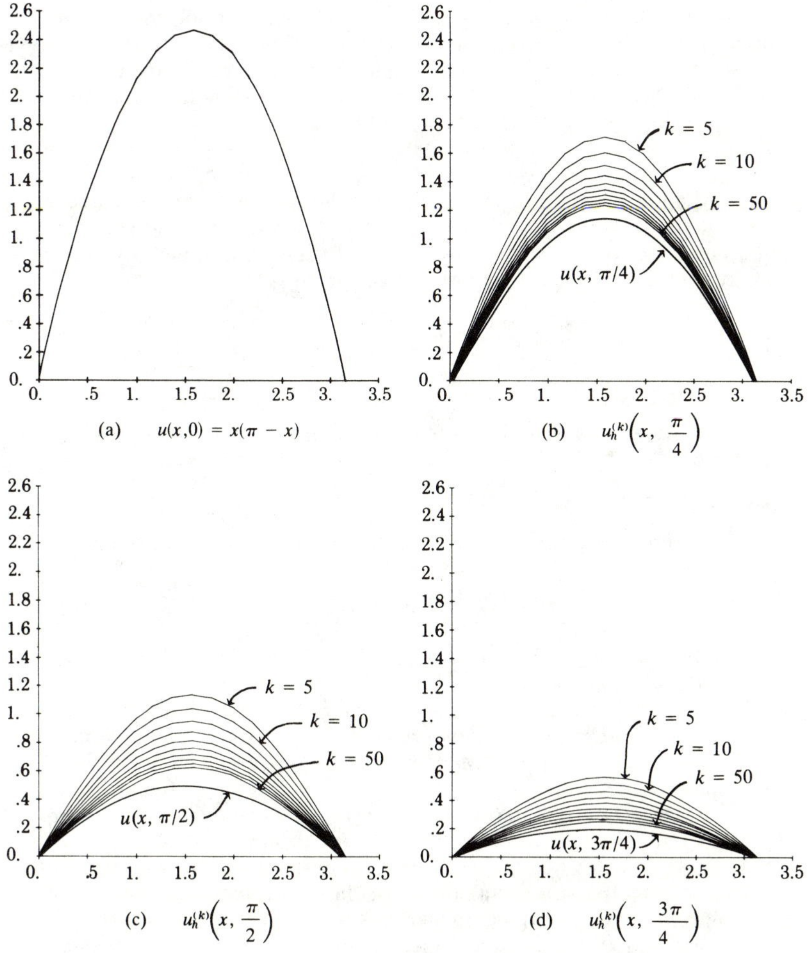

(a) $u(x,0) = x(\pi - x)$

(b) $u_h^{(k)}\left(x, \frac{\pi}{4}\right)$

(c) $u_h^{(k)}\left(x, \frac{\pi}{2}\right)$

(d) $u_h^{(k)}\left(x, \frac{3\pi}{4}\right)$

Fig. 15.2

$$x = mh, \qquad 0 \leqq y \leqq \pi; \qquad m = 1, \dots, 15.$$

It is easy to see that the resulting initial approximation is

(15.20) $$u_h^{(0)}(mh, nh) = \frac{h}{\pi} m(\pi - mh)(\pi - nh); \qquad m, n = 1, \dots, 15.$$

Aided by a digital computer, we now apply the successive approximation scheme described above, starting with the initial approximation (15.20) and computing the values of $u_h^{(k)}$ for $k = 1, 2, \dots, 50$. Partial results of these computations are displayed in graphical form in Figure 15.2. This figure displays the graphs of $u_h^{(k)}(x, y)$, as a function of x, for the four values $y = 0, \pi/4, \pi/2, 3\pi/4$. The graphs are in the form of polygonal curves formed by joining the values of $u_h^{(k)}$ at neighboring nodes with straight line segments. Figure 15.2(b), for example, displays the successive approximations $u_h^{(k)}(x, \pi/4)$ for the values $k = 5, 10, \dots, 50$. The heavy curve in this figure displays the values $u(x, \pi/4)$ of the exact series solution (15.19) of problem (15.17), (15.18), computed with an accuracy of two decimal places.

One would surmise from the graphs that the successive approximations $u_h^{(k)}$ will not converge to u. Of course this is true since $u(x, y)$ is generally different from $u_h(x, y)$ for $h > 0$. The error between $u_h^{(k)}$ and u which is apparent in the graphs is composed of two main parts:

$$u_h^{(k)}(x, y) - u_h(x, y) \text{ (the successive approximation error)}$$

and

$$u_h(x, y) - u(x, y) \text{ (the discretization error).}$$

There are also some roundoff errors present in the calculations themselves, of course.

Problems

15.1. Prove the error estimate (15.7). [*Hint:* From Taylor's formula,

$$v(E) = v(P) + hv_x(P) + \frac{h^2}{2} v_{xx}(P) + \frac{h^3}{3!} v_{xxx}(P) + \frac{h^4}{4!} v_{xxxx}(\bar{E}),$$

where $\bar{E}$ is a point on the segment PE. Write similar formulas for $v(N)$, $v(W)$ and $v(S)$.]

15.2. Prove the discrete mean value property (15.10).

15.3. Prove the *discrete maximum principle:* Let u_h be a function defined at all interior and boundary nodes and suppose u_h possesses the mean value property (15.10). Then, if h is sufficiently small, u_h assumes its maximum (and minimum) value at a boundary node and only at a boundary node unless u_h is identically constant.

15.4. Show that the system (15.9) has a unique solution [*Hint:* Assume $\mathbf{b} = \mathbf{0}$. If $A\mathbf{u}_h = \mathbf{0}$ has a non-zero solution $\mathbf{u}_h$, then $\mathbf{u}_h$ must have a maximum non-zero component, which may be assumed to be positive. From the discrete mean value property (15.10) conclude that the values of u_h must be positive at certain boundary nodes. From the special form of $\mathbf{b}$, reach a contradiction.]

15.5. Obtain the series solution (15.19) of the Dirichlet problem (15.17), (15.18) using the method of separation of variables and Fourier series.

15.6. Show that the indexing scheme for the interior nodes in Example 15.1 can be given explicitly by assigning the index $i = m + 15(n - 1)$ to the node $P = (mh, nh)$; $m, n = 1, \dots, 15$, i.e.

$$P_{m+15(n-1)} = (mh, nh).$$

15.7. Show that equations (15.8) for the nodes P_1 and P_2 of Example 15.1 are, respectively,

$$h^{-2}[u_h(P_2) + u_h(P_{16}) + h(\pi - h) - 4u_h(P_1)] = 0$$
$$h^{-2}[u_h(P_3) + u_h(P_{17}) + u_h(P_1) + 2h(\pi - 2h) - 4u_h(P_2)] = 0$$

where $h = 2^{-4}\pi$.

15.8. Show that the formula for obtaining the $(k + 1)$st successive approximation at node P_{17} in Example 15.1 is given by

$$u_h^{(k+1)}(P_{17}) = \frac{1}{4}[u_h^{(k)}(P_{18}) + u_h^{(k)}(P_{32}) + u_h^{(k+1)}(P_{16}) + u_h^{(k+1)}(P_2)].$$

16. The Neumann Problem

We recall the statement of the (interior) Neumann problem given in Section 4. Let Ω be a bounded domain in R^n with smooth boundary $\partial\Omega$, and let $\mathbf{n} = \mathbf{n}(x)$ be the outward unit normal vector to $\partial\Omega$ at the point x. Let f be a given function defined and continuous on $\partial\Omega$. Find a function $u \in C^0(\bar{\Omega})$ which is harmonic in Ω and whose outer normal derivative $\partial u/\partial n$ on $\partial\Omega$ is equal to f, i.e.

$$\nabla^2 u = 0 \quad \text{in} \quad \Omega, \tag{16.1}$$

$$\frac{\partial u}{\partial n}(x) = f(x), \qquad x \in \partial\Omega. \tag{16.2}$$

We have imposed rather strong requirements on $\partial\Omega$ and f in order to keep the discussion at an elementary level. In fact, to get simple proofs of the two theorems below, we impose the additional assumption that $u \in C^1(\bar{\Omega})$.

It is easy to see that if u is a solution of (16.1), (16.2) and c is any constant, then $u + c$ is also a solution. The following theorem asserts that although there is no uniqueness of solution for the Neumann problem, there is uniqueness up to an additive arbitrary constant.

Theorem 16.1. Any two solutions of the Neumann problem (16.1), (16.2) can differ only by a constant.

Proof. The difference $\tilde{u} = u_1 - u_2$ of any two solutions must satisfy (16.1) and (16.2) with $f(x) \equiv 0$. From the first Green's identity (equation (1.10) of Chapter VI) with $u = w = \tilde{u}$ we have

$$\int_\Omega |\nabla\tilde{u}|^2 dx = 0,$$

from which it follows that $\tilde{u}$ must be constant in Ω.

Just as easily we can prove that (16.1), (16.2) can have a solution only if the data f satisfy the condition

$$\int_{\partial\Omega} f(x)d\sigma = 0. \tag{16.3}$$

Theorem 16.2. Condition (16.3) is necessary for the existence of a solution of the Neumann problem (16.1), (16.2).

Proof. If u is a solution of (16.1), (16.2), then using the Divergence Theorem we must have

$$0 = \int_{\Omega} \nabla^2 u dx = \int_{\Omega} \nabla\cdot\nabla u dx = \int_{\partial\Omega} \nabla u\cdot\mathbf{n} d\sigma = \int_{\partial\Omega} \frac{\partial u}{\partial n} d\sigma = \int_{\partial\Omega} f(x)d\sigma.$$

The physical interpretation of u as the steady state temperature in Ω with given heat flux on $\partial\Omega$ explains the necessity of condition (16.3), which requires that the total heat flux across $\partial\Omega$ must be zero. If this were not the case, there would be a net change in the heat energy contained in Ω and such a change is impossible under steady state conditions.

Example 16.1. Consider the Neumann problem for the unit disc in R^2

$$\nabla^2 u = 0 \qquad 0 \leqq r < 1, \qquad -\pi \leqq \theta \leqq \pi, \tag{16.4}$$

$$\frac{\partial u}{\partial r}(1, \theta) = f(\theta), \qquad -\pi \leqq \theta \leqq \pi. \tag{16.5}$$

The necessary condition (16.3) for the existence of a solution is, in this case,

$$\int_{-\pi}^{\pi} f(\theta)d\theta = 0. \tag{16.6}$$

The problem can be solved by the method used in Section 7 for solving the Dirichlet problem for the disc. We look for a solution in the form of a superposition of the harmonic functions (7.3),

$$u(r, \theta) = \frac{A_0}{2} + \sum_{n=1}^{\infty} r^n(A_n \cos n\theta + B_n \sin n\theta). \tag{16.7}$$

The boundary condition (16.5) will be satisfied if

$$f(\theta) = \sum_{n=1}^{\infty} (nA_n \cos n\theta + nB_n \sin n\theta), \qquad -\pi \leqq \theta \leqq \pi. \tag{16.8}$$

Assuming for simplicity that f is continuous and has sectionally continuous derivative on the boundary of the disc, f can be represented by its Fourier series

$$f(\theta) = \frac{a_0}{2} + \sum_{n=1}^{\infty} (a_n \cos n\theta + b_n \sin n\theta), \qquad -\pi \leqq \theta \leqq \pi.$$

Since f must also satisfy condition (16.6), the constant term $a_0/2$ will be zero (why?). Therefore (16.8) will be satisfied if

$$(16.9) \qquad A_n = \frac{a_n}{n} = \frac{1}{n\pi}\int_{-\pi}^{\pi} f(\theta)\cos n\theta\, d\theta, \qquad n = 1, 2, \dots,$$

$$(16.10) \qquad B_n = \frac{b_n}{n} = \frac{1}{n\pi}\int_{-\pi}^{\pi} f(\theta)\sin n\theta\, d\theta, \qquad n = 1, 2, \dots .$$

We conclude that the solution of (16.4), (16.5) is given by (16.7) with coefficients $A_n, B_n; n = 1, 2, \dots$, given by (16.9) and (16.10), and A_0 being arbitrary. The solution has been determined up to an arbitrary constant, in accordance with Theorem 16.1.

We will not go any further into the study of the Neumann problem. We close this section by summarizing the basic results for the two-dimensional problem. It can be shown, using the method of inversion with respect to a circle, that Theorems 16.1 and 16.2 are also valid for the Neumann problem for an exterior domain Ω, provided that the complement of Ω contains a disc and u is required to satisfy the condition at infinity that u remains bounded. Moreover, using the methods of potential theory, it can be shown that condition (16.3) is sufficient as well as necessary for the existence of solution of either the interior or exterior Neumann problems. In fact, if Ω is a simply connected domain, the Neumann problem for Ω can be transformed to a Dirichlet problem for Ω provided condition (16.3) is satisfied. The method of transformation exploits the properties of conjugate harmonic functions, in particular the property described in Problem 14.7 (see Problem 16.2).

Problems

16.1. Solve the Neumann problem (16.4), (16.5), if $f(\theta) = |\theta| - (\pi/2)$, $-\pi \leqq \theta \leqq \pi$.

16.2. Let Ω be a simply connected domain in R^2 with smooth boundary and suppose that $u(x, y)$ is a solution of the Neumann problem (16.1), (16.2) satisfying condition (16.3). Let $v(x, y)$ be conjugate harmonic to $u(x, y)$ and suppose that both $u$and v are in $C^1(\bar{\Omega})$. Use Problem 14.7 to show that

$$(16.10) \qquad v(x, y) = v(x_0, y_0) + \int_{(x_0,y_0)}^{(x,y)} f(s)ds, \; (x, y) \in \partial\Omega,$$

where (x_0, y_0) is any fixed point of $\partial\Omega$ and the integration is carried along $\partial\Omega$ in the counterclockwise direction. Show that (16.10) defines v on $\partial\Omega$ as a continuous single valued function. Thus v is a solution of a Dirichlet problem. Once $v(x, y)$ is found, the solution $u(x, y)$ of the original Neumann problem is determined as the conjugate harmonic of $v(x, y)$. Note that this method determines u up to an arbitrary additive constant.

16.3. Use the method described in Problem 16.2 to find the solution of

$$\nabla^2 u = 0 \qquad 0 \leqq r < 1, \qquad 0 \leqq \theta \leqq 2\pi$$

$$\frac{\partial u}{\partial r}(1, \theta) = \cos 2\theta, \qquad 0 \leqq \theta \leqq 2\pi.$$

16.4. Use the Taylor expansion

$$\log(1 - z) = -\sum_{n=1}^{\infty} \frac{z^n}{n}, \qquad |z| < 1$$

to sum the series (16.7), with coefficients A_n, B_n given by (16.9), (16.10), and obtain the integral representation

$$u(r, \theta) = \frac{A_0}{2} - \frac{1}{2\pi} \int_0^{2\pi} f(\phi) \log [1 - r^2 - 2r \cos (\theta - \phi)] d\phi$$

for the solution of the Neumann problem (16.4), (16.5).

References for Chapter VII

1. Courant, R., and Hilbert, D.: *Methods of Mathematical Physics,* New York: Interscience Publishers, Inc., 1953.
2. Kellogg, O. D.: *Foundations of Potential Theory,* New York: Dover Publications, Inc., 1953.
3. Petrovskii, I G.: *Partial Differential Equations,* Philadelphia: W. B. Saunders Co., 1967.
4. Churchill, R. V.: *Fourier Series and Boundary Value Problems,* New York: McGraw Hill Book Co., 1963.
5. Tolstov, G. P.: *Fourier Series,* Englewood Cliffs, N. J.: Prentice Hall Inc., 1962.
6. Taylor, A. E.: *Advanced Calculus,* Boston: Ginn and Co., 1955.
7. Churchill, R. V.: *Complex Variables and Applications,* New York: McGraw Hill Book Co., 1960.
8. Kober, H.: *Dictionary of Conformal Representation,* New York: Dover Press, 1952.
9. Ahlfors, L. V.: *Complex Analysis,* New York: McGraw Hill Book Co., Inc., 1953.
10. Forsythe, G. E., and Wasow, W. R.: *Finite-Difference Methods for Partial Differential Equations,* New York: John Wiley & Sons, 1960.
11. Smith, G. D.: *Numerical Solutions of Partial Differential Equations,* New York: Oxford University Press, 1965.

CHAPTER VIII

The wave equation

In this chapter we present a detailed study of the wave equation, which is the most important example of a linear partial differential equation of hyperbolic type. In Section 1 we describe some simple solutions of the equation, including solutions known as plane waves and spherical waves which physically represent traveling waves with plane or spherical profiles. Sections 2 through 6 are devoted to the study of the initial value problem. In Section 2 this problem is carefully defined and illustrated by representative physical problems in one, two and three space dimensions. In Section 3 we derive an inequality which must be satisfied by every solution of the wave equation, and which involves certain integrals representing the energy of the physical wave described by the solution. The method of derivation is known as the energy method and the inequality is known as the domain of dependence inequality. It follows from the inequality that the value of the solution of the initial value problem at any given point of space-time depends only on the values of the initial data on a certain portion of space known as the domain of dependence. The details are given in Section 4, where the uniqueness of solution of the initial value problem and the conservation of the total energy of the solution are also proved. In Section 5 it is shown that Kirchhoff's formula gives the solution of the initial value problem in three space dimensions. The method of descent (going from three to two and one dimensions) is then used to obtain the corresponding solution in two and one space dimensions. In Section 6 the properties of the solution of the initial value problem in various space dimensions are discussed. It is shown that in three space dimensions the so called Huygens' principle holds, a consequence of which is the sharp propagation of signals. In two space dimensions Huygens' principle does not hold and consequently propagating signals always have a decaying trailing edge. This phenomenon is known as diffusion of waves. Sections 7 through 10 are devoted to the study of the initial-boundary value problem. In Section 7 the uniqueness of solution is proved by the energy method, and the reflection of waves by plane boundaries is discussed. The vibrations of a string and of a rectangular membrane are studied in Sections 7 and 8, respectively. The method of solution involves separation of variables, solution of an associated eigenvalue problem, and Fourier series expansions of the initial data. This method is generalized in Section 10, where the properties of an eigenvalue problem for the Laplacian operator are summarized, and the fundamental theorem concerning the representation of functions by eigenfunction expansions is stated. This general method is once more illustrated by applying it to the study of vibrations of a circular membrane.

1. Some Solutions of the Wave Equation. Plane and Spherical Waves

The wave equation

$$\frac{\partial^2 u}{\partial x_1^2} + \dots + \frac{\partial^2 u}{\partial x_n^2} - \frac{\partial^2 u}{\partial t^2} = 0 \tag{1.1}$$

is the simplest and most important partial differential equation of hyperbolic type. We have seen in Chapter V that every second order hyperbolic equation may be reduced by a transformation of coordinates to a canonical form which, at least at any given point, has principal part the same as that of the wave equation. This would seem to indicate that most important properties of hyperbolic equations are shared by the wave equation. It turns out that, with some exceptions, this is actually true. Therefore, for the mathematician who is interested in the theory of hyperbolic equations, a careful study of the wave equation is essential. For the physical scientist, engineer and applied mathematician, the reason for studying the wave equation is obvious since this equation describes many wave propagation and vibration phenomena (see Chapter VI, Section 4).

A solution of the wave equation (1.1) is a function of $n + 1$ variables $x_1, \dots, x_n$ and t. The variables $x_1, \dots, x_n$ are called the space variables and t is called the time variable. We will use the notation $x = (x_1, \dots, x_n)$ where x is a point in the n-dimensional space R^n known as the x-space. Equation (1.1) is known as the n-dimensional wave equation where n indicates the num-

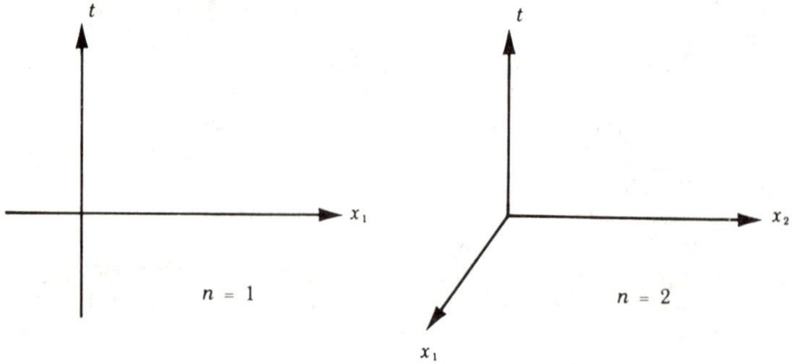

Fig. 1.1

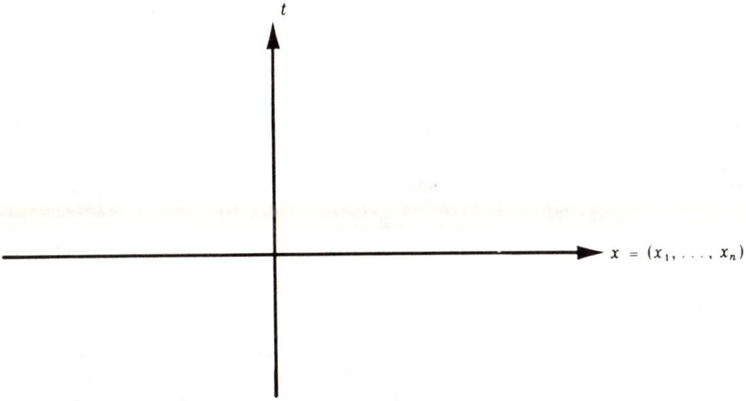

Fig. 1.2

ber of space variables. We emphasize that u is a function defined in the space R^{n+1} known as the (x, t)-space. For $n = 1$ and $n = 2$ we can draw figures of the (x, t)-space as shown in Figure 1.1. However for $n \geqq 3$ this is no longer possible. For this reason it is sometimes convenient to represent the (x, t)-space by representing the x-space by one axis as shown in Figure 1.2.

In this section we describe some solutions of the wave equation.

Let $\xi_x = (\xi_1, \ldots, \xi_n)$ be a unit vector in R^n so that

$$\xi_1^2 + \cdots + \xi_n^2 = 1. \tag{1.2}$$

Then for each fixed t and each constant c, the equation

$$\xi_1 x_1 + \cdots + \xi_n x_n - t = c \tag{1.3}$$

represents a plane in the x-space R^n. The vector ξ_x is normal to this plane and as t increases, the plane travels in the direction of ξ_x with speed 1. (To see this consider the special case $\xi = (1, 0, \ldots, 0)$.) Now, let $F(y)$ be a C^2 function of the single variable y. Then it is easy to verify that

$$u(x_1, \ldots, x_n, t) = F(\xi_1 x_1 + \cdots + \xi_n x_n - t) \tag{1.4}$$

is a solution of the wave equation (1.1) (see Problem 1.1). The value of u on the traveling plane (1.3) is constant and equal to $F(c)$. For this reason, solutions of the form (1.4) are known as *plane waves*. We illustrate plane wave solutions of the wave equation with examples for $n = 1$, 2 and 3.

For $n = 1$, equation (1.1) is the one-dimensional wave equation

$$\frac{\partial^2 u}{\partial x_1^2} - \frac{\partial^2 u}{\partial t^2} = 0. \tag{1.5}$$

Condition (1.2) is $\xi_1{}^2 = 1$ and the only possible values of ξ_1 are $\xi_1 = +1$ and $\xi_1 = -1$. The corresponding traveling planes are, in this case, traveling points in R^1 given by

(1.6)
$$x_1 - t = c$$

and

(1.7)
$$x_1 + t = c,$$

respectively. (1.6) describes a point traveling in the positive x_1 direction with speed 1, while (1.7) describes a point traveling in the negative x_1 direction with speed 1. The corresponding plane wave solutions are

(1.8)
$$u(x_1, t) = F(x_1 - t)$$

and

(1.9)
$$u(x_1, t) = G(x_1 + t)$$

where F and G are arbitrary C^2 functions of a single variable. Solutions of the form (1.8) represent waves traveling in the positive x_1 direction with speed 1. Let us suppose for example that $F(y)$ is the "blip" function shown in Figure 1.3. Figure 1.4 shows the graph of the solution (1.8) as a function of x_1 for several values of t. Recalling that the one-dimensional wave equation (1.5) describes the motions of a stretched string which lies along the x_1-axis when in equilibrium, the drawings in Figure 1.4 may be viewed as photographs, at the indicated instants of time, of a wavelet traveling along the string. Figure 1.5 shows the (x_1, t)-plane on which the solution (1.8) is defined. The shaded strip indicates the region where $u(x_1, t) \neq 0$. Solutions of the form (1.9) represent waves traveling in the negative x_1 direction. The student should draw figures corresponding to Figures 1.4 and 1.5 for (1.9) taking $G(y) = F(y)$ as shown in Figure 1.3 (see Problem 1.2).

For $n = 2$ equation (1.1) is the two-dimensional wave equation

(1.10)
$$\frac{\partial^2 u}{\partial x_1^2} + \frac{\partial^2 u}{\partial x_2^2} - \frac{\partial^2 u}{\partial t^2} = 0.$$

There are infinitely many unit vectors $\xi_x = (\xi_1, \xi_2)$ in R^2. For example $\xi_x = (1/\sqrt{2}, 1/\sqrt{2})$ is such a vector and the corresponding traveling "plane" is, in this case, a traveling line in R^2 given by

(1.11)
$$\frac{1}{\sqrt{2}} x_1 + \frac{1}{\sqrt{2}} x_2 - t = c.$$

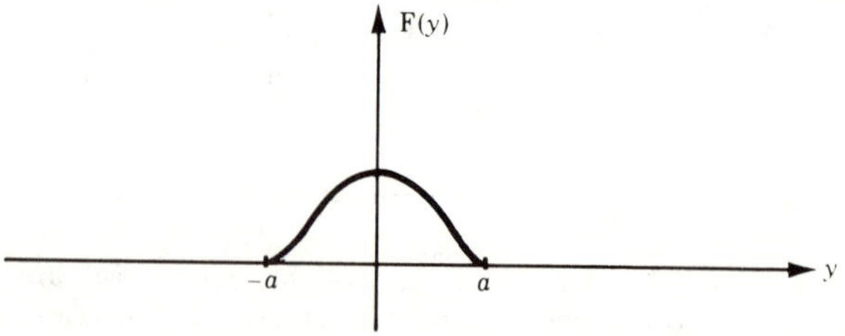

Fig. 1.3

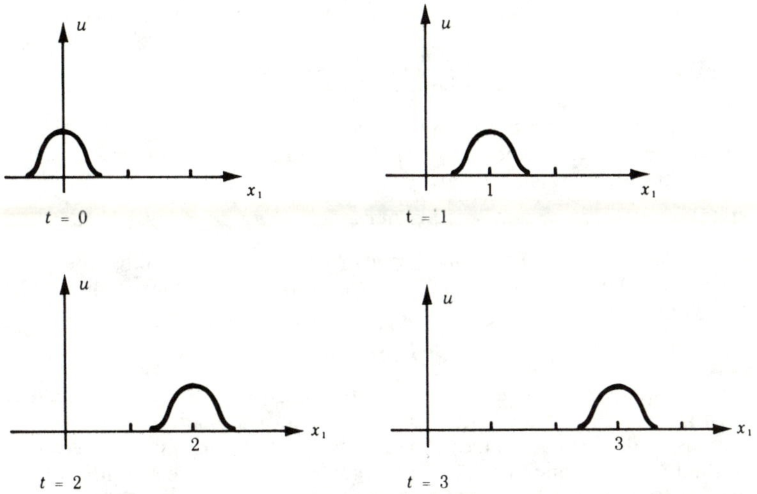

Fig. 1.4

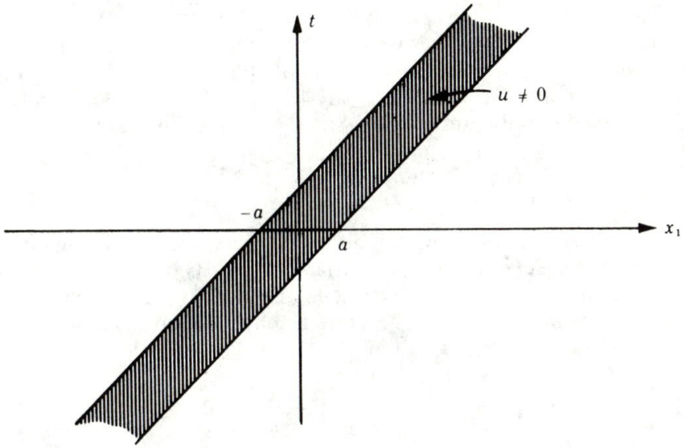

Fig. 1.5

(1.11) describes a line in the x-space R^2 traveling in the direction of its normal $(1/\sqrt{2}, 1/\sqrt{2})$ with speed 1. The corresponding plane wave solutions of (1.10) are

$$u(x_1, x_2, t) = F\left(\frac{1}{\sqrt{2}} x_1 + \frac{1}{\sqrt{2}} x_2 - t\right) \tag{1.12}$$

where F is an arbitrary C^2 function of a single variable. Further study of this example is indicated in Problem 1.3.

For $n = 3$ equation (1.1) is the three-dimensional wave equation

$$\frac{\partial^2 u}{\partial x_1^2} + \frac{\partial^2 u}{\partial x_2^2} + \frac{\partial^2 u}{\partial x_3^2} - \frac{\partial^2 u}{\partial t^2} = 0. \tag{1.13}$$

Again there are infinitely many unit vectors $\xi_x = (\xi_1, \xi_2, \xi_3)$ in R^3. For example one such vector is $\xi_x = (1, 0, 0)$ and the corresponding traveling planes are given by

$$x_1 - t = c. \tag{1.14}$$

(1.14) describes a plane in the x-space R^3 traveling in the direction of its normal (1, 0, 0) with speed 1. The corresponding plane wave solutions of (1.13) are

$$u(x_1, x_2, x_3, t) = F(x_1 - t). \tag{1.15}$$

Further study of this example is indicated in Problem 1.4.

In Chapter V, Section 7, we have seen that every solution of the one-dimensional wave equation can be expressed as the sum of two plane waves of the form (1.8) and (1.9). However, for $n \geqq 2$ there are solutions of the wave equation which are not finite sums of plane waves. Among these solutions there is another class of simple solutions known as *spherical waves.* A spherical wave is a solution of the wave equation whose value (for each fixed t) is constant on spheres in the x-space centered at the origin. In order to study spherical waves we introduce spherical coordinates in the x-space. The Laplacian operator in terms of spherical coordinates in R^n is given by formula (2.3) of Chapter VII. Hence, the wave equation (1.1), using spherical coordinates in the x-space R^n, takes the form

$$\frac{1}{r^{n-1}}\frac{\partial}{\partial r}\left(r^{n-1}\frac{\partial u}{\partial r}\right) + \frac{1}{r^2}\Lambda_n u - \frac{\partial^2 u}{\partial t^2} = 0, \tag{1.16}$$

where Λ_n is a second order partial differential operator involving differentiations with respect to the angular variables only. By definition, a spherical wave is a solution of (1.16) which depends only on r and t and does not depend on angular variables. Therefore a spherical wave is a function $u = u(r, t)$ which satisfies the equation

$$\frac{1}{r^{n-1}}\frac{\partial}{\partial r}\left(r^{n-1}\frac{\partial u}{\partial r}\right) - \frac{\partial^2 u}{\partial t^2} = 0. \tag{1.17}$$

Equation (1.17) is known as the *equation of spherical waves.*

For $n = 3$ the equation of spherical waves is

$$\frac{1}{r^2}\frac{\partial}{\partial r}\left(r^2\frac{\partial u}{\partial r}\right) - \frac{\partial^2 u}{\partial t^2} = 0, \qquad (n = 3). \tag{1.18}$$

This equation can be simplified by introducing the new dependent variable $w = ru$. Substituting into (1.18), the equation for w is found to be

$$\frac{\partial^2 w}{\partial r^2} - \frac{\partial^2 w}{\partial t^2} = 0 \tag{1.19}$$

which is the one-dimensional wave equation. We already know that every solution of (1.19) can be expressed as the sum of solutions of the form $F(r - t)$ and $G(r + t)$, where F and G are C^2 functions of a single variable. Therefore every solution of (1.18) can be expressed as the sum of solutions of the form

$$u(r, t) = \frac{F(r - t)}{r} \tag{1.20}$$

and

$$u(r, t) = \frac{G(r + t)}{r}. \tag{1.21}$$

Solutions of the three-dimensional wave equation of the form (1.20) represent expanding (or outgoing) spherical waves, while solutions of the form (1.21) represent contracting (or incoming) spherical waves. The speed of expansion or contraction of these spherical waves is 1.

For $n = 2$ equation (1.17) is

$$\frac{1}{r}\frac{\partial}{\partial r}\left(r\frac{\partial u}{\partial r}\right) - \frac{\partial^2 u}{\partial t^2} = 0 \qquad (n = 2). \tag{1.22}$$

Equation (1.22) is also known as the *equation of cylindrical waves* (see Problem 1.7). This equation is not as easy to solve as (1.18). Solutions may be obtained by the method of separation of variables described below (see Problems 1.8 and 1.9).

Let us apply the method of separation of variables to the wave equation. We look for solutions of (1.1) of the form

$$u(x, t) = v(x)T(t) \tag{1.23}$$

where $v(x)$ is a function of the space variables and $T(t)$ is a function of t only. Substituting (1.23) into (1.1) and separating the variables we obtain

$$\frac{\dfrac{\partial^2 v(x)}{\partial x_1^2} + \dots + \dfrac{\partial^2 v(x)}{\partial x_n^2}}{v(x)} = \frac{T''(t)}{T(t)}. \tag{1.24}$$

Since the left-hand side is a function of x only, while the right-hand side is a function of t only, the two functions must be constant functions and in fact equal to the same constant. Hence (1.24) is equivalent to

$$\frac{\dfrac{\partial^2 v(x)}{\partial x_1^2} + \dots + \dfrac{\partial^2 v(x)}{\partial x_n^2}}{v(x)} = \frac{T''(t)}{T(t)} = -\mu,$$

or to the pair of equations

$$\frac{\partial^2 v}{\partial x_1^2} + \dots + \frac{\partial^2 v}{\partial x_n^2} + \mu v = 0 \tag{1.25}$$

$$T'' + \mu T = 0 \tag{1.26}$$

where μ is some constant. Equation (1.25) is an important equation of elliptic type and is known as the *reduced wave equation.* Equation (1.26) is a simple ordinary differential equation.

Let us consider first the case $\mu > 0$, say $\mu = \omega^2, \omega > 0$. In this case (1.26) has two linearly independent solutions

$$\cos \omega t, \sin \omega t$$

which are the real and imaginary parts of the complex valued solution $e^{i\omega t}$. Solutions of the reduced wave equation (1.25) can be found by further separation of variables as was done for Laplace's equation in Chapter VII, Section 2. We do this here for $n = 3$ and leave the case $n = 2$ for the problems. In terms of spherical coordinates equation (1.25) becomes

$$\frac{1}{r^2}\frac{\partial}{\partial r}\left(r^2\frac{\partial v}{\partial r}\right) + \frac{1}{r^2}\Lambda_3 v + \omega^2 v = 0. \tag{1.27}$$

We look for solutions of this equation in the form

$$v(r, \theta, \phi) = R(r)Y(\theta, \phi). \tag{1.28}$$

Substitution of (1.28) into (1.27) and separation of variables yields

$$\frac{\frac{1}{r^2}\frac{d}{dr}\left(r^2\frac{dR}{dr}\right) + \omega^2 R}{\frac{1}{r^2}R} = -\frac{\Lambda_3 Y}{Y} = \gamma$$

where γ is the separation constant. The separated equations are

$$\Lambda_3 Y + \gamma Y = 0 \tag{1.29}$$

$$r^2R'' + 2rR' + (\omega^2 r^2 - \gamma)R = 0. \tag{1.30}$$

As discussed in Chapter VII, Section 2, equation (1.29) has smooth solutions only when γ is equal to one of the values

$$\gamma_k = k(k+1), \qquad k = 0, 1, 2, \ldots . \tag{1.31}$$

For each such γ_k there are $2k + 1$ linearly independent solutions of (1.29) denoted by

$$Y_k^{(\ell)}(\theta, \phi), \qquad \ell = 1, 2, \ldots, 2k+1, \tag{1.32}$$

which are known as the Laplace spherical harmonics. For each γ_k, equation (1.30) becomes

$$r^2R'' + 2rR' + [\omega^2 r^2 - k(k+1)]R = 0. \tag{1.33}$$

In terms of the new dependent variable

$$w = r^{1/2}R \tag{1.34}$$

equation (1.33) becomes

$$r^2w'' + rw' + \left[\omega^2 r^2 - \left(k + \frac{1}{2}\right)^2\right] w = 0. \tag{1.35}$$

Equation (1.35) is *Bessel's equation* of order $k + 1/2$ with parameter ω. It has two linearly independent solutions (Bessel's functions),

(1.36) $$J_{k+1/2}(\omega r) \quad \text{and} \quad J_{-(k+1/2)}(\omega r),$$

where

(1.37) $$J_\nu(z) = \sum_{m=0}^{\infty} (-1)^m \frac{(z/2)^{\nu+2m}}{m!\Gamma(\nu + m + 1)}$$

with Γ denoting the gamma function. The solutions (1.36) are distinguished by their behavior at the origin; $J_{k+1/2}(\omega r)$ behaves like $r^{k+1/2}$ near $r = 0$ while $J_{-(k+1/2)}(\omega r)$ behaves like $r^{-(k+1/2)}$ near $r = 0$. Consequently, for each $k = 0, 1, 2, \ldots$, equation (1.33) has two linearly independent solutions,

(1.38) $$r^{-1/2}J_{k+1/2}(\omega r) \quad \text{and} \quad r^{-1/2}J_{-(k+1/2)}(\omega r)$$

which behave, respectively, like r^k and r^{-k-1} near $r = 0$. The corresponding product solutions of the reduced wave equation (1.27) are

(1.39) $$\begin{cases} r^{-1/2}J_{k+1/2}(\omega r)\, Y_k^{(\ell)}(\theta, \phi) \quad \text{and} \quad r^{-1/2}J_{-(k+1/2)}(\omega r)\, Y_k^{(\ell)}(\theta, \phi), \\ k = 0, 1, 2, \ldots ; \qquad \ell = 1, 2, \ldots , 2k + 1. \end{cases}$$

Thus, by the method of separation of variables we have found the following collection of solutions for the three-dimensional wave equation,

(1.40) $$r^{-1/2} \begin{Bmatrix} J_{k+1/2}(\omega r) \\ J_{-(k+1/2)}(\omega r) \end{Bmatrix} Y_k^{(\ell)}(\theta, \phi) \begin{Bmatrix} \cos \omega t \\ \sin \omega t \end{Bmatrix}; \quad k = 0, 1, 2, \ldots ; \qquad \ell = 1, 2, \ldots , 2k + 1.$$

Because of the harmonic dependence on t, these solutions are referred to as oscillatory solutions of the three-dimensional wave equation.

Nonoscillatory solutions correspond to the case $\mu < 0$, say $\mu = -\omega^2$, $\omega > 0$. Two linearly independent solutions of the time equation are, in this case, $e^{\omega t}$ and $e^{-\omega t}$ and the equations corresponding to (1.27), (1.29) and (1.30) are obtained by replacing ω^2 by $-\omega^2$. The equation corresponding to (1.35) is

(1.41) $$r^2w'' + rw' - \left[\omega^2r^2 + \left(k + \frac{1}{2}\right)^2\right] w = 0.$$

Equation (1.41) is Bessel's equation with purely imaginary argument (see Watson)[1]. It has two linearly independent solutions (Bessel's functions),

(1.42) $$I_{k+1/2}(\omega r) \quad \text{and} \quad I_{-(k+1/2)}(\omega r),$$

where

(1.43) $$I_\nu(z) = \sum_{m=0}^{\infty} \frac{(z/2)^{\nu+2m}}{m!\Gamma(\nu+m+1)}.$$

Problems

1.1. Show that (1.4) is a solution of the wave equation (1.1).

1.2. Draw the figures corresponding to Figures 1.4 and 1.5 for the solution (1.9), taking $G(y) = F(y)$ as shown in Figure 1.3.

1.3. Let $F(y)$ be the blip function shown in Figure 1.3. In the (x_1, x_2) plane draw the lines where (1.12) has its maximum value and the strips where $u(x_1, x_2, t) \neq 0$ for $t = 0, 1, 2, 5$.

1.4. Let $F(y)$ be the blip function shown in Figure 1.3. In the (x_1, x_2, x_3)-space draw the planes where (1.15) has its maximum value and the slabs where $u(x_1, x_2, x_3, t) \neq 0$ for $t = 0, 1, 2, 5$.

1.5. Show that in the (x, t)-space R^{n+1} the planes (1.3) are characteristic with respect to the wave equation (1.1). Also show that the plane wave solutions (1.4) are constant on these planes.

1.6. Derive (1.19) from (1.18).

1.7. In R^3 introduce cylindrical coordinates

$$x_1 = \rho \cos \theta, \qquad x_2 = \rho \sin \theta, \qquad x_3 = z,$$

and show that the three-dimensional wave equation becomes

$$\text{(1.44)} \qquad \frac{1}{\rho}\frac{\partial}{\partial \rho}\left(\rho \frac{\partial u}{\partial \rho}\right) + \frac{1}{\rho^2}\frac{\partial^2 u}{\partial \theta^2} + \frac{\partial^2 u}{\partial z^2} - \frac{\partial^2 u}{\partial t^2} = 0.$$

If u does not depend on θ and z, equation (1.44) becomes equation (1.22) (with $r = \rho$). Why is (1.22) called the equation of cylindrical waves?

1.8. Apply the method of separation of variables to the reduced wave equation (1.25) for $n = 2$. Consider only the case $\mu > 0$, say $\mu = \omega^2$, $\omega > 0$. Introduce polar coordinates and look for solutions of the form $v(r, \theta) = R(r)\Theta(\theta)$ with the periodicity conditions $\Theta(0) = \Theta(2\pi)$, $\Theta'(0) = \Theta'(2\pi)$. Show that the equation for R is Bessel's equation of order n with parameter ω,

$$r^2R'' + rR' + (\omega^2 r^2 - n^2)R = 0,$$

where $n = 0, 1, 2, \ldots$. Derive the following solutions for the two-dimensional wave equation,

$$\begin{Bmatrix} J_n(\omega r) \\ Y_n(\omega r) \end{Bmatrix} \begin{Bmatrix} \cos n\theta \\ \sin n\theta \end{Bmatrix} \begin{Bmatrix} \cos \omega t \\ \sin \omega t \end{Bmatrix}, \quad n = 0, 1, 2, \ldots$$

where J_n is Bessel's function of order n and Y_n is Bessel's function of the second kind of order n (see Watson)[1].

1.9. From the solutions of the two-dimensional wave equation obtained in Problem 1.8, determine those which are independent of θ. These are solutions of the equation of cylindrical waves (1.22).

1.10. The solutions (1.40) with $k = 0$ are spherically symmetric ($Y_0^{(1)}(\theta, \phi) = 1$) and hence they must be of the form (1.20) or (1.21). Show that the combinations

$$[r^{-1/2}J_{1/2}(\omega r) \pm ir^{-1/2}J_{-1/2}(\omega r)]e^{i\omega t}$$

are of this form. [*Hint:* Compare the series expansions of $\cos z$ and sin z with those of $J_{1/2}(z)$ and $J_{-1/2}(z)$ and use the fact that $\Gamma(m + 1/2) = (2m)!\sqrt{\pi}/(4^m m!)$ to obtain the formulas

$$J_{1/2}(z) = \sqrt{\frac{2}{\pi z}}\sin z \text{ and } J_{-1/2}(z) = \sqrt{\frac{2}{\pi z}}\cos z.]$$

2. The Initial Value Problem

The basic problem associated with the wave equation

$$\frac{\partial^2 u}{\partial x_1^2} + \ldots + \frac{\partial^2 u}{\partial x_n^2} - \frac{\partial^2 u}{\partial t^2} = 0, \tag{2.1}$$

is the initial value problem, or Cauchy problem. This problem asks for the solution $u(x_1, \ldots, x_n, t)$ of (2.1) which satisfies the given initial conditions

$$u(x_1, \ldots, x_n, 0) = \phi(x_1, \ldots, x_n), \tag{2.2}$$

$$\frac{\partial u}{\partial t}(x_1, \ldots, x_n, 0) = \psi(x_1, \ldots, x_n). \tag{2.3}$$

The functions ϕ and ψ, known as the initial data, are given functions which are defined in the x-space at $t = 0$. The plane $t = 0$ in the (x, t)-space R^{n+1} is known as the initial surface or initial manifold of the problem.

Although we are mainly concerned with the cases $n = 1, n = 2$ and $n = 3$, much of the discussion will be carried out for general n. However, even when the discussion is for general n, it is convenient to visualize the problem under consideration for the special case $n = 2$. For $n = 2$, the initial value problem asks for the solution $u(x_1, x_2, t)$ of the two-dimensional wave equation

$$\frac{\partial^2 u}{\partial x_1^2} + \frac{\partial^2 u}{\partial x_2^2} - \frac{\partial^2 u}{\partial t^2} = 0 \tag{2.4}$$

satisfying the initial conditions

$$u(x_1, x_2, 0) = \phi(x_1, x_2) \tag{2.5}$$

$$\frac{\partial u}{\partial t}(x_1, x_2, 0) = \psi(x_1, x_2). \tag{2.6}$$

The solution is to be defined in the three-dimensional (x_1, x_2, t)-space. The initial conditions (2.5) and (2.6) are specified on the initial surface $t = 0$ which, in this case, is the (x_1, x_2)-plane.

Our purpose is to show that, under certain conditions on the initial data ϕ and ψ, the initial value problem (2.1)–(2.3) is well-posed. We also want to find a formula for the solution of the problem. We already know from the Cauchy-Kovalevsky theorem that if ϕ and ψ are analytic at the origin of the x-space, then the problem has a unique solution which is defined and analytic in a neighborhood of the origin of the (x, t)-space. Here, however, we want to enlarge the scope of our study beyond this limited result. We want to solve the problem "in the large"; i.e., for all (x, t) in R^{n+1} and not just in a neighborhood of the origin. Also we want to eliminate the assumption of analyticity on ϕ and ψ.

Concerning our aim to solve the initial value problem in the large, it should be noted that it is enough to solve the problem only in the upper half-space $t \geqq 0$. The reason for this is that the problem for the lower half-space $t \leqq 0$ can be reduced to the problem for the upper half-space by making the transformation $t' = -t$ and noting that under this transformation the wave equation (2.1) remains unchanged. (This is not true for the heat equation!) In what follows we will only study the "forward" problem which asks for the solution of the initial value problem (2.1)–(2.3) for $t \geqq 0$. This of course is the problem that most often arises in physics.

It should also be noted that the initial value problem with initial surface the plane $t = t^0$ can be immediately reduced to the problem (2.1), (2.2), (2.3) with initial surface $t = 0$ by making the transformation $t' = t - t^0$. Obviously, under this transformation the wave equation remains unchanged.

We have already seen in problems of previous chapters of this book that the initial value problem for the one-dimensional wave equation,

$$\frac{\partial^2 u}{\partial x^2} - \frac{\partial^2 u}{\partial t^2} = 0, \qquad x \in R^1, \qquad t > 0, \tag{2.7}$$

$$u(x, 0) = \phi(x), \qquad x \in R^1, \tag{2.8}$$

$$\frac{\partial u}{\partial t}(x, 0) = \psi(x), \qquad x \in R^1, \tag{2.9}$$

is a well-posed problem and that its solution is given by the formula

$$u(x, t) = \frac{1}{2}[\phi(x + t) + \phi(x - t)] + \frac{1}{2}\int_{x-t}^{x+t} \psi(\tau)d\tau. \tag{2.10}$$

Existence, uniqueness and the derivation of formula (2.10) (assuming that $\phi \in C^2(R^1)$ and $\psi \in C^1(R^1)$) were the subject of Problem 4.1 of Chapter VI. Indeed, existence can be proved by simply verifying that (2.10) satisfies (2.7)–(2.9). Uniqueness of the solution of the problem follows from the fact that the general solution of (2.7) is $u(x, t) = F(x - t) + G(x + t)$ and from the fact that the initial conditions (2.8) and (2.9) uniquely determine the functions F and G and yield (in a unique way) formula (2.10) for the solution of the problem. For the initial value problem with $n > 1$ we cannot follow this simple procedure to prove existence and uniqueness and obtain a formula for the solution since we do not have a simple formula for the general solution of the wave equation in more than one space variable. Instead, what we will do is first prove uniqueness for general n by using a method known as the energy method. Then we will write down the formula for the solution of the problem for $n = 2$ and $n = 3$ and will verify by direct computation that it satisfies the wave equation and the initial conditions. This will prove existence. Finally we will prove continuous dependence on the initial data using the formula for the solution of the problem.

In Section 4 of Chapter VI we described some physical phenomena which are governed by the wave equation. In studying the initial value problem for the wave equation it is useful to keep in mind an associated physical problem. The problem of determining the small vibrations of an infinite stretched string, given its initial displacement and velocity, leads to an initial value problem for the one-dimensional wave equation. The problem of determining the small vibrations of an infinite stretched membrane, given its initial displacement and velocity, leads to an initial value problem for the two-dimensional wave equation. Finally, the problem of determining the propagation of sound (here u represents the deviation from ambient pressure) in an infinite atmosphere, given the initial distribution of u and $\partial u/\partial t$, leads to an initial value problem for the three-dimensional wave equation. One may object to all these examples as being unrealistic since there are no infinite strings, membranes or atmospheres. However, as we will see, the presence of boundaries in the x-space (corresponding to boundaries of a string, membrane or atmosphere) affects $u(x, t)$ only when $t \geqq T$, where $T = T(x)$ depends on the distance of x from the boundary. This is due to the fact that disturbances propagate with finite speed. Thus, when boundaries are present, solving the initial value problem is still relevant but only over finite intervals of time.

Problems

2.1. Consider the initial value problem (2.7)–(2.9) for the one-dimensional wave equation.

(a) Suppose that $\phi = \psi = 0$ in some interval of length ℓ. Up to what time t can you be sure that $u = 0$ at the center of the interval?

(b) Suppose that $\phi = \psi = 0$ outside the interval $|x| \leqq 1$. Up to what time can you be sure that $u = 0$ at (i) $x = 3$, (ii) $x = 10$, and (iii) $x = -5$?

2.2. Write down the formula for the solution of the initial value problem

$$u_{xx} - \frac{1}{c^2} u_{tt} = 0, \qquad x \in R^1, \qquad t > 0,$$

$$u(x, 0) = \phi(x), \qquad u_t(x, 0) = \psi(x), \qquad x \in R^1.$$

Then answer the questions of Problem 2.1 for this problem.

2.3. Consider the initial value problem (2.7)–(2.9) and suppose that $\phi = \psi = 0$ outside the interval $|x| \leqq \ell$. Show that at any fixed point x^0, u is eventually constant. More precisely, show that there are numbers $T = T(x^0)$ and U such that $u(x^0, t) = U$ for $t > T(x^0)$. Find $T(x^0)$ and U.

2.4. Consider the initial value problem (2.7)–(2.9).

(a) Find $u(\pi, t)$ and $u(-\pi, t)$ for $t = 0, \pi/2, \pi, 3\pi/2, 2\pi$, if

$$\phi(x) = \begin{cases} \cos x & \text{for } \ |x| < \dfrac{\pi}{2} \\ 0 & \text{for } \ |x| \geqq \dfrac{\pi}{2} \end{cases}$$

$$\psi(x) = 0 \quad \text{for} \quad x \in R^1.$$

(b) Follow the instructions of part (a) with ϕ and ψ interchanged.

[Use formula (2.10). Ignore the fact that the initial data are not smooth at $x = \pm\pi/2$. We will discuss this difficulty later on.]

3. The Domain of Dependence Inequality. The Energy Method

In this section we derive an important inequality which is satisfied by solutions of the wave equation. This inequality, known as an *energy inequality,* yields immediately the uniqueness of solution of the initial value problem. It also yields the important fact that the value of the solution at a point depends only on the values of the initial data on a finite part of the initial manifold. For this reason, the inequality is called the *domain of dependence inequality*.

We will first explain the inequality and then we will give its proof. The method of proof is known as the *energy method.* The applications and explanation of the use of the word energy will be given in the next section.

Let us first recall some facts concerning characteristic directions and characteristic cones for the wave equation (see Example 2.10 of Chapter V). If $\boldsymbol{\nu} = (\nu_1, \ldots, \nu_n, \nu_t)$ is a unit vector in R^{n+1} describing a characteristic direction, then by definition,

$$\nu_1^2 + \cdots + \nu_n^2 - \nu_t^2 = 0. \tag{3.1}$$

Moreover, since $\boldsymbol{\nu}$ is a unit vector, we must have

$$\nu_t = \pm\frac{1}{\sqrt{2}}. \tag{3.2}$$

Thus the normal to a characteristic surface must always make a 45° angle with the t-axis. Now let $(x^0, t^0) = (x_1{}^0, \ldots, x_n{}^0, t^0)$ be a fixed point in R^{n+1}. The equation

$$(x_1 - x_1^0)^2 + \cdots + (x_n - x_n^0)^2 - (t - t^0)^2 = 0 \tag{3.3}$$

describes a double conical surface with apex at (x^0, t^0), axis parallel to the t-axis and generators making a 45° angle with the t-axis. The surface (3.3) is characteristic with respect to the wave equation and is known as the *characteristic cone with apex at* (x^0, t^0). The upper part, that is, the part for which $t \geqq t^0$, is known as the *forward characteristic cone with apex at* (x^0, t^0), while the lower part ($t \leqq t^0$) is known as the *backward characteristic cone with apex at* (x^0, t^0). (See Fig. 3.1 where the forward and backward cones are shown for $n = 2$.)

Consider now the backward characteristic cone with apex at $(x^0, t^0) \in R^{n+1}$. The surface and interior of this cone are given by the inequalities

$$(x_1 - x_1^0)^2 + \cdots + (x_n - x_n^0)^2 - (t - t^0)^2 \leqq 0, \qquad t \leqq t^0. \tag{3.4}$$

Let T be any number $\leqq t^0$ and consider the plane

$$t = T \tag{3.5}$$

in R^{n+1}. The plane (3.5) is actually the x-space R^n at $t = T$. The part of this plane cut off by the backward characteristic cone with apex at (x^0, t^0) is the

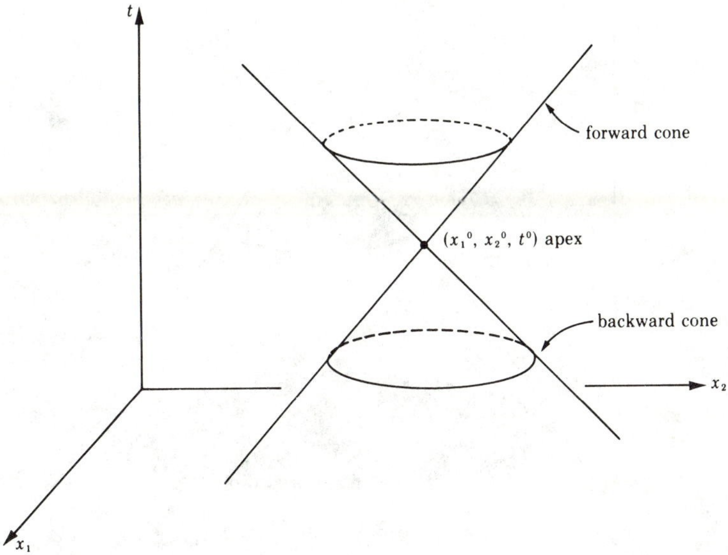

Fig. 3.1

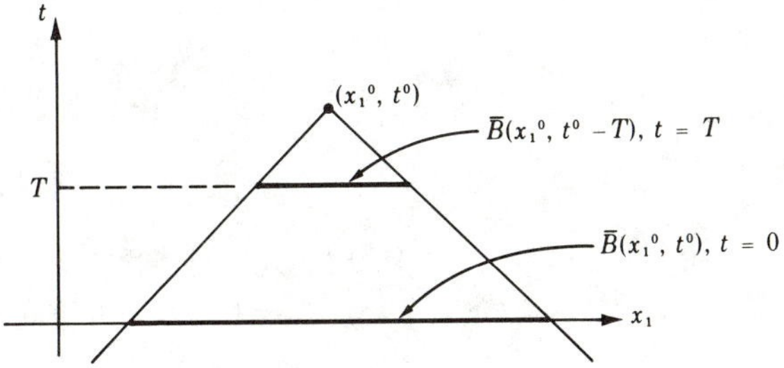

Fig. 3.2

set of points (x, t) satisfying both (3.4) and (3.5), and is therefore given by

$$(3.6) \qquad (x_1 - x_1^0)^2 + \cdots + (x_n - x_n^0)^2 \leqq (t^0 - T)^2, \qquad t = T.$$

This set is the closed ball $\bar{B}(x^0, t^0 - T)$ (with center at x^0 and radius $t^0 - T$) in the x-space R^n at $t = T$. Figures 3.2, 3.3 and 3.4 illustrate these cut-off balls for $n = 1, n = 2$ and general n, respectively. For $n = 1$, 2 and 3 the balls are, respectively, intervals, discs and solid spheres. In Figure 3.4, the x-space is represented by one axis and consequently the balls are represented by intervals.

Theorem 3.1. *(Domain of dependence inequality.)* Let (x^0, t^0) be a point in R^{n+1} with $t^0 > 0$, and let Ω be the conical domain in R^{n+1} bounded by the backward characteristic cone with apex at (x^0, t^0) and by the plane $t = 0$.

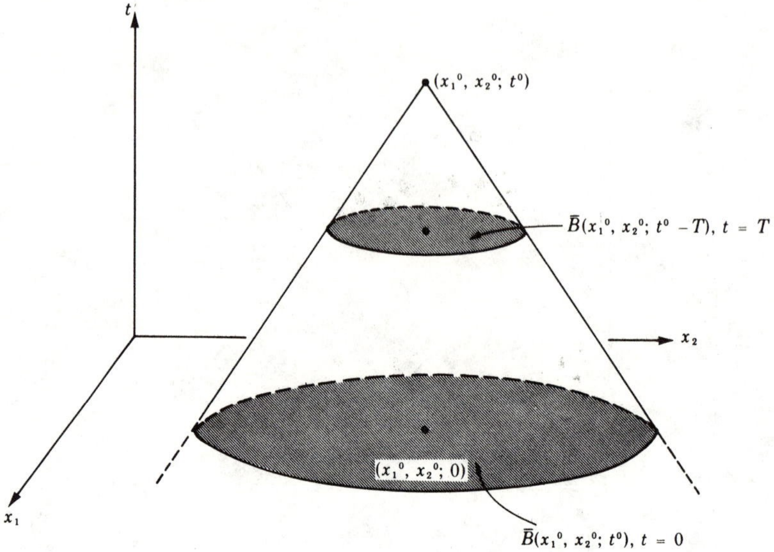
t
$(x_1^0, x_2^0; t^0)$
$\bar{B}(x_1^0, x_2^0; t^0 - T)$, $t = T$
x_2
$(x_1^0, x_2^0; 0)$
x_1
$\bar{B}(x_1^0, x_2^0; t^0)$, $t = 0$

Fig. 3.3

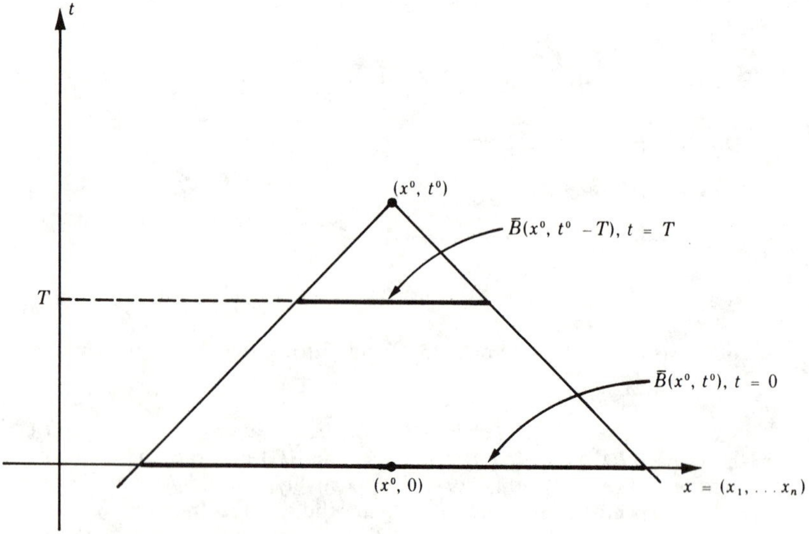
t
(x^0, t^0)
$\bar{B}(x^0, t^0 - T)$, $t = T$
T
$\bar{B}(x^0, t^0)$, $t = 0$
$(x^0, 0)$
$x = (x_1, \ldots x_n)$

Fig. 3.4

Suppose that the function u is in $C^2(\bar{\Omega})$ and satisfies the wave equation in Ω. Then for any number T, $0 \leqq T \leqq t^0$, the following inequality is satisfied:

$$(3.7)\quad \int_{\bar{B}(x^0,\, t^0-T)} (u_{x_1}^2 + \cdots + u_{x_n}^2 + u_t^2)\big|_{t=T}\, dx \leqq \int_{\bar{B}(x^0,\, t^0)} (u_{x_1}^2 + \cdots + u_{x_n}^2 + u_t^2)\big|_{t=0}\, dx.$$

Note that the regions of integration in the left and right sides of inequality (3.7) are, respectively, the balls cut off from the planes $t = T$ and $t = 0$ by the backward characteristic cone with apex at (x^0, t^0) as shown in Figures 3.2–3.4, and the corresponding integrands, which are the sums of the squares of the first derivatives of u, are evaluated, respectively, at $t = T$ and $t = 0$.

Proof of Theorem 3.1. We give the proof for $n = 2$, although the proof for general n is not really different (see Problem 3.2). The method of proof, known as the *energy method,* is based on the differential identity

$$(3.8)\quad 2u_t(u_{x_1x_1} + u_{x_2x_2} - u_{tt}) = (2u_tu_{x_1})_{x_1} + (2u_tu_{x_2})_{x_2} - (u_{x_1}^2 + u_{x_2}^2 + u_t^2)_t$$

which is easy to verify (see Problem 3.1). Let Ω_T be the part of Ω below the plane $t = T$ (see Fig. 3.5) and let C_T be the part of the backward characteristic cone with apex at $(x_1{}^0, x_2{}^0; t^0)$ which lies between the planes $t = 0$ and $t = T$. Integrating (3.8) over Ω_T and using the fact that u satisfies the wave equation in Ω, we obtain

$$(3.9)\quad 0 = \int_{\Omega_T} [(2u_tu_{x_1})_{x_1} + (2u_tu_{x_2})_{x_2} - (u_{x_1}^2 + u_{x_2}^2 + u_t^2)_t]\, dx_1dx_2dt.$$

Now the intergrand in (3.9) is the divergence of the vector field $(2u_tu_{x_1}, 2u_tu_{x_2}, -(u_{x_1}^2 + u_{x_2}^2 + u_t{}^2))$ in the (x_1, x_2, t)-space. Applying the divergence theorem to (3.9) we obtain

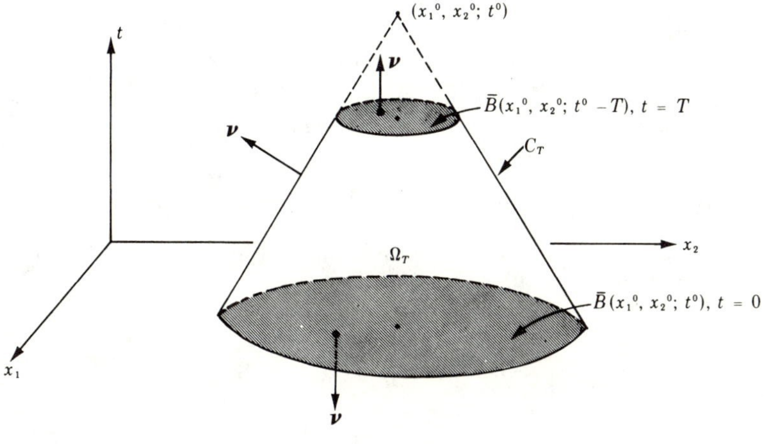

Fig. 3.5

$$(3.10)\qquad 0 = \int_{\partial\Omega_T} [2u_t u_{x_1}\nu_1 + 2u_t u_{x_2}\nu_2 - (u_{x_1}^2 + u_{x_2}^2 + u_t^2)\nu_t]\, d\sigma$$

where $\partial\Omega_T$ is the boundary of Ω_T, $d\sigma$ is the element of surface on $\partial\Omega_T$ and $\boldsymbol{\nu} = (\nu_1, \nu_2, \nu_t)$ is the unit normal vector on $\partial\Omega_T$ in the direction exterior to Ω_T. Since the boundary $\partial\Omega_T$ consists of three parts, the top and bottom discs and the conical surface C_T, we split the integral in (3.10) into the sum of three integrals. On the top disc, $x \in \bar{B}(x_1{}^0, x_2{}^0; t^0 - T)$, $t = T$ and $\boldsymbol{\nu} = (0, 0, 1)$. Hence the integral on the top disc is

$$(3.11)\qquad -\int_{\bar{B}(x_1^0,\, x_2^0;\, t^0 - T)} (u_{x_1}^2 + u_{x_2}^2 + u_t^2)\,\big|_{t=T}\, dx_1 dx_2.$$

On the bottom disc, $x \in \bar{B}(x_1{}^0, x_2{}^0; t^0)$, $t = 0$ and $\boldsymbol{\nu} = (0, 0, -1)$. Hence the integral on the bottom disc is

$$(3.12)\qquad \int_{\bar{B}(x_1^0,\, x_2^0;\, t^0)} (u_{x_1}^2 + u_{x_2}^2 + u_t^2)\,\big|_{t=0}\, dx_1 dx_2.$$

On the conical surface C_T the exterior normal $\boldsymbol{\nu} = (\nu_1, \nu_2, \nu_t)$ defines a characteristic direction and hence must satisfy the relations

$$(3.13)\qquad \nu_1^2 + \nu_2^2 = \nu_t^2, \qquad \nu_t = \frac{1}{\sqrt{2}}.$$

Hence the integral on C_T is

$$\int_{C_T} [2u_t u_{x_1}\nu_1 + 2u_t u_{x_2}\nu_2 - (u_{x_1}^2 + u_{x_2}^2 + u_t^2)\nu_t] d\sigma$$

$$= \sqrt{2}\int_{C_T} [2u_t u_{x_1}\nu_t\nu_1 + 2u_t u_{x_2}\nu_t\nu_2 - (u_{x_1}^2 + u_{x_2}^2 + u_t^2)\nu_t^2] d\sigma$$

$$= \sqrt{2}\int_{C_T} [2u_t u_{x_1}\nu_t\nu_1 + 2u_t u_{x_2}\nu_t\nu_2 - u_{x_1}^2\nu_t^2 - u_{x_2}^2\nu_t^2 - u_t^2\nu_1^2 - u_t^2\nu_2^2] d\sigma,$$

where both relations in (3.13) were used. Now, recognizing the integrand in the last integral as the negative of the sum of two squares, the integral on C_T is

$$(3.14)\qquad -\sqrt{2}\int_{C_T} [(u_{x_1}\nu_t - u_t\nu_1)^2 + (u_{x_2}\nu_t - u_t\nu_2)^2]\, d\sigma.$$

Since the integral in (3.10) is equal to the sum of the integrals (3.11), (3.12) and (3.14), equation (3.10) becomes

$$0 = -\int_{\bar{B}(x_1^0, x_2^0; t^0 - T)} (u_{x_1}^2 + u_{x_2}^2 + u_t^2)\,\big|_{t=T}\, dx_1 dx_2 + \int_{\bar{B}(x_1^0, x_2^0; t^0)} (u_{x_1}^2 + u_{x_2}^2 + u_t^2)\,\big|_{t=0}\, dx_1 dx_2 - \sqrt{2}\int_{C_T} [(u_{x_1}\nu_t - u_t\nu_1)^2 + (u_{x_2}\nu_t - u_t\nu_2)^2]\, d\sigma.$$

Finally, since the third quantity in this equation is always $\leqq 0$ (why?), we obtain the inequality

$$\int_{\bar{B}(x_1^0, x_2^0; t^0 - T)} (u_{x_1}^2 + u_{x_2}^2 + u_t^2)|_{t=T}\, dx_1 dx_2 \leqq \int_{\bar{B}(x_1^0, x_2^0; t^0)} (u_{x_1}^2 + u_{x_2}^2 + u_t^2)|_{t=0}\, dx_1 dx_2 \tag{3.15}$$

which is inequality (3.7) for $n = 2$.

Problems

3.1. The differential identity (3.8) expresses the product of $2u_t$ and the left side of the wave equation as the divergence of a vector field in (x_1, x_2, t)-space.

(a) Verify (3.8) by performing the indicated differentiations on the right side to obtain the left side.

(b) Derive (3.8) by expressing each term on the left side as the difference of two derivatives [*Hint:* $2u_t u_{x_1 x_1} = (2u_t u_{x_1})_{x_1} - 2u_{tx_1} u_{x_1} = (2u_t u_{x_1})_{x_1} - (u_{x_1}^2)_t$.]

3.2. Prove Theorem 3.1 for general n. First verify the differential identity corresponding to (3.8) for general n,

$$\begin{aligned} &2u_t(u_{x_1 x_1} + \cdots + u_{x_n x_n} - u_{tt}) \\ &\quad = (2u_t u_{x_1})_{x_1} + \cdots + (2u_t u_{x_n})_{x_n} - (u_{x_1}^2 + \cdots + u_{x_n}^2 + u_t^2)_t. \end{aligned} \tag{3.16}$$

3.3. Let ∇ stand for the "differentiation vector" in the x-space only,

$$\nabla = \left(\frac{\partial}{\partial x_1}, \ldots, \frac{\partial}{\partial x_n}\right)$$

(a) Show that the differential identity (3.16) can be written in the form

$$2u_t(\nabla^2 u - u_{tt}) = \nabla\cdot(2u_t \nabla u) - (|\nabla u|^2 + u_t^2)_t. \tag{3.17}$$

(b) Show that the domain of dependence inequality (3.7) can be written in the form

$$\int_{\bar{B}(x^0, t^0 - T)} (|\nabla u|^2 + u_t^2)|_{t=T}\, dx \leqq \int_{\bar{B}(x^0, t^0)} (|\nabla u|^2 + u_t^2)\,|_{t=0}\, dx. \tag{3.18}$$

3.4. The p.d.e.

$$\frac{\partial^2 u}{\partial x_1^2} + \cdots + \frac{\partial^2 u}{\partial x_n^2} - \frac{\partial^2 u}{\partial t^2} - q(x)u = 0 \tag{3.19}$$

arises in the study of wave propagation in a nonhomogeneous medium. The function $q(x)$ depends only on the space variables and is non-negative. Under the assumptions of Theorem 3.1, with u satisfying (3.19) instead of the wave equation, derive the domain of dependence inequality

$$\begin{aligned}(3.20)\qquad &\int_{\bar{B}(x^0,t^0-T)} (u_{x_1}^2 + \cdots + u_{x_n}^2 + u_t^2 + qu^2)\,|_{t=T}\, dx \\ &\leqq \int_{\bar{B}(x^0,t^0)} (u_{x_1}^2 + \ldots + u_{x_n}^2 + u_t^2 + qu^2)\,|_{t=0}\, dx.\end{aligned}$$

4. Uniqueness in the Initial Value Problem. Domain of Dependence and Range of Influence. Conservation of Energy

In this section we apply the domain of dependence inequality to the question of uniqueness in the initial value problem,

$$(4.1)\qquad \frac{\partial^2 u}{\partial x_1^2} + \cdots + \frac{\partial^2 u}{\partial x_n^2} - \frac{\partial^2 u}{\partial t^2} = 0, \qquad x \in R^n, t > 0,$$

$$(4.2)\qquad u(x, 0) = \phi(x),\ u_t(x, 0) = \psi(x), \qquad x \in R^n.$$

Theorem 4.1. Let (x^0, t^0) be a point in R^{n+1} with $t^0 > 0$ and let Ω be the conical domain in R^{n+1} bounded by the backward characteristic cone with apex at (x^0, t^0) and by the plane $t = 0$. Let u be a function in $C^2(\bar{\Omega})$ satisfying the wave equation in Ω and suppose that u and u_t vanish on the base of Ω, i.e.

$$u(x, 0) = u_t(x, 0) = 0 \text{ for } x \in \bar{B}(x^0, t^0).$$

Then u vanishes in $\bar{\Omega}$.

Proof. Since $u(x, 0) = 0$ for $x \in \bar{B}(x^0, t^0)$, we also have $u_{x_1}(x, 0) = \cdots = u_{x_n}(x, 0) = 0$ for $x \in \bar{B}(x^0, t^0)$. Hence the integral on the right side of the domain of dependence inequality (3.7) is equal to zero. Since the integral on the left side is non-negative, it must also be equal to zero, i.e.

$$(4.3)\qquad \int_{\bar{B}(x^0, t^0-T)} (u_{x_1}^2 + \cdots + u_{x_n}^2 + u_t^2)\,|_{t=T}\, dx = 0$$

for every T, $0 \leqq T \leqq t^0$. Now the integrand in (4.3) is continuous and non-negative and hence it must be equal to zero. We conclude that

$$u_{x_1}(x, T) = \cdots = u_{x_n}(x, T) = u_t(x, T) = 0$$

for $x \in \bar{B}(x^0, t^0 - T)$, $0 \leqq T \leqq t^0$. This means that all first order partial derivatives of $u(x, t)$ vanish in Ω and hence u must be constant in Ω. By continuity, u must be constant in $\bar{\Omega}$ and since $u = 0$ on the base of $\bar{\Omega}$, $u = 0$ everywhere in $\bar{\Omega}$.

Theorem 4.1 yields immediately the following uniqueness result.

Corollary 4.1. Let the point (x^0, t^0) and the conical domain Ω be as described in Theorem 4.1. Suppose that there are two functions u_1 and u_2 which are in $C^2(\bar{\Omega})$, satisfy the wave equation in Ω and on the base of Ω, $u_1 = u_2$ and $\partial u_1/\partial t = \partial u_2/\partial t$. Then $u_1 \equiv u_2$ in $\bar{\Omega}$.

Proof. Let $\tilde{u} = u_1 - u_2$. It is easy to check that $\tilde{u}$ satisfies all the assumptions of Theorem 4.1. Hence $\tilde{u} \equiv 0$ in $\bar{\Omega}$, or $u_1 \equiv u_2$ in $\bar{\Omega}$.

Corollary 4.2. Let u_1 and u_2 be solutions of the initial value problem (4.1), (4.2) which are in C^2 for $x \in R^n$ and $t \geqq 0$. Then $u_1 \equiv u_2$.

While Corollary 4.2 asserts the uniqueness of solutions of the initial value problem, Corollary 4.1 gives the more precise information that knowing u and u_t only on the base of the conical domain Ω uniquely determines u in $\bar{\Omega}$. In particular, the values of u and u_t on the base of Ω uniquely determine the value of u at the apex of Ω. We state this precisely in the following corollary.

Corollary 4.3. Let the point (x^0, t^0) and the conical domain Ω be as described in Theorem 4.1 and let u be a function in $C^2(\bar{\Omega})$ satisfying the wave equation in Ω. Then the value of u at (x^0, t^0) is uniquely determined by the values of u and u_t on the base of Ω.

According to Corollary 4.3, the value of the solution u of the initial value problem (4.1), (4.2) at a point (x^0, t^0) (with $t^0 > 0$) is uniquely determined by the values of the initial data ϕ and ψ on the part of the initial surface $t = 0$ cut off by the backward characteristic cone with apex at (x^0, t^0). For this reason, this part of the initial surface is known as the *domain of dependence* of the solution at the point (x^0, t^0). The values of the initial data outside the domain of dependence do not affect the value of the solution at (x^0, t^0). As we saw in Section 3, the domain of dependence of the solution at (x^0, t^0) is the closed ball $\bar{B}(x^0, t^0)$ at $t = 0$ (see Figs. 3.2–3.4).

Let us now change our point of view and consider a point $(x^1, 0)$ on the initial surface $t = 0$ and ask the following question: At what points of the upper half (x, t)-space $t > 0$ is the value of the solution of the initial value problem influenced by the values of the initial data at $(x^1, 0)$? To study this question let us consider the region R consisting of all points which lie on or inside the forward characteristic cone with apex at $(x^1, 0)$. Then it is easy to see that the domain of dependence of the solution at any point of the upper half (x, t)-space $t > 0$ which is outside of R does not contain the point $(x^1, 0)$. Consequently, the value of the solution at points outside of R is not influenced by the values of the initial data at $(x^1, 0)$. It follows that the value of the solution may be influenced by the values of the initial data at $(x^1, 0)$ only at points of R. For this reason R is known as the *range of influence* of the initial data at the point $(x^1, 0)$. Figure 4.1 illustrates the domain of dependence and range of influence for $n = 2$.

Since we already know that the solution of the initial value problem for the one-dimensional wave equation is given by formula (2.10), the results of this section for $n = 1$ could have been obtained by direct inspection of this formula. Indeed the value of the solution at (x^0, t^0) is given by

$$\text{(4.4)} \quad u(x^0, t^0) = \frac{1}{2}[\phi(x^0 + t^0) + \phi(x^0 - t^0)] + \frac{1}{2}\int_{x^0-t^0}^{x^0+t^0} \psi(\tau)d\tau$$

$$= \frac{1}{2}[u(x^0 + t^0, 0) + u(x^0 - t^0, 0)] + \frac{1}{2}\int_{x^0-t^0}^{x^0+t^0} \frac{\partial u}{\partial t}(\tau, 0)d\tau.$$

The conical domain Ω is, in this case, the triangle PAB shown in Figure 4.2

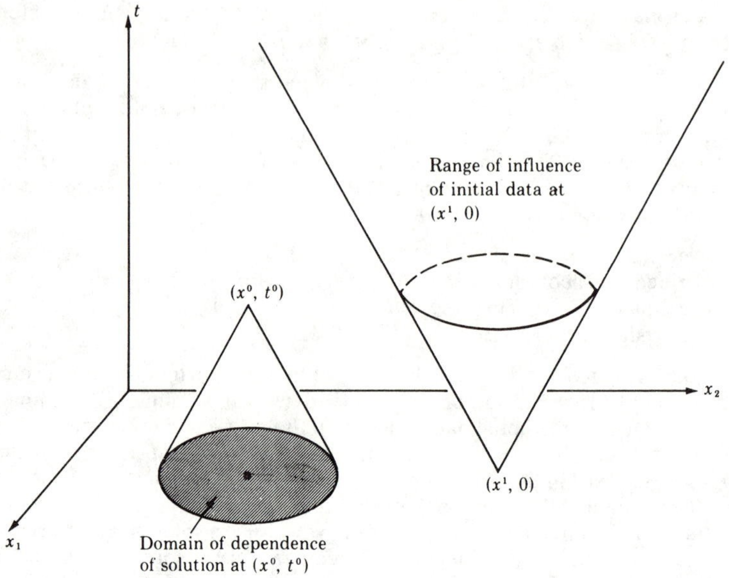

Fig. 4.1

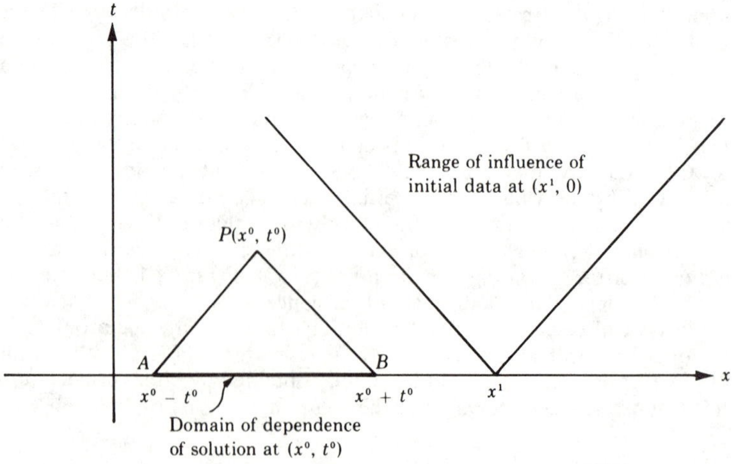

Fig. 4.2

and the base of Ω is the interval AB on the x-axis from $x^0 - t^0$ to $x^0 + t^0$. It is clear from (4.4) that the value of u at (x^0, t^0) depends only on the values of u and u_t on the interval AB and this interval is the domain of dependence of the solution at (x^0, t^0). It is also easy to see that the range of influence of the data at a point $(x^1, 0)$ on the initial line $t = 0$ is the triangular region described by

$$|x - x^1| \leqq t, t \geqq 0,$$

and shown in Figure 4.2.

We conclude this section with a discussion of an important law of wave propagation known as conservation of energy. If D is a region in the x-space R^n, the integral

$$E(u; D, T) = \frac{1}{2}\int_D (u_{x_1}^2 + \cdots + u_{x_n}^2 + u_t^2)\,|_{t=T}\, dx \tag{4.5}$$

is known as the *energy* of $u(x, t)$ contained in the region D at time $t = T$. The reason for the use of the word energy is that if a solution $u(x, t)$ of the wave equation describes a wave propagation phenomenon, the integral in (4.5) is precisely what physicists call the energy of the wave in the region D at time $t = T$. The domain of dependence inequality (3.7) is known as an energy inequality and the method of its derivation is known as the energy method because the integrals involved are energies. In fact, (3.7) can be written as

$$E(u; \bar{B}(x^0, t^0 - T), T) \leqq E(u; \bar{B}(x^0, t^0), 0). \tag{4.6}$$

Theorem 4.2. *(Conservation of energy.)* If u is the solution of the initial value problem (4.1), (4.2) which is in C^2 for $x \in R^n$ and $t \geqq 0$, and if the initial data ϕ and ψ vanish outside some ball $B(0, R)$ in R^n, then the energy of u contained in the whole space R^n remains constant (independent of t); i.e., for every $T \geqq 0$,

$$\frac{1}{2}\int_{R^n} (u_{x_1}^2 + \cdots + u_{x_n}^2 + u_t^2)\,|_{t=T}\, dx = \frac{1}{2}\int_{R^n} (u_{x_1}^2 + \cdots + u_{x_n}^2 + u_t^2)\,|_{t=0}\, dx \tag{4.7}$$

or

$$E(u; R^n, T) = E(u; R^n, 0). \tag{4.8}$$

The proof of this theorem is left for the problems.

Problems

4.1. Suppose that u is a C^2 solution of the wave equation (4.1) and that at $t = 0$, $u = u_t = 0$ in a ball of radius R. Up to what time t can you be sure that $u = 0$ at the center of the ball?

4.2. Suppose that u is a C^2 solution of the two-dimensional wave equation and that at $t = 0$, $u = u_t = 0$ outside the disc $x_1^2 + x_2^2 \leqq 1$. Up to what time can you be sure that $u = 0$ at the points (a) $(x_1, x_2) = (5, 0)$, (b) $(x_1, x_2) = (0, 10)$, and (c) $(x_1, x_2) = (2, 3)$?

4.3. Let (x^0, t^0) be a point in R^{n+1}. Suppose that u satisfies the wave equation inside the backward characteristic cone C with apex at (x^0, t^0) and u is in C^2 on and inside C. Show that for any T, $T < t^0$, the value of u at (x^0, t^0) is uniquely determined by the values of u and u_t on the part of the plane $t = T$ cut off by C.

4.4. Consider the initial value problem (4.1)-(4.2) for $n = 2$. Let D be a region in the initial plane $t = 0$ and let D' be the part of the upper half (x, t)-space $t > 0$ filled by cones with axes parallel to the t-axis, generators making a 45° angle with the t-axis and bases contained in the region D. Then, according to Corollary 4.3, the values of the initial data in the region D uniquely determine the value of u in D'. What is D' if D is a square?

4.5. Show that if u is the solution of the initial value problem (4.1), (4.2) and if the initial data ϕ and ψ vanish outside the ball $\bar{B}(0, R)$ in R^n, then $u(x, t)$ must vanish outside the set of points defined by the inequalities

$$r \leqq R + t, t \geqq 0, \tag{4.9}$$

where $r = |x|$. The set (4.7) may be called the range of influence of the data in the ball $\bar{B}(0, R)$. Sketch the set (4.9) for $n = 1$ and $n = 2$.

4.6. Prove Theorem 4.2 for $n = 2$. [*Hint:* Use the energy method. Integrate the differential identity (3.8) over the cylindrical region

$$r \leqq R + 2T, 0 \leqq t \leqq T.$$

Use the divergence theorem and the assertion in Problem 4.5 to obtain the equality (4.7).]

4.7. The p.d.e.

$$\frac{\partial^2 u}{\partial x_1^2} + \cdots + \frac{\partial^2 u}{\partial x_n^2} - \frac{\partial^2 u}{\partial t^2} = f(x, t) \tag{4.10}$$

describes the forced vibrations of a homogeneous n-dimensional body. Except for a constant multiplicative factor, the function $f(x, t)$ describes the density of the externally applied force at the point x of the body at time t. Prove the uniqueness of solution of the initial value problem for the equation (4.10) with initial conditions (4.2). Specifically, prove Corollaries 4.1 and 4.2 with equation (4.10) replacing the wave equation (4.1).

4.8. Do Problem 4.7 with the equation

$$\frac{\partial^2 u}{\partial x_1^2} + \cdots + \frac{\partial^2 u}{\partial x_n^2} - \frac{\partial^2 u}{\partial t^2} - q(x)u = f(x, t) \tag{4.11}$$

instead of equation (4.10). [*Hint:* Use Problem 3.4.]

5. Solution of the Initial Value Problem. Kirchhoff's Formula. The Method of Descent

In this section we will establish the existence of solution of the initial value problem for the wave equation in two and three space variables. We will do this by writing down a formula for $u(x, t)$ and verifying by direct computation that it satisfies the wave equation and the initial conditions. Since we have already proved the uniqueness of solution, this formula must then give the solution of the problem. We will not go into the method of derivation of the formula for the solution.

We first consider the initial value problem for the three-dimensional wave equation,

$$\frac{\partial^2 u}{\partial x_1^2} + \frac{\partial^2 u}{\partial x_2^2} + \frac{\partial^2 u}{\partial x_3^2} - \frac{\partial^2 u}{\partial t^2} = 0; \qquad x \in R^3, \qquad t > 0 \tag{5.1}$$

$$u(x, 0) = \phi(x), \qquad x \in R^3, \tag{5.2}$$

$$\frac{\partial u}{\partial t}(x, 0) = \psi(x), \qquad x \in R^3. \tag{5.3}$$

We are looking for the solution $u(x, t)$ of this problem which is in C^2 for $x \in R^3$ and $t \geqq 0$. As we will see from the formula for the solution, u will satisfy this smoothness requirement provided that the initial data ϕ and ψ satisfy the smoothness conditions $\phi \in C^3(R^3)$ and $\psi \in C^2(R^3)$. From now on we assume that ϕ and ψ satisfy these conditions.

First we state a lemma which reduces the problem of finding the solution of (5.1), (5.2), (5.3) to the problem of finding the solution of (5.1) satisfying the special initial conditions,

$$u(x, 0) = 0, \qquad x \in R^3 \tag{5.4}$$

$$\frac{\partial u}{\partial t}(x, 0) = p(x), \qquad x \in R^3. \tag{5.5}$$

Lemma 5.1. Let u_p be the solution of the initial value problem (5.1), (5.4), (5.5) and assume that u_p is in C^3 for $x \in R^3$ and $t \geqq 0$. Then $v = \partial u_p/\partial t$ satisfies (5.1) and the initial conditions,

$$u(x, 0) = p(x), \qquad x \in R^3, \tag{5.6}$$

$$\frac{\partial u}{\partial t}(x, 0) = 0, \qquad x \in R^3, \tag{5.7}$$

and v is in C^2 for $x \in R^3$ and $t \geqq 0$.

Proof. That v satisfies (5.1) follows from the more general and obvious fact that any derivative of a solution of a homogeneous p.d.e. with constant coefficients is also a solution of the equation. Now, since u_p satisfies (5.5),

$$v(x, 0) = \frac{\partial}{\partial t} u_p(x, 0) = p(x).$$

Moreover,

$$\frac{\partial v}{\partial t}(x, 0) = \frac{\partial^2 u_p}{\partial t^2}(x, 0) = \left(\frac{\partial^2 u_p}{\partial x_1^2} + \frac{\partial^2 u_p}{\partial x_2^2} + \frac{\partial^2 u_p}{\partial x_3^2}\right)\Big|_{t=0} = 0.$$

Here we used the fact that u_p satisfies (5.1) and the fact that u_p vanishes for $t = 0$; hence all its space derivatives also vanish at $t = 0$.

Lemma 5.1 and the principle of superposition yield immediately

Lemma 5.2. In the notation of Lemma 5.1, the solution of the initial value problem (5.1), (5.2), (5.3), which is in C^2 for $x \in R^3$ and $t \geqq 0$, is given by

$$u = \frac{\partial u_\phi}{\partial t} + u_\psi \tag{5.8}$$

provided that u_ϕ is in C^3 and u_ψ is in C^2 for $x \in R^3$ and $t \geqq 0$.

According to Lemma 5.2, in order to solve the initial value problem (5.1), (5.2), (5.3), it is enough to know the solution of the special initial value problem (5.1), (5.4), (5.5).

Lemma 5.3. Suppose that $p \in C^k(R^3)$ where k is any integer $\geqq 2$. Then the solution of the initial value problem (5.1), (5.4), (5.5) is given by the formula

$$u_p(x, t) = \frac{1}{4\pi t} \int_{S(x,\, t)} p(x')d\sigma_t \tag{5.9}$$

and u_p is in C^k for $x \in R^3$ and $t \geqq 0$.

Formula (5.9) is known as *Kirchhoff's formula*. In (5.9), $S(x, t)$ denotes the surface of the sphere in R^3 with center at the point x and radius t, $d\sigma_t$ is the element of surface on $S(x, t)$ and x' is the variable point of integration on $S(x, t)$. It is sometimes useful to rewrite formula (5.9) by introducing new variables of integration. We set

$$x' = x + t\alpha$$

or

$$x_1' = x_1 + t\alpha_1,\ x_2' = x_2 + t\alpha_2,\ x_3' = x_3 + t\alpha_3$$

where $\alpha = (\alpha_1, \alpha_2, \alpha_3)$ is a unit vector in the direction of $x' - x$ (see Fig. 5.1). As x' varies over the sphere $S(x, t)$, α varies over the unit sphere $S(0, 1)$. Denoting by $d\sigma_1$ the element of surface on $S(0, 1)$ and using the fact that $d\sigma_t = t^2 d\sigma_1$, Kirchhoff's formula (5.9) takes the form

$$u_p(x, t) = \frac{t}{4\pi} \int_{S(0,1)} p(x + t\alpha)d\sigma_1. \tag{5.10}$$

Remember that in the integral in (5.10) the variable point of integration is α. Still another way of writing Kirchhoff's formula is in terms of the mean value (average) $M[p, S(x, t)]$ of the function p over the sphere $S(x, t)$,

$$M[p, S(x, t)] = \frac{1}{4\pi t^2} \int_{S(x,t)} p(x')d\sigma_t = \frac{1}{4\pi} \int_{S(0,1)} p(x + t\alpha)d\sigma_1. \tag{5.11}$$

Kirchhoff's formula then becomes

$$u_p(x, t) = tM[p, S(x, t)]. \tag{5.12}$$

Proof of Lemma 5.3. First we prove that u_p satisfies the initial condi-

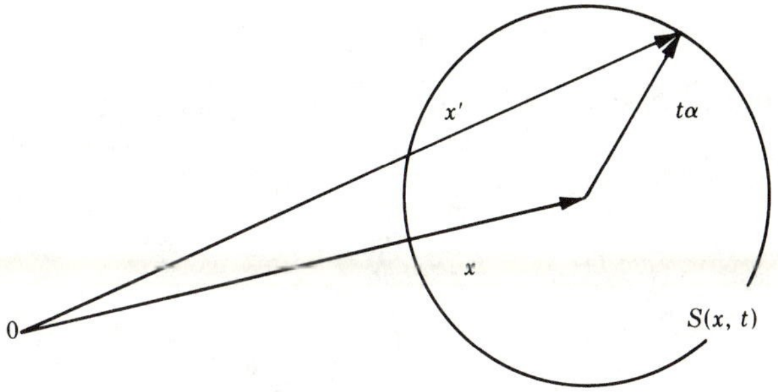

Fig. 5.1

tions (5.4) and (5.5). Using the continuity of p it is easy to show (Problem 5.1) that

$$\lim_{t\to 0^+} M[p, S(x, t)] = p(x). \tag{5.13}$$

Hence, it follows from (5.12) that

$$u_p(x, t) \to 0 \text{ as } t \to 0^+$$

and condition (5.4) is satisfied. Next, differentiating (5.10) with respect to t we obtain

$$\frac{\partial u_p}{\partial t}(x, t) = \frac{1}{4\pi}\int_{S(0,1)} p(x + t\alpha)d\sigma_1 + \frac{t}{4\pi}\int_{S(0,1)} \nabla p(x + t\alpha)\cdot\alpha d\sigma_1. \tag{5.14}$$

Here we used the chain rule,

$$\frac{\partial}{\partial t} p(x + t\alpha) = \frac{\partial}{\partial t} p(x_1 + t\alpha_1, x_2 + t\alpha_2, x_3 + t\alpha_3)$$

$$= D_1 p(x + t\alpha)\alpha_1 + D_2 p(x + t\alpha)\alpha_2 + D_3 p(x + t\alpha)\alpha_3 = \nabla p(x + t\alpha)\cdot\alpha$$

where $D_i p$ denotes the partial derivative of p with respect to its ith variable. By (5.13) the first term in (5.14) approaches $p(x)$ as $t \to 0^+$. Since the derivatives of p are continuous, the integral in the second term of (5.14) remains bounded as $t \to 0^+$ and hence the second term in (5.14) approaches 0 as $t \to 0^+$. Thus

$$\frac{\partial u_p}{\partial t}(x, t) \to p(x) \text{ as } t \to 0^+$$

and the initial condition (5.5) is satisfied.

In order to prove that u_p satisfies the wave equation we rewrite equation (5.14) as follows,

$$\frac{\partial u_p}{\partial t}(x, t) = \frac{1}{t} u_p(x, t) + \frac{1}{4\pi t}\int_{S(x,\, t)} \nabla p(x')\cdot\alpha d\sigma_t.$$

Now, using the fact that α is the exterior unit normal vector to $S(x, t)$ and applying the Divergence Theorem, we obtain

$$(5.15)\qquad \frac{\partial u_p}{\partial t}(x, t) = \frac{1}{t}\, u_p(x, t) + \frac{1}{4\pi t}\int_{B(x,\, t)} \nabla^2 p(x')dx'$$

where $B(x, t)$ is the ball in R^3 with center at x and radius t, x' is the variable point of integration in $B(x, t)$ and dx' is the element of volume. Differentiation of (5.15) with respect to t yields

$$\frac{\partial^2 u_p}{\partial t^2}(x, t) = -\frac{1}{t^2}\, u_p(x, t) + \frac{1}{t}\,\frac{\partial u_p}{\partial t}(x, t)$$
$$-\frac{1}{4\pi t^2}\int_{B(x,t)} \nabla^2 p(x')dx' + \frac{1}{4\pi t}\,\frac{\partial}{\partial t}\int_{B(x,t)} \nabla^2 p(x')dx'.$$

Substitution from (5.15) for $\partial u_p/\partial t$ and simplification yields

$$\frac{\partial^2 u_p}{\partial t^2}(x, t) = \frac{1}{4\pi t}\,\frac{\partial}{\partial t}\int_{B(x,t)} \nabla^2 p(x')dx'.$$

Now, it is an easy exercise to compute the derivative on the right side of this equation (Problem 5.2) and obtain

$$\frac{\partial^2 u_p}{\partial t^2}(x, t) = \frac{1}{4\pi t}\int_{S(x,t)} \nabla^2 p(x')d\sigma_t,$$

or

$$(5.16)\qquad \frac{\partial^2 u_p}{\partial t^2}(x, t) = \frac{t}{4\pi}\int_{S(0,1)} \nabla^2 p(x + t\alpha)d\sigma_1.$$

Finally, applying the Laplacian with respect to x on equation (5.10) we obtain

$$(5.17)\qquad \nabla^2 u_p(x, t) = \frac{t}{4\pi}\int_{S(0,1)} \nabla^2 p(x + t\alpha)d\sigma_1.$$

Comparing (5.16) and (5.17) we see that u_p satisfies the wave equation (5.1).

The assertion of the lemma concerning the smoothness of u_p follows easily from a study of the formula (5.10). The proof of the lemma is now complete.

The solution of the initial value problem (5.1), (5.2), (5.3) is now obtained by combining Lemmas 5.2 and 5.3.

Theorem 5.1. Suppose that $\phi \in C^3(R^3)$ and $\psi \in C^2(R^3)$. Then the solution of the initial value problem (5.1), (5.2), (5.3) for the three-dimensional wave equation is given by

$$(5.18)\qquad u(x, t) = \frac{1}{4\pi t}\int_{S(x,t)} \psi(x')d\sigma_t + \frac{\partial}{\partial t}\left[\frac{1}{4\pi t}\int_{S(x,t)} \phi(x')d\sigma_t\right]$$

and the solution is in C^2 for $x \in R^3$ and $t \geqq 0$.

We consider next the initial value problem for the two-dimensional wave equation

$$\frac{\partial^2 u}{\partial x_1^2} + \frac{\partial^2 u}{\partial x_2^2} - \frac{\partial^2 u}{\partial t^2} = 0; \qquad x \in R^2, \qquad t > 0 \tag{5.19}$$

$$u(x, 0) = \phi(x), \qquad x \in R^2, \tag{5.20}$$

$$\frac{\partial u}{\partial t}(x, 0) = \psi(x), \qquad x \in R^2. \tag{5.21}$$

The solution of this problem can be obtained from the solution of the three-dimensional problem by a method known as the *method of descent.* This method consists of regarding the initial data $\phi(x) = \phi(x_1, x_2)$ and $\psi(x) = \psi(x_1, x_2)$ as functions defined in the three dimensional space R^3 which do not depend on the third variable x_3. If these functions are substituted into Kirchhoff's formula, they yield solutions of the three-dimensional wave equation which do not depend on the variable x_3, and consequently they are actually solutions of the two-dimensional wave equation (5.19). If $p(x) = p(x_1, x_2)$ in Kirchhoff's formula (5.9), it can be shown (Problem 5.3) that

$$\begin{aligned} u_p(x, t) &= \frac{1}{4\pi t} \int_{S(x_1, x_2, x_3; t)} p(x_1', x_2') d\sigma_t \\ &= \frac{1}{4\pi t} \int_{S(x_1, x_2, 0; t)} p(x_1', x_2') d\sigma_t \\ &= \frac{1}{2\pi} \int_{\bar{B}(x_1, x_2; t)} \frac{p(x_1', x_2')}{\sqrt{t^2 - (x_1' - x_1)^2 - (x_2' - x_2)^2}}\, dx_1'\, dx_2' \end{aligned} \tag{5.22}$$

and $u_p(x, t) = u_p(x_1, x_2; t)$. In (5.22) $\bar{B}(x_1, x_2; t)$ is the ball (disc) in R^2 with center at (x_1, x_2) and radius t. Thus, by "descending" from three dimensions to two dimensions we obtain the solution of the two-dimensional initial value problem (5.19), (5.20), (5.21) given in the following theorem.

Theorem 5.2. Suppose that $\phi \in C^3(R^2)$ and $\psi \in C^2(R^2)$. Then the solution of the initial value problem (5.19), (5.20), (5.21) for the two-dimensional wave equation is given by

$$\begin{aligned} u(x_1, x_2; t) = &\frac{1}{2\pi} \int_{\bar{B}(x_1, x_2; t)} \frac{\psi(x_1', x_2')}{\sqrt{t^2 - (x_1' - x_1)^2 - (x_2' - x_2)^2}}\, dx_1'\, dx_2' \\ &+ \frac{\partial}{\partial t}\left[\frac{1}{2\pi} \int_{\bar{B}(x_1, x_2; t)} \frac{\phi(x_1', x_2')}{\sqrt{t^2 - (x_1' - x_1)^2 - (x_2' - x_2)^2}}\, dx_1'\, dx_2'\right] \end{aligned} \tag{5.23}$$

and the solution is in C^2 for $(x_1, x_2) \in R^2$ and $t \geqq 0$.

The method of descent can be also used to obtain the solution of the one-dimensional initial value problem

$$\frac{\partial^2 u}{\partial x_1^2} - \frac{\partial^2 u}{\partial t^2} = 0; \qquad x_1 \in R^1, \qquad t > 0, \tag{5.24}$$

$$u(x_1, 0) = \phi(x_1), \qquad x_1 \in R^1, \tag{5.25}$$

$$\frac{\partial u}{\partial t}(x_1, 0) = \psi(x_1), \qquad x_1 \in R^1, \tag{5.26}$$

from the solution of the three-dimensional problem. In this case we must "descend two dimensions." If $p(x) = p(x_1)$ in Kirchhoff's formula (5.9), it can be shown (Problem 5.4) that

$$
\begin{aligned}
u_p(x, t) &= \frac{1}{4\pi t} \int_{S(x_1, x_2, x_3; t)} p(x_1')\,d\sigma_t \\
&= \frac{1}{4\pi t} \int_{S(x_1, 0, 0; t)} p(x_1)\,d\sigma_t = \frac{1}{2} \int_{x_1 - t}^{x_1 + t} p(x_1')\,dx_1'
\end{aligned}
\tag{5.27}
$$

and $u_p(x, t) = u_p(x_1, t)$.

Theorem 5.3. Suppose that $\phi \in C^2(R^1)$ and $\psi \in C^1(R^1)$. Then the solution of the initial value problem (5.24), (5.25), (5.26) for the one-dimensional wave equation is given by

$$
\begin{aligned}
u(x_1, t) &= \frac{1}{2} \int_{x_1 - t}^{x_1 + t} \psi(x_1')\,dx_1' + \frac{\partial}{\partial t}\left[\frac{1}{2} \int_{x_1 - t}^{x_1 + t} \phi(x_1')\,dx_1'\right] \\
&= \frac{1}{2} \int_{x_1 - t}^{x_1 + t} \psi(x_1')\,dx_1' + \frac{1}{2}[\phi(x_1 + t) + \phi(x_1 - t)]
\end{aligned}
\tag{5.28}
$$

and the solution is in C^2 for $x_1 \in R^1$ and $t \geqq 0$.

The solution (5.28) is, of course, the solution that we obtained earlier for the one-dimensional problem.

Problems

5.1. Prove (5.13) for any continuous function p.

5.2. Show that

$$\frac{\partial}{\partial t} \int_{B(x, t)} \nabla^2 p(x')\,dx' = \int_{S(x, t)} \nabla^2 p(x')\,d\sigma_t.$$

[*Hint:* Introduce spherical coordinates.]

5.3. Derive (5.22). [*Hint:* Project $d\sigma_t$ onto the (x_1', x_2')-plane and derive the relation

$$d\sigma_t = \frac{t}{\sqrt{t^2 - (x_1' - x_1)^2 - (x_2' - x_2)^2}}\,dx_1'\,dx_2'.]$$

5.4. Derive (5.27).

5.5. Consider the initial value problem

$$\frac{\partial^2 u}{\partial x_1^2} + \cdots + \frac{\partial^2 u}{\partial x_n^2} - \frac{1}{c^2}\frac{\partial^2 u}{\partial t^2} = 0; \qquad x \in R^n, t > 0$$

$$u(x, 0) = \phi(x), \qquad \frac{\partial u}{\partial t}(x, 0) = \psi(x), \; x \in R^n.$$

Write down the formulas for the solution of this problem for $n = 1, 2, 3$.

6. Discussion of the Solution of the Initial Value Problem. Huygens' Principle. Diffusion of Waves

First, let us complete the proof that the initial value problem is well-posed. It remains to show that the solution depends continuously on the initial data. This follows easily from an examination of the formulas for the solution. For $n = 1$, the solution is given by (5.28) and involves the initial data and integrals of the initial data. Hence, if the data change by a small amount, the solution will also change by a small amount. For $n = 2$ and $n = 3$ the solution is given by (5.23) and (5.18), respectively, and involves integrals and time derivatives of integrals of the initial data. Hence, if the data and the first order derivatives of the data change by a small amount, the solution will also change by a small amount. Note that for $n = 2$ and $n = 3$, the data as well as their first order derivatives are required to change by a small amount. For $n > 3$, derivatives of higher order (depending on n) of the data must be required to change by a small amount.

Next, let us recall the result obtained in Section 4 concerning the domain of dependence of the solution of the initial value problem at the point (x, t): $u(x,t)$ depends on the values of the initial data on the part of the initial surface $t = 0$ cut off by the backward characteristic cone with apex at (x, t). This part of the initial surface is the closed ball $\bar{B}(x, t)$ in the x-space R^n with center at x and radius t. An examination of the formulas (5.18), (5.23) and (5.28) shows that indeed, for $n = 1, 2, 3$, the value of $u(x, t)$ depends on the values of the initial data in $\bar{B}(x, t)$. However a striking phenomenon occurs in the case $n = 3$: $u(x, t)$ depends only on the data (and their derivatives) over the boundary $S(x, t)$ of the ball $\bar{B}(x, t)$. This phenomenon was first discovered by Huygens and is known as *Huygens' principle.* While formula (5.18) shows that Huygens' principle holds for $n = 3$, formulas (5.23) and (5.28) show that Huygens' principle does not hold for $n = 1$ and $n = 2$. Indeed, for $n = 1$ and $n = 2$, $u(x, t)$ depends on the values of the data

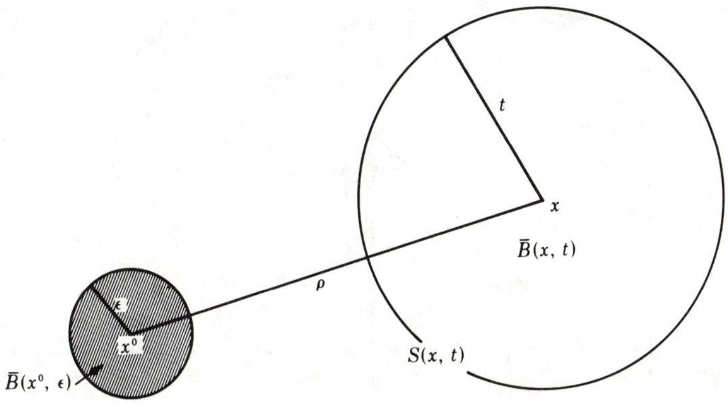

Fig. 6.1. x-space R^n at time t

over the whole ball $\bar{B}(x, t)$. In general it can be shown that Huygens' principle holds for every odd $n \geqq 3$ and does not hold for any even n.

In order to understand the implications of Huygens' principle let us consider the initial value problem for the n-dimensional wave equation with initial data vanishing everywhere except in a small ball $\bar{B}(x^0, \epsilon)$ with center at the point x^0 of R^n and radius ϵ. Let x be a fixed point in R^n outside of $\bar{B}(x^0, \epsilon)$ and let us study the values of $u(x, t)$ for $t \geqq 0$ (see Fig. 6.1). If $n = 3$ (or n is odd and $\geqq 3$), Huygens' principle holds and $u(x, t)$ is given in terms of integrals of the data and their derivatives over $S(x, t)$. Therefore $u(x, t) = 0$ for all t for which $S(x, t)$ does not intersect $\bar{B}(x^0, \epsilon)$. If ρ is the distance between x and x^0, $S(x, t)$intersects $\bar{B}(x^0, t)$ only when t is in the interval $\rho - \epsilon \leqq t \leqq \rho + \epsilon$. Consequently, $u(x, t) = 0$ for $t < \rho - \epsilon$ and again for $t > \rho + \epsilon$ while $u(x, t)$ may be non-zero only in the interval $\rho - \epsilon \leqq t \leqq \rho + \epsilon$. If $n = 2$ (or $n = 1$, or n is even), Huygens' principle does not hold and $u(x, t)$ is given in terms of integrals of the data over $\bar{B}(x, t)$. Since $\bar{B}(x, t)$ intersects $\bar{B}(x^0, \epsilon)$ for all $t \geqq \rho - \epsilon$, it follows that $u(x, t) = 0$ for $t < \rho - \epsilon$, while $u(x, t)$ may be non-zero for all $t \geqq \rho - \epsilon$. Figure 6.2 illustrates all this discussion in the (x, t)-space. Note that the history of $u(x, t)$ at the point x is described along the line passing through x and parallel to the t-axis. Figures 6.3 (a) and (b) show the regions of the upper-half (x, t)-space where $u(x, t)$ may be non-zero if the data are not zero only in the ball $B(x^0, \epsilon)$.

It is clear from the above discussion that a disturbance which is initially confined in a small ball $\bar{B}(x^0, \epsilon)$ of the three-dimensional space gives rise to an expanding spherical wave having a leading and a trailing edge. Indeed,

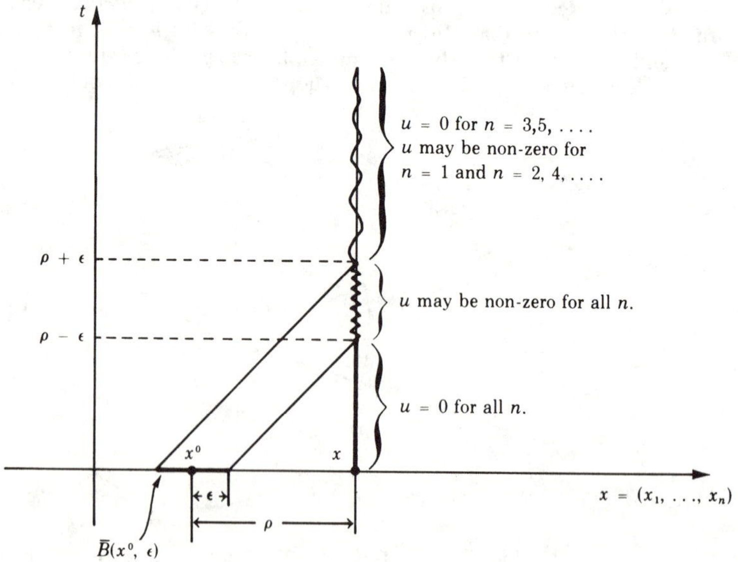

Fig. 6.2

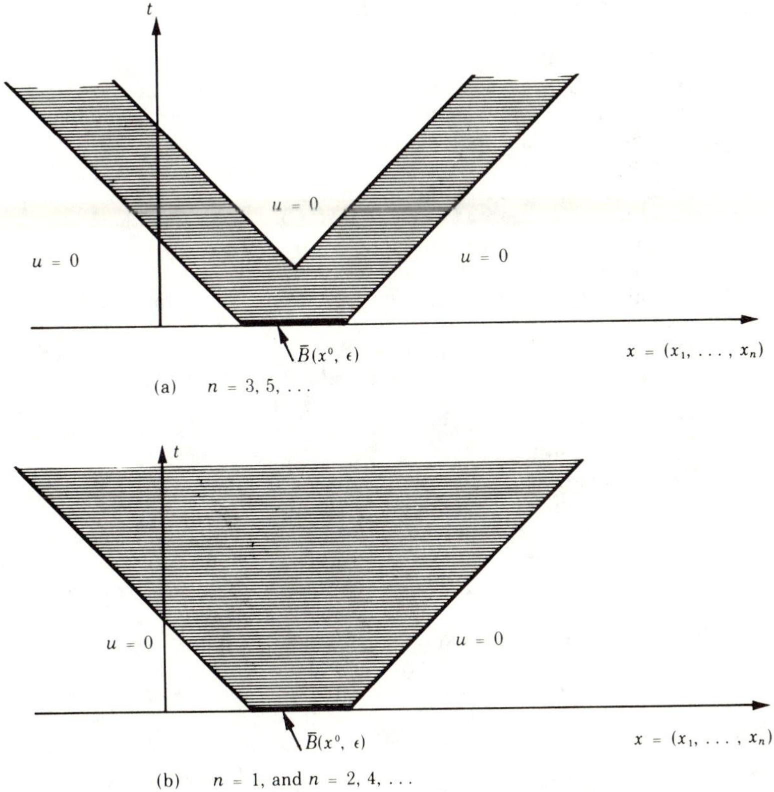

Fig. 6.3. u may be non-zero in the shaded regions

at each instant t ($t > \epsilon$) the disturbance is confined in a shell having outer boundary the surface $S(x^0, t + \epsilon)$ (leading edge of the wave) and inner boundary the surface $S(x^0, t - \epsilon)$ (trailing edge of the wave). In contrast, a disturbance which is initially confined in a small ball $\bar{B}(x^0, \epsilon)$ of the two-dimensional space gives rise to an expanding wave having a leading edge but (generally) no trailing edge. At each instant t the disturbance may be present everywhere inside $S(x^0, t + \epsilon)$. Nevertheless at each fixed point x the disturbance tends to zero with time, as is easy to see from formula (5.23). Thus, while a disturbance initially confined in a finite region of the three-dimensional space gives rise to an expanding wave with sharp leading and trailing edges, a disturbance initially confined in a finite region of the two-dimensional space gives rise to a wave with a sharp leading edge but a slowly decaying trailing edge. This phenomenon of slowly decaying trailing edges is known as *diffusion of waves.* Since diffusion of waves cannot occur when Huygens' principle holds, it follows that there is no diffusion of waves for odd $n \geqq 3$, while there is diffusion of waves for even n. The case $n = 1$ is special in that there may not be any decaying trailing edge at all. At each fixed x, $u(x, t)$ may be constant for all $t \geqq |x - x_0| + \epsilon$ (see Problem 2.3 and also the end of this section).

It is interesting to consider Huygens' principle and the diffusion of waves in relation to physical phenomena that are governed by the wave equation. Assuming that the propagation of sound in the atmosphere is governed by the three-dimensional wave equation, a sound disturbance which is initially confined in a finite region of the space gives rise to a wave with sharp leading and trailing edges. At any point of the atmosphere, the sound is "heard" during a finite interval of time and no trace of the sound is left over after this interval. (Think of clapping your hands once.) On the other hand, assuming that the propagation of waves in a two-dimensional medium, such as a stretched membrane or a liquid surface, is governed by the two-dimensional wave equation, a disturbance which is initiated in a finite region gives rise to a wave which leaves behind it a slowly decaying trace. (Think of dropping a rock in a pond.) Actually the trace that is left behind the wave decays rapidly because of the presence of friction which is not taken into account in the derivation of the wave equation.

We discuss next the possible validity of the formulas for the solution of the initial value problem when boundaries are present or when u is known to satisfy the wave equation only in some region D of the x-space which is not necessarily the whole of R^n. Suppose, for example, that we want to study the vibrations of a finite stretched string which occupies the segment AB of the x-axis as shown in Figure 4.2. Then for x in this segment and all $t > 0$, u must satisfy the one-dimensional wave equation. According to Corollary 4.3, the value of u at any point in the triangle PAB is uniquely determined by the initial conditions $u(x, 0) = \phi(x)$ and $u_t(x, 0) = \psi(x)$ for x in the interval AB. Now, it is easy to see that as long as (x, t) is in the triangle PAB, formula (5.28) defines a function which satisfies the wave equation and the initial conditions. Hence, for (x, t) in the triangle PAB, $u(x, t)$ must be given by formula (5.28). In order to determine $u(x, t)$ for larger t, i.e., for t for which (x, t) lies above the triangle PAB, we must take into account the fact that the string is finite and make use of the boundary conditions that are specified at the ends of the string. This leads to the study of an initial-boundary value problem which is the subject of the last part of this chapter. However, as long as (x, t) is in the triangle PAB, the presence of the boundary points of the string has no effect on $u(x, t)$, and $u(x, t)$ is given by formula (5.28). The situation is similar for $n > 1$. If u is required to satisfy the n-dimensional wave equation for x in some domain D of the x-space R^n and $t > 0$, and if u is required to satisfy the initial conditions $u(x, 0) = \phi(x)$ and $u_t(x, 0) = \psi(x)$ for $x \in D$, then, for every $x \in D$, as long as the part of the initial surface $t = 0$ cut off by the backward characteristic cone with apex at (x, t) is contained in D, $u(x, t)$ is given by formula (5.23) for $n = 2$ and by formula (5.18) for $n = 3$ (see also Problem 4.4). Briefly, for every $x \in D$, $u(x, t)$ is given by the formulas for the solution of the initial value problem, as long as t is less than the distance of x from the boundary of D.

Let us consider now the possibility that the initial data ϕ and ψ are not as smooth as required by Theorems 5.1–5.3. For simplicity we limit the discussion to the case $n = 1$, although similar comments are valid for all n. According to Theorem 5.3, if $\phi \in C^2(R^1)$ and $\psi \in C^1(R^1)$, the solution of the initial value problem (5.24), (5.25), (5.26) is given by formula (5.28), which for convenience we reproduce here (with $x = x_1$),

$$u(x, t) = \frac{1}{2}[\phi(x + t) + \phi(x - t)] + \frac{1}{2}\int_{x-t}^{x+t} \psi(x')dx'. \tag{5.28}$$

Moreover, this solution is twice continuously differentiable in the upper-half (x, t)-plane $t \geqq 0$. Suppose now that ϕ and ψ are only sectionally continuous, i.e., continuous for all $x \in R^1$ except at isolated points where they may have finite jumps. Then formula (5.28) still defines a function $u(x, t)$ which, however, is no longer twice continuously differentiable but only sectionally continuous, i.e., continuous everywhere in the upper-half (x, t)-plane $t \geqq 0$, except along certain curves where it may have finite jumps. Since such a function $u(x, t)$ may not even be differentiable, it may not be a solution of the wave equation in the (classical) sense that u_{xx} and u_{tt} exist and are equal. In order to avoid this kind of difficulty, the concept of a *generalized solution* of a partial differential equation was developed. It is beyond the scope of his book to go into an explanation of this concept. It suffices here to say that if ϕ and ψ are sectionally continuous, then (5.28) defines a generalized solution of the wave equation and, in fact, a generalized solution of the initial value problem (5.24)–(5.26). As a simple illustration, suppose that $\psi(x) \equiv 0$ and $\phi(x)$ is equal to the *Heaviside function:*

$$H(x) = \begin{cases} 0 & \text{for} \quad x < 0 \\ 1 & \text{for} \quad x \geqq 0. \end{cases} \tag{6.1}$$

Then the (generalized) solution of the initial value problem is given by

$$u(x, t) = \frac{1}{2} H(x + t) + \frac{1}{2} H(x - t) \tag{6.2}$$

and is shown in Figure 6.4. Note that the solution has a jump discontinuity along the characteristic lines $x - t = 0$ and $x + t = 0$ and that the jump across each characteristic is constant. This illustrates the general phenomenon that a discontinuity in $u(x, 0)$ at some point of the initial line gives rise to discontinuities in $u(x, t)$ along the characteristics emanating from that point.

We conclude this section with some simple examples of the initial value problem for the one-dimensional wave equation. Suppose first that $\psi(x) \equiv 0$. Then the solution (5.28) of the problem becomes

$$u(x, t) = \frac{1}{2} \phi(x + t) + \frac{1}{2} \phi(x - t) \tag{6.3}$$

and consists of two traveling waves. As we saw in Section 1, $\phi(x + t)/2$ represents a wave traveling to the left with speed 1, while $\phi(x - t)/2$ represents a wave traveling to the right with speed 1. If, for example, ϕ is the cosine pulse

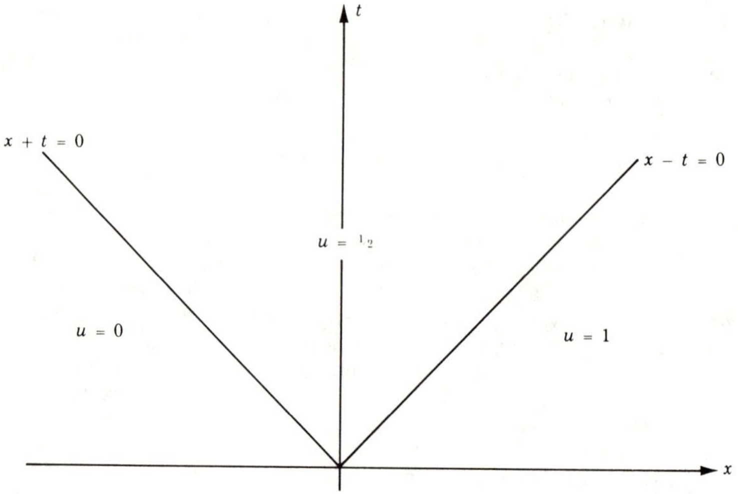

Fig. 6.4

$$(6.4)\qquad \phi(x) = \begin{cases} \cos \pi x & \text{for } |x| \leqq \dfrac{1}{2}, \\ 0 & \text{for } |x| > \dfrac{1}{2}, \end{cases}$$

then the solution (6.3) consists of two cosine pulses each having amplitude one-half the amplitude of the initial pulse, one traveling to the left and the other to the right. The initial pulse splits into two equal pulses traveling in opposite directions. Figure 6.5 shows the graph of $u(x, t)$ as a function of x at different instances of time.

Next, suppose that $\phi(x) \equiv 0$. Then the solution (5.28) becomes

$$(6.5)\qquad u(x, t) = \frac{1}{2} \int_{x-t}^{x+t} \psi(x')dx'.$$

It is easy to see that if $\psi(x)$ vanishes outside some interval I of the x-axis, then at each fixed point x, the interval of integration $(x - t, x + t)$ in (6.5) eventually covers I and $u(x, t)$ becomes constant (see Problem 2.3). If, for example,

$$(6.6)\qquad \psi(x) = \begin{cases} 1 & \text{for } |x| \leqq 1 \\ 0 & \text{for } |x| > 1 \end{cases}$$

then at each fixed x, $u(x, t)$ eventually becomes equal to the constant

$$\frac{1}{2} \int_{-1}^{1} 1 \, dx' = 1.$$

Figure 6.6 shows the graph of $u(x, t)$ as a function of x at different instances of time.

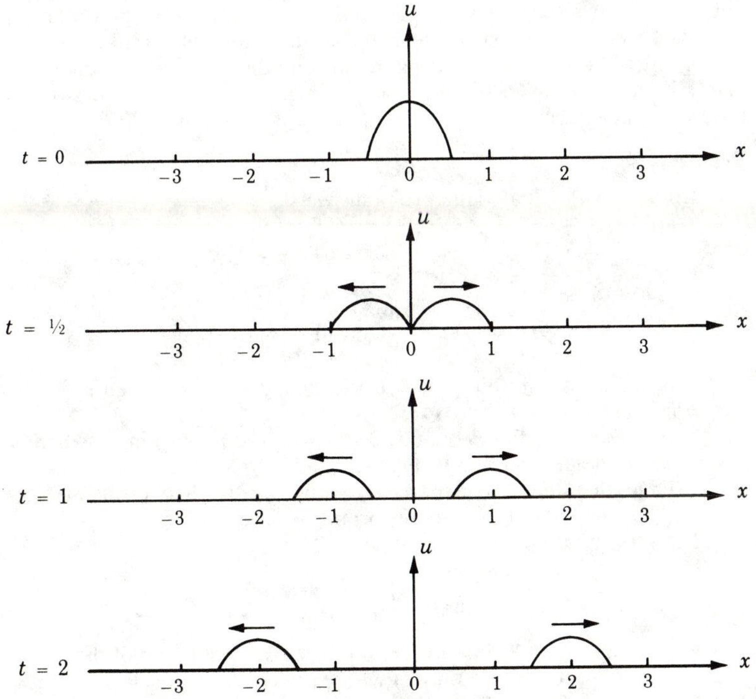

Fig. 6.5

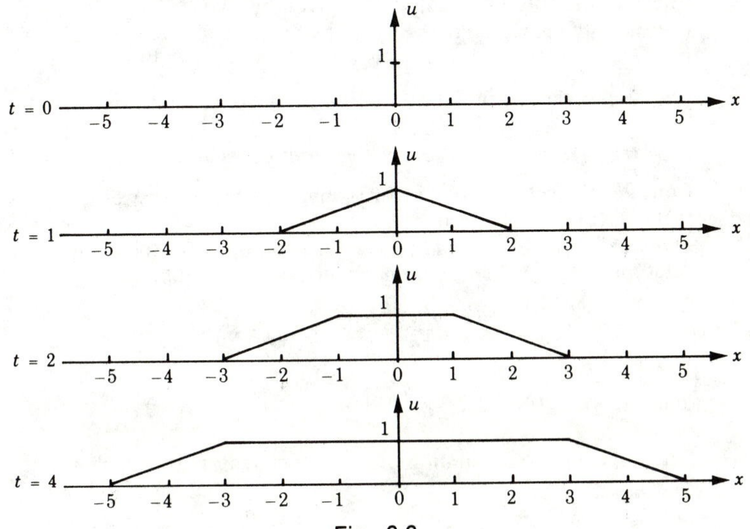

Fig. 6.6

The above two examples show that if an "infinite" stretched string is subjected to an initial displacement pulse, the pulse splits into two equal pulses which travel along the string in opposite directions. If, on the other hand, the string is subjected to an initial velocity pulse (a kick), the string is raised to a new level.

Problems

6.1. Consider the initial value problem

$$\frac{\partial^2 u}{\partial x_1^2} + \cdots + \frac{\partial^2 u}{\partial x_n^2} - \frac{1}{c^2}\frac{\partial^2 u}{\partial t^2} = 0; \qquad x \in R^n, \qquad t > 0,$$

$$u(x, 0) = \phi(x), \qquad \frac{\partial u}{\partial t}(x, 0) = \psi(x), \qquad x \in R^n,$$

with ϕ and ψ vanishing outside the ball $\bar{B}(0, 1)$ with center at the origin and radius 1.

(a) Take $c = 1$. For $n = 1, 2, 3$ describe the regions in R^n where u may be different than zero when $t = 1$, $t = 2$ and $t = 10$.

(b) Do the same as in part (a) with $c \neq 1$. (See Problem 5.5.) Interpret c as the speed of propagation of waves.

6.2. Let $\chi_1(x)$ be the function defined in R^n by

$$\chi_1(x) = \begin{cases} 1 & \text{for } |x| \leqq 1 \\ 0 & \text{for } |x| > 1 \end{cases}$$

(a) For $n = 1, 2, 3$, find $u(0, t)$ if $u(x, t)$ is the solution of the initial value problem (4.1), (4.2) with $\phi(x) \equiv 0$ and $\psi(x) = \chi_1(x)$.

(b) For $n = 1, 2, 3$, find $u(0, t)$ if $u(x, t)$ is the solution of the initial value problem (4.1), (4.2) with $\phi(x) \equiv \chi_1(x)$ and $\psi(x) \equiv 0$.

6.3. Consider the solution (6.3) with ϕ given by (6.4).

(a) Indicate the region in the upper half (x, t)-plane where $u(x, t) \neq 0$

(b) Draw the graph of $u(x, t)$ as a function of t ($t \geqq 0$) at $x = 2$.

(c) Discuss the discontinuities in the partial derivatives $\frac{\partial u}{\partial x}(x, t)$ and $\frac{\partial u}{\partial t}(x, t)$. Where do these discontinuities occur in the (x, t)-plane?

6.4. Consider the solution (6.5) with ψ given by (6.6). Answer questions (a), (b), (c) of Problem 6.3.

6.5. *Duhamel's principle.* Let $u(x, t)$ be the solution of the following initial value problem for the nonhomogeneous wave equation,

$$(6.7) \qquad \frac{\partial^2 u}{\partial x_1^2} + \cdots + \frac{\partial^2 u}{\partial x_n^2} - \frac{\partial^2 u}{\partial t^2} = f(x, t); \qquad x \in R^n, \qquad t > 0,$$

$$(6.8) \qquad u(x, 0) = 0, \qquad \frac{\partial u}{\partial t}(x, 0) = 0, \qquad x \in R^n.$$

Let $v(x, t; \tau)$ be the solution of the associated "pulse problem"

$$(6.9) \qquad \frac{\partial^2 v}{\partial x_1^2} + \ldots + \frac{\partial^2 v}{\partial x_n^2} - \frac{\partial^2 v}{\partial t^2} = 0; \qquad x \in R^n, \qquad t > \tau,$$

(6.10) $v(x, \tau; \tau) = 0, \quad \frac{\partial}{\partial t} v(x, \tau; \tau) = -f(x, \tau), \qquad x \in R^n.$

(Note that in this problem the initial conditions are prescribed at $t = \tau$.) Show that

(6.11) $$u(x, t) = \int_0^t v(x, t; \tau) d\tau.$$

6.6. Use Duhamel's principle and Kirchhoff's formula to show that the solution of the initial value problem (6.7), (6.8) with $n = 3$ is given by

(6.12) $$u(x, t) = -\frac{1}{4\pi} \int_{\bar{B}(x,t)} \frac{f(x', t - r)}{r} dx_1' dx_2' dx_3'$$

where

$$r = | x - x' | = [(x_1 - x_1')^2 + (x_2 - x_2')^2 + (x_3 - x_3')^2]^{1/2}$$

and $\bar{B}(x, t)$ is the ball in R^3 with center at x and radius t.

6.7. Find the solution of the initial value problem

$$\frac{\partial^2 u}{\partial x_1^2} + \frac{\partial^2 u}{\partial x_2^2} + \frac{\partial^2 u}{\partial x_3^2} - \frac{\partial^2 u}{\partial t^2} = f(x, t); \qquad x \in R^3, \qquad t > 0$$

$$u(x, 0) = \phi(x), \qquad \frac{\partial u}{\partial t}(x, 0) = \psi(x), \qquad x \in R^3.$$

6.8. Continuous dependence on data for the initial value problem may be formulated in an energy sense, which is often appropriate in engineering applications. In this setting, changes in the initial data and solution are measured in terms of energy. Let u and v be two solutions of (4.1), (4.2) which satisify the conditions of Theorem 4.2. Show that $E(u - v; R^n, T) = E(u - v; R^n, 0)$ and formulate a statement of continuous dependence on initial data in terms of energy.

7. Wave Propagation in Regions with Boundaries. Uniqueness of Solution of the Initial-Boundary Value Problem. Reflection of Waves

We turn now to the study of the initial-boundary value problem for the wave equation. As we discussed in Section 4 of Chapter VI, this problem arises in the study of wave propagation in regions with boundaries. Important examples are vibrations of bounded bodies such as finite strings and membranes, and propagation of acoustical or electromagnetic waves in the presence of finite or infinite reflectors. (Traditionally, one speaks of vibrations in a finite body and of wave propagation in an infinite body although the two phenomena are in fact the same.) The general problem is the following: Let Ω be a domain in R^n. Find a function $u(x, t)$ defined for $x \in \bar{\Omega}$ and $t \geqq 0$, and satisfying the wave equation

(7.1) $$u_{x_1x_1} + \dots + u_{x_nx_n} - u_{tt} = 0 \quad \text{for} \quad x \in \Omega \quad \text{and} \quad t > 0,$$

the initial conditions

(7.2) $$u(x, 0) = \phi(x), \qquad u_t(x, 0) = \psi(x) \quad \text{for} \quad x \in \bar{\Omega},$$

and one of the boundary conditions

(7.3) $$u(x, t) = 0 \quad \text{for} \quad x \in \partial\Omega \quad \text{and} \quad t \geqq 0,$$

or

(7.4) $$\frac{\partial u}{\partial n}(x, t) = 0 \quad \text{for} \quad x \in \partial\Omega \quad \text{and} \quad t \geqq 0,$$

where the differentiation in (7.4) is in the direction of the exterior normal to $\partial\Omega$. Note that the solution is required to satisfy only one of the boundary conditions (7.3) or (7.4). Note also that we limit our study to homogeneous boundary conditions.

Figure 7.1 shows a graphical representation of the initial-boundary value problem (7.1), (7.2), (7.3) for $n = 1$ with Ω being a finite interval (a, b). Figure 7.2 shows the same problem for $n = 2$ with Ω a bounded domain in R^2.

For simplicity in proving the results of this section, we assume that the desired solution of the initial-boundary value problem (7.1), (7.2), (7.3) or (7.1), (7.2), (7.4) is of class C^2 for $x \in \bar{\Omega}$ and $t \geqq 0$, and that the domain Ω is bounded and normal in the sense that the divergence theorem is applicable. We first prove the following result on conservation of energy, from which follows easily the uniqueness of solution of the initial-boundary value problem.

Theorem 7.1. *(Conservation of energy.)* Let Ω be a bounded domain in R^n and suppose that $u(x, t)$ is a solution of the wave equation (7.1) satisfying one of the boundary conditions (7.3) or (7.4). Suppose also that

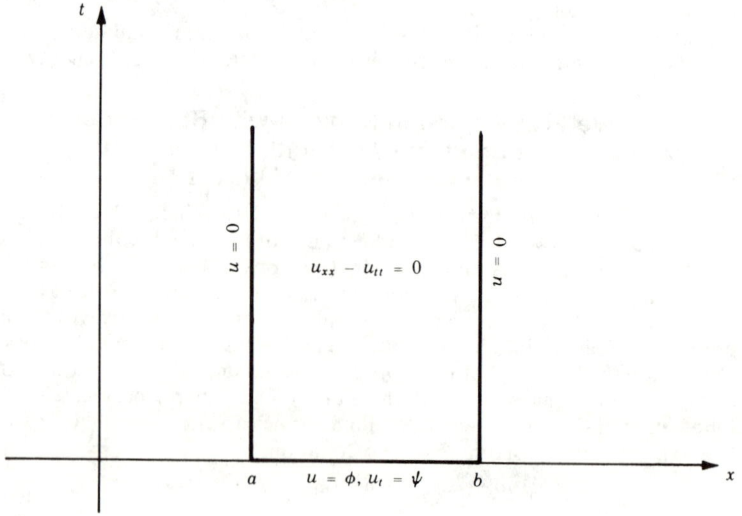

Fig. 7.1

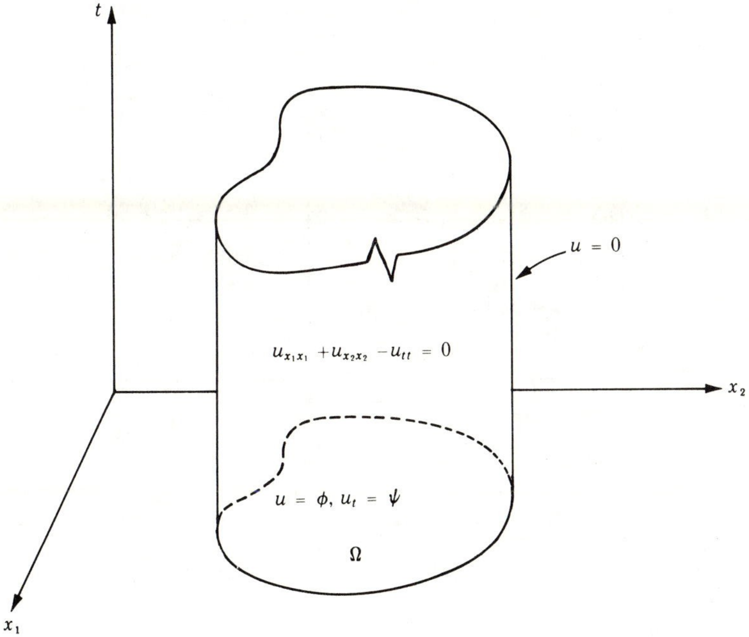

Fig. 7.2

$u(x, t)$ is in C^2 for $x \in \bar{\Omega}$ and $t \geqq 0$. Then the energy of u contained in Ω remains constant (independent of t), i.e., for every $T \geq 0$,

$$(7.5)\qquad \int_\Omega (u_{x_1}^2 + \cdots + u_{x_n}^2 + u_t^2)\,|_{t=T}\,dx = \int_\Omega (u_{x_1}^2 + \cdots + u_{x_n}^2 + u_t^2)\,|_{t=0}\,dx,$$

or, in the notation (4.5),

$$(7.6)\qquad E(u; \Omega, T) = E(u; \Omega, 0).$$

Proof. We give the proof for $n = 2$ since it is easier to visualize the geometry in this case. (The proof for any $n \geqq 1$ is not really different.) Let Ω^T be the cylindrical domain in the (x_1, x_2, t)-space consisting of the points (x_1, x_2, t) for which $(x_1, x_2) \in \Omega$ and $0 < t < T$ (see Fig. 7.3). We integrate the differential identity (3.8) over Ω^T. Since u satisfies the wave equation we have

$$(7.7)\qquad 0 = \iiint_{\Omega^T} [(2u_tu_{x_1})_{x_1} + (2u_tu_{x_2})_{x_2} - (u_{x_1}^2 + u_{x_2}^2 + u_t^2)_t]dx_1dx_2dt.$$

The integrand in this equation is the divergence of the vector field $(2u_tu_{x_1}, 2u_tu_{x_2}, -(u_{x_1}^2 + u_{x_2}^2 + u_t^2))$. Applying the divergence theorem we obtain

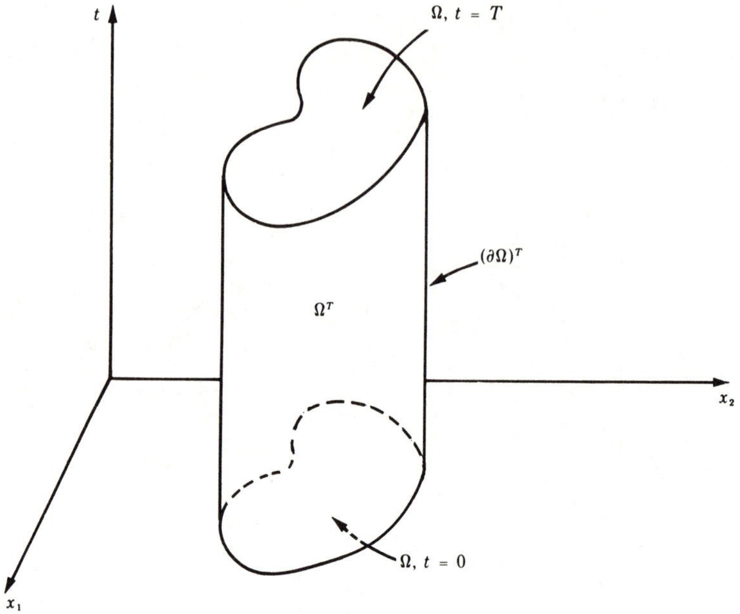

Fig. 7.3

$$(7.8) \qquad 0 = \iint_{\partial\Omega^T} [2u_t u_{x_1}\nu_1 + 2u_t u_{x_2}\nu_2 - (u_{x_1}^2 + u_{x_2}^2 + u_t^2)\nu_t]d\sigma$$

where $\partial\Omega^T$ is the boundary of Ω^T, $d\sigma$ the element of surface on $\partial\Omega^T$ and $\boldsymbol{\nu} = (\nu_1, \nu_2, \nu_t)$ the unit normal vector on $\partial\Omega^T$ in the direction exterior to Ω^T. Since the boundary $\partial\Omega^T$ consists of three parts, the top and bottom of the cylinder and the lateral cylindrical surface, we split the integral in (7.8) into three integrals. On the top of the cylinder, $(x_1, x_2) \in \Omega, t = T, \boldsymbol{\nu} = (0, 0, 1)$ and the integral is

$$(7.9) \qquad -\iint_{\Omega} (u_{x_1}^2 + u_{x_2}^2 + u_t^2)\,|_{t=T}\, dx_1 dx_2.$$

On the bottom, $(x_1, x_2) \in \Omega, t = 0, \boldsymbol{\nu} = (0, 0, -1)$ and the integral is

$$(7.10) \qquad \iint_{\Omega} (u_{x_1}^2 + u_{x_2}^2 + u_t^2)\,|_{t=0}\, dx_1 dx_2.$$

On the cylindrical surface, $(x_1, x_2) \in \partial\Omega, 0 \le t \le T, \boldsymbol{\nu} = (n_1, n_2, 0)$, where $\mathbf{n} = (n_1, n_2)$ is the exterior normal to $\partial\Omega$, and the integral is

$$(7.11) \qquad \int_0^T \int_{\partial\Omega} (2u_t u_{x_1} n_1 + 2u_t u_{x_2} n_2) ds dt = \int_0^T \int_{\partial\Omega} 2u_t \frac{\partial u}{\partial n}\, ds dt,$$

where ds is the element of length on the boundary curve $\partial\Omega$. Since the integral in (7.8) is the sum of the integrals (7.9), (7.10), and (7.11), we have

$$
(7.12)\qquad
\begin{aligned}
0 = &-\iint_{\Omega} (u_{x_1}^2 + u_{x_2}^2 + u_t^2)\,|_{t=T}\, dx_1 dx_2 \\
&+ \iint_{\Omega} (u_{x_1}^2 + u_{x_2}^2 + u_t^2)\,|_{t=0}\, dx_1 dx_2 + \int_0^T \int_{\partial\Omega} 2u_t \frac{\partial u}{\partial n}\, ds dt.
\end{aligned}
$$

If u satisfies the boundary condition (7.4), the last integral in (7.12) is obviously zero. If u satisifies (7.3), then $u_t = 0$ on $\partial\Omega$ for $t \geqq 0$ (reason: for $(x_1, x_2) \in \partial\Omega$, $u(x_1, x_2, t)$ is constant $(=0)$ independent of t) and the last integral in (7.12) is again zero. Hence, in either case, (7.12) yields the conclusion (7.5) of the theorem for $n = 2$.

If u is required to satisfy the initial condition (7.2) in addition to the assumptions of Theorem 7.1, then the constant energy of u in Ω can be computed in terms of the initial data. Equation (7.5) becomes in this case,

$$
(7.13)\qquad \int_{\Omega} (u_{x_1}^2 + \ldots + u_{x_n}^2 + u_t^2)\,|_{t=T}\, dx = \int_{\Omega} (\phi_{x_1}^2 + \ldots + \phi_{x_n}^2 + \psi^2) dx.
$$

Theorem 7.2. Let Ω be a bounded domain in R^n and suppose that $u(x, t)$ is a solution of either one of the initial-boundary value problems (7.1), (7.2), (7.3) or (7.1), (7.2), (7.4). Suppose also that $u(x, t)$ is of class C^2 for $x \in \bar{\Omega}$ and $t \geqq 0$. If the initial data ϕ and ψ vanish in Ω, then u must vanish for every (x, t) with $x \in \bar{\Omega}$ and $t \geqq 0$.

Proof. Since ϕ vanishes in Ω, the derivatives of ϕ must also vanish in Ω. Therefore, from (7.13),

$$
(7.14)\qquad \int_{\Omega} (u_{x_1}^2 + \ldots + u_{x_n}^2 + u_t^2)\,|_{t=T}\, dx = 0
$$

for every $T \geqq 0$. Since the integrand in (7.14) is continuous and nonnegative, it must be zero. We conclude that for $x \in \Omega$ and $t \geqq 0$, all first order partial derivatives of u must vanish and therefore u must be constant. By continuity, u must be constant in the closed region of points (x, t) with $x \in \bar{\Omega}$ and $t \geqq 0$, and since $u(x, 0) = 0$ for $x \in \bar{\Omega}$, the constant value of u in this region must be zero.

Corollary 7.1. *(Uniqueness.)* There can be at most one solution of the initial-boundary value problem (7.1), (7.2), (7.3) or (7.1), (7.2), (7.4).

Proof. The difference of any two solutions satisfies the assumptions of Theorem 7.2.

The above results on conservation of energy and uniqueness for the initial-boundary value problem are valid even when the domain Ω is unbounded, provided that the initial data vanish outside some finite ball $\bar{B}(0, R)$ in R^n. The necessary modification of the proof of conservation of energy is based on the observation that, in spite of the presence of boundaries, the range of influence of the initial data in the ball $\bar{B}(0, R)$ is the set of points in (x, t)-space for which

$$x \in \Omega, \qquad |x| \leqq R + t, \qquad t \geqq 0. \tag{7.15}$$

Since the initial data are assumed to vanish outside $\bar{B}(0, R)$, the solution u must vanish outside the set (7.15). Consequently, in order to prove conservation of energy, we integrate the differential identity (3.8) over the cylindrical domain in the (x, t)-space consisting of those points for which $x \in \Omega$ and $|x| < R + 2T$ and $0 < t < T$. (See also Problems 4.5 and 4.6.)

At this point, instead of addressing ourselves to the questions of existence and continuous dependence of solution on the initial data for the general initial-boundary value problem, we will solve specific problems which frequently arise in applications. In the remainder of this section we will solve some problems for which Ω is infinite and has plane boundaries. We will do this by introducing imaginary data in the complement of Ω, thereby reducing the problem to the initial value problem for the wave equation which we already know how to solve. In the next two sections we will solve the initial-boundary value problems for a finite string and for a rectangular membrane using the method of separation of variables and Fourier series. These two examples will serve as models for explaining the method of solution of the general problem which is briefly outlined in the last section of the chapter.

Example 7.1. *(Semi-infinite string with fixed end.)* Solve the problem

$$\begin{cases} u_{xx} - u_{tt} = 0; & 0 < x < \infty, \quad 0 < t \\ u(x, 0) = \phi(x), \quad u_t(x, 0) = \psi(x); & 0 \leqq x < \infty \\ u(0, t) = 0; \; 0 \leqq t. \end{cases} \tag{7.16}$$

Let $\tilde{\phi}_0$ and $\tilde{\psi}_0$ be the odd extensions of the initial data ϕ and ψ to the whole x-axis,

$$\tilde{\phi}_0(x) = \begin{cases} \phi(x) & \text{if } x > 0 \\ -\phi(-x) & \text{if } x < 0 \end{cases} \qquad \tilde{\psi}_0(x) = \begin{cases} \psi(x) & \text{if } x > 0 \\ -\psi(-x) & \text{if } x < 0 \end{cases}$$

and let $\tilde{u}(x, t)$ be the solution of the initial value problem

$$\tilde{u}_{xx} - \tilde{u}_{tt} = 0; \qquad -\infty < x < \infty, \qquad 0 < t$$

$$\tilde{u}(x, 0) = \tilde{\phi}_0(x), \qquad \tilde{u}_t(x, 0) = \tilde{\psi}_0(x); \quad -\infty < x < \infty.$$

We know that

$$\tilde{u}(x, t) = \frac{1}{2}[\tilde{\phi}_0(x + t) + \tilde{\phi}_0(x - t)] + \frac{1}{2}\int_{x-t}^{x+t} \tilde{\psi}_0(x')dx'$$

Since the functions $\tilde{\phi}_0$ and $\tilde{\psi}_0$ are odd, it is easy to see that $\tilde{u}(0, t) = 0$ for all $t \geqq 0$. Moreover, when $x > 0$, $\tilde{u}(x, t)$ satisfies the wave equation and the initial conditions of problem (7.16). Therefore the desired solution of (7.16) is the function $\tilde{u}(x, t)$ restricted to $x \geqq 0$,

$$u(x, t) = \frac{1}{2}[\tilde{\phi}_0(x+t) + \tilde{\phi}_0(x-t)] + \frac{1}{2}\int_{x-t}^{x+t} \tilde{\psi}_0(x')dx'; \tag{7.17}$$

$$0 \leqq x, \qquad 0 \leqq t.$$

As an illustration of this general solution let us suppose that ψ is identically zero and ϕ is a cosine pulse centered at $x = 3$,

$$\phi(x) = \begin{cases} \cos \pi (x-3), & \text{for } |x-3| < \frac{1}{2} \\ 0, & \text{for } 0 < x < 2.5 \text{ and } 3.5 < x < \infty . \end{cases} \tag{7.18}$$

The solution (7.17) (which is a generalized solution in this case since the initial data are not C^2) consists of two pulses, each having amplitude one-half the amplitude of the initial pulse, one traveling to the left and the other to the right. When the pulse traveling to the left reaches the fixed point $x = 0$, it is reflected there and then travels to the right with changed sign. Figure 7.4 shows the graph of u as a function of x at different instances of time. The solution is easiest to understand if imaginary data are drawn on the negative x-axis. Thus, the $t = 0$ part of the figure shows the graph of the

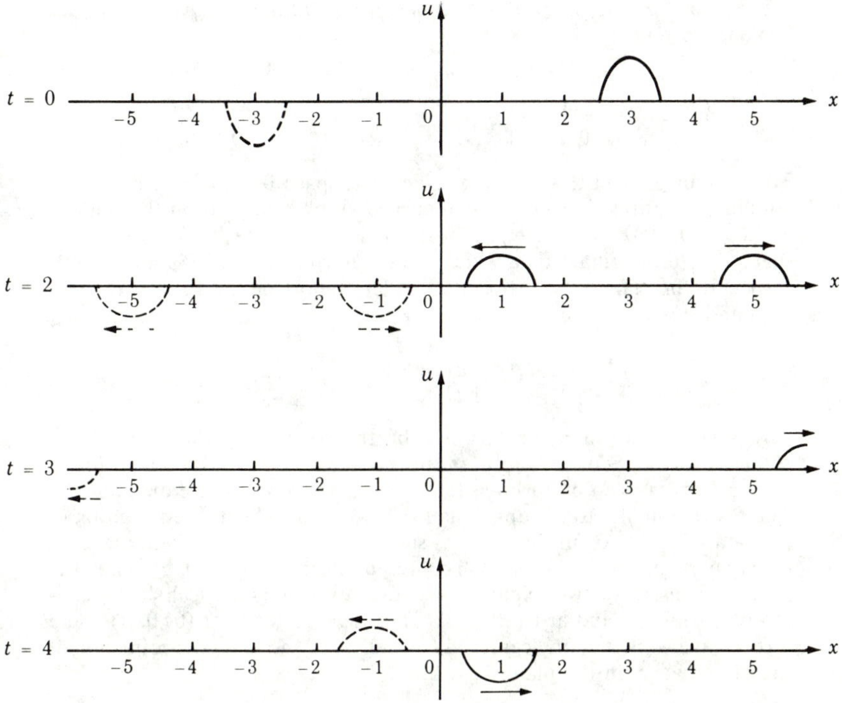

Fig. 7.4

odd extension $\tilde{\phi}_0$ of the initial pulse ϕ. The reflected pulse should be interpreted as originating from the imaginary initial pulse on the negative x-axis.

Example 7.2. *(Semi-infinite string with free end.)* If the end $x = 0$ of a string is free to move up or down, the slope u_x at the free end must be zero. We must solve the problem

$$(7.19)\qquad \begin{cases} u_{xx} - u_{tt} = 0; & 0 < x < \infty, \quad 0 < t \\ u(x, 0) = \phi(x), \quad u_t(x, 0) = \psi(x); & 0 \leqq x < \infty \\ u_x(0, t) = 0; & 0 \leqq t. \end{cases}$$

A method similar to the one used in Example 7.1 yields the solution

$$(7.20)\qquad u(x, t) = \frac{1}{2}[\tilde{\phi}_e(x + t) + \tilde{\phi}_e(x - t)] + \frac{1}{2}\int_{x-t}^{x+t} \tilde{\psi}_e(x')dx'$$

where $\tilde{\phi}_e$ and $\tilde{\psi}_e$ are the even extensions of ϕ and ψ to the whole x-axis,

$$\tilde{\phi}_e(x) = \begin{cases} \phi(x) & \text{if } x > 0 \\ \phi(-x) & \text{if } x < 0 \end{cases} \qquad \tilde{\psi}_e(x) = \begin{cases} \psi(x) & \text{if } x > 0 \\ \psi(-x) & \text{if } x < 0. \end{cases}$$

Example 7.3. *(Reflection of acoustical waves from a hard plane surface.)* Solve the problem

$$(7.21)\qquad \begin{cases} u_{x_1x_1} + u_{x_2x_2} + u_{x_3x_3} - u_{tt} = 0; & x \in R^3 \text{ with } 0 < x_3,\ 0 < t, \\ u(x, 0) = \phi(x),\ u_t(x, 0) = \psi(x); & x \in R^3 \text{ with } 0 \leqq x_3, \\ u_{x_3}(x_1, x_2, 0, t) = 0; & -\infty < x_1,\ x_2 < \infty,\ 0 \leq t. \end{cases}$$

The domain Ω is in this case the upper-half space $0 < x_3$. The boundary condition requires that the normal derivative of u be zero on the boundary $x_3 = 0$ of Ω. To solve the problem we introduce in the lower-half space $x_3 < 0$ imaginary data which are the mirror images of ϕ and ψ with respect to the plane $x_3 = 0$. Specifically, let ϕ and ψ be the even extensions of ϕ and ψ with respect to x_3, i.e.,

$$\tilde{\phi}(x_1, x_2, x_3) = \begin{cases} \phi(x_1, x_2, x_3) & \text{if } x_3 > 0 \\ \phi(x_1, x_2, -x_3) & \text{if } x_3 < 0, \end{cases}$$

with a similar formula for $\tilde{\psi}$. Let $\tilde{u}$ be the solution of the initial value problem for the wave equation in the whole (x_1, x_2, x_3)-space with initial data $\tilde{\phi}$ and $\tilde{\psi}$ (given by formula (5.18) with $\tilde{\phi}$ and $\tilde{\psi}$ in place of ϕ and ψ). The restriction of $\tilde{u}$ to the upper-half space $x_3 \geqq 0$ is the desired solution of problem (7.21). As an illustration, suppose that the initial data are zero everywhere except in a small ball centered at the point $(0, 0, a)$. Then the solution consists of two expanding spherical waves (spherical shells) centered at $(0, 0, a)$ and at $(0, 0, -a)$. The wave centered at $(0, 0, a)$ is the direct wave while the one centered at $(0, 0, -a)$ is actually the reflection of the direct wave on the plane $x_3 = 0$.

Problems

7.1. Consider the initial-boundary value problems for the telegraph equation

(7.22) $u_{xx} - u_{tt} - au_t - bu = 0; \quad 0 < x < L, \quad 0 < t,$

(7.23) $u(x, 0) = \phi(x) \quad u_t(x, 0) = \psi(x); \quad 0 \leqq x \leqq L$

(7.24) $u(0, t) = 0, \quad u(L, t) = 0; \quad 0 \leqq t$

or

(7.25) $u_x(0, t) = 0, \quad u_x(L, t) = 0; \quad 0 \leqq t.$

Assume that a and b are nonnegative constants.

(a) Multiply (7.22) by $2u_t$ and derive the differential identity

$$(2u_t u_x)_x - (u_x^2 + u_t^2 + bu^2)_t - 2au_t^2 = 0.$$

(b) Prove that if u satisfies (7.22) and either one of the boundary conditions (7.24) or (7.25), then

$$\int_0^L (u_x^2 + u_t^2 + bu^2)\,|_{t=T}\, dx \leqq \int_0^L (u_x^2 + u_t^2 + bu^2)\,|_{t=0}\, dx.$$

(c) State and prove a uniqueness theorem for the above initial-boundary value problems.

7.2 State and prove conservation of energy and uniqueness theorems for the initial-boundary value problems for equation (3.19). Note that the energy in this case has an extra term.

7.3. State and prove uniqueness of solution of the initial-boundary value problems for the wave equation (4.10) with a forcing term.

7.4. (a) Show that if u satisfies the wave equation (7.1) and the mixed boundary condition

(7.26) $$\frac{\partial u}{\partial n} + \alpha u = 0; \quad x \in \partial\Omega, \quad t \geqq 0,$$

where $\alpha = \alpha(x)$ is a nonnegative continuous function defined for $x \in \partial\Omega$, then

$$\int_\Omega (u_{x_1}^2 + \cdots + u_{x_n}^2 + u_t^2)\,|_{t=T}\, dx + \int_{\partial\Omega} \alpha u^2\,|_{t=T}\, ds$$

$$= \int_\Omega (u_{x_1}^2 + \cdots + u^2{}_{x_n} + u_t^2)\,|_{t=0}\, dx + \int_{\partial\Omega} \alpha u^2\,|_{t=0}\, ds$$

where ds is the element of surface on $\partial\Omega$.

(b) Prove the uniqueness of solution of the initial-boundary value problem (7.1), (7.2), (7.26).

7.5. Verify by direct substitution that (7.20) satisfies (7.19). Discuss the solution (7.20) when ψ is identically zero and ϕ is the cosine pulse (7.18). In particular draw a figure analogous to Figure 7.4.

7.6. Consider the initial-boundary value problem for the two-dimensional wave equation in the first quadrant

$$u_{xx} + u_{yy} - u_{tt} = 0; \quad x > 0, \quad y > 0, \quad t > 0$$

$$u(x, y, 0) = \phi(x, y), \quad u_t(x, y, 0) = \psi(x, y); \quad x \geqq 0, \quad y \geqq 0$$

$$u(0, y, t) = u(x, 0, t) = 0; \quad x \geqq 0, \quad y \geqq 0, \quad t \geqq 0.$$

Introduce appropriate imaginary data in the other three quadrants to reduce the problem to an initial value problem. Discuss the solution if the initial data are zero everywhere except in a small circle in the first quadrant.

7.7. Consider the initial-boundary value problem for the three-dimensional wave equation in a domain between two parallel planes,

$$u_{x_1x_1} + u_{x_2x_2} + u_{x_3x_3} - u_{tt} = 0; \qquad x \in R^3 \text{ with } 0 < x_3 < a,\ 0 < t$$

$$u(x, 0) = \phi(x),\ u_t(x, 0) = \psi(x) \qquad x \in R^3 \text{ with } 0 \leqq x_3 \leqq a$$

$$u(x_1, x_2, 0, t) = u(x_1, x_2, a, t) = 0, \qquad -\infty < x_1, x_2 < \infty,\ 0 \leqq t.$$

Introduce an infinite sequence of imaginary data to reduce the problem to an initial value problem. Discuss the solution if the initial data are zero everywhere except in a small ball.

7.8. Suppose that the boundary of the domain Ω consists of two pieces, $\partial\Omega = S_1 + S_2$ and suppose u is required to satisfy the boundary condition

$$u(x, t) = 0 \quad \text{for} \quad x \in S_1,$$

(7.27) $$\frac{\partial u}{\partial n}(x, t) = 0 \quad \text{for} \quad x \in S_2; \qquad 0 \leq t.$$

Show that conservation of energy and uniqueness results are still valid for the problem (7.1), (7.2), (7.27).

8. The Vibrating String

Consider a taut string of length L with both ends fixed, such as the string of a guitar. In order to determine the vibrations of the string we must solve the initial-boundary value problem

(8.1) $$u_{xx} - u_{tt} = 0; \qquad 0 < x < L, \qquad 0 < t,$$

(8.2) $$u(x, 0) = \phi(x), \qquad u_t(x, 0) = \psi(x); \qquad 0 \leqq x \leqq L,$$

(8.3) $$u(0, t) = 0, \qquad u(L, t) = 0; \qquad 0 \leqq t.$$

We know that this problem can have at most one solution. We will find the solution by using separation of variables and Fourier series. The method of solution consists of two steps. First, using separation of variables we find an infinite collection of functions which satisfy the wave equation (8.1) and the boundary conditions (8.3). Then we superpose these functions and use Fourier series representations of the initial data ϕ and ψ to satisfy the initial conditions (8.2). The resulting solution of (8.1), (8.2), (8.3) will be in the form of an infinite series.

We begin by looking for solutions of (8.1) in the form

(8.4) $$u(x, t) = X(x)T(t).$$

Substituting into (8.1) and separating the variables we find

$$\frac{X''}{X} = \frac{T''}{T} = -\lambda,$$

where $-\lambda$ is the separation constant. The minus sign is for later convenience. We conclude that a function $u(x, t)$ of the form (8.4) will be a solution of (8.1) if $X(x)$ and $T(t)$ satisfy the ordinary differential equations

(8.5) $$X'' + \lambda X = 0, \qquad 0 < x < L$$

(8.6) $$T'' + \lambda T = 0, \qquad 0 < t.$$

It is also easy to see that (except for the trivial and uninteresting case $u(x, t) \equiv 0$) a function of the form (8.4) will satisfy the boundary conditions (8.3) if

(8.7) $$X(0) = 0, \qquad X(L) = 0.$$

We are faced then with finding nontrivial solutions $X(x)$ of the boundary value problem (8.5), (8.7). Note that this problem always has the trivial solution $X(x) \equiv 0$. It turns out that for most values of λ, the trivial solution is the only solution of the problem. To see this it is convenient to consider the three cases $\lambda < 0$, $\lambda = 0$ and $\lambda > 0$ separately, since in each of these cases the general solution of equation (8.5) has a different form.

If $\lambda < 0$, the general solution of (8.5) is

$$X(x) = C_1 e^{\sqrt{-\lambda}\, x} + C_2 e^{-\sqrt{-\lambda}\, x}.$$

To satisfy the boundary conditions (8.7) the constants must be such that

$$C_1 + C_2 = 0,$$

$$C_1 e^{\sqrt{-\lambda}L} + C_2 e^{-\sqrt{-\lambda}L} = 0.$$

The only solution of this system of equations for C_1, C_2 is the trivial one, $C_1 = C_2 = 0$. Hence the only solution of (8.5), (8.7) is the trivial solution.

If $\lambda = 0$, the general solution of (8.5) is

$$X(x) = C_1 + C_2 x.$$

To satisfy the boundary conditions (8.7) we must have

$$C_1 = 0,$$
$$C_1 + C_2 L = 0,$$

so that $C_1 = C_2 = 0$ and again the trivial solution is the only solution of (8.5), (8.7).

Finally, if $\lambda > 0$ the general solution of (8.5) is

$$X(x) = C_1 \cos \sqrt{\lambda} x + C_2 \sin \sqrt{\lambda} x$$

and the boundary conditions (8.7) will be satisfied if

$$C_1 = 0,$$

$$C_1 + C_2 \sin \sqrt{\lambda} L = 0.$$

This system of equations for C_1 and C_2 has nontrivial solutions, namely $C_1 = 0$ and C_2 arbitrary, if and only if

$$\sin \sqrt{\lambda}L = 0, \tag{8.8}$$

or

$$\sqrt{\lambda}L = k\pi,$$

where k is an integer. Since λ is assumed to be positive, we conclude that the only values of λ for which the problem (8.5), (8.7) has nontrivial solutions are

$$\lambda_k = \frac{k^2\pi^2}{L^2}, \qquad k = 1, 2, \ldots . \tag{8.9}$$

For each value λ_k of λ, problem (8.5), (8.7) has infinitely many solutions, namely

$$X_k(x) = D_k \sin\frac{k\pi x}{L}, \qquad k = 1, 2, \ldots, \tag{8.10}$$

where D_k is an arbitrary constant. (Note that these infinitely many solutions corresponding to a given λ_k differ only by a multiplicative constant, i.e., they are linearly dependent.)

The problem of finding nontrivial solutions of (8.5), (8.7) is an example of a general problem known as the *eigenvalue problem* or *Sturm-Liouville problem*. The values of λ for which the problem has nontrivial solutions are known as *eigenvalues* and the corresponding nontrivial solutions are called *eigenfunctions*. The eigenvalues of (8.5), (8.7) are given by (8.9) and the corresponding eigenfunctions by (8.10).

Let us return now to the problem of finding nontrivial solutions of the wave equation (8.1) satisfying the boundary conditions (8.3). The general solution of equation (8.6) for T, with λ being an eigenvalue λ_k, is

$$T_k(t) = A_k \cos\frac{k\pi t}{L} + B_k \sin\frac{k\pi t}{L}, \qquad k = 1, 2, \ldots, \tag{8.11}$$

where A_k and B_k are arbitrary constants. Substituting (8.10) and (8.11) into (8.4) we obtain the infinite collection of functions

$$u_k(x, t) = \sin\frac{k\pi x}{L}\left(A_k \cos\frac{k\pi t}{L} + B_k \sin\frac{k\pi t}{L}\right), \; k = 1, 2, \ldots, \tag{8.12}$$

each of which satisfies the wave equation (8.1) and the boundary conditions (8.3). (In (8.12) the constant D_k has been incorporated into the arbitrary constants A_k and B_k.)

Each of the functions $u_k(x, t)$ (with reasonable values of A_k and B_k) represents a possible motion of the string known as the *kth mode of vibration* or *kth harmonic*. Any one of these modes can be excited if the proper initial conditions prevail. Indeed, if the initial conditions are

$$u(x, 0) = A_k \sin\frac{k\pi x}{L}, \qquad u_t(x, 0) = B_k\frac{k\pi}{L} \sin\frac{k\pi x}{L}, \; 0 < x < L, \tag{8.13}$$

for some positive integer k, then the solution of the initial-boundary value problem (8.1), (8.13), (8.3) is the function $u_k(x, t)$ given by (8.12), and the string will vibrate in its kth mode. With more general initial conditions however, it is reasonable to expect that many, possibly all, modes of vibration will be excited at the same time. This suggests that we try to solve the initial-boundary value problem (8.1), (8.2), (8.3) by superposition of the functions $u_k(x, t)$.

Let us suppose then that the solution of the problem (8.1), (8.2), (8.3) can be written in the form of the infinite series

$$(8.14) \qquad u(x, t) = \sum_{k=1}^{\infty} \sin \frac{k\pi x}{L}\left(A_k \cos \frac{k\pi t}{L} + B_k \sin \frac{k\pi t}{L}\right),$$

where the constants A_k and B_k are to be determined. These constants must be assigned values such that (8.14) will satisfy the wave equation (8.1), the initial conditions (8.2) and the boundary conditions (8.3). We note first that if the constants A_k and B_k go to zero like k^{-4} as $k \to \infty$, i.e., if there is a constant $M > 0$ such that

$$(8.15) \qquad |A_k| \leqq Mk^{-4}, \qquad |B_k| \leqq Mk^{-4}, \qquad k = 1, 2, \dots,$$

then the series in (8.14), as well as the series obtained by termwise differentiation twice with respect to x or t, all converge uniformly for $0 \leqq x \leqq L$ and $t \geqq 0$. It follows that if the coefficients A_k, B_k satisfy the condition (8.15), then the function $u(x, t)$ defined by (8.14) is of class C^2 for $0 \leqq x \leqq L$ and $t \geqq 0$, and $u(x, t)$ satisfies the wave equation (8.1) (since each term of the series does) and the boundary conditions (8.3). In order that $u(x, t)$ satisfy the initial conditions (8.2), the constants A_k and B_k must be chosen so that

$$(8.16) \qquad \phi(x) = \sum_{k=1}^{\infty} A_k \sin \frac{k\pi x}{L}, \qquad 0 \leqq x \leqq L,$$

$$(8.17) \qquad \psi(x) = \sum_{k=1}^{\infty} \frac{k\pi}{L} B_k \sin \frac{k\pi x}{L}, \qquad 0 \leqq x \leqq L.$$

Thus the constants A_k and $(k\pi/L)B_k$, $k = 1, 2, \dots$, must be the coefficients of the Fourier sine series representations of the functions $\phi(x)$ and $\psi(x)$, respectively, on the interval $0 \leqq x \leqq L$. From formulas (8.40) of section 8, Chapter VII, these constants are given by

$$(8.18) \qquad A_k = \frac{2}{L}\int_0^L \phi(x) \sin \frac{k\pi x}{L}\, dx, \qquad k = 1, 2, \dots,$$

$$(8.19) \qquad B_k = \frac{L}{k\pi}\frac{2}{L}\int_0^L \psi(x) \sin \frac{k\pi x}{L}\, dx, \qquad k = 1, 2, \dots.$$

Moreover, it follows from an obvious modification of Theorem 8.2 of Chapter VII (see also Problem 8.13 of Chapter VII) that the representations (8.16) and (8.17) are valid in the sense of absolute and uniform convergence provided that the initial data satisfy the following conditions:

(1) ϕ and ψ are continuous and have sectionally continuous derivatives for $0 \leqq x \leqq L$; and (2) $\phi(0) = \psi(0) = \phi(\pi) = \psi(\pi) = 0$ (Problem 8.1). Since $|\sin k\pi t/L| \leq 1$ and $|\cos k\pi t/L| \leq 1$, the conditions (1) and (2) also guarantee that the series in (8.14) converges absolutely and uniformly for $0 \leqq x \leqq L$ and $t \leqq 0$ to the continuous function $u(x, t)$, which obviously satisfies the boundary condition (8.3) and the first of the initial conditions (8.2). This function u will also satisfy the second initial condition and the wave equation itself if the coefficients A_k and B_k satisfy condition (8.15). For condition (8.15) to hold it is necessary to impose additional conditions on the initial data ϕ and ψ (see Problem 8.2). In practice, however, we only check that ϕ and ψ satisfy the above conditions (1) and (2) and view (8.14) as the generalized solution of the problem (8.1), (8.2), (8.3).

Example 8.1. A guitar string of length L is pulled at the midpoint by the amount a and then released. Find the motion of the string.

We must solve the problem (8.1), (8.2), (8.3) with $\psi \equiv 0$ and

$$\phi(x) = \frac{2a}{L} \cdot \begin{cases} x, & \text{for } 0 \leqq x \leqq \dfrac{L}{2} \\ L - x, & \text{for } \dfrac{L}{2} \leqq x \leqq L. \end{cases} \tag{8.20}$$

The solution is given by (8.14). Obviously $B_k = 0$ for all k, while the A_k are the Fourier sine series coefficients of the function $\phi(x)$. Except for the multiplicative constant $2a/L$, $\phi(x)$ is the function whose Fourier sine series we obtained in Chapter VII, formula (8.41). Therefore, the motion of the string is given by

$$u(x, t) = \frac{8a}{\pi^2} \sum_{k=1}^{\infty} \frac{\sin(k\pi/2)}{k^2} \sin\frac{k\pi x}{L} \cos\frac{k\pi t}{L}. \tag{8.21}$$

By noting that the terms with even k are zero and setting $k = 2n + 1$, $n = 0, 1, \ldots$, we obtain the alternate formula

$$u(x, t) = \frac{8a}{\pi^2} \sum_{n=0}^{\infty} \frac{(-1)^n}{(2n+1)^2} \sin\frac{(2n+1)\pi x}{L} \cos\frac{(2n+1)\pi t}{L}, \tag{8.22}$$

which shows clearly that the resulting motion of the string is, in this case, the superposition of odd harmonics.

Let us examine more closely formula (8.12) for the kth harmonic or mode of vibration of the string. In order to relate more closely to the physical parameters of the string, let us return to the original time variable by replacing t with ct,

$$u_k(x, t) = \sin\frac{k\pi x}{L}\left(A_k \cos\frac{k\pi ct}{L} + B_k \sin\frac{k\pi ct}{L}\right), \tag{8.23}$$

where $c = \sqrt{T/\rho}$, T being the tension in the string and ρ its density (mass per unit length). Using a trigonometric identity, we rewrite (8.23) in the form

$$u_k(x, t) = E_k \sin \frac{k\pi x}{L} \cos k\omega(t - t_k), \tag{8.24}$$

where E_k and t_k are new arbitrary constants and

$$\omega = \frac{\pi c}{L} = \frac{\pi}{L} \sqrt{\frac{T}{\rho}}. \tag{8.25}$$

From (8.24) we see that when the string vibrates in its kth mode, each point of the string (fixed x) performs a harmonic motion with circular frequency (radians per second) $k\omega$ and amplitude proportional to $\sin(k\pi x/L)$. When $k = 1$ the string vibrates in what is known as its *fundamental mode,*

$$u_1(x, t) = E_1 \sin \frac{\pi x}{L} \cos \omega(t - t_1), \tag{8.26}$$

with *fundamental frequency* (cycles per second)

$$f = \frac{\omega}{2\pi} = \frac{1}{2L} \sqrt{\frac{T}{\rho}}. \tag{8.27}$$

The frequency of the kth harmonic is k times the fundamental frequency f. When we listen to a string which has been plucked in an arbitrary manner, we generally hear all of the harmonics, but their various frequencies are all integral multiples of the fundamental frequency f. This property of vibrating strings (which, as we will see, is not shared by vibrating membranes) is the reason why strings are capable of producing musical tones. Furthermore, the presence of the various harmonics imparts a characteristic known as "timbre" to the musical tone produced by a string, in contrast to a "pure tone" (produced, for example, by an electronic instrument) which is composed of a vibration of a single frequency. In practice, since the amplitude and energy of each harmonic decreases with increasing k, we are mainly hearing the lower harmonics. Moreover, under special initial conditions, some of the harmonics may be altogether absent. For instance, in Example 8.1 all even harmonics are missing because the initial displacement of the string is symmetric with respect to its middle $x = L/2$. It is even possible to eliminate some of the harmonics, and consequently raise the pitch of the emitted tone, by lightly touching one's finger at certain points of the string after the string has been plucked. To see this, consider Figure 8.1, which shows the graphs of the amplitudes (up to a multiplicative constant) of the first four harmonics. The points where the amplitude is zero, and which therefore remain stationary, are known as *nodes*. The fundamental mode has only two nodes, the ends $x = 0$ and $x = L$. The second harmonic has three nodes, the ends and the midpoint $x = L/2$. The third harmonic has four nodes, etc. Suppose now that after the player has plucked the string, he then gently touches the midpoint of the string with his finger. The effect of this action is to force the midpoint into a node and thus eliminate all harmonics except those which have the midpoint as a node. In particular, the fundamental mode will disappear and the lowest emitted frequency will be that of the second harmonic, namely $2f$. Thus the pitch of the emitted musical tone will be doubled.

Formula (8.27) shows the dependence of the fundamental frequency of a vibrating string on its length, tension and density. It is not difficult to see how these three factors enter into the construction, tuning and playing of a guitar (Problem 8.4).

As another illustration of the method of separation of variables and Fourier series, let us study the vibrations of a string of length L with both ends free to slide on frictionless vertical posts. Instead of (8.3), the boundary conditions in this case are

$$u_x(0, t) = 0, \quad u_x(L, t) = 0; \quad 0 \leqq t. \tag{8.28}$$

We must solve the initial-boundary value problem (8.1), (8.2), (8.28). After separation of variables, the eigenvalue problem to be solved is

$$\begin{cases} X'' + \lambda X = 0 \\ X'(0) = X'(L) = 0. \end{cases} \tag{8.29}$$

In this case $\lambda = 0$ is an eigenvalue of the problem with constants as the corresponding eigenfunctions. The sequence of eigenvalues is

$$\lambda_0 = 0; \quad \lambda_k = \frac{k^2\pi^2}{L^2}, \quad k = 1, 2, \ldots \tag{8.30}$$

and the corresponding eigenfunctions are

$$X_0(x) = D_0; \quad X_k(x) = D_k \cos\frac{k\pi x}{L}, \quad k = 1, 2, \ldots \tag{8.31}$$

where $D_0, D_1, D_2, \ldots$ are arbitrary constants. The general solution of the T equation (8.6), when λ is an eigenvalue, is

$$\begin{aligned} T_0(t) &= \frac{A_0}{2} + \frac{B_0}{2}t; \\ T_k(t) &= A_k \cos\frac{k\pi t}{L} + B_k \sin\frac{k\pi t}{L}, \quad k = 1, 2, \ldots \end{aligned} \tag{8.32}$$

where $A_0, A_1, A_2, \ldots, B_0, B_1, B_2, \ldots$ are arbitrary constants. Each of the functions

$$\begin{cases} u_0(x, t) = \dfrac{A_0}{2} + \dfrac{B_0}{2}t \\ u_k(x, t) = \cos\dfrac{k\pi x}{L}\left(A_k \cos\dfrac{k\pi t}{L} + B_k \sin\dfrac{k\pi t}{L}\right), \quad k = 1, 2, \ldots \end{cases} \tag{8.33}$$

satisfies the wave equation (8.1) and the boundary conditions (8.28). The functions $u_k(x, t)$, with $k = 1, 2, \ldots$, represent the various modes of vibration of the string, while $u_0(x, t)$ represents a translation (rigid motion) which is possible since the ends of the string are assumed to be free, allowing the string to slide up or down. Superposition of the functions (8.33) yields

$$u(x, t) = \frac{A_0}{2} + \frac{B_0}{2}t + \sum_{k=1}^{\infty} \cos\frac{k\pi x}{L}\left(A_k \cos\frac{k\pi t}{L} + B_k \sin\frac{k\pi t}{L}\right). \tag{8.34}$$

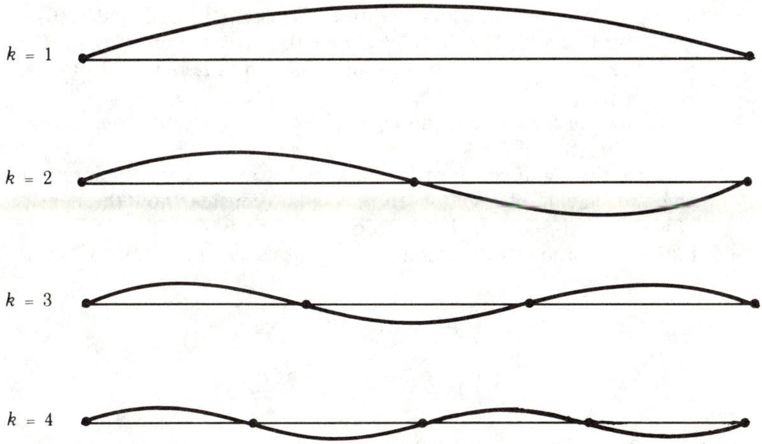

Fig. 8.1

In order for (8.34) to satisfy the initial conditions (8.2) the constants A_k and B_k must be chosen so that

$$\phi(x) = \frac{A_0}{2} + \sum_{k=1}^{\infty} A_k \cos\frac{k\pi x}{L}, \qquad 0 \leqq x \leqq L, \tag{8.35}$$

$$\psi(x) = \frac{B_0}{2} + \sum_{k=1}^{\infty} \frac{k\pi}{L} B_k \cos\frac{k\pi x}{L}, \qquad 0 \leqq x \leqq L. \tag{8.36}$$

Thus the constants $A_0, A_1, A_2, \ldots$, and $B_0, (\pi/L)B_1, (2\pi/L)B_2, \ldots$ must be the coefficients of the Fourier cosine series expansions of the functions $\phi(x)$ and $\psi(x)$, respectively, on the interval $0 \leqq x \leqq L$. From formulas (8.39) of Section 8, Chapter VII, these constants are given by

$$A_k = \frac{2}{L}\int_0^L \phi(x)\cos\frac{k\pi x}{L}\,dx, \qquad k = 0, 1, 2, \ldots \tag{8.37}$$

$$\begin{aligned} B_0 &= \frac{2}{L}\int_0^L \psi(x)dx; \\ B_k &= \frac{L}{k\pi}\frac{2}{L}\int_0^L \psi(x)\cos\frac{k\pi x}{L}\,dx, \qquad k = 1, 2, \ldots . \end{aligned} \tag{8.38}$$

The solution of the initial-boundary value problem (8.1), (8.2), (8.28) is given by the series (8.34), where the coefficients in this series are given by formulas (8.37) and (8.38).

Problems

8.1. Prove the assertion following formula (8.15).
8.2. Show that A_k and B_k will satisfy condition (8.15) if the initial data ϕ

and ψ satisfy the following conditions (a) $\phi \in C^4([0, L])$; $\phi^{(p)}(0) = \phi^{(p)}(\pi) = 0, p = 0, 1, 2, 3$. (b) $\psi \in C^3([0, L])$; $\psi^{(p)}(0) = \psi^{(p)}(\pi) = 0, p = 0, 1, 2$. [*Hint:* Use an obvious modification of Lemma 8.2 of Chapter VII.]

8.3. How are the constants E_k and t_k in (8.24) related to the constants A_k and B_k in (8.23)?

8.4. Discuss the significance of formula (8.27) in the construction, tuning and playing of a guitar. In particular, consider how the density ρ is made large on the lower frequency strings.

8.5 Find the solution of the initial value problem (8.1), (8.2), (8.3) with $\phi(x) \equiv 0$ and

$$\psi(x) = \begin{cases} x & \text{for } 0 \leqq x \leqq \dfrac{L}{2} \\ L - x & \text{for } \dfrac{L}{2} \leqq x \leq L. \end{cases}$$

8.6. A string of length L with both ends fixed, such as the string of a piano, is struck at its midportion, the impact thus imparting to the string an initial velocity $\psi(x)$ in the shape of the rectangular pulse given by the right side of equation (8.42) of Chapter VII. Find the motion of the string.

8.7. A string of length π with fixed ends is displaced into a parabolic arc and then released from rest ($u_t(x, 0) = 0$). Assuming that

$$u(x, 0) = bx(\pi - x), \quad 0 \leqq x \leqq \pi,$$

where b is a small constant, derive the formula for the motion of the string,

$$u(x, t) = \frac{4b}{\pi} \sum_{n=1}^{\infty} \frac{1 - (-1)^n}{n^3} \sin nx \cos nt.$$

8.8. Solve the problem (8.1), (8.2), (8.3) with initial data

$$\phi(x) = a \sin \frac{2\pi x}{L}, \quad \psi(x) = b \sin \frac{5\pi x}{L}, \quad 0 \leqq x \leqq L,$$

where a and b are small constants.

8.9. Show that the solution of problem (8.1), (8.2), (8.3) with $L = \pi$ and initial data

$$\phi(x) = \frac{2a}{\pi} \begin{cases} x & \text{for } 0 \leqq x \leqq \dfrac{\pi}{2} \\ \pi - x & \text{for } \dfrac{\pi}{2} \leqq x \leqq \pi \end{cases} ; \psi(x) = bx(\pi - x), 0 \leqq x \leqq \pi$$

is given by $u(x, t) =$

$$\frac{8a}{\pi^2} \sum_{n=1}^{\infty} \frac{\sin(n\pi/2)}{n^2} \sin nx \cos nt + \frac{4b}{\pi} \sum_{n=1}^{\infty} \frac{1 - (-1)^n}{n^4} \sin nx \sin nt.$$

8.10. Supply all the details of the method of separation of variables and Fourier series for solving the initial-boundary value problem for the string with free ends, which was outlined at the end of this section.

8.11. Consider the following initial-boundary value problem for the non-homogeneous wave equation

$$(8.39)\quad u_{xx} - u_{tt} = f(x, t); \qquad 0 < x < L, \qquad 0 < t,$$

$$(8.40)\qquad u(x, 0) = 0, \qquad u_t(x, 0) = 0; \qquad 0 \leqq x \leqq L,$$

$$(8.3)\qquad u(0, t) = 0, \qquad u(L, t) = 0, \qquad 0 \leqq t;$$

which arises in the study of forced vibrations of a string with zero initial conditions. Use a modification of Duhamel's principle (Problem 6.5) to obtain the following formula for the solution

$$(8.41)\quad u(x, t) = \sum_{k=1}^{\infty} \left[\int_0^t f_k(\tau) \sin \frac{k\pi(t - \tau)}{L}\, d\tau \right] \sin \frac{k\pi x}{L}$$

where

$$(8.42)\quad f_k(\tau) = -\frac{2}{k\pi} \int_0^L f(x, \tau) \sin \frac{k\pi x}{L}\, dx, \qquad k = 1, 2, \ldots .$$

What is the solution of problem (8.39), (8.2), (8.3)? [*Remark:* In deriving (8.41), do not worry about series convergence or the validity of the interchange of summation and integration. If $f(x, t)$ is assumed to be sufficiently smooth, it is easy to show that (8.41) will converge and will satisfy (8.39), (8.40), (8.3). When a formula such as (8.41) for the solution of a problem is obtained in this carefree manner, this formula is sometimes referred to as the *formal solution* of the problem.]

8.12. If $f(x, t) = F(t) \sin(\pi x/L)$, use Problem 8.11 to derive the following formula for the solution of problem (8.39), (8.40), (8.3),

$$u(x, t) = -\frac{L}{\pi} \sin \frac{\pi x}{L} \int_0^t F(\tau) \sin \frac{\pi(t - \tau)}{L}\, d\tau.$$

9. Vibrations of a Rectangular Membrane

Consider a stretched membrane which is fastened to a rectangular frame of length a and width b. In order to study its vibrations we must solve the initial-boundary value problem

$$(9.1)\quad u_{xx} + u_{yy} - u_{tt} = 0; \qquad 0 < x < a, \qquad 0 < y < b, \qquad 0 < t,$$

$$(9.2)\quad \begin{cases} u(x, y, 0) = \phi(x, y), \\ u_t(x, y, 0) = \psi(x, y); \end{cases} \qquad 0 \leqq x \leqq a, \qquad 0 \leqq y \leqq b,$$

$$(9.3)\quad \begin{cases} u(0, y, t) = 0, \quad u(a, y, t) = 0; \quad 0 \leqq y \leqq b, \quad 0 \leqq t \\ u(x, 0, t) = 0, \quad u(x, b, t) = 0; \quad 0 \leqq x \leqq a, \quad 0 \leqq t. \end{cases}$$

Our method of solution will be the same as that used for the vibrating string. First, using separation of variables we will obtain a collection of functions satisfying the wave equation (9.1) and the boundary conditions (9.3). Then we will form a superposition of these functions which will also satisfy the initial conditions (9.2).

We begin by looking for solutions of (9.1) which have the form

$$u(x, y, t) = v(x, y)T(t). \tag{9.4}$$

By the usual method of separation of variables we find that in order for a function u of the form (9.4) to satisfy (9.1), the functions v and T must satisfy the equations

$$v_{xx} + v_{yy} + \lambda v = 0; \qquad 0 < x < a, \qquad 0 < y < b \tag{9.5}$$

$$T'' + \lambda T = 0; \qquad 0 < t. \tag{9.6}$$

Moreover, (except for the trivial and uninteresting case $u \equiv 0$), in order for (9.4) to satisfy the boundary conditions (9.3), $v(x, y)$ must satisfy the boundary conditions

$$\begin{cases} v(0, y) = 0, & v(a, y) = 0; \quad 0 \leqq y \leqq b \\ v(x, 0) = 0, & v(x, b) = 0; \quad 0 \leqq x \leqq a. \end{cases} \tag{9.7}$$

We are faced then with the boundary value problem (9.5), (9.7) for the elliptic partial differential equation (9.5). This is an eigenvalue problem. We must find the values of λ (eigenvalues) for which the problem has nontrivial solutions (eigenfunctions).

It turns out that the eigenvalue problem (9.5), (9.7) can be solved by the method of separation of variables. (This is due to the simplicity of the rectangular domain for the problem.) Let us look for solutions of (9.5) of the form

$$v(x, y) = X(x)Y(y). \tag{9.8}$$

Substitution of (9.8) into (9.5) and separation of variables yields

$$\frac{Y''}{Y} + \lambda = -\frac{X''}{X}.$$

Since the left side is a function of y only while the right side is a function of x only, each side must be equal to the same constant, say μ. Hence

$$X'' + \mu X = 0, \qquad 0 < x < a \tag{9.9}$$

$$Y'' + \nu Y = 0, \qquad 0 < y < b \tag{9.10}$$

where $\nu = \lambda - \mu$. The boundary conditions (9.7) lead to the boundary conditions for $X(x)$ and $Y(y)$,

$$X(0) = 0, \qquad X(a) = 0 \tag{9.11}$$

$$Y(0) = 0, \qquad Y(b) = 0. \tag{9.12}$$

Problems (9.9), (9.11) and (9.10), (9.12) are eigenvalue problems of the type that we solved in the previous section. Their eigenvalues and corresponding eigenfunctions are

$$\mu_m = \frac{m^2\pi^2}{a^2}, \qquad X_m(x) = D_m \sin\frac{m\pi x}{a}; \qquad m = 1, 2, \ldots$$

$$\nu_n = \frac{n^2\pi^2}{b^2}, \qquad Y_n(y) = E_n \sin\frac{n\pi y}{b}; \qquad n = 1, 2, \ldots .$$

Since $\lambda = \mu + \nu$, we conclude that the eigenvalue problem (9.5), (9.7) has a double sequence of eigenvalues

$$\text{(9.13)} \qquad \lambda_{mn} = \pi^2\left(\frac{m^2}{a^2} + \frac{n^2}{b^2}\right); \qquad m, n, = 1, 2, \ldots ,$$

with corresponding eigenfunctions

$$\text{(9.14)} \quad v_{mn}(x, y) = D_{mn} \sin\frac{m\pi x}{a} \sin\frac{n\pi y}{b}; \qquad m, n = 1, 2, \ldots ,$$

where D_{mn} are arbitrary constants.

When λ is one of the eigenvalues (9.13), the general solution of the T equation (9.6) is

$$\text{(9.15)} \quad T_{mn}(t) = A_{mn} \cos\sqrt{\lambda_{mn}}\, t + B_{mn} \sin\sqrt{\lambda_{mn}}\, t; \qquad m, n = 1, 2, \ldots .$$

Substitution of (9.14) and (9.15) into (9.4) yields the double sequence of functions

$$\text{(9.16)} \quad u_{mn}(x, y, t) = \sin\frac{m\pi x}{a} \sin\frac{n\pi y}{b} (A_{mn} \cos\sqrt{\lambda_{mn}}\, t + B_{mn} \sin\sqrt{\lambda_{mn}}\, t); \qquad m, n = 1, 2, \ldots ,$$

each of which satisfies the wave equation (9.1) and the boundary conditions (9.3). If the constants A_{mn} and B_{mn} tend to zero sufficiently rapidly as $m, n \to \infty$, then the superposition

$$\text{(9.17)} \quad u(x, t) = \sum_{m,n=1}^{\infty} \sin\frac{m\pi x}{a} \sin\frac{n\pi y}{b} \cdot (A_{mn} \cos\sqrt{\lambda_{mn}}\, t + B_{mn} \sin\sqrt{\lambda_{mn}}\, t)$$

will also satisfy (9.1) and (9.3). In order for (9.17) to satisfy the initial conditions (9.2), the constants A_{mn} and B_{mn} must be chosen so that

$$\text{(9.18)} \qquad \phi(x, y) = \sum_{m,n=1}^{\infty} A_{mn} \sin\frac{m\pi x}{a} \sin\frac{n\pi y}{b}$$

$$\text{(9.19)} \qquad \psi(x, y) = \sum_{m,n=1}^{\infty} \sqrt{\lambda_{mn}} B_{mn} \sin\frac{m\pi x}{a} \sin\frac{n\pi y}{b} .$$

Here we need a theorem asserting that a function of two variables which is defined in a rectangle $0 \leqq x \leqq a$, $0 \leqq y \leqq b$ and satisfies certain conditions can be represented by a *double Fourier sine series*. Such a theorem can be found for example in the book of Tolstov.[2] If the representation (9.18) is assumed to be valid in the sense of uniform convergence on the rectangle $0 \leqq x \leqq a$, $0 \leqq y \leqq b$, then it is easy to obtain the following formula for the coefficients,

$$(9.20) \quad A_{mn} = \frac{4}{ab} \int_0^a \int_0^b \phi(x, y) \sin \frac{m\pi x}{a} \sin \frac{n\pi y}{b} \, dxdy; \qquad m, n = 1, 2, \ldots .$$

The formula for the coefficients B_{mn} is obvious.

Let us discuss now the eigenvalue problem (9.5), (9.7) and the associated modes of vibration of the membrane. For simplicity we consider only square membranes, $a = b$. Concerning the eigenvalue problem (9.5), (9.7) we make the following important observation. When $m = n$ there is only one linearly independent eigenfunction

$$(9.21) \qquad v_{mm}(x, y) = \sin \frac{m\pi x}{a} \sin \frac{m\pi y}{a}$$

corresponding to the eigenvalue

$$(9.22) \qquad \lambda_{mm} = \frac{\pi^2}{a^2} 2m^2.$$

However, when $m \neq n$ there are two linearly independent eigenfunctions

$$(9.23) \quad v_{mn}(x, y) = \sin \frac{m\pi x}{a} \sin \frac{n\pi y}{a}, \qquad v_{nm}(x, y) = \sin \frac{n\pi x}{a} \sin \frac{m\pi y}{a}$$

corresponding to the same eigenvalue

$$(9.24) \qquad \lambda_{mn} = \lambda_{nm} = \frac{\pi^2}{a^2} (m^2 + n^2).$$

Thus, the situation here is in contrast to that of the one-dimensional eigenvalue problem (8.5), (8.7) for which there is only one linearly independent eigenfunction corresponding to each eigenvalue. Now, a mode of vibration of the membrane corresponding to the eigenvalue λ_{mn} is given by

$$u_{mn}(x, y, t) = v_{mn}(x, y)[A_{mn} \cos \sqrt{\lambda_{mn}}\, t + B_{mn} \sin \sqrt{\lambda_{mn}}\, t]$$

where v_{mn} is an eigenfunction corresponding to λ_{mn}. For simplicity let us take $B_{mn} = 0$ (the effect of this is to pick a convenient value for the phase of the motion) and let us incorporate the arbitrary constant A_{mn} into v_{mn}. Also, let us return to the real time variable by replacing t by ct where $c = \sqrt{T/\rho}$, T being the tension and ρ the density of the membrane. Then, a mode of vibration corresponding to λ_{mn} is given by

$$(9.25) \qquad u_{mn}(x, y, t) = v_{mn}(x, y) \cos \sqrt{\lambda_{mn}}\, ct,$$

where v_{mn} is any eigenfunction corresponding to λ_{mn}. From (9.25) it follows that each point of the membrane performs a simple harmonic motion with frequency

$$f_{mn} = \frac{c\sqrt{\lambda_{mn}}}{2\pi} = \frac{c}{2a}\sqrt{m^2 + n^2} \tag{9.26}$$

and amplitude equal to the value of the eigenfunction v_{mn} at that point. The fundamental (lowest) frequency of the membrane is

$$f_{11} = \frac{c}{2a}\sqrt{2}. \tag{9.27}$$

From (9.26) and (9.27) it follows that, in contrast to a vibrating string, the frequencies of the higher modes of vibration of a square membrane are not integral multiples of the fundamental frequency. When a drum is struck in an arbitrary manner, many modes of vibration are simultaneously present, but the various frequencies that we hear are not integral multiples of the same frequency. This accounts for the lack of musical quality of a tone emitted by a drum. It is possible however, at least in theory, to excite only a single mode of vibration given by (9.25). Remembering that $v_{mn}(x, y)$ represents the amplitude of motion of the points of the membrane, the points (x, y) satisfying

$$v_{mn}(x, y) = 0 \tag{9.29}$$

form curves known as *nodal curves.* The points of the membrane lying on nodal curves of v_{mn} remain stationary. Clearly the nodal curves of each of the two eigenfunctions (9.23) are straight lines. However, since linear combinations of the eigenfunctions (9.23) are also eigenfunctions corresponding to the same eigenvalue (9.24), the nodal curves corresponding to these eigenfunctions can assume many complex forms, as illustrated in Figure 9.1.

Problems

9.1. Derive formula (9.20) for the coefficients of the double Fourier sine series representation (9.18) of the function ϕ under the assumption of

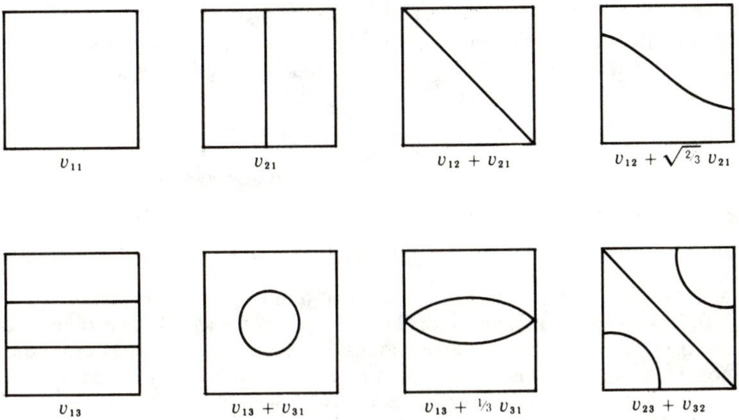

Fig. 9.1

uniform convergence of the series.

9.2. Show that the nodal curves of modes of vibration of a square membrane corresponding to eigenvalues of the form (9.22) are always straight lines.

9.3. Discuss the lowest frequency emitted by a vibrating square membrane in terms of its physical parmeters.

10. Vibrations in Finite Regions. The General Method of Separation of Variables and Eigenfunction Expansions. Vibrations of a Circular Membrane

The problems of a vibrating string and of a vibrating rectangular membrane are special cases of a general problem of vibrations in finite regions. The study of this problem requires the solution of the following initial-boundary value problem for the wave equation in a bounded domain Ω of R^n with piecewise smooth boundary: Find the function $u(x, t)$ defined for $x \in \bar{\Omega}$ and $t \geqq 0$, and satisfying the wave equation

$$\nabla^2 u - u_{tt} = 0, \quad \text{for} \quad x \in \Omega \quad \text{and} \quad t > 0, \tag{10.1}$$

the initial conditions

$$u(x, 0) = \phi(x), \qquad u_t(x, 0) = \psi(x), \quad \text{for} \quad x \in \bar{\Omega} \tag{10.2}$$

and the boundary condition

$$u(x, t) = 0, \quad \text{for} \quad x \in \partial\Omega \quad \text{and} \quad t \geqq 0. \tag{10.3}$$

The method of separation of variables and Fourier series which was employed in the last two sections can be generalized (at least in theory) to obtain a formula for the solution of (10.1), (10.2), (10.3). The formula will involve series expansions of the initial data ϕ and ψ in terms of the eigenfunctions of an associated eigenvalue problem for the Laplacian operator.

The method begins by looking for functions $u(x, t)$ of the form

$$u(x, t) = v(x)T(t)$$

which satisfy the wave equation (10.1) and the boundary condition (10.3). By the usual arguments of the method of separation of variables, $v(x)$ must satisfy

$$\nabla^2 v + \lambda v = 0 \quad \text{for} \quad x \in \Omega \tag{10.4}$$

$$v = 0 \quad \text{for} \quad x \in \partial\Omega \tag{10.5}$$

and $T(t)$ must be a solution of the o.d.e.

$$T'' + \lambda T = 0 \quad \text{for} \quad t > 0 \tag{10.6}$$

where $-\lambda$ is the separation constant. The boundary value problem (10.4), (10.5) is an eigenvalue problem. The values of λ for which the problem has nontrivial solutions are the eigenvalues of the problem. The corresponding nontrivial solutions are the eigenfunctions. Just as in the case of the Dirichlet problem for Laplace's equation, a solution of (10.4), (10.5) is required to be in $C^2(\Omega) \cap C^0(\bar{\Omega})$.

We summarize here the properties of the eigenvalues and eigenfunctions of the eigenvalue problem (10.4), (10.5):

1. The eigenvalues are all positive.

2. If v_1 and v_2 are eigenfunctions corresponding to the eigenvalues λ_1 and λ_2 respectively, and if $\lambda_1 \neq \lambda_2$, then

$$\int_\Omega v_1(x)v_2(x)dx = 0. \tag{10.7}$$

Any two functions v_1, v_2 satisfying (10.7) are said to be *orthogonal.* Accordingly, property 2 states that eigenfunctions corresponding to distinct eigenvalues are orthogonal.

3. The eigenvalues form a countable set of numbers $\{\lambda_n: n = 1, 2, ...\}$. Moreover $\lambda_n \to \infty$ as $n \to \infty$.

4. Each eigenvalue has a finite number of linearly independent eigenfunctions corresponding to it. This number is called the *multiplicity* of the eigenvalue. By a process known as *orthogonalization,* it is possible to form linear combinations of these eigenfunctions which are mutually orthogonal. The number of mutually orthogonal eigenfunctions thus formed is equal to the multiplicity of the eigenvalue.

5. Every eigenfunction belongs to $C^1(\bar{\Omega})$ and is analytic in Ω.

Assuming property 5, properties 1 and 2 can be easily proved using Green's identities described in Section 1 of Chapter VI. The proof of property 1 is the subject of Problem 1.4 of Chapter VI. The proof of property 2 is outlined in Problem 10.1. The proofs of properties 3 and 4 and the proof of the eigenfunction expansion theorem stated below involve the theory of integral equations and can be found, for example, in the book of Vladimirov,[3] §24.6. The proof of analyticity of an eigenfunction in Ω is similar to the proof of the corresponding property for harmonic functions. That an eigenfunction belongs to $C^1(\bar{\Omega})$ can be proved by the methods of potential theory (see, for example, Sobolev,[4] Lecture 15). It should be noted that the proofs of some of these results may require additional smoothness assumptions for the boundary of Ω.

On the basis of the above stated properties, the eigenvalues of (10.4), (10.5) may be listed in order of increasing magnitude

$$0 < \lambda_1 \leqq \lambda_2 \leqq \lambda_3 \leqq \dots \leqq \lambda_n \leqq \dots ; \qquad \lambda_n \to \infty \quad \text{as} \quad n \to \infty, \tag{10.8}$$

with an eigenvalue being listed in this sequence as many times as its multiplicity. The corresponding eigenfunctions

$$v_1, v_2, v_3, \dots, v_n, \dots \tag{10.9}$$

are chosen so that they are mutually orthogonal, i.e.

$$\int_\Omega v_m(x)v_n(x)dx = 0, \qquad m \neq n. \tag{10.10}$$

The following theorem concerning the representation of functions by means of eigenfunction expansions is of fundamental importance.

Theorem 10.1. *(Eigenfunction expansion theorem.)* Suppose that $f \in C^2(\bar{\Omega})$ and $f(x) = 0$ for $x \in \partial\Omega$. Then f can be represented by a series involving the eigenfunctions (10.9) of the eigenvalue problem (10.4), (10.5),

$$f(x) = \sum_{n=1}^{\infty} a_n v_n(x), \qquad x \in \bar{\Omega}, \tag{10.11}$$

with the series converging absolutely and uniformly to f on $\bar{\Omega}$. The coefficients in (10.11) are given by

$$a_n = \frac{\int_{\Omega} f(x) v_n(x) dx}{\int_{\Omega} [v_n(x)]^2 dx} \tag{10.12}$$

Example 10.1. The eigenvalue problem for the "one-dimensional Laplacian,"

$$v'' + \lambda v = 0, \qquad 0 < x < L$$

$$v(0) = v(L) = 0,$$

was solved in Section 8. The eigenvalues are

$$\lambda_n = \frac{n^2\pi^2}{L^2}, \qquad n = 1, 2, \dots ,$$

with corresponding eigenfunctions

$$v_n(x) = \sin\frac{n\pi x}{L}, \qquad n = 1, 2, \dots .$$

Each eigenvalue has multiplicity 1. The orthogonality relation (10.10) is, in this case,

$$\int_0^L \sin\frac{m\pi x}{L}\sin\frac{n\pi x}{L}\, dx = 0, \qquad m \neq n,$$

and can be proved immediately. The eigenfunction expansion theorem is, in this case, the Fourier sine series representation theorem (see Theorem 8.2 and Problem 8.13, Chapter VII).

Example 10.2. The eigenvalue problem for the two-dimensional Laplacian

$$v_{xx} + v_{yy} + \lambda v = 0; \qquad 0 < x < a, \qquad 0 < y < a$$

$$v(0, y) = v(a, y) = 0, \qquad 0 \leqq y \leqq a$$

$$v(x, 0) = v(x, a) = 0, \qquad 0 \leqq x \leqq a$$

was solved in Section 9. It is convenient in this case to use a double indexing of the eigenvalues and eigenfunctions. The eigenvalues are

$$\lambda_{mn} = \frac{\pi^2}{a^2}(m^2 + n^2); \qquad m, n = 1, 2, \ldots$$

with corresponding eigenfunctions

$$v_{mn} = \sin\frac{m\pi x}{a}\sin\frac{n\pi y}{a}; \qquad m, n = 1, 2, \ldots .$$

The eigenvalues λ_{mm}, $m = 1, 2, \ldots$ have multiplicity 1. The eigenvalues λ_{mn}, $m \neq n; m, n = 1, 2, \ldots$, have multiplicity 2 since $\lambda_{mn} = \lambda_{nm}$ and the eigenfunctions

$$\sin\frac{m\pi x}{a}\sin\frac{n\pi y}{a}, \ \sin\frac{n\pi x}{a}\sin\frac{m\pi y}{a}$$

are linearly independent and both correspond to λ_{mn}. The orthogonality relation (10.10) is, in this case,

$$\int_0^a\int_0^a \sin\frac{m\pi x}{a}\sin\frac{n\pi y}{a}\sin\frac{m'\pi x}{a}\sin\frac{n'\pi y}{a}\,dxdy = 0, \quad (m, n) \neq (m', n'); \qquad m, n, m', n' = 1, 2, \ldots \tag{10.13}$$

(Problem 10.3). The eigenfunction expansion theorem is, in this case, the representation theorem of a function of two variables by a double Fourier sine series which was mentioned in Section 9. We have

$$f(x, y) = \sum_{m,n=1}^{\infty} a_{mn}\sin\frac{m\pi x}{a}\sin\frac{n\pi y}{a}; \qquad 0 \leqq x \leqq a,\ 0 \leqq y \leqq a, \tag{10.14}$$

where

$$a_{mn} = \frac{4}{a^2}\int_0^a\int_0^a f(x, y)\sin\frac{m\pi x}{a}\sin\frac{n\pi y}{a}\,dxdy. \tag{10.15}$$

Let us return now to our original initial-boundary value problem (10.1), (10.2), (10.3). The general solution of the T equation (10.6), with λ being one of the eigenvalues λ_n of the eigenvalue problem (10.4), (10.5), is

$$T_n(t) = A_n\cos\sqrt{\lambda_n}\,t + B_n\sin\sqrt{\lambda_n}\,t, \qquad n = 1, 2, \ldots . \tag{10.16}$$

Multiplying the eigenfunction $v_n(x)$ by the functions $T_n(t)$ we obtain the sequence of functions

$$u_n(x, t) = v_n(x)[A_n\cos\sqrt{\lambda_n}\,t + B_n\sin\sqrt{\lambda_n}\,t], \qquad n = 1, 2, \ldots \tag{10.17}$$

each of which satisfies the wave equation (10.1) and the boundary condition (10.3). Each of the functions (10.17) represents a possible mode of vibration in the domain Ω. If the constants A_n, B_n tend to zero sufficiently rapidly as $n \to \infty$, then the superposition

$$u(x, t) = \sum_{n=1}^{\infty} v_n(x)[A_n\cos\sqrt{\lambda_n}\,t + B_n\sin\sqrt{\lambda_n}\,t] \tag{10.18}$$

will also satisfy (10.1) and (10.3). In order for (10.18) to satisfy the initial conditions (10.2) the constants A_n and B_n must be chosen so that

$$\phi(x) = \sum_{n=1}^{\infty} A_n v_n(x) \tag{10.19}$$

$$\psi(x) = \sum_{n=1}^{\infty} \sqrt{\lambda_n} B_n v_n(x). \tag{10.20}$$

Thus A_n and $\sqrt{\lambda_n} B_n$, $n = 1, 2, \ldots$ must be the coefficients of the eigenfunction expansions of the initial data ϕ and ψ respectively. From Theorem 10.1,

$$A_n = \frac{\int_\Omega \phi(x) v_n(x) dx}{\int_\Omega [v_n(x)]^2 dx}, \quad B_n = \frac{\int_\Omega \psi(x) v_n(x) dx}{\sqrt{\lambda_n} \int_\Omega [v_n(x)]^2 dx} \tag{10.21}$$

We conclude that, under appropriate smoothness conditions on ϕ and ψ, the solution of the initial-boundary value problem (10.1), (10.2), (10.3) is given by the series (10.18) with coefficients given by (10.21).

Example 10.3. *(Vibrations of a circular membrane.)* For the study of vibrations of a circular membrane of radius a it is convenient to use polar coordinates. The initial-boundary value problem to be solved is

$$\frac{1}{r}(ru_r)_r + \frac{1}{r^2} u_{\theta\theta} - u_{tt} = 0; \quad 0 \leqq r < a, \ -\pi \leqq \theta \leqq \pi, \quad 0 < t, \tag{10.22}$$

$$\begin{cases} u(r, \theta, 0) = \phi(r, \theta), \\ u_t(r, \theta, 0) = \psi(r, \theta); \end{cases} \quad 0 \leqq r \leqq a, \quad -\pi \leqq \theta \leqq \pi \tag{10.23}$$

$$u(a, \theta, t) = 0; \quad -\pi \leqq \theta \leqq \pi, \quad 0 \leqq t. \tag{10.24}$$

We separate the space and time variables by looking for solutions of (10.22) and (10.24) of the form

$$u(r, \theta, t) = v(r, \theta) T(t). \tag{10.25}$$

The function v must satisfy the eigenvalue problem (10.4), (10.5) which in polar coordinates has the form

$$\frac{1}{r}(rv_r)_r + \frac{1}{r^2} v_{\theta\theta} + \lambda v = 0; \quad 0 < r < a, \ -\pi \leqq \theta \leqq \pi \tag{10.26}$$

$$v(a, \theta) = 0, \quad -\pi \leqq \theta \leqq \pi \tag{10.27}$$

while the function T must satisfy equation (10.6).

The simplicity of the domain $\Omega = \{(r, \theta): 0 \leqq r < a, -\pi \leqq \theta \leqq \pi\}$ allows us to solve (10.26), (10.27) by the method of separation of variables. We look for nontrivial solutions $v(r, \theta)$ of (10.26), (10.27) which are in $C^2(\Omega) \cap C^0(\bar{\Omega})$ and have the form

$$v(r, \theta) = R(r)\Theta(\theta). \tag{10.28}$$

Following the usual procedure of separation of variables we find that Θ and R must satisfy, respectively,

$$\Theta'' + \mu\Theta = 0, \qquad -\pi \leqq \theta \leqq \pi \tag{10.29}$$

$$\Theta(\pi) = \Theta(-\pi), \qquad \Theta'(\pi) = \Theta'(-\pi) \tag{10.30}$$

and

$$r^2R'' + rR' + (\lambda r^2 - \mu)R = 0, \qquad 0 < r < a \tag{10.31}$$

$$|R(0)| < \infty, \qquad R(a) = 0, \tag{10.32}$$

where μ is the separation constant. Conditions (10.30) and the first of conditions (10.32) follow from the smoothness requirements on v.

Problem (10.29), (10.30) is an eigenvalue problem. Conditions (10.30) are known as periodic boundary conditions. The eigenvalues of the problem are

$$\mu_n = n^2, \qquad n = 0, 1, 2, \ldots \tag{10.33}$$

with corresponding linearly independent eigenfunctions

$$\Theta_0(\theta) = 1, \qquad \Theta_n(\theta) = \cos n\theta,\ \sin n\theta; \qquad n = 1, 2, \ldots . \tag{10.34}$$

Thus, the eigenvalue 0 is of multiplicity 1, while all the others are of multiplicity 2.

For each fixed value n^2 of μ, problem (10.31), (10.32) is an eigenvalue problem of still another type: it has nontrivial solutions (eigenfunctions) only for certain values of λ (eigenvalues). Equation (10.31) with $\mu = n^2$,

$$r^2R'' + rR' + (\lambda r^2 - n^2)R = 0, \tag{10.35}$$

is *Bessel's equation* of order n with parameter λ (see Section 1 of this chapter). The details of finding the eigenvalues and eigenfunctions of (10.35), (10.32) are tedious and we do not present them here (see, for example, Churchill,[5] Chapter 8). The functions $J_n(x)$, known as *Bessel's functions of the first kind of order n,* have been extensively studied and their values have been tabulated (see Equation (1.37)). Each $J_n(x)$ vanishes for an infinite sequence of positive values of x tending to $+\infty$. The eigenvalues of (10.35), (10.32) are the positive roots $\lambda = \lambda_{nj}; j = 1, 2, \ldots ,$ of the equation

$$J_n(\sqrt{\lambda}a) = 0, \tag{10.36}$$

and the corresponding eigenfunctions are

$$R_{nj}(r) = J_n(\sqrt{\lambda_{nj}}\, r); \quad n = 0, 1, 2, \ldots ; \qquad j = 1, 2, \ldots . \tag{10.37}$$

Let us return now to our original eigenvalue problem (10.26), (10.27). Substituting (10.34) and (10.37) into (10.28) we obtain the eigenfunctions

$$v_{nj} = J_n(\sqrt{\lambda_{nj}}\, r)\cos n\theta, \qquad J_n(\sqrt{\lambda_{nj}}\, r)\sin n\theta; \quad n = 0, 1, 2, \ldots ; \qquad j = 1, 2, \ldots , \tag{10.38}$$

corresponding to the eigenvalues

(10.39) $$\lambda_{nj}; \quad n = 0, 1, 2, \dots; \quad j = 1, 2, \dots .$$

The eigenvalues λ_{0j} are of multiplicity 1. All the others are of multiplicity 2. As a special case of the eigenfunction expansion Theorem 10.1, the following expansion theorem can be proved: Suppose that $f(r, \theta)$ is C^2 for $0 \leqq r \leqq a$, $-\pi \leqq \theta \leqq \pi$, and $f(a, \theta) = 0$ for $-\pi \leqq \theta \leqq \pi$. Then f can be represented by a series involving the eigenfunctions (10.38) of the eigenvalue problem (10.26), (10.27),

(10.40) $$f(r, \theta) = \sum_{n=0}^{\infty} \sum_{j=1}^{\infty} [A_{nj} J_n (\sqrt{\lambda_{nj}}\, r) \cos n\theta + B_{nj} J_n(\sqrt{\lambda_{nj}}\, r) \sin n\theta]$$

with the series converging absolutely and uniformly to f. For each $j = 1, 2, \dots$ the coefficients are given by

$$A_{0j} = \frac{1}{\pi a^2 [J_1(\sqrt{\lambda_{nj}}\, a)]^2} \int_{-\pi}^{\pi} \int_0^a f(r, \theta) J_0(\sqrt{\lambda_{0j}}\, r) r\,dr\,d\theta$$

(10.41) $$A_{nj} = \frac{2}{\pi a^2 [J_{n+1}(\sqrt{\lambda_{0j}}\, a)]^2} \int_{-\pi}^{\pi} \int_0^a f(r, \theta) J_n (\sqrt{\lambda_{nj}}\, r) \cos n\theta \, r\,dr\,d\theta,$$

$$B_{nj} = \frac{2}{\pi a^2 [J_{n+1}(\sqrt{\lambda_{nj}}\, a)]^2} \int_{-\pi}^{\pi} \int_0^a f(r, \theta) J_n (\sqrt{\lambda_{nj}}\, r) \sin n\theta \, r\,dr\,d\theta,$$

Finally, let us return to the initial-boundary value problem (10.22), (10.23), (10.24). The T equation (10.6), with λ being one of the eigenvalues (10.39), has two linearly independent solutions

(10.42) $$T_{nj} = \cos \sqrt{\lambda_{nj}}\, t, \ \sin \sqrt{\lambda_{nj}}\, t.$$

Multiplying (10.38) and (10.42), we obtain the collection of functions

(10.43) $$\begin{cases} J_n(\sqrt{\lambda_{nj}}r) \cos n\theta \cos\sqrt{\lambda_{nj}}t, \ J_n(\sqrt{\lambda_{nj}}r) \sin n\theta \cos \sqrt{\lambda_{nj}}t, \\ J_n(\sqrt{\lambda_{nj}}r) \cos n\theta \sin \sqrt{\lambda_{nj}}t, \ J_n(\sqrt{\lambda_{nj}}r) \sin n\theta \sin \sqrt{\lambda_{nj}}t; \\ \qquad n = 0, 1, 2, \dots; j = 1, 2, \dots \end{cases}$$

each of which satisfies (10.22) and (10.24). It is left now as an exercise to form a superposition of (10.43) satisfying, in addition, the initial conditions (10.23) under appropriate conditions on ϕ and ψ (Problem 10.6).

Problems

10.1. Prove that eigenfunctions of (10.4), (10.5) corresponding to distinct eigenvalues are orthogonal. [*Hint:* Prove (10.7), where $\nabla^2 u_1 + \lambda_1 u_1 = 0$, $\nabla^2 u_2 + \lambda_2 u_2 = 0$. Multiply the first of these equations by u_2, the second by u_1, subtract, integrate over Ω, use Green's second identity and the fact that u_1 and u_2 vanish on $\partial\Omega$.]

10.2. Assuming that the representation (10.11) is valid in the sense of uniform convergence, derive formula (10.12) for the computation of the coefficients.

10.3. Prove the orthogonality relation (10.13).

10.4. Under the assumptions of Theorem 10.1 prove the "Parseval relation"

$$\sum_{n=1}^{\infty} a_n^2 \int_\Omega [v_n(x)]^2 dx = \int_\Omega [f(x)]^2 dx. \tag{10.44}$$

10.5. Is there an expansion theorem in terms of the eigenfunctions of problem (10.29), (10.30)?

10.6. Write out the series solution of the initial boundary value problem (10.22), (10.23), (10.24).

10.7. Solve problem (10.1), (10.2), (10.3), if Ω is the rectangular parallelepiped in R^3,

$$0 < x < a, \qquad 0 < y < b, \qquad 0 < z < c.$$

10.8. Consider the "forced vibrations" problem in a domain Ω of R^n

$$\nabla^2 u - u_{tt} = f(x, t); \qquad x \in \Omega, \qquad t > 0 \tag{10.45}$$

$$u(x, 0) = 0, \qquad u_t(x, 0) = 0, \qquad x \in \bar{\Omega} \tag{10.46}$$

$$u(x, t) = 0; \qquad x \in \partial\Omega, \qquad t \geqq 0. \tag{10.3}$$

Use Duhamel's principle to obtain the (formal) solution

$$u(x, t) = \sum_{n=1}^{\infty} \left[\int_0^t f_n(\tau) \sin \sqrt{\lambda_n}(t - \tau) d\tau \right] v_n(x) \tag{10.47}$$

where

$$f_n(\tau) = -\frac{1}{\sqrt{\lambda_n}} \frac{\int_\Omega f(x, \tau) v_n(x) dx}{\int_\Omega [v_n(x)]^2 dx}, \quad n = 1, 2, \ldots, \tag{10.48}$$

and λ_n, $v_n(x)$; $n = 1, 2, \ldots$ are, respectively, the eigenvalues (10.8) and corresponding orthogonal eigenfunctions (10.9) of the eigenvalue problem (10.4), (10.5). What is the solution of problem (10.45), (10.2), (10.3)? [See remark in Problem 8.11.]

10.9. In Problem 10.8, suppose that the forcing function is

$$f(x, t) = F(x) \sin \omega t.$$

(a) If ω^2 is not equal to any of the eigenvalues λ_n, $n = 1, 2, \ldots$, derive the following formula for the solution of (10.45), (10.46), (10.3),

$$u(x, t) = \sum_{n=1}^{\infty} \frac{a_n}{\sqrt{\lambda_n}} \left[\frac{\sqrt{\lambda_n} \sin \omega t - \omega \sin \sqrt{\lambda_n}\, t}{\omega^2 - \lambda_n} \right] v_n(x) \tag{10.49}$$

where the a_n, $n = 1, 2, \ldots$, are the coefficients of the expansion of

$F(x)$ in terms of the eigenfunctions $v_n(x)$, $n = 1, 2, \ldots$.

(b) If $\omega^2 = \lambda_j$ where λ_j is one of the eigenvalues, show that the solution is given by (10.49) with the jth term in the series replaced by

$$\frac{a_j}{2\sqrt{\lambda_j}}\left(t \cos\sqrt{\lambda_j}\, t - \frac{1}{\sqrt{\lambda_j}} \sin \sqrt{\lambda_{ij}}\, t\right) v_j(x).$$

(This is the case of resonance.)

References for Chapter VIII

1. Watson, G. N.: *A Treatise on the Theory of Bessel Functions, New York:* Cambridge University Press, 1966.
2. Tolstov, G. P.: *Fourier Series,* Englewood Cliffs, N. J.: Prentice Hall Inc., 1962.
3. Vladimirov, V. S.: *Equations of Mathematical Physics,* New York: Marcel Dekker, Inc., 1971.
4. Sobolev, S. L.: *Partial Differential Equations of Mathematical Physics,* Elmsford, N.Y.: Pergamon Press, 1965.
5. Churchill, R. V.: *Fourier Series and Boundary Value Problems,* Ed. 2, New York: McGraw Hill Book Co., 1963.

CHAPTER IX

The heat equation

The heat equation is the most important example of a linear partial differential equation of parabolic type. The first two sections of this chapter are devoted to the study of the initial-boundary value problem for the heat equation in one space variable. The study of heat conduction in a finite rod leads to this problem. In Section 1 we prove a maximum-minimum theorem from which follow immediately the uniqueness of solution of the problem and the continuous dependence of the solution on the initial and boundary data. In Section 2 we obtain the solution of the problem by the method of separation of variables and Fourier series. Problems with a variety of boundary conditions are also considered. Section 3 is devoted to the study of the initial value problem for the one-dimensional heat equation. In Section 4 we give a summary of the basic results concerning the initial and initial-boundary value problems for the heat equation in more than one space dimensions. An application to transistor theory is presented in Section 5.

1. Heat Conduction in a Finite Rod. The Maximum-Minimum Theorem and Its Consequences

Let us consider the problem of determining the temperature distribution in a cylindrical rod of length L which is made of homogeneous, isotropic material and has insulated cylindrical surface. If initially the temperature does not vary over cross-sections perpendicular to the center-line, the same will be true at later times. Consequently, if the center line occupies the portion of the x-axis from $x = 0$ to $x = L$, the temperature u will be a function of x and t only and will satisfy the one-dimensional heat equation (see Chapter VI, Section 2). Assuming that the initial temperature distribution and the temperature at the ends $x = 0$ and $x = L$ for $t \geqq 0$ are known, we are faced with the initial-boundary value problem

(1.1) $$u_t - u_{xx} = 0; \qquad 0 < x < L, \qquad 0 < t$$

(1.2) $$u(x, 0) = \phi(x); \qquad 0 \leqq x \leqq L$$

(1.3) $$u(0, t) = f_1(t), \qquad u(L, t) = f_2(t); \qquad 0 \leqq t.$$

The functions ϕ, f_1 and f_2 are assumed to be continuous and satisfy the compatibility conditions

$$\phi(0) = f_1(0), \qquad \phi(L) = f_2(0).$$

A precise statement of the problem is: Find a function $u(x, t)$ defined and continuous in the closed strip $0 \leqq x \leqq L$, $0 \leqq t$ and satisfying the heat equation (1.1) in the interior of the strip, the initial condition (1.2) at the bottom $t = 0$ and the boundary conditions (1.3) at the two vertical sides $x = 0$ and $x = L$ (see Fig. 1.1).

In this section we will prove a maximum-minimum theorem from which follow immediately the uniqueness and continuous dependence of solution on the initial and boundary data for the problem (1.1), (1.2), (1.3). In the next section we will use the method of separation of variables and Fourier series to solve the problem in the case of homogeneous boundary conditions ($f_1 \equiv f_2 \equiv 0$).

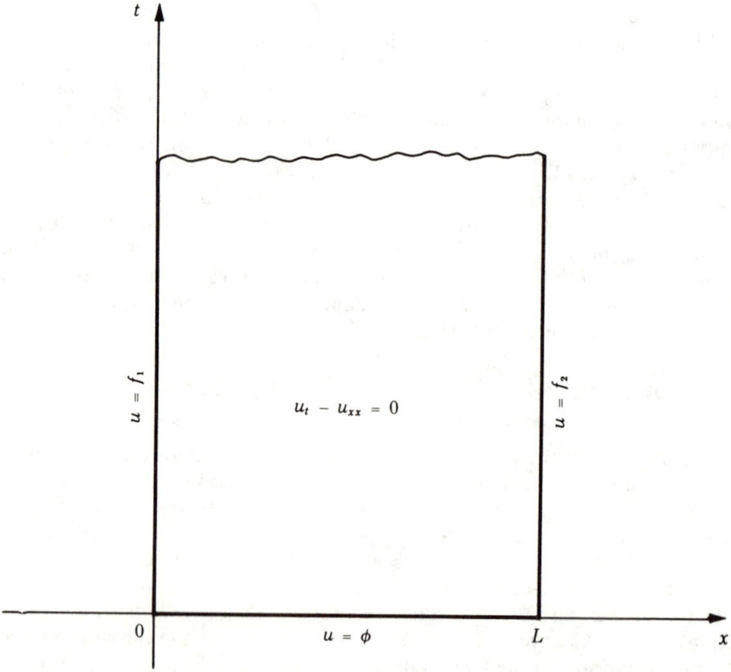

Fig. 1.1

Theorem 1.1. (*Maximum-minimum theorem.*) Let T be any number >0. Suppose that the function $u(x, t)$ is continuous in the closed rectangle R,

$$R: 0 \leqq x \leqq L, \qquad 0 \leqq t \leqq T,$$

and satisfies the heat equation (1.1) for $0 < x < L$ and $0 < t \leqq T$. Then u attains its maximum and minimum values on the lower base $t = 0$ or on the vertical sides $x = 0$, $x = L$ of R.

Proof. We prove the maximum part of the theorem first. The proof will be by contradiction. Let M be the maximum value of u in R and suppose that, contrary to the assertion of the theorem, the maximum value of u on the lower base and vertical sides of R is $M - \epsilon$, where $\epsilon > 0$. Let (x_0, t_0) be a point where u attains its maximum in R, $u(x_0, t_0) = M$. We must have $0 < x_0 < L$ and $t_0 > 0$. Let us introduce the auxiliary function

$$v(x, t) = u(x, t) + \frac{\epsilon}{4L^2}(x - x_0)^2.$$

On the lower base and vertical sides of R,

$$v(x, t) \leqq M - \epsilon + \frac{\epsilon}{4} = M - \frac{3\epsilon}{4},$$

while $v(x_0, t_0) = M$. Therefore, the maximum value of v in R is not attained on the lower base and vertical sides of R. Let (x_1, t_1) be a point where v attains its maximum. We must have $0 < x_1 < L$ and $0 < t_1 \leqq T$. At (x_1, t_1), v must satisfy the necessary conditions for a maximum, namely, $v_t = 0$ if $t_1 < T$ or $v_t \geqq 0$ if $t_1 = T$, and $v_{xx} \leqq 0$. Hence, at (x_1, t_1),

$$v_t - v_{xx} \geqq 0.$$

On the other hand

$$v_t - v_{xx} = u_t - u_{xx} - \frac{\epsilon}{2L^2} < 0.$$

We have reached a contradiction, and the maximum part of the theorem is proved. The minimum assertion is reduced immediately to the maximum assertion by considering the function $w = -u$ and noting that w has a maximum value where u has a minimum. Since w satisfies all the assumptions of the theorem, it must attain its maximum value on the lower base and vertical sides of R. Consequently u must attain its minimum value there.

Corollary 1.1. *(Uniqueness.)* There is at most one solution of the initial-boundary value problem (1.1), (1.2), (1.3).

Proof. Let u be the difference of any two solutions of the problem. Then u satisfies (1.1), (1.2), (1.3) with zero initial and boundary data. From Theorem 1.1, if T is any number > 0, $u(x, t) = 0$ for $0 \leqq x \leqq L$ and $0 \leqq t \leqq T$. Since T is arbitrary, $u(x, t) = 0$ in the whole strip $0 \leqq x \leqq L$, $0 \leqq t$.

Corollary 1.2. *(Continuous dependence on data.)* The solution of the initial-boundary value problem (1.1), (1.2), (1.3) depends continuously on the initial and boundary data of the problem in the following sense: Let u and $\tilde{u}$ be the solutions of the problem with data ϕ, f_1, f_2 and $\tilde{\phi}$, $\tilde{f}_1$, $\tilde{f}_2$, respectively. Let T and ϵ be any positive numbers. If

$$\max_{0 \leqq x \leqq L} |\phi(x) - \tilde{\phi}(x)| \leqq \epsilon,$$

$$\max_{0 \leqq t \leqq T} |f_1(t) - \tilde{f}_1(t)| \leqq \epsilon, \qquad \max_{0 \leqq t \leqq T} |f_2(t) - \tilde{f}_2(t)| \leqq \epsilon,$$

then

$$\max_{\substack{0 \leqq x \leqq L \\ 0 \leqq t \leqq T}} |u(x, t) - \tilde{u}(x, t)| \leqq \epsilon.$$

The proof is left as an exercise.

The following consequence of the maximum-minimum theorem agrees with physical intuition. Suppose that when $t = 0$, the maximum temperature in the rod is U and that at all subsequent times $t > 0$, the temperature at the ends of the rod is $\leqq U$. Then at no point in the rod will the temperature ever rise above U. This makes sense physically in view of the well known fact that heat flows from hot to cool.

Just as in the case of the wave equation, the heat equation is invariant under t-translation, that is, the form of the equation does not change if the variable t is replaced by $t' = t - t_0$ where t_0 is any fixed number. Consequently, if the initial time in a problem is the time $t = t_0$, the above translation will reduce the problem to one with initial time $t = 0$. An important consequence of this is that the maximum-minimum theorem is valid with R being any rectangle of the form

$$R: 0 \leqq x \leqq L, \qquad t_0 \leqq t \leqq T;$$

u must attain its maximum and minimum values on the lower base $t = t_0$ or on the vertical sides $x = 0$, $x = L$ of R. (See also Problem 1.4.)

In contrast with the wave equation, the heat equation is not invariant under reversal of time. Indeed, under the substitution $t' = -t$ equation (1.1) becomes

$$u_{t'} + u_{xx} = 0$$

which differs from the heat equation in an essential manner. As a consequence, it turns out that, in general, the initial-boundary value problem for the heat equation cannot be solved "backward" in time. (If the temperature at the ends of a rod is always known, then knowledge of the temperature in the rod at a given instant determines the temperature at subsequent times but does not, in general, determine the temperature at previous times.) We will see the reason for this in the next section. Here, we should point out that the maximum-minimum theorem is "one-way" in time (see Problem 1.5). In physical terms this property of the heat equation is described by saying that heat conduction is an irreversible process.

Finally, we mention an obvious modification of the initial-boundary value problem (1.1), (1.2), (1.3) which is finite in time. If the boundary data functions $f_1(t)$ and $f_2(t)$ are known only over a finite interval of time $0 \leqq t \leqq$

T, then one can only ask for the determination of $u(x, t)$ in the finite rectangle $0 \leqq x \leqq L$, $0 \leqq t \leqq T$. Uniqueness and continuous dependence on the data for this problem follow immediately from the maximum-minimum theorem.

A word of caution: the letter t in the heat equation (1.1) does not represent physical time. Remember that in Chapter VI, equation (2.7) was simplified to equation (2.8) by changing the time scale: The quantity $(k/c\rho)t$, with k, c, ρ being physical constants and t time, was replaced by t' and then the prime was dropped. Thus, in order to return to physical parameters the letter t in all formulas of this chapter should be replaced by $(k/c\rho)t$.

Problems

1.1. Prove Corollary 1.2.

1.2. At time $t = 0$ the temperature in a rod of length L is everywhere zero. During the time interval $0 \leqq t \leqq T_1$ the temperature at the two ends of the rod is kept equal to zero. When $t > T_1$ the temperature at the ends is gradually raised to 100°. Find the temperature $u(x, t)$ in the rod during the time interval $0 \leqq t \leqq T_1$.

1.3. Solve the initial-boundary value problem

$$u_t - u_{xx} = 0; \qquad 0 < x < L, \qquad t > 0$$

$$u(x, 0) = a + (b - a)\frac{x}{L}, \qquad 0 \leqq x \leqq L$$

$$u(0, t) = a, \qquad u(L, t) = b; \qquad 0 \leqq t,$$

where a and b are constants.

1.4. Let t_0 be any number less than T and suppose that $u(x, t)$ is continuous in the rectangle R: $0 \leqq x \leqq L$, $t_0 \leqq t \leqq T$, satisfies the heat equation in the interior of R and at the top $t = T$, and vanishes on the lower base $t = t_0$ and vertical sides $x = 0$, $x = L$ of R. Show that $u(x, t) = 0$ for $0 \leqq x \leqq L$. Conclude that at any time T, the temperature distribution in a finite rod is uniquely determined by the temperature distribution at any previous time t_0 and by the temperature at the two ends during the time interval $[t_0, T]$.

1.5. Explain how the argument in the proof of Theorem 1.1 would fail to prove a maximum-minimum theorem with the roles of the top and bottom of the rectangle R reversed.

1.6. Suppose that $u(x, t)$ is of class C^2 in the closed strip $0 \leqq x \leqq L$, $0 \leqq t$ and satisfies the heat equation in the interior of the strip and one of the two boundary conditions on each vertical side of the strip

$$u(0, t) = 0, \quad \text{or} \quad u_x(0, t) = 0; \qquad 0 \leqq t,$$

$$u(L, t) = 0, \quad \text{or} \quad u_x(L, t) = 0; \qquad 0 \leqq t.$$

Show that for any $T \geqq 0$,

$$\int_0^L u^2 \big|_{t=T}\, dx \leqq \int_0^L u^2 \big|_{t=0}\, dx.$$

[*Hint:* Derive the differential identity

$$2u(u_t - u_{xx}) = (u^2)_t - (2uu_x)_x + 2u_x^2$$

and integrate over the rectangle $0 \leqq x \leqq L$, $0 \leqq t \leqq T$.]

1.7. Use the result of Problem 1.6 to prove uniqueness of solution for each of the following initial-boundary value problems, assuming the solution is of class C^2 in the strip $0 \leqq x \leqq L$, $0 \leqq t$:

$$u_t - u_{xx} = 0; \qquad 0 < x < L, \qquad 0 < t$$

$$u(x, 0) = \phi(x); \qquad 0 \leqq x \leqq L$$

(a) $$u(0, t) = f_1(t), \qquad u(L, t) = f_2(t); \qquad 0 \leq t,$$

or (b) $$u_x(0, t) = f_1(t), \qquad u_x(L, t) = f_2(t); \qquad 0 \leqq t,$$

or (c) $$u(0, t) = f_1(t), \qquad u_x(L, t) = f_2(t); \qquad 0 \leqq t.$$

2. Solution of the Initial-Boundary Value Problem for the One-Dimensional Heat Equation

Consider a cylindrical rod of length L with insulated cylindrical surface and center-line occupying the interval $[0, L]$ of the x-axis. Suppose that when $t = 0$ the temperature in the rod is a known function of x and that for all $t \geqq 0$ the temperature at the two ends of the rod is kept equal to 0. In order to determine the temperature distribution in the rod for $t \geqq 0$ we must solve the initial-boundary value problem

(2.1) $$u_t - u_{xx} = 0; \qquad 0 < x < L, \qquad 0 < t,$$

(2.2) $$u(x, 0) = \phi(x); \qquad 0 \leqq x \leqq L,$$

(2.3) $$u(0, t) = 0, \qquad u(L, t) = 0; \qquad 0 \leqq t.$$

The function ϕ is assumed to be continuous and vanish at the ends $x = 0$ and $x = L$. From the previous section, we know that this problem can have at most one solution which, if it exists, must depend continuously on the initial data $\phi(x)$. In this section we will obtain the solution of the problem using separation of variables and Fourier series. The method of solution is exactly the same as the one used in Section 8 of Chapter VIII for solving the initial-boundary value problem for the one-dimensional wave equation.

A function of the form

(2.4) $$u(x, t) = X(x)T(t)$$

will satisfy the heat equation (2.1) and the boundary conditions (2.3) if $X(x)$ satisfies

(2.5) $$X'' + \lambda X = 0, \qquad 0 < x < L$$

(2.6) $$X(0) = 0, \qquad X(L) = 0$$

and $T(t)$ satisfies

(2.7) $$T' + \lambda T = 0, \qquad 0 < t$$

where $-\lambda$ is the separation constant. We have already solved the eigenvalue problem (2.5), (2.6). The eigenvalues are

$$\lambda_n = \frac{n^2\pi^2}{L^2}, \quad n = 1, 2, \ldots \tag{2.8}$$

with corresponding eigenfunctions

$$X_n(x) = A_n \sin\frac{n\pi x}{L}, \qquad n = 1, 2, \ldots . \tag{2.9}$$

The general solution of the T equation (2.7) with λ being an eigenvalue λ_n is

$$T_n(t) = B_n e^{-\frac{n^2\pi^2}{L^2}t} \qquad n = 1, 2, \ldots . \tag{2.10}$$

Substituting (2.9) and (2.10) into (2.4) we obtain the infinite sequence of functions

$$u_n(x, t) = C_n \sin\frac{n\pi x}{L} e^{-\frac{n^2\pi^2}{L^2}t} \qquad n = 1, 2, \ldots , \tag{2.11}$$

each of which satisfies the heat equation (2.1) and the boundary conditions (2.3). Hopefully, the initial condition (2.2) will be satisfied by a superposition of the functions (2.11),

$$u(x, t) = \sum_{n=1}^{\infty} C_n \sin\frac{n\pi x}{L} e^{-\frac{n^2\pi^2}{L^2}t} . \tag{2.12}$$

Let us postpone for a moment the consideration of questions of convergence and differentiability of the series in (2.12). In order to satisfy (2.2) the constants in (2.12) must be chosen so that

$$\phi(x) = \sum_{n=1}^{\infty} C_n \sin\frac{n\pi x}{L}, \tag{2.13}$$

i.e., the C_n must be the coefficients of the Fourier sine series representation of the function $\phi(x)$ on the interval $[0, L]$. From formula (8.40) of Chapter VII,

$$C_n = \frac{2}{L}\int_0^L \phi(x) \sin\frac{n\pi x}{L}\,dx. \tag{2.14}$$

Let us assume now that ϕ satisfies the following conditions: (1) ϕ is continuous and has a sectionally continuous derivative on $[0, L]$, and (2) $\phi(0) = \phi(L) = 0$. Then we know (Problem 8.13 of Chapter VII) that the representation (2.13) with coefficients (2.14) is valid in the sense of absolute and uniform convergence on the interval $[0, L]$. Since

$$0 < e^{-\frac{n^2\pi^2}{L^2}t} \leqq 1 \quad \text{for all} \quad t \geqq 0,$$

the series in (2.12) with coefficients (2.14) converges absolutely and uniformly for $0 \leqq x \leqq L$ and $t \geqq 0$. Therefore the function $u(x, t)$ defined by this series is continuous in the closed strip $0 \leqq x \leqq L$, $t \geqq 0$. Obviously $u(x, t)$ satisfies the boundary conditions (2.3). It remains to show that $u(x, t)$ satisfies the heat equation in the interior of the strip. To do this it suffices to show that all the series obtained by termwise differentiation of the series in (2.12), once with respect to t or twice with respect to x, converge uniformly in any closed strip of the form $0 \leqq x \leqq L$, $t_0 \leqq t$ with t_0 being any positive number, however small. But this follows from the Weierstrass M-test using the fact that for any $t_0 > 0$,

$$\frac{n^2\pi^2}{L^2} e^{-\frac{n^2\pi^2}{L^2} t} \leqq 1 \quad \text{for all} \quad t \geqq t_0,$$

provided n is sufficiently large. We have shown that the desired solution of the initial-boundary value problem (2.1), (2.2), (2.3) is given by the series (2.12) with coefficients given by (2.14).

Example 2.1. If the initial temperature in the rod is given by the sawtooth function

$$\phi(x) = \frac{2U}{L} \cdot \begin{cases} x & \text{for} \quad 0 \leqq x \leqq \dfrac{L}{2} \\ L - x & \text{for} \quad \dfrac{L}{2} \leqq x \leqq L, \end{cases} \tag{2.15}$$

where U is a constant, and if the ends are always kept at temperature zero, we find, using the Fourier sine series representation (8.41) of Chapter VII, that the temperature in the rod will be given by

$$u(x, t) = \frac{8U}{\pi^2} \sum_{n=0}^{\infty} \frac{(-1)^n}{(2n+1)^2} \sin\frac{(2n+1)\pi x}{L} e^{-\frac{(2n+1)^2\pi^2}{L^2} t} \tag{2.16}$$

We now state and prove a fundamental property of the solution of the initial value problem (2.1), (2.2), (2.3).

Theorem 2.1. The solution $u(x, t)$ of the initial-boundary value problem (2.1), (2.2), (2.3) for the heat equation is of class C^∞ when $t > 0$; i.e., the partial derivatives of all orders with respect to x and t are continuous at every point (x, t) with $t > 0$.

Proof. Since $u(x, t)$ is given by the series (2.12) it suffices to show that all series obtained from (2.12) by termwise differentiation any number of times with respect to x or t converge uniformly when $t \geqq t_0$ with t_0 being any positive number, however small. But this follows from the Weierstrass M-test by noting that the absolute value of a typical term of any such series is bounded by

$$|C_n| \frac{n^k\pi^k}{L^k} e^{-\frac{n^2\pi^2}{L^2} t_0} \tag{2.17}$$

when $t \geqq t_0$. Since the C_n are bounded, the series with terms (2.17) converges for any value of the integer k.

Let us look for example at the solution (2.16) of Example 2.1. The derivative u_x of this solution is not continuous at the point $(x, t) = (L/2, 0)$. In fact, since $u(x, 0) = \phi(x)$ where ϕ is given by (2.15), the initial temperature distribution has a corner at $x = L/2$. However, as soon as t becomes positive, this corner disappears and u becomes infinitely smooth.

Theorem 2.1 is a special instance of a general result which asserts that any solution of the heat equation is of class C^∞ in the interior of its domain of definition.

It should be noted that in proving Theorem 2.1, we only used the fact that the coefficients C_n are bounded. Thus, if the initial temperature distribution $\phi(x)$ had jump discontinuities, its Fourier coefficients would still be bounded and the solution (2.12) would still be C^∞ for $t > 0$. Of course, in such a case, (2.12) would not be continuous in the closed strip $0 \leqq x \leqq L, t \geqq 0$, but it would still be the solution of the problem (2.1), (2.2), (2.3) in a more general sense.

We can use Theorem 2.1 to prove that, in general, the initial-boundary value problem for the heat equation cannot be solved "backward" in time, not even over a very small time interval. Indeed, suppose that for some $\epsilon > 0$, $u(x, t)$ is a continuous function in the closed rectangle $0 \leqq x \leqq L, -\epsilon \leqq t \leqq 0$ and satisfies

$$u_t - u_{xx} = 0; \qquad 0 < x < L, \qquad -\epsilon < t < 0$$

$$u(0, t) = 0, \qquad u(L, t) = 0; \qquad -\epsilon \leqq t \leqq 0$$

$$u(x, 0) = \phi(x), \qquad 0 \leqq x \leqq L,$$

where $\phi(x)$ is continuous. Let

$$\psi(x) = u(x, -\epsilon).$$

Now solve the forward problem with initial line $t = -\epsilon$ and initial data $\psi(x)$ on this line. By uniqueness, the solution of this forward problem must be equal to $u(x, t)$ in the rectangle $0 \leqq x \leqq L, -\epsilon \leqq t \leqq 0$. From Theorem 2.1, the solution of the forward problem must be C^∞ for all t greater than the initial time $t = -\epsilon$. In particular, $u(x, 0)$ must be C^∞. This is a contradiction since $u(x, 0) = \phi(x)$, unless of course $\phi(x)$ were C^∞ to begin with, and not just continuous. Incidentally, it can be shown that even if the "backward" problem were solvable, the solution would not depend continuously on the initial data.

The initial-boundary value problem with nonhomogeneous boundary conditions

$$u(0, t) = f_1(t), \qquad u(L, t) = f_2(t); \qquad 0 \leqq t \tag{2.18}$$

can be reduced to the problem with homogeneous boundary conditions (2.3) provided that one can find some solution $u_1(x, t)$ of the heat equation (2.1) satisfying the boundary conditions (2.18). $u_1(x, t)$ need not satisfy any particular initial condition. Indeed, it is easy to verify that the function

$$u(x, t) = u_1(x, t) + u_2(x, t) \tag{2.19}$$

will be the solution of the nonhomogeneous problem (2.1), (2.2), (2.18) provided that $u_2(x, t)$ is the solution of the homogeneous problem (2.1), (2.3) with initial condition

(2.20) $$u_2(x, 0) = \phi(x) - u_1(x, 0) \qquad 0 \leqq x \leqq L.$$

Problems 2.2 and 2.3 can be solved by this method.

Let us consider now the problem of determining the temperature distribution in a rod with insulated ends. We must solve the initial-boundary value problem for the heat equation (2.1) with initial condition (2.2) and boundary conditions

(2.21) $$u_x(0, t) = 0, \qquad u_x(L, t) = 0, \qquad 0 \leqq t.$$

The proof of uniqueness of solution (under rather stringent conditions) was outlined in Problems 1.6 and 1.7. Using the method of separation of variables and Fourier series the solution of the problem is found to be

(2.22) $$u(x, t) = \frac{a_0}{2} + \sum_{k=1}^{\infty} a_k \cos\frac{k\pi x}{L} e^{-\frac{k^2\pi^2}{L^2}t}$$

where a_k are the coefficients of the Fourier cosine series representation of the initial temperature $\phi(x)$ on the interval $[0, L]$,

(2.23) $$a_k = \frac{2}{L}\int_0^L \phi(x) \cos\frac{k\pi x}{L}\, dx, \qquad k = 0, 1, 2, \ldots .$$

Finally, let us consider the problem of determining the temperature distribution in a rod with one end kept at temperature zero and the other end insulated. We must solve the initial-boundary value problem for (2.1) with initial condition (2.2) and boundary conditions

(2.24) $$u(0, t) = 0, \qquad u_x(L, t) = 0; \qquad 0 \leqq t.$$

After separation of variables, the eigenvalue problem to be solved is

(2.25) $$X'' + \lambda X = 0, \qquad 0 < x < L$$

(2.26) $$X(0) = 0, \qquad X'(L) = 0.$$

It is easy to check that there are no non-positive eigenvalues of this problem. When $\lambda > 0$, the general solution of (2.25) is

$$X(x) = A_1 \cos\sqrt{\lambda}x + A_2 \sin\sqrt{\lambda}x,$$

and the boundary conditions (2.26) will be satisfied if

$$A_1 = 0, \qquad A_2\sqrt{\lambda} \cos\sqrt{\lambda}L = 0.$$

The eigenvalues and corresponding eigenfunctions are

(2.27) $$\lambda_n = \frac{(2n-1)^2\pi^2}{4L^2}, \qquad n = 1, 2, \ldots ,$$

(2.28) $$X_n(x) = A_n \sin\frac{(2n-1)\pi x}{2L}, \qquad n = 1, 2, \dots .$$

The general solution of the T equation (2.7) with λ being the eigenvalue λ_n is

(2.29) $$T_n(t) = B_n e^{-\frac{(2n-1)^2\pi^2}{4L^2}t}, \qquad n = 1, 2, \dots .$$

Multiplying (2.28) and (2.29) we obtain the sequence of functions

(2.30) $$u_n(x, t) = C_n \sin\frac{(2n-1)\pi x}{2L} e^{-\frac{(2n-1)^2\pi^2}{4L^2}t}, \qquad n = 1, 2, \dots,$$

each of which satisfies the heat equation (2.1) and the boundary conditions (2.24). Hopefully, the initial condition (2.2) will be satisfied by a superposition of these functions,

(2.31) $$u(x, t) = \sum_{n=1}^{\infty} C_n \sin\frac{(2n-1)\pi x}{2L} e^{-\frac{(2n-1)^2\pi^2}{4L^2}t}.$$

In order to satisfy (2.2) the constants C_n must be chosen so that

(2.32) $$\phi(x) = \sum_{n=1}^{\infty} C_n \sin\frac{(2n-1)\pi x}{2L}.$$

Without using any general result on representations of functions by series of eigenfunctions, we can show that the representation (2.32) is possible and obtain a formula for the coefficients using only standard Fourier series theory. The presence of $2L$ rather than L in (2.32) suggests that we extend the function ϕ, which is defined on $[0, L]$, to a function defined on $[0, 2L]$, and represent the extended function by a Fourier sine series on the interval $[0, 2L]$. It turns out that extending ϕ so that the resulting extension is symmetric with respect to $x = L$ gives the desired result. (Such an extension is also suggested by the observation that a temperature distribution which is symmetric about $x = L$ should result in no heat transfer across $x = L$.) Let

(2.33) $$\tilde{\phi}(x) = \begin{cases} \phi(x) & \text{for } 0 \leqq x \leqq L \\ \phi(2L - x) & \text{for } L \leqq x \leqq 2L. \end{cases}$$

The Fourier sine series representation of $\tilde{\phi}$ on $[0, 2L]$ is

(2.34) $$\tilde{\phi}(x) = \sum_{n=1}^{\infty} b_n \sin\frac{n\pi x}{2L}, \qquad 0 \leqq x \leqq 2L,$$

where

(2.35) $$b_n = \frac{2}{2L}\int_0^{2L} \tilde{\phi}(x) \sin\frac{n\pi x}{2L}\, dx.$$

It is easy to show (Problem 2.8) that, because of (2.33),

$$(2.36)\qquad b_{2n-1} = \frac{2}{L}\int_0^L \tilde{\phi}(x)\sin\frac{(2n-1)\pi x}{2L}\,dx, \qquad b_{2n} = 0;$$
$$n = 1, 2, \ldots .$$

Since $\tilde{\phi}(x) = \phi(x)$ for $0 \leqq x \leqq L$ we conclude that (under appropriate conditions on ϕ) the representation (2.32) is valid with

$$(2.37)\qquad C_n = \frac{2}{L}\int_0^L \phi(x)\sin\frac{(2n-1)\pi x}{2L}\,dx, \qquad n = 1, 2, \ldots .$$

We conclude that the solution of the initial-boundary value problem (2.1), (2.2), (2.24) is given by (2.31) with coefficients given by (2.37).

Problems

2.1. The initial temperature distribution in a rod of length π is given by

$$\phi(x) = Ux(\pi - x), \qquad 0 \leqq x \leqq \pi,$$

and both ends of the rod are always kept at zero temperature. Find the temperature in the rod for $t \geqq 0$.

2.2. Consider a rod of length L and suppose that the end $x = 0$ is always kept at temperature a while the end $x = L$ is kept at temperature b. If the initial temperature in the rod is given by the function $\phi(x)$, where $\phi(0) = a$ and $\phi(L) = b$, find the formula for the temperature in the rod for $t \geqq 0$. What is the limit of the temperature distribution as $t \to \infty$?

2.3. Solve the initial boundary value problem for the heat equation (2.1) with initial condition (2.2) and boundary conditions

$$u(0, t) = a + ct, \qquad u(L, t) = b + ct; \qquad 0 \leqq t,$$

where $\phi(0) = a$ and $\phi(L) = b$. [*Hint:* Consider the function $a + (b - a)x/L + ct + (c/2)x(x - L)$.]

2.4. Use separation of variables and Fourier series to derive the solution (2.22) of the problem (2.1), (2.2), (2.21).

2.5. Find the limit as $t \to \infty$ of the solution of problem (2.1), (2.2), (2.3) and of problem (2.1), (2.2), (2.21). Explain the answers in physical terms.

2.6. If both ends of a rod are insulated and the initial temperature is everywhere equal to 100°, what will the future temperature be?

2.7. Find the temperature distribution in a rod of length L with insulated ends if the initial temperature distribution in the rod is given by

$$\phi(x) = \begin{cases} 0 & \text{for } 0 \leqq x < \dfrac{L-a}{2} \\ U & \text{for } \dfrac{L-a}{2} < x < \dfrac{L+a}{2} \\ 0 & \text{for } \dfrac{L+a}{2} < x \leqq L \end{cases}$$

where U is a constant.

2.8. Derive formulas (2.36) for the coefficients of the Fourier sine series expansion (2.34) over the interval $[0, 2L]$ for a function $\tilde{\phi}(x)$ satisfying (2.33).

2.9. Prove the following form of Duhamel's principle for the heat equation: Let $u(x, t)$ be the solution of the following initial-boundary value problem for the nonhomogeneous heat equation

$$(2.38) \qquad u_t - u_{xx} = f(x, t); \qquad 0 < x < L, \qquad 0 < t$$

$$(2.39) \qquad u(x, 0) = 0, \qquad 0 \leqq x \leqq L$$

$$(2.3) \qquad u(0, t) = 0, \qquad u(L, t) = 0, \qquad 0 \leqq t.$$

For each $\tau \geqq 0$, let $v(x, t; \tau)$ be the solution of the associated "pulse problem"

$$v_t - v_{xx} = 0; \qquad 0 < x < L, \qquad \tau < t$$

$$v(x, \tau; \tau) = f(x, \tau), \qquad 0 \leqq x \leqq L$$

$$v(0, t; \tau) = 0, \qquad v(L, t; \tau) = 0, \qquad \tau \leqq t.$$

(Note that in this problem the initial time is $t = \tau$.) Show that

$$u(x, t) = \int_0^t v(x, t; \tau) d\tau.$$

2.10. Apply Duhamel's principle described in Problem 2.9 to obtain the following formula for the solution of problem (2.38), (2.39), (2.3),

$$u(x, t) = \sum_{n=1}^{\infty} \left[\int_0^t f_n(\tau) e^{-\frac{n^2\pi^2}{L^2}(t-\tau)} d\tau \right] \sin \frac{n\pi x}{L},$$

where

$$f_n(\tau) = \frac{2}{L} \int_0^L f(x, \tau) \sin \frac{n\pi x}{L} dx.$$

In particular, if

$$f(x, t) = F(t) \sin \frac{\pi x}{L}$$

show that

$$u(x, t) = \left(\int_0^t F(\tau) e^{\frac{\pi^2}{L^2}\tau} d\tau \right) \sin \frac{\pi x}{L} e^{-\frac{\pi^2}{L^2}t}$$

2.11. Find the solution of the initial-boundary value problem (2.38), (2.2), (2.3).

3. The Initial Value Problem for the One-Dimensional Heat Equation

The problem of determining the temperature distribution in an infinitely long cylinder with insulated cylindrical surface leads to the initial value problem for the heat equation

$$(3.1) \qquad u_t - u_{xx} = 0; \qquad -\infty < x < \infty, \qquad 0 < t,$$

$$(3.2) \qquad u(x, 0) = \phi(x), \qquad -\infty < x < \infty.$$

The given function ϕ is assumed to be continuous for $-\infty < x < \infty$, while the desired solution $u(x, t)$ is to be continuous in the closed upper-half space $-\infty < x < \infty$, $0 \leqq t$. If we impose no other conditions on $u(x, t)$, problem (3.1), (3.2) has infinitely many solutions. A demonstration of this fact can be found, for example, in the book of A. Friedman,[1] pp. 30–31, where nontrivial solutions of the heat equation vanishing on the initial line $t = 0$ are constructed. These solutions are unbounded, however. It turns out that imposing the condition that $u(x, t)$ must be bounded guarantees the uniqueness of solution of (3.1), (3.2).

Initial Value Problem

Let $\phi(x)$ be continuous and bounded for $-\infty < x < \infty$. Find a function $u(x, t)$ which is continuous and bounded in the closed upper half plane $-\infty < x < \infty$, $0 \leqq t$, and satisfies the heat equation (3.1) and the initial condition (3.2).

As we will see, this problem is well-posed. The uniqueness and continuous dependence of the solution on the initial data follow immediately from the following extreme value theorem.

Theorem 3.1. *(Extreme value theorem.)* Suppose that $u(x, t)$ is continuous and bounded in the closed upper-half plane $-\infty < x < \infty$, $0 \leqq t$, and satisfies the heat equation (3.1). Let

$$M = \sup_{-\infty<x<\infty} u(x, 0), \qquad m = \inf_{-\infty<x<\infty} u(x, 0).$$

Then

$$m \leqq u(x, t) \leqq M,$$

for every (x, t) with $-\infty < x < \infty$, $0 \leqq t$.

Note that $u(x, 0)$ may or may not attain its maximum or minimum values at any point on the infinite interval $-\infty < x < \infty$. This is why we must use the supremum and infimum of $u(x, 0)$ over the interval $-\infty < x < \infty$ rather than the maximum and minimum values of $u(x, 0)$. A proof of Theorem 3.1, which is based on the maximum-minimum Theorem 1.1 for finite rectangles, can be found in the book of Petrovskii,[2] page 345. The proofs of the following corollaries are left for the problems.

Corollary 3.1. *(Uniqueness.)* There is at most one solution of the initial value problem (3.1), (3.2) which is continuous and bounded for $-\infty < x < \infty$, $0 \leqq t$.

Corollary 3.2. A solution of the initial value problem (3.1), (3.2) which is continuous and bounded for $-\infty < x < \infty$, $0 \leqq t$, depends continuously on the initial data $\phi(x)$.

The formula for the solution of the initial value problem (3.1), (3.2), given in the following theorem, can be derived using Fourier integral representations, which are natural extensions of Fourier series representations to the case of (nonperiodic) functions defined over infinite intervals. (See, for example, Churchill,[3] §65–66.)

Theorem 3.2. Let $\phi(x)$ be continuous and bounded for $-\infty < x < \infty$. The solution $u(x, t)$ of the initial value problem (3.1), (3.2), which is continuous and bounded for $-\infty < x < \infty$, $0 \leqq t$, is given by

$$(3.3) \qquad u(x, t) = \begin{cases} \dfrac{1}{2\sqrt{\pi}} \displaystyle\int_{-\infty}^{+\infty} \frac{1}{\sqrt{t}} e^{-\frac{(x-\xi)^2}{4t}} \phi(\xi) d\xi, & \text{when } t > 0 \\ \phi(x), & \text{when } t = 0. \end{cases}$$

Proof. Remembering that the function ϕ is assumed to be bounded, it is easy to see that for every (x, t) with $t > 0$, the integral in (3.3) converges. The same is true with all the integrals obtained by differentiation under the integral sign any number of times with respect to x or t. (This is due to the presence of the negative exponential which goes to zero very rapidly as $\xi \to \pm\infty$.) In fact all these integrals converge uniformly in a neighborhood of any point (x, t) with $t > 0$. Consequently $u(x, t)$ is C^∞ in the open upper-half plane $-\infty < x < \infty, t > 0$. Now, for each ξ, $-\infty < \xi < \infty$, the integrand in (3.3) satisfies the heat equation when $t > 0$ (Problem 3.3). By the principle of superposition (formula (9.10) of Chapter V), the function $u(x, t)$ defined by the integral in (3.3) satisfies the heat equation (3.1) when $t > 0$. By definition, $u(x, t)$ also satisfies the initial condition (3.2).

Next, we show that $u(x, t)$ is bounded. Since ϕ is bounded, it is enough to show that $u(x, t)$ is bounded for $t > 0$. Suppose that

$$|\phi(x)| \leqq M \quad \text{for} \quad -\infty < x < \infty.$$

Then, when $t > 0$,

$$|u(x, t)| \leqq \frac{M}{2\sqrt{\pi}} \int_{-\infty}^{\infty} \frac{1}{\sqrt{t}} e^{-\frac{(x-\xi)^2}{4t}} d\xi = \frac{M}{\sqrt{\pi}} \int_{-\infty}^{\infty} e^{-\eta^2} d\eta = M.$$

In the next to the last equality the change of variable $\eta = (\xi - x)/(2\sqrt{t})$ was used. The last equality is the well known formula

$$(3.4) \qquad \frac{1}{\sqrt{\pi}} \int_{-\infty}^{\infty} e^{-\eta^2} d\eta = 1.$$

To complete the proof of the theorem we must show that $u(x, t)$ is continuous in the closed upper-half plane $t \geqq 0$. Since we have already shown that $u(x, t)$ is C^∞ in the open half-plane $t > 0$, it is only necessary to show that $u(x, t)$ is continuous at every point $(x_0, 0)$ of the x-axis; i.e.,

$$(3.5) \qquad \lim_{\substack{(x,t)\to(x_0,0) \\ t>0}} \left[\frac{1}{2\sqrt{\pi}} \int_{-\infty}^{\infty} \frac{1}{\sqrt{t}} e^{-\frac{(x-\xi)^2}{4t}} \phi(\xi) d\xi \right] = \phi(x_0).$$

The proof of of (3.5) is similar to the proof of (10.19) of Chapter VII and we leave it as a (more difficult) exercise (Problem 3.4).

It can be shown that if the initial temperature $\phi(x)$ is bounded and continuous except for a finite number of points where it has finite jumps, then the bounded solution of the initial value problem (3.1), (3.2) is still given by the integral in (3.3) when $t > 0$. Again, the integral defines a C^∞ function when $t > 0$. Moreover, for every point x_0 where $\phi(x)$ is continuous, the integral approaches $\phi(x_0)$ as $(x, t) \to (x_0, 0)$.

Example 3.1. Find the solution of the initial value problem (3.1), (3.2) if the initial temperature is given by

$$\phi(x) = \begin{cases} U_1 & \text{for } x < 0 \\ U_2 & \text{for } x > 0. \end{cases}$$

When $t > 0$, the solution is given by

$$(3.6) \qquad u(x, t) = \frac{U_1}{2\sqrt{\pi}} \int_{-\infty}^{0} \frac{1}{\sqrt{t}} e^{-\frac{(x-\xi)^2}{4t}} d\xi + \frac{U_2}{2\sqrt{\pi}} \int_{0}^{\infty} \frac{1}{\sqrt{t}} e^{-\frac{(x-\xi)^2}{4t}} d\xi.$$

After some manipulation involving changes of variables of integration, (3.6) is simplified to

$$(3.7) \qquad u(x, t) = \frac{1}{2}(U_1 + U_2) + (U_2 - U_1)\frac{1}{\sqrt{\pi}} \int_0^{\frac{x}{2\sqrt{t}}} e^{-\eta^2} d\eta.$$

It is easy to see, using formula (3.4), that

$$(3.8) \qquad \lim_{(x,t)\to(x_0,0)} u(x, t) = \begin{cases} U_1 & \text{if } x_0 < 0 \\ U_2 & \text{if } x_0 > 0 \end{cases}$$

Note also that

$$u(0, t) = \frac{1}{2}(U_1 + U_2), \qquad 0 < t$$

and hence

$$\lim_{t\to 0} u(0, t) = \frac{1}{2}(U_1 + U_2).$$

We close by noting the following consequence of formula (3.3) for the solution of the initial value problem (3.1), (3.2). Suppose that the initial temperature $\phi(x)$ is zero everywhere except over a small interval (a, b) of the x-axis where ϕ is positive. When $t > 0$ the temperature at any point x is given by

$$u(x, t) = \frac{1}{2\sqrt{\pi t}} \int_a^b e^{-\frac{(x-\xi)^2}{4t}} \phi(\xi) d\xi.$$

Since the integrand is positive, it follows that at any point x (no matter how far x is from the interval (a, b)) the temperature becomes positive as soon as t becomes positive. The physical meaning of this is that heat is conducted with infinite speed. Of course we know from physical observation that this conclusion is false. The explanation of this paradox lies in the fact that the assumptions under which the equation of heat conduction was derived are not exactly verified in nature. Nevertheless, experimental measurements show that the heat equation gives a good approximate description of the process of heat conduction.

Problems

3.1. Prove Corollary 3.1.
3.2. Prove Corollary 3.2.
3.3. Prove that for each ξ, $-\infty < \xi < \infty$, the integrand in (3.3) satisfies the heat equation when $t > 0$.
3.4. Prove (3.5).
3.5. Use Theorem 3.2 to show that

$$\frac{1}{2\sqrt{\pi}} \int_{-\infty}^{\infty} \frac{1}{\sqrt{t}} e^{-\frac{(x-\xi)^2}{4t}} d\xi = 1,$$

and derive formula (3.4).
3.6. Derive (3.7) from (3.6).
3.7. Derive (3.8) from (3.7).
3.8 The *error function*

$$\operatorname{erf}(z) = \frac{2}{\sqrt{\pi}} \int_0^z e^{-\eta^2} d\eta$$

occurs frequently in the theory of probability and there exist extensive tables of its values. In particular, its value at ∞, $\operatorname{erf}(\infty) = 1$, yields formula (3.4). Express the solution (3.7) of Example 3.1 in terms of the error function.
3.9. The study of the temperature distribution in a semi-infinite rod with its end kept at zero temperature or being insulated leads to the following initial-boundary value problems

(3.9) $$u_t - u_{xx} = 0; \qquad 0 < x < \infty, \qquad 0 < t$$

(3.10) $$u(x, 0) = \phi(x), \qquad 0 \leqq x < \infty$$

(3.11) $$u(0, t) = 0, \qquad 0 \leqq t$$

or

(3.12) $$u_x(0, t) = 0, \qquad 0 \leqq t.$$

The desired solution is to be continuous and bounded in the closed first quadrant $0 \leqq x < \infty$, $0 \leqq t < \infty$. Problems (3.9), (3.10), (3.11), and (3.9), (3.10), (3.12) can be reduced to problem (3.1), (3.2) by extending ϕ over the whole x-axis to an odd or even function (see Examples 7.1 and 7.2 of Chapter VIII).

(a) Derive the following formula for the solution of (3.9), (3.10), (3.11) when $t > 0$,

(3.13) $$u(x, t) = \frac{1}{2\sqrt{\pi}} \int_0^{\infty} \frac{1}{\sqrt{t}} \left[e^{-\frac{(x-\xi)^2}{4t}} - e^{-\frac{(x+\xi)^2}{4t}} \right] \phi(\xi) d\xi.$$

(b) Derive the following formula for the solution of (3.9), (3.10), (3.12) when $t > 0$,

$$(3.14) \qquad u(x, t) = \frac{1}{2\sqrt{\pi}} \int_0^\infty \frac{1}{\sqrt{t}} \left[e^{-\frac{(x-\xi)^2}{4t}} + e^{-\frac{(x+\xi)^2}{4t}} \right] \phi(\xi) d\xi .$$

(c) Verify directly that (3.13) and (3.14) satisfy (3.11) and (3.12) respectively.

3.10. The temperature in a semi-infinite rod is initially constant, equal to U_0. If the temperature at the end $x = 0$ of the rod is kept equal to zero, derive the formula

$$u(x, t) = U_0 \operatorname{erf}\left(\frac{x}{2\sqrt{t}}\right)$$

for the temperature in the rod at any time $t > 0$. What happens to the temperature in the rod as $t \to \infty$? [Use the results of Problems 3.8 and 3.9.]

3.11. The temperature in a semi-infinite rod is initially zero. If the end $x = 0$ is kept at constant temperature U_0 derive the formula

$$u(x, t) = U_0 \left[1 - \operatorname{erf}\left(\frac{x}{2\sqrt{t}}\right)\right]$$

for the temperature in the rod.

3.12. If

$$\phi(x) = \begin{cases} 0 & \text{for} \quad 0 < x < L \\ 1 & \text{for} \quad L < x < \infty \end{cases}$$

derive the following formula for the solution of (3.9), (3.10), (3.11)

$$u(x, t) = \frac{1}{2}\left[\operatorname{erf}\left(\frac{L + x}{2\sqrt{t}}\right) - \operatorname{erf}\left(\frac{L - x}{2\sqrt{t}}\right)\right].$$

3.13. (a) Formulate and prove Duhamel's principle for solving the initial value problem for the nonhomogeneous heat equation

$$(3.15) \qquad u_t - u_{xx} = f(x, t); \qquad -\infty < x < \infty, \qquad 0 < t$$

$$(3.16) \qquad u(x, 0) = 0, \qquad -\infty < x < \infty.$$

(b) Apply Duhamel's principle to find the following (formal) solution of (3.15), (3.16)

$$(3.17) \qquad u(x, t) = \int_0^t \int_{-\infty}^\infty \frac{1}{2\sqrt{\pi(t - \tau)}} e^{-\frac{(x-\xi)^2}{4(t-\tau)}} f(\xi, \tau) d\xi d\tau.$$

3.14. Verify that the function

$$u(x, t) = Cxt^{-\frac{3}{2}} e^{-\frac{x^2}{4t}}, \qquad -\infty < x < \infty, \qquad 0 < t,$$

where C is a constant, satisfies the heat equation (3.1) and assumes zero initial values in the sense that for each x, $-\infty < x < \infty$,

$$\lim_{t \to 0^+} u(x, t) = 0.$$

How do you reconcile this example with Corollary 3.1 which asserts uniqueness of the problem (3.1), (3.2) with $\phi = 0$?

4. Heat Conduction in More than One Space Dimension

The results of our study of the heat equation in one space variable can be generalized in a straightforward way to the heat equation in any number of space variables.

First, let us consider the problem of heat conduction in three-dimensional space. The initial value problem asks for a function $u(x, t) = u(x_1, x_2, x_3, t)$ which is continuous and bounded for $x \in R^3$ and $t \geqq 0$, and which satisfies the heat equation

$$(4.1) \qquad u_t - (u_{x_1x_1} + u_{x_2x_2} + u_{x_3x_3}) = 0; \qquad x \in R^3, \qquad t > 0,$$

and the initial condition

$$(4.2) \qquad u(x, 0) = \phi(x), \qquad x \in R^3.$$

The given function ϕ is continuous and bounded in R^3 and describes the initial temperature distribution in space. Problem (4.1), (4.2) is well-posed and its solution is given by

$$(4.3) \quad u(x, t) = \begin{cases} \left(\dfrac{1}{2\sqrt{\pi t}}\right)^3 \displaystyle\int_{R^3} \exp\left[-\dfrac{(x_1 - \xi_1)^2 + (x_2 - \xi_2)^2 + (x_3 - \xi_3)^2}{4t}\right] \\ \qquad \qquad \phi(\xi_1, \xi_2, \xi_3) d\xi_1 d\xi_2 d\xi_3, \quad \text{for} \quad t < 0 \\ \phi(x), \quad \text{for} \quad t = 0. \end{cases}$$

This solution is C^∞ for $x \in R^3$ and $t > 0$. If ϕ is bounded but only piecewise continuous, then the bounded solution of the initial value problem (4.1), (4.2) is still given by the integral in (4.3) when $t > 0$. Again, this integral defines a C^∞ function when $t > 0$. Moreover, at every point $x^0 \in R^3$ where ϕ is continuous, the integral approaches $\phi(x^0)$ as $(x, t) \to (x^0, 0)$.

Next, let us look at the problem of heat conduction in a bounded body with piecewise smooth boundary if the temperature at the boundary is kept equal to zero. The study of this problem leads to the following initial-boundary value problem in a bounded domain Ω of R^n,

$$(4.4) \qquad u_t - \nabla^2 u = 0; \qquad x \in \Omega, \qquad t > 0$$

$$(4.5) \qquad u(x, 0) = \phi(x), \qquad x \in \bar{\Omega}$$

$$(4.6) \qquad u(x, t) = 0; \qquad x \in \partial\Omega, \qquad t \geqq 0.$$

The uniqueness and continuous dependence of the solution on the initial data follows from a generalization of the maximum-minimum theorem to n space dimensions (see Vladimirov[4]). Problem (4.4), (4.5), (4.6) can be solved by the method of separation of variables and eigenfunction expansions, which was described in Section 10 of Chapter VIII in connection with the corresponding problem for the wave equation. The (formal) solution is given by

$$u(x, t) = \sum_{n=1}^{\infty} a_n v_n(x) e^{-\lambda_n t}, \tag{4.7}$$

where $\lambda_1, \lambda_2, \dots$ are the eigenvalues (10.8) and $v_1(x), v_2(x), \dots$ are the corresponding eigenfunctions (10.9) of the eigenvalue problem (10.4), (10.5) of Chapter VIII while $a_1, a_2, \dots$ are the coefficients of the eigenfunction expansion of ϕ,

$$a_n = \frac{\int_\Omega \phi(x) v_n(x) dx}{\int_\Omega [v_n(x)]^2 dx}, \qquad n = 1, 2, \dots. \tag{4.8}$$

If ϕ is such that the coefficients (4.8) are bounded, it can be shown, using the properties of the eigenvalues and eigenfunctions, that the function $u(x, t)$ defined by (4.7) is in C^∞ for $x \in \Omega$ and $t > 0$ and satisfies the heat equation (4.4) and the boundary condition (4.6). If ϕ satisfies the stronger conditions that $\phi \in C^2(\bar{\Omega})$ and $\phi(x) = 0$ for $x \in \partial\Omega$, then it follows from the eigenfunction expansion Theorem 10.1 of Chapter VIII that (4.7) also satisfies the initial condition (4.5). In fact, in this case the function $u(x, t)$ defined by (4.7) is the solution of (4.4), (4.5), (4.6) in the strict sense that $u(x, t)$ is continuous for $x \in \bar{\Omega}$ and $t \geqq 0$. However, if ϕ has jump discontinuities but is such that its coefficients (4.8) are bounded, (4.7) is still the solution of (4.4), (4.5), (4.6) in a more general sense.

In Section 2 we applied the above method of solution of the initial-boundary value problem to the case of the heat equation in one space variable. We give here a further illustration of the method by applying it to the study of heat conduction in a rectangular plate.

Example 4.1. Consider a rectangular plate of length a and width b and suppose that the two surfaces of the plate are insulated while its four edges are kept at zero temperature. If at $t = 0$ the temperature in the plate does not vary across its thickness the same will be true at later times. To determine the history of the temperature distribution in the plate we must solve the initial-boundary value problem

$$u_t - (u_{xx} + u_{yy}) = 0; \qquad 0 < x < a,\ 0 < y < b, \qquad 0 < t \tag{4.9}$$

$$u(x, y, 0) = \phi(x, y); \qquad 0 \leqq x \leqq a, \qquad 0 \leqq y \leqq b, \tag{4.10}$$

$$\begin{cases} u(0, y, t) = 0, & u(a, y, t) = 0; \quad 0 \leqq y \leqq b, \quad 0 \leqq t \\ u(x, 0, t) = 0, & u(x, b, t) = 0; \quad 0 \leqq x \leqq a, \quad 0 \leqq t. \end{cases} \tag{4.11}$$

Using some of the results of Chapter VIII, Section 9, it is easy to obtain the following solution of (4.9), (4.10), (4.11),

$$u(x, t) = \sum_{m,n=1}^{\infty} A_{mn} \sin\frac{m\pi x}{a} \sin\frac{n\pi y}{b} \exp\left[-\pi^2\left(\frac{m^2}{a^2} + \frac{n^2}{b^2}\right)t\right], \tag{4.12}$$

where

$$(4.13)\quad A_{mn} = \frac{4}{ab}\int_0^a\int_0^b \phi\,(x, y)\sin\frac{m\pi x}{a}\sin\frac{n\pi y}{b}\,dxdy;\ m, n = 1, 2, \ldots .$$

Example 4.2. Consider the following initial-boundary value problem for the heat equation in a bounded domain Ω of R^n with nonhomogeneous but time-independent boundary condition,

$$(4.14)\quad \begin{cases} u_t - \nabla^2 u = 0; & x \in \Omega, \quad t > 0 \\ u(x, 0) = \phi(x), & x \in \bar{\Omega} \\ u(x, t) = f(x); & x \in \partial\Omega, \quad t \geqq 0. \end{cases}$$

This problem can be solved by the following method. Let $v(x)$ be the solution of the Dirichlet problem

$$(4.15)\quad \begin{cases} \nabla^2 v = 0, & x \in \Omega \\ v(x) = f(x), & x \in \partial\Omega \end{cases}$$

and let $w(x, t)$ be the solution of the initial-boundary value problem for the heat equation with homogeneous boundary condition

$$(4.16)\quad \begin{cases} w_t - \nabla^2 w = 0; & x \in \Omega, \quad t > 0 \\ w(x, 0) = \phi(x) - v(x), & x \in \Omega \\ w(x, t) = 0; & x \in \partial\Omega, \quad t \geqq 0. \end{cases}$$

Then it is easy to see that

$$(4.17)\quad u(x, t) = v(x) + w(x, t)$$

is the solution of (4.14). It is left as an exercise (Problem 4.9) to show that as $t \to \infty$, $u(x, t)$ tends to the steady state temperature distribution $v(x)$, i.e.,

$$(4.18)\quad \lim_{t\to\infty} u(x, t) = v(x), \qquad x \in \bar{\Omega}.$$

Problems

4.1. Verify that (4.3) satisfies the heat equation (4.1) when $t > 0$.

4.2. If initially the temperature in space is constant, equal to U_0, then obviously the bounded solution of (4.1), (4.2) is $u(x, t) = U_0$. Verify that formula (4.3) gives this solution.

4.3. Find the solution of (4.1), (4.2) if

$$\phi(x) = \begin{cases} U_1 & \text{for} \quad x_3 < 0 \\ U_2 & \text{for} \quad x_3 > 0. \end{cases}$$

4.4. State the well posed initial value problem for the heat equation in two space variables and write down its solution.

4.5. Derive the solution (4.12) of problem (4.9), (4.10), (4.11).

4.6. Formulate and solve the problem described in Example 4.1 for a circular plate of radius a instead of a rectangular plate. [Use polar coordinates and Example 10.3 of Chapter VIII.]

4.7. *Cooling of a spherical body.* The temperature in a spherical body of radius a is initially constant, equal to U_0. At all subsequent time the boundary of the solid is kept at zero temperature. Obviously the temperature u in the solid will depend only on the distance r from the center of the body and the time t, i.e., $u = u(r, t)$. Show that

$$u(r, t) = \frac{2aU_0}{\pi} \sum_{n=1}^{\infty} \frac{(-1)^{n+1}}{nr} \sin \frac{n\pi r}{a} e^{-\frac{n^2\pi^2}{a^2}t}.$$

4.8. Solve Problem 4.7 if the surface of the spherical body is insulated instead of being kept at zero temperature.

4.9. Prove (4.18).

4.10. *Heating of a spherical body.* The temperature in a spherical body of radius a is initially zero. At all subsequent time the boundary of the body is kept at constant temperature U_1. Use the result of Problem 4.7 to find the temperature $u(r, t)$ in the body.

4.11. Formulate and find a formula for the solution of the initial-boundary value problem for the heat equation in the half-space of points (x_1, x_2, x_3) with $x_3 > 0$, if
(a) the boundary $x_3 = 0$ is kept at zero temperature,
(b) the boundary $x_3 = 0$ is insulated.

4.12. Consider the initial-boundary value problem for the nonhomogeneous heat equation in a bounded domain Ω of R^n (there are heat sources in Ω),

$$u_t - \nabla^2 u = f(x, t); \quad x \in \Omega, \quad t > 0 \tag{4.19}$$

$$u(x, 0) = 0, \quad x \in \bar{\Omega} \tag{4.20}$$

$$u(x, t) = 0; \quad x \in \partial\Omega, \quad t \geqq 0. \tag{4.6}$$

Generalize Duhamel's principle described in Problem 2.9 and derive the following formula for the solution of (4.19), (4.20), (4.6)

$$u(x, t) = \sum_{n=1}^{\infty} \left[\int_0^t f_n(\tau) e^{-\lambda_n(t-\tau)} d\tau\right] v_n(x), \tag{4.21}$$

where $\lambda_1, \lambda_2, \ldots$ are the eigenvalues (10.8) and $v_1(x), v_2(x), \ldots$ are the corresponding eigenfunctions of the eigenvalue problem (10.4), (10.5) of Chapter VIII, and

$$f_n(\tau) = \frac{\int_\Omega f(x, \tau) v_n(x) dx}{\int_\Omega [v_n(x)]^2 dx}. \tag{4.22}$$

4.13. Find a formula for the solution of (4.19), (4.5), (4.6).

4.14. Describe a method of solution of the problem

$$\begin{aligned} u_t - \nabla^2 u &= F(x, t); \quad x \in \Omega, \quad t > 0 \\ u(x, 0) &= \phi(x), \quad x \in \Omega, \\ u(x, t) &= f(x), \quad x \in \partial\Omega, \quad t \geqq 0. \end{aligned}$$

4.15. Let $u(x, t)$ be the solution of the heat conduction problem

$$\begin{aligned} u_t - \nabla^2 u &= 0, \qquad x \in \Omega, \qquad 0 < t \\ u(x, 0) &= \phi(x), \qquad x \in \bar{\Omega} \\ u(x, t) &= 0, \qquad x \in \partial\Omega, \qquad 0 \le t \end{aligned}$$

where Ω is a bounded domain in R^3 with smooth boundary $\partial\Omega$. Consider a portion Γ of the boundary surface $\partial\Omega$. Show that the total amount of heat flowing across Γ, i.e.,

$$Q = \int_0^\infty \left[\iint_\Gamma \frac{\partial u}{\partial n} \, d\sigma \right] dt,$$

is given by

$$Q = -\int_\Omega \phi(x) h(x) dx$$

where h is the solution of the Dirichlet problem

$$\nabla^2 h = 0, \qquad x \in \Omega,$$

$$h(x) = \begin{cases} 1, & x \in \Gamma \\ 0, & x \in \partial\Omega - \Gamma. \end{cases}$$

h is called the *harmonic measure* of the surface Γ. [*Hint:* Apply Green's identity to u and h].

5. An Application to Transistor Theory

As we mentioned in Chapter VI, the heat equation arises in diffusion processes other than the diffusion of heat. The diffusion of charges in transistors is one phenomenon in which the heat equation occurs. For example, the calculation of certain electrical parameters of a *p-n-p* transistor with an exponentially graded base requires the solution of the following initial-boundary value problem (see Lindmayer and Wrigley[5]):

(5.1) $$p_t - D\left(p_{xx} - \frac{\eta}{L} p_x\right) = 0; \qquad 0 < x < L, \qquad 0 < t,$$

(5.2) $$p(x, 0) = \frac{KL}{D} \left[\frac{1 - e^{-\eta\left(1 - \frac{x}{L}\right)}}{\eta} \right], \qquad 0 \le x \le L,$$

(5.3) $$p(0, t) = p(L, t) = 0, \qquad 0 \le t.$$

The function $p(x, t)$ describes very closely the concentration of excess holes (positive charge carriers) at time t and position x in the base of the transistor, which occupies the interval $0 \le x \le L$. The constants K, D and η are all positive, with $\eta = 0$ being permitted as a limiting case.

We will first construct a series solution of the problem (5.1), (5.2), (5.3) and then use this solution to compute the reclaimable charge, a quantity which is required in certain transistor calculations. This example demonstrates that series solutions can be used to obtain information of practical value.

By following the technique outlined in Problem 5.1, the equation (5.1) is found to have solutions of the form

(5.4) $$p(x, t) = e^{\frac{\eta x}{2L} - \frac{D\eta^2}{4L^2}t} u(x, t),$$

where u satisfies the heat equation

(5.5) $$u_t - Du_{xx} = 0; \qquad 0 < x < L, \qquad 0 < t.$$

It is easily seen from (5.2), (5.3) and (5.4) that u must satisfy the initial condition

(5.6) $$u(x, 0) = \phi(x) \equiv \frac{KL}{D\eta} e^{-\frac{\eta x}{2L}} [1 - e^{-\eta\left(1 - \frac{x}{L}\right)}], \qquad 0 \le x \le L,$$

and the boundary conditions

(5.7) $$u(0, t) = u(L, t) = 0, \qquad 0 \le t.$$

The initial-boundary value problem (5.5), (5.6), (5.7) was solved in Section 2. Remembering to replace t by Dt in the formula (2.12) for the solution (see the last paragraph of Section 1), we find

$$u(x, t) = \sum_{n=1}^{\infty} C_n \sin \frac{n\pi x}{L} e^{-\frac{n^2\pi^2 D}{L^2}t},$$

where the constants C_n are the Fourier sine series coefficients of the function $\phi(x)$ given by (5.6). The coefficients C_n are easily computed (Problem 5.2), with the result

(5.8) $$\phi(x) = \frac{2KL}{D} \frac{(1 - e^{-\eta})}{\eta} \sum_{n=1}^{\infty} \frac{n\pi}{n^2\pi^2 + \eta^2/4} \sin \frac{n\pi x}{L}.$$

The concentration p is found to be

(5.9) $$p(x, t) = \frac{2\pi KL}{D} \frac{(1 - e^{-\eta})}{\eta} e^{\frac{\eta x}{2L}} \sum_{n=1}^{\infty} \frac{n}{n^2\pi^2 + \eta^2/4} \sin \frac{n\pi x}{L} \cdot \exp\left[-\frac{D}{L^2}\left(n^2\pi^2 + \frac{\eta^2}{4}\right)t\right].$$

For fixed x, $0 \le x \le L$, the function $p(x, t)$ given by (5.9) describes the manner in which the concentration of excess holes at position x collapses as $t \to \infty$. This collapse of hole concentration gives rise to a current $I(t)$, the emitter discharge current, which is given by

(5.10) $$I(t) = I_p \frac{D}{K} p_x(x, t) |_{x=0}, \qquad 0 < t,$$

where I_p is a constant.

Because of the presence of the exponential time factors in the series (5.9), the derivative $p_x(0, t)$, $0 < t$, can be computed by termwise differentiation of this series (see Section 2). In this way we obtain

$$(5.11)\quad I(t) = 2I_p \frac{(1 - e^{-\eta})}{\eta} \sum_{n=1}^{\infty} \frac{n^2\pi^2}{n^2\pi^2 + \eta^2/4} \exp\left[-\frac{D}{L^2}\left(n^2\pi^2 + \frac{\eta^2}{4}\right)t\right]$$

so long as $t > 0$. Observe that the series (5.11) does not converge for $t = 0$.

A physical quantity of interest in transistor design is the reclaimable charge Q, defined by the improper integral

$$(5.12)\qquad Q = \int_{0^+}^{\infty} I(t)dt = \lim_{\substack{\epsilon\to 0^+ \\ R\to\infty}} \int_{\epsilon}^{R} I(t)dt.$$

We will derive a simple, closed form expression for Q. First, the improper integral (5.12) for Q can be easily evaluated by substituting the series (5.11) for $I(t)$ into (5.12) and integrating termwise (Problem 5.3). This produces a series expression for Q,

$$(5.13)\qquad Q = \frac{2I_pL^2}{D}\frac{(1 - e^{-\eta})}{\eta} \sum_{n=1}^{\infty} \left[\frac{n\pi}{n^2\pi^2 + \eta^2/4}\right]^2.$$

The numerical series appearing in (5.13) can be expressed in closed form by an appropriate application of Parseval's relation. In fact, the terms of the series in (5.13) are precisely the squares of the Fourier sine coefficients of the function

$$\psi(x) = \frac{D}{2KL}\frac{\eta}{1 - e^{-\eta}}\phi(x),$$

which is easily seen from (5.8). By applying Parseval's relation to the function $\psi(x)$ and using the expression (5.6) for $\phi(x)$, it is found that (Problem 5.4)

$$(5.14)\qquad \sum_{n=1}^{\infty} \left[\frac{n\pi}{n^2\pi^2 + \eta^2/4}\right]^2 = \frac{2}{L}\int_0^L [\psi(x)]^2\, dx = \frac{1}{2}\,\frac{\dfrac{\sinh\eta}{\eta} - 1}{\cosh\eta - 1}.$$

It now follows from (5.13) and (5.14) that Q may be written in the convenient form

$$Q = I_p \frac{L^2}{D}\frac{(1 - e^{-\eta})}{\eta}\left[\frac{\dfrac{\sinh\eta}{\eta} - 1}{\cosh\eta - 1}\right].$$

This equation for Q is used in the study of transistor design (see Lindmayer and Wrigley,[5] Chapter 4, §4).

Problems

5.1. Consider the parabolic equation

$$v_t - k(v_{xx} + av_x + bv) = 0$$

where a, b and k are real constants and $k > 0$. This equation can be reduced to the heat equation by introducing the new dependent variable u through the relation

$$v(x, t) = e^{\alpha x+\beta t}\, u(x, t).$$

Show that if

$$\alpha = -\frac{a}{2} \quad \text{and} \quad \beta = k(\alpha^2 + a\alpha + b)$$

then u must satisfy the heat equation

$$u_t - ku_{xx} = 0.$$

5.2. Obtain the Fourier series expansion (5.8).

5.3. Carry out the derivation of (5.13) from (5.11) and (5.12). First compute $\int_\epsilon^R I(t)dt$ for $0 < \epsilon < R < \infty$, and then perform the limiting operations. Justify all steps of the computation.

5.4. (a) Show that for an odd sectionally continuous function $f(x)$, $-L < x < L$, Parseval's relation (8.34) of Chapter VIII can be written in the form

$$\frac{2}{L}\int_0^L |f(x)|^2\, dx = \sum_{n=1}^{\infty} b_n^2.$$

(b) Derive (5.14).

References for Chapter IX

1. Friedman, A: *Partial Differential Equations of Parabolic Type,* Englewood Cliffs, N.J.: Prentice Hall, Inc., 1964.
2. Petrovskii, I. G.: *Partial Differential Equations,* Philadelphia: W. B. Saunders Co., 1967.
3. Churchill, R. V.: *Fourier Series and Boundary Value Problems,* Ed. 2, New York: McGraw Hill Book Co., 1963.
4. Vladimirov, V. S.: *Equations of Mathematical Physics,* New York: Marcel Dekker, Inc., 1971.
5. Lindmayer, J., and Wrigley, C. Y.: *Fundamentals of Semiconductor Devices,* Princeton, N.J.: Van Nostrand Co., 1965.

CHAPTER X

Systems of first order linear and quasi-linear equations

In this chapter we study systems of linear and quasi-linear partial differential equations of first order. Such systems arise in many areas of mathematics, engineering and the physical sciences. In Section 1 we describe five examples and then introduce a useful matrix notation. In Section 2 we define linear hyperbolic systems in two independent variables and then show how the introduction of new unknowns reduces such systems to a canonical form in which the principal part of the ith equation involves only the ith unknown. In Section 3 we define the characteristic curves of a linear hyperbolic system and observe that the principal part of the ith equation of the system in canonical form is actually the ordinary derivative of the ith unknown along the ith characteristic curve. This observation is the basis of the method of characteristics for solving the initial value problem for the system. An application to the system governing electrical transmission lines is discussed. In Section 4 we present a brief discussion of general quasi-linear hyperbolic systems. Finally in Section 5 we study in some detail a quasi-linear hyperbolic system of two equations in two unknowns which governs the one-dimensional isentropic flow of an inviscid gas.

1. Examples of Systems. Matrix Notation

Systems of first order partial differential equations arise in many areas of mathematics, engineering, and the physical sciences. We begin this section by describing a few important examples.

The *Cauchy-Riemann equations*

$$(1.1)\qquad \begin{cases} \dfrac{\partial u}{\partial x} - \dfrac{\partial v}{\partial y} = 0 \\[2mm] \dfrac{\partial u}{\partial y} + \dfrac{\partial v}{\partial x} = 0 \end{cases}$$

play an important role in the study of analytic functions of a complex variable (see Churchill,[1] Chapter 2). Equations (1.1) form a system of two equations in the two unknown functions $u = u(x, y)$ and $v = v(x, y)$ of the two independent variables x and y.

In electrical engineering, the study of transmission lines leads to the equations

$$(1.2)\qquad \begin{cases} L\dfrac{\partial I}{\partial t} + \dfrac{\partial E}{\partial x} + RI = 0 \\[2ex] C\dfrac{\partial E}{\partial t} + \dfrac{\partial I}{\partial x} + GE = 0 \end{cases}$$

(see Sokolnikoff and Redheffer,[2] p. 465). The variable x denotes the directed distance along the transmission line from some fixed point on the line; t denotes time. The electrical properties of the line are assumed to be known and depend only on x. $C = C(x)$ is the capacitance to ground per unit length, $L = L(x)$ is the inductance per unit length, $R = R(x)$ is the resistance per unit length and $G = G(x)$ is the conductance to ground per unit length. $I = I(x, t)$ and $E = E(x, t)$ are, respectively, the current and potential at the point x of the line at time t. Equations (1.2) form a system of two equations in the two unknowns I and E and two independent variables x and t.

The study of fluid dynamics leads to many systems of first order partial differential equations. The form of the system that governs a particular fluid flow depends on the assumptions made for that flow. For example, the two-dimensional motion of a perfect inviscid fluid is governed by *Euler's equations of motion*,

$$(1.3)\qquad \begin{cases} \dfrac{\partial u}{\partial t} + u\dfrac{\partial u}{\partial x} + v\dfrac{\partial u}{\partial y} + \dfrac{1}{\rho}\dfrac{\partial p}{\partial x} = 0 \\[2ex] \dfrac{\partial v}{\partial t} + u\dfrac{\partial v}{\partial x} + v\dfrac{\partial v}{\partial y} + \dfrac{1}{\rho}\dfrac{\partial p}{\partial y} = 0 \\[2ex] \dfrac{\partial \rho}{\partial t} + \dfrac{\partial}{\partial x}(\rho u) + \dfrac{\partial}{\partial y}(\rho v) = 0 \\[2ex] \dfrac{\partial}{\partial t}\left(\dfrac{p}{\rho^\gamma}\right) + u\dfrac{\partial}{\partial x}\left(\dfrac{p}{\rho^\gamma}\right) + v\dfrac{\partial}{\partial y}\left(\dfrac{p}{\rho^\gamma}\right) = 0. \end{cases}$$

This is a system of four partial differential equations in the four unknown functions u, v, p and ρ and three independent variables x, y and t. $u = u(x, y, t)$ and $v = v(x, y, t)$ are the x and y components of the velocity of the fluid at the point (x, y) at time t, while $p = p(x, y, t)$ and $\rho = \rho(x, y, t)$ are the pressure and density of the fluid. γ is a constant greater than 1 known as the ratio of specific heats. A derivation of (1.3) can be found in most books on fluid mechanics or in Garabedian,[3] Chapter 14.

In Section 4 of this chapter we will study in some detail the one-dimensional *isentropic* (or homentropic) flow of an inviscid gas. This flow is governed by the equations

$$(1.4)\qquad \begin{cases} \dfrac{\partial u}{\partial t} + u\dfrac{\partial u}{\partial x} + \dfrac{c^2}{\rho}\dfrac{\partial \rho}{\partial x} = 0 \\[2ex] \dfrac{\partial \rho}{\partial t} + u\dfrac{\partial \rho}{\partial x} + \rho\dfrac{\partial u}{\partial x} = 0. \end{cases}$$

Here $u = u(x, t)$ and $\rho = \rho(x, t)$ are the velocity and density of the gas at position x and time t while $c = c(\rho)$ is the local speed of sound, which is assumed to be a known function of ρ. (Again, see Garabedian,[3] Chapter 14 for a derivation.) Equations (1.4) form a system of two equations in the two unknowns u and ρ and two independent variables x and t. If the assumption of constant entropy is dropped, the flow is said to be *nonisentropic* (or nonhomentropic) and is governed by the system of three equations in the three unknowns u, ρ, p,

$$(1.5)\qquad \begin{cases} \dfrac{\partial u}{\partial t} + u\dfrac{\partial u}{\partial x} + \dfrac{1}{\rho}\dfrac{\partial p}{\partial x} = 0 \\[2ex] \dfrac{\partial \rho}{\partial t} + u\dfrac{\partial \rho}{\partial x} + \rho\dfrac{\partial u}{\partial x} = 0 \\[2ex] \dfrac{\partial p}{\partial t} + u\dfrac{\partial p}{\partial x} + c^2\rho\dfrac{\partial u}{\partial x} = 0. \end{cases}$$

Here u, ρ and c denote the same quantities as in (1.4), while $p = p(x, t)$ is the pressure of the gas at position x and time t (see Becker,[4] Chapter 3).

In this book we limit our study of systems to those which consist of m equations in m unknowns and only two independent variables, and which are of the form

$$(1.6)\qquad \frac{\partial u_i}{\partial t} + a_{i1}\frac{\partial u_1}{\partial x} + a_{i2}\frac{\partial u_2}{\partial x} + \cdots + a_{im}\frac{\partial u_m}{\partial x} + b_i = 0, \qquad i = 1, 2, \ldots, m.$$

Each of the unknowns $u_i = u_i(x, t)$, $i = 1, 2, \ldots, m$, is assumed to be a function of the two independent variables x and t, while the coefficients a_{ij}, b_i may be functions of the unknowns as well as of x and t,

$$a_{ij} = a_{ij}(x, t, u_1, \ldots, u_m), \quad b_i = b_i(x, t, u_1, \ldots, u_m); \qquad i, j = 1, 2, \ldots, m.$$

We have chosen to use the letter t rather than y since this variable usually indicates time in physical applications. Note carefully that the system (1.6) has a special form with respect to the variable t: the ith equation contains the derivative with respect to t of u_i only, and this derivative has unit coefficient.

It is easy to see that the systems (1.1), (1.2), (1.4) and (1.5) are special cases of the general system (1.6). For example (1.4) has the form (1.6) with $m = 2, u_1 = u, u_2 = \rho$, and

$$a_{11} = u_1, \qquad a_{12} = c^2(u_2)/u_2, \qquad a_{21} = u_1, \qquad a_{22} = u_2, \qquad b_1 = b_2 = 0.$$

The classification of the first order system (1.6) according to linearity is similar to the classification of single first order equations (see Chapter III,

Section 1). If the coefficients a_{ij} actually depend on the unknowns $u_1, \dots,$ u_m, the system is called *quasi-linear.* If the a_{ij} do not depend on $u_1, \dots, u_m$, and the b_i depend linearly on $u_1, \dots, u_m$, the system is said to be *linear.* If the a_{ij} are independent of $u_1, \dots, u_m$, and the b_i depend on $u_1, \dots, u_m$ but not linearly, the system is called *almost linear.* Systems (1.1) and (1.2) are linear while (1.4) and (1.5) are quasi-linear. Note that the derivatives of the unknowns appear linearly in (1.6). A system of first order equations in which the derivatives appear nonlinearly is said to be *nonlinear.*

The system (1.6) can be written in a more convenient and compact form by introducing matrix notation. Let **u** and **b** denote the column vectors

$$\mathbf{u} = \begin{bmatrix} u_1 \\ u_2 \\ \cdot \\ \cdot \\ \cdot \\ u_m \end{bmatrix}, \qquad \mathbf{b} = \begin{bmatrix} b_1 \\ b_2 \\ \cdot \\ \cdot \\ \cdot \\ b_m \end{bmatrix}$$

and A the $m \times m$ matrix

$$A = [a_{ij}]_{i,j=1,\dots,m} = \begin{bmatrix} a_{11} & \cdot & \cdot & \cdot & a_{1m} \\ \cdot & \cdot & \cdot & \cdot & \cdot \\ \cdot & \cdot & \cdot & \cdot & \cdot \\ a_{m1} & \cdot & \cdot & \cdot & a_{mm} \end{bmatrix}$$

Of course we are dealing here with vector and matrix functions; each component or entry is a function. Remembering the definition of matrix multiplication and that differentiation means differentiation of each component or entry, it is easy to see that the system (1.6) can be written in the form

$$\text{(1.7)} \qquad \frac{\partial \mathbf{u}}{\partial t} + A(x, t, \mathbf{u}) \frac{\partial \mathbf{u}}{\partial x} + \mathbf{b}(x, t, \mathbf{u}) = \mathbf{0}.$$

In (1.7) we display the variables of A and **b** to remind us that they may depend not only on the independent variables x and t but also on the unknown **u**.

As an example, the system (1.4) can be written in the form

$$\text{(1.8)} \qquad \frac{\partial}{\partial t}\begin{bmatrix} u \\ \rho \end{bmatrix} + \begin{bmatrix} u & c^2/\rho \\ \rho & u \end{bmatrix} \frac{\partial}{\partial x}\begin{bmatrix} u \\ \rho \end{bmatrix} = \begin{bmatrix} 0 \\ 0 \end{bmatrix}$$

which is a special case of (1.7) with

$$\mathbf{u} = \begin{bmatrix} u \\ \rho \end{bmatrix}, \qquad A = \begin{bmatrix} u & c^2/\rho \\ \rho & u \end{bmatrix}, \qquad \mathbf{b} = \begin{bmatrix} 0 \\ 0 \end{bmatrix}.$$

We close this section with the definition of solution of the first order system (1.7). Since **u** is m-dimensional, $A(x, t, \mathbf{u})$ and $\mathbf{b}(x, t, \mathbf{u})$ are functions of $m + 2$ variables. We assume that these functions are defined in some domain $\tilde{\Omega}$ in R^{m+2}. Naturally, a matrix or vector function is said to be of class C^k in some domain if each of its entries or components are of class C^k in that domain.

Definition 1.1. A *solution* of the system (1.7) in a domain Ω of R^2 is a (vector) function $\mathbf{u} = \mathbf{u}(x, t)$ which is defined and C^1 in Ω and is such that the following two conditions are satisfied:

(i) For every $(x, t) \in \Omega$, the point $(x, t, \mathbf{u}(x, t))$ is in the domain $\tilde{\Omega}$ of A and $\mathbf{b}$.

(ii) When $\mathbf{u} = \mathbf{u}(x, t)$ is substituted into (1.7), the resulting (vector) equation is an identity in (x, t) for all (x, t) in Ω.

Problems

1.1. Write the systems (1.1), (1.2) and (1.5) in matrix notation.

1.2. Consider the telegraph equation of Example 2.5 of Chapter V,

$$u_{xx} - \frac{1}{c^2} u_{tt} + au_t + bu = 0.$$

Let $u_1 = u, u_2 = u_x, u_3 = u_t$ and show that u_1, u_2, u_3 must satisfy the system of three equations,

$$\begin{aligned} &\frac{\partial u_1}{\partial t} - u_3 = 0 \\ &\frac{\partial u_2}{\partial t} - \frac{\partial u_3}{\partial x} = 0 \\ &\frac{\partial u_3}{\partial t} - c^2\left(\frac{\partial u_2}{\partial x} + au_3 + bu_1\right) = 0. \end{aligned} \tag{1.9}$$

Write this system in matrix form. This example illustrates how a higher order equation may be reduced to a system of first order equations.

2. Linear Hyperbolic Systems. Reduction to Canonical Form

We study in this section linear and almost linear systems of m first order equations in m unknowns and two independent variables of the form

$$\frac{\partial u_i}{\partial t} + \sum_{j=1}^{m} a_{ij}(x, t)\frac{\partial u_j}{\partial x} + b_i(x, t, u_1, \dots, u_m) = 0; \qquad i = 1, 2, \dots m. \tag{2.1}$$

In obvious matrix notation the system (2.1) has the form

$$\frac{\partial \mathbf{u}}{\partial t} + A(x, t)\frac{\partial \mathbf{u}}{\partial x} + \mathbf{b}(x, t, \mathbf{u}) = \mathbf{0}. \tag{2.2}$$

We assume that the coefficient matrix $A(x, t)$ is of class C^1 in the domain of the (x, t)-plane under consideration.

Just as in the case of a single partial differential equation, it turns out that most of the important properties of solutions of the system (2.2) depend only on its principal part $\mathbf{u}_t + A(x, t)\mathbf{u}_x$. Since this principal part is completely characterized by the coefficient matrix $A(x, t)$, this matrix plays a fundamental role in the study of (2.2). There are two important classes of systems of the form (2.2) which are defined in terms of properties of the matrix $A(x, t)$. Recall that an eigenvalue of A is a root $\lambda = \lambda(x, t)$ of the characteristic equation

$$\text{(2.3)} \qquad \det|A - \lambda I| = \begin{vmatrix} a_{11} - \lambda & a_{12} & \cdots & a_{1m} \\ a_{21} & a_{22} - \lambda & \cdots & a_{2m} \\ \cdot & \cdot & & \cdot \\ \cdot & \cdot & & \cdot \\ \cdot & \cdot & & \cdot \\ a_{m1} & a_{m2} & \cdots & a_{mm} - \lambda \end{vmatrix} = 0,$$

and that for each eigenvalue λ of A, there is at least one nontrivial column vector $\mathbf{p} = \mathbf{p}(x, t)$ such that

$$\text{(2.4)} \qquad A\mathbf{p} = \lambda\mathbf{p}.$$

The vector $\mathbf{p}$ is known as an eigenvector of A corresponding to the eigenvalue λ.

Definition 2.1. The system (2.2) is said to be *elliptic* at the point (x, t) if $A(x, t)$ has no real eigenvalues. (2.2) is said to be elliptic in a domain Ω of R^2 if it is elliptic at every point of Ω.

The system of Cauchy-Riemann equations (1.1) is elliptic in R^2. Indeed, the matrix A for this system is

$$A = \begin{bmatrix} 0 & 1 \\ -1 & 0 \end{bmatrix}$$

and the characteristic equation (2.3) is

$$\lambda^2 + 1 = 0,$$

which has no real roots. In this book we will not discuss elliptic systems any further.

Definition 2.2. The system (2.2) is said to be *hyperbolic* at the point (x, t), if $A(x, t)$ has m real and distinct eigenvalues $\lambda_1(x, t), \lambda_2(x, t), \ldots, \lambda_m(x, t)$. (2.2) is said to be hyperbolic in a domain Ω of R^2 if it is hyperbolic at every point of Ω.

Since the eigenvalues $\lambda_k(x, t)$, $k = 1, 2, \ldots, m$, are assumed to be distinct, it follows from a well-known theorem of linear algebra that the corresponding eigenvectors

$$\text{(2.5)} \qquad \mathbf{p}_k(x, t) = \begin{bmatrix} p_{1k}(x, t) \\ p_{2k}(x, t) \\ \cdot \\ \cdot \\ \cdot \\ p_{mk}(x, t) \end{bmatrix}; \qquad k = 1, 2, \ldots, m,$$

are linearly independent. Remember that the eigenvector $\mathbf{p}_k$ corresponding to the eigenvalue λ_k is determined only up to a multiplicative constant. Any multiple of $\mathbf{p}_k$ is also an eigenvector corresponding to λ_k. (*Caution:* Usually hyperbolicity is defined by requiring only that A has m real eigenvalues and m linearly independent eigenvectors. If A has m real and distinct eigenvalues, the system (2.2) is sometimes said to be hyperbolic in the narrow sense or strictly hyperbolic. For simplicity we will stick to Definition 2.2.)

Example 2.1. Let a and b be positive constants and consider the system

$$\begin{cases} \dfrac{\partial u_1}{\partial t} + a\dfrac{\partial u_2}{\partial x} = 0 \\ \dfrac{\partial u_2}{\partial t} + b\dfrac{\partial u_1}{\partial x} = 0. \end{cases} \tag{2.6}$$

The matrix form of this system is

$$\frac{\partial}{\partial t}\begin{bmatrix} u_1 \\ u_2 \end{bmatrix} + \begin{bmatrix} 0 & a \\ b & 0 \end{bmatrix}\frac{\partial}{\partial x}\begin{bmatrix} u_1 \\ u_2 \end{bmatrix} = \begin{bmatrix} 0 \\ 0 \end{bmatrix}$$

so that

$$A = \begin{bmatrix} 0 & a \\ b & 0 \end{bmatrix}.$$

The characteristic equation is

$$\lambda^2 - ab = 0.$$

If we set $c_0 = \sqrt{ab}$, the eigenvalues of A are $\lambda_1 = c_0$ and $\lambda_2 = -c_0$. The system (2.6) is hyperbolic in the entire (x, t)-plane. The eigenvectors of A corresponding to the eigenvalues λ_1 and λ_2 are easily found using (2.4). They are

$$\mathbf{p}_1 = \begin{bmatrix} \sqrt{a} \\ \sqrt{b} \end{bmatrix}, \qquad \mathbf{p}_2 = \begin{bmatrix} \sqrt{a} \\ -\sqrt{b} \end{bmatrix}. \tag{2.7}$$

We will now show that if the system (2.2) is hyperbolic at a point (x, t), it is possible to obtain a very simple canonical form of the system in a neighborhood of (x, t) by introducing new unknowns. We make use of the eigenvalues and eigenvectors of the matrix A. Let Λ be the $m \times m$ diagonal matrix with diagonal entries the eigenvalues of A, and let P be the $m \times m$ matrix with columns the corresponding eigenvectors (2.5) of A,

$$\Lambda = \begin{bmatrix} \lambda_1 & 0 & \dots & 0 \\ 0 & \lambda_2 & \dots & 0 \\ \cdot & \cdot & & \cdot \\ \cdot & \cdot & & \cdot \\ \cdot & \cdot & & \cdot \\ 0 & 0 & \dots & \lambda_m \end{bmatrix}, \qquad P = \begin{bmatrix} p_{11} & p_{12} & \cdots & p_{1m} \\ p_{21} & p_{22} & \cdots & p_{2m} \\ \cdot & \cdot & & \cdot \\ \cdot & \cdot & & \cdot \\ \cdot & \cdot & & \cdot \\ p_{m1} & p_{m2} & \cdots & p_{mm} \end{bmatrix}. \tag{2.8}$$

Using an implicit function theorem and the assumption that A is C^1, it can be shown that the matrices Λ and P are C^1 in some neighborhood of (x, t) (see Petrovskii,[5] §7). Consequently, since the eigenvalues of A are assumed to be distinct at (x, t), they remain distinct in some neighborhood U of (x, t), and the columns of P are linearly independent in U. From linear algebra, P is nonsingular in U and if P^{-1} denotes its inverse, then

(2.9) $$P^{-1}AP = \Lambda \quad \text{in} \quad U.$$

We introduce now the new unknown $\mathbf{v}$ by the relation

(2.10) $$\mathbf{v} = P^{-1}\mathbf{u}.$$

Then

(2.11) $$\mathbf{u} = P\mathbf{v}$$

and

(2.12) $$\frac{\partial \mathbf{u}}{\partial t} = P\frac{\partial \mathbf{v}}{\partial t} + \frac{\partial P}{\partial t}\mathbf{v}, \qquad \frac{\partial \mathbf{u}}{\partial x} = P\frac{\partial \mathbf{v}}{\partial x} + \frac{\partial P}{\partial x}\mathbf{v}.$$

Substituting (2.11) and (2.12) into (2.2) we obtain

$$P\frac{\partial \mathbf{v}}{\partial t} + AP\frac{\partial \mathbf{v}}{\partial x} + \frac{\partial P}{\partial t}\mathbf{v} + A\frac{\partial P}{\partial x}\mathbf{v} + \mathbf{b}(x, t, P\mathbf{v}) = \mathbf{0}.$$

Finally, multiplying this equation from the left by P^{-1} and using (2.9) we obtain the desired canonical form in the neighborhood U of (x, t),

(2.13) $$\frac{\partial \mathbf{v}}{\partial t} + \Lambda(x, t)\frac{\partial \mathbf{v}}{\partial x} + \mathbf{c}(x, t, \mathbf{v}) = \mathbf{0},$$

where

(2.14) $$\mathbf{c} = P^{-1}\left(\frac{\partial P}{\partial t} + A\frac{\partial P}{\partial x}\right)\mathbf{v} + P^{-1}\mathbf{b}(x, t, P\mathbf{v}).$$

The simplicity of the canonical form (2.13) becomes apparent if we write it out in component form,

(2.15) $$\frac{\partial v_i}{\partial t} + \lambda_i(x, t)\frac{\partial v_i}{\partial x} + c_i(x, t, v_1, \ldots, v_m) = 0, \qquad i = 1, 2, \ldots, m.$$

The principal part of the ith equation involves only the ith unknown v_i.

Example 2.1 *(continued)*. Since the system (2.6) is hyperbolic in the whole of R^2 and has constant coefficients, the transformation to canonical form is valid in the entire (x, t)-plane. We have

$$\Lambda = \begin{bmatrix} c_0 & 0 \\ 0 & -c_0 \end{bmatrix}, \qquad P = \begin{bmatrix} \sqrt{a} & \sqrt{a} \\ \sqrt{b} & -\sqrt{b} \end{bmatrix}$$

The new unknown $\mathbf{v} = P^{-1}\mathbf{u}$ must satisfy the canonical form of system (2.6),

$$\frac{\partial \mathbf{v}}{\partial t} + \Lambda\frac{\partial \mathbf{v}}{\partial x} = \mathbf{0},$$

or

(2.16) $$\begin{cases} \dfrac{\partial v_1}{\partial t} + c_0\dfrac{\partial v_1}{\partial x} = 0 \\[2ex] \dfrac{\partial v_2}{\partial t} - c_0\dfrac{\partial v_2}{\partial x} = 0. \end{cases}$$

The canonical form (2.16) is, in this case, particularly simple. Each equation involves only one unknown and can be easily solved by the methods of Chapter III or Chapter V. The general solution of (2.16) is

$$(2.17)\qquad \begin{aligned} v_1(x, t) &= f(x - c_0 t) \\ v_2(x, t) &= g(x + c_0 t) \end{aligned}$$

where f and g are arbitrary C^1 functions of a single variable. Now, returning to the unknowns u_1 and u_2 by using $\mathbf{u} = P\mathbf{v}$, we obtain the general solution of (2.6)

$$(2.18)\qquad \begin{aligned} u_1(x, t) &= \sqrt{a}[f(x - c_0 t) + g(x + c_0 t)] \\ u_2(x, t) &= \sqrt{b}[f(x - c_0 t) - g(x + c_0 t)]. \end{aligned}$$

Example 2.2. Consider the system

$$(2.19)\qquad \begin{cases} \dfrac{\partial u_1}{\partial t} + \dfrac{\partial u_1}{\partial x} + x^2 \dfrac{\partial u_2}{\partial x} = 0 \\ \dfrac{\partial u_2}{\partial t} + t^2 \dfrac{\partial u_1}{\partial x} + \dfrac{\partial u_2}{\partial x} = 0. \end{cases}$$

The matrix form of this system is

$$\frac{\partial}{\partial t}\begin{bmatrix} u_1 \\ u_2 \end{bmatrix} + \begin{bmatrix} 1 & x^2 \\ t^2 & 1 \end{bmatrix} \frac{\partial}{\partial x}\begin{bmatrix} u_1 \\ u_2 \end{bmatrix} = \begin{bmatrix} 0 \\ 0 \end{bmatrix},$$

so that

$$A = \begin{bmatrix} 1 & x^2 \\ t^2 & 1 \end{bmatrix}.$$

The characteristic equation is

$$(\lambda - 1)^2 - x^2 t^2 = 0,$$

and the eigenvalues of A are

$$\lambda_1 = 1 + |xt|, \qquad \lambda_2 = 1 - |xt|.$$

The eigenvalues are distinct and the system is hyperbolic at every point of the (x, t)-plane except at points of the x- and t-axes. Let us confine our attention to the domain Ω consisting of the open first quadrant $x > 0$, $t > 0$. We know that we can transform (2.19) to canonical form in a neighborhood of every point of Ω. Actually in this case we can do so in the whole of Ω. We leave it to the reader to show that in Ω,

$$\Lambda = \begin{bmatrix} 1 + xt & 0 \\ 0 & 1 - xt \end{bmatrix}, \qquad P = \begin{bmatrix} x & x \\ t & -t \end{bmatrix}, \quad P^{-1} = \frac{1}{2xt}\begin{bmatrix} t & x \\ t & -x \end{bmatrix},$$

and that the new unknown $\mathbf{v} = P^{-1}\mathbf{u}$ must satisfy the canonical form of system (2.19) in Ω,

$$(2.20)\quad \frac{\partial}{\partial t}\begin{bmatrix} v_1 \\ v_2 \end{bmatrix} + \begin{bmatrix} 1 + xt & 0 \\ 0 & 1 - xt \end{bmatrix} \frac{\partial}{\partial x}\begin{bmatrix} v_1 \\ v_2 \end{bmatrix} + \frac{1}{2xt}\begin{bmatrix} t + x + xt^2 & t - x + xt^2 \\ t - x - xt^2 & t + x - xt^2 \end{bmatrix}\begin{bmatrix} v_1 \\ v_2 \end{bmatrix} = \begin{bmatrix} 0 \\ 0 \end{bmatrix}.$$

Problems

2.1. Write the Cauchy-Riemann equations (1.1) in matrix form and obtain the characteristic equation $\lambda^2 + 1 = 0$.

2.2. Verify the matrix product differentiation rule (2.12).

2.3. Supply all of the computational details in Example 2.2.

2.4. Show that if a solution (u_1, u_2) of system (2.6) is C^2, then each of the unknowns u_1, u_2 must satisfy the wave equation,

$$\frac{\partial^2 u}{\partial t^2} - ab\,\frac{\partial^2 u}{\partial x^2} = 0.$$

2.5. The study of long gravity waves on the surface of a fluid in a channel leads to the linear system of two equations in the two unknowns v and ζ,

$$(2.21)\quad \begin{cases} \dfrac{\partial v}{\partial t} + g\,\dfrac{\partial \zeta}{\partial x} = 0 \\[2ex] \dfrac{\partial \zeta}{\partial t} + \dfrac{S_0}{b}\dfrac{\partial v}{\partial x} + \dfrac{S_0'}{b}\,v = 0. \end{cases}$$

Here b and g are positive constants, $S_0 = S_0(x)$ is a given positive function representing the equilibrium cross-sectional area of the fluid in the channel and $S_0' = dS_0/dx$ (see Landau and Lifshitz,[6] §13, for details and derivation).

(a) Show that the system (2.21) is hyperbolic in the whole (x, t)-plane.

(b) Introduce the new unknowns v_1, v_2 related to v and ζ by

$$\begin{bmatrix} v \\ \zeta \end{bmatrix} = \begin{bmatrix} \sqrt{g} & \sqrt{g} \\ \sqrt{\dfrac{S_0}{b}} & -\sqrt{\dfrac{S_0}{b}} \end{bmatrix}\begin{bmatrix} v_1 \\ v_2 \end{bmatrix}$$

to obtain the canonical form

$$(2.22)\quad \frac{\partial}{\partial t}\begin{bmatrix} v_1 \\ v_2 \end{bmatrix} + \begin{bmatrix} \sqrt{\dfrac{gS_0}{b}} & 0 \\ 0 & -\sqrt{\dfrac{gS_0}{b}} \end{bmatrix}\frac{\partial}{\partial x}\begin{bmatrix} v_1 \\ v_2 \end{bmatrix} + \frac{S_0'}{4\sqrt{S_0}}\sqrt{\frac{g}{b}}\begin{bmatrix} 3v_1 + v_2 \\ -(v_1 + 3v_2) \end{bmatrix} = \begin{bmatrix} 0 \\ 0 \end{bmatrix}.$$

(c) Find the general solution of (2.21) if the channel cross-sectional area is constant; i.e., $S_0' = 0$. (Compare with (2.6).)

3. The Method of Characteristics for Linear Hyperbolic Systems. Application to Electrical Transmission Lines

Consider the (almost) linear system (2.1) of m equations in m unknowns and two independent variables which can be written in the matrix form

$$\frac{\partial \mathbf{u}}{\partial t} + A(x, t)\frac{\partial \mathbf{u}}{\partial x} + \mathbf{b}(x, t, \mathbf{u}) = \mathbf{0}. \tag{3.1}$$

The coefficient matrix $A(x, t)$ is assumed to be C^1 in the domain of interest. Just as in the case of a single equation, the definition of characteristic curves of system (3.1) involves the principal part $D_t + A(x, t)D_x$ of the differential operator appearing in (3.1). This definition is stated in Problem 3.1 where the derivation of the following more direct definition is outlined.

Definition 3.1. A *characteristic curve* of system (3.1) is a curve in the (x, t)-plane given by $x = x(t)$, where $x(t)$ is a solution of the differential equation

$$\frac{dx}{dt} = \lambda(x, t), \tag{3.2}$$

with $\lambda(x, t)$ being an eigenvalue of the coefficient matrix $A(x, t)$.

More briefly, a characteristic curve (or simply a characteristic) of (3.1) is a solution curve of (3.2) with $\lambda(x, t)$ being an eigenvalue of $A(x, t)$. Since the system (3.1) is said to be elliptic if $A(x, t)$ has no real eigenvalues, it follows that elliptic systems, just like elliptic equations, have no characteristic curves.

Let us suppose from now on that the system (3.1) is hyperbolic in a domain Ω of the (x, t)-plane. Then, in a neighborhood of any point $(\bar{x}, \bar{t})$ of Ω, the matrix $A(x, t)$ has m distinct eigenvalues $\lambda_1(x, t), \lambda_2(x, t), \dots, \lambda_m(x, t)$ which are C^1 in that neighborhood since $A(x, t)$ is assumed to be C^1 in Ω. It follows from Theorem 4.1 of Chapter I that there are exactly m distinct characteristic curves of (3.1) passing through $(\bar{x}, \bar{t})$, each curve corresponding to an eigenvalue of $A(x, t)$. The characteristic curve corresponding to the eigenvalue $\lambda_i(x, t)$, is the solution curve of the initial value problem,

$$\frac{dx}{dt} = \lambda_i(x, t), \qquad x(\bar{t}) = \bar{x}. \tag{3.3}$$

Since the eigenvalues are distinct, the characteristic curves are never tangent. Moreover, at a point where an eigenvalue is zero, the corresponding characteristic is parallel to the t-axis. However, a characteristic curve of (3.1) can never be parallel to the x-axis. (This of course is due to the fact that (3.1) is special with respect to the variable t.)

A curve C in the (x, t)-plane is said to be characteristic at a point (x^0, t^0) with respect to the system (3.1) if there is a characteristic curve of (3.1) which is tangent to C at (x^0, t^0). Note that the x-axis is nowhere characteristic with respect to (3.1).

Example 3.1. Consider the system (2.6) of Example 2.1. The eigenvalues of A are $\lambda_1 = c_0$ and $\lambda_2 = -c_0$. The characteristics corresponding to λ_1 are the solution curves of

$$\frac{dx}{dt} = c_0,$$

which are the lines

$$x = c_0 t + c_1.$$

The characteristics corresponding to λ_2 are the solution curves of

$$\frac{dx}{dt} = -c_0$$

which are the lines

$$x = -c_0 t + c_2.$$

In this example the characteristics are straight lines because the coefficient matrix A is constant. To each eigenvalue corresponds a one-parameter family of characteristic lines, and through each point of the (x, t)-plane pass two characteristic lines, one from each of these families.

In the previous section we saw that the introduction of new unknowns makes it possible to transform the hyperbolic system (3.1) to its canonical form

$$\frac{\partial \mathbf{v}}{\partial t} + \Lambda(x, t)\frac{\partial \mathbf{v}}{\partial x} + \mathbf{c}(x, t, \mathbf{v}) = \mathbf{0}, \tag{3.4}$$

in a neighborhood of any point of Ω. $\Lambda(x, t)$ is the diagonal matrix of eigenvalues of $A(x, t)$, given by (2.8). Clearly, the characteristic curves of a hyperbolic system remain invariant under transformation of the system to its canonical form: the characteristic curves of the canonical form (3.4) are the same as those of the original system (3.1). The component form of (3.4),

$$\frac{\partial v_i}{\partial t} + \lambda_i(x, t)\frac{\partial v_i}{\partial x} + c_i(x, t, \mathbf{v}) = 0, \qquad i = 1, 2, \ldots, m, \tag{3.5}$$

shows that the principal part of the ith equation of a hyperbolic system in canonical form involves only the ith unknown v_i.

We now make the following observation which is of fundamental importance in the study of hyperbolic systems: Along a characteristic curve corresponding to the eigenvalue λ_i, the principal part of the ith equation in (3.5) is actually an ordinary derivative of the ith unknown v_i. Indeed, let C_i be a characteristic curve corresponding to λ_i. By definition, C_i is given by $x = x_i(t)$ where $x_i(t)$ is a solution of (3.2) with $\lambda = \lambda_i$, i.e.,

$$\frac{dx_i(t)}{dt} = \lambda_i(x_i(t), t). \tag{3.6}$$

Let $V_i(t)$ be the function $v_i(x, t)$ restricted to C_i,

(3.7) $$V_i(t) = v_i(x, t)\,|_{(x,t)\in C_i} = v_i(x_i(t), t).$$

Then

$$\frac{dV_i(t)}{dt} = \frac{\partial v_i}{\partial t}(x_i(t), t) + \frac{\partial v_i}{\partial x}(x_i(t), t)\frac{dx_i(t)}{dt},$$

and using (3.6),

(3.8) $$\frac{dV_i(t)}{dt} = \frac{\partial v_i}{\partial t}(x_i(t), t) + \lambda_i(x_i(t), t)\frac{\partial v_i}{\partial x}(x_i(t), t).$$

The right hand side of (3.8) is the principal part of the ith equation in (3.5) restricted to the characteristic curve C_i. It follows that the ith equation in (3.5), when restricted to a characteristic curve corresponding to λ_i, is the ordinary differential equation,

(3.9) $$\frac{dV_i}{dt} + c_i(x_i(t), t, \mathbf{v}(x_i(t), t)) = 0.$$

Example 3.1 *(continued).* The canonical form of system (2.6),

(2.16) $$\begin{aligned} \frac{\partial v_1}{\partial t} + c_0\frac{\partial v_1}{\partial x} &= 0 \\ \frac{\partial v_2}{\partial t} - c_0\frac{\partial v_2}{\partial x} &= 0 \end{aligned}$$

was derived in the previous section. Along the characteristics $x = c_0t + c_1$ corresponding to $\lambda_1 = c_0$, the first equation in (2.16) is

$$\frac{dV_1}{dt} = 0 \quad \text{where} \quad V_1(t) = v_1(c_0t + c_1, t).$$

Along the characteristics $x = -c_0t + c_2$ corresponding to $\lambda_2 = -c_0$, the second equation in (2.16) is

$$\frac{dV_2}{dt} = 0, \quad \text{where} \quad V_2(t) = v_2(-c_0t + c_2, t).$$

The above observation, that the equations of a hyperbolic system in canonical form are ordinary differential equations along the corresponding characteristic curves, leads to two important results concerning the initial value problem for the system: (i) the characteristic curves are exceptional for the initial value problem, and (ii) the characteristic curves can be used to solve the problem. The initial value problem for a system of first order equations in two independent variables is analogous to the initial value problem for a single first order equation. It asks for a solution of the system which has given values on a given curve in R^2.

Initial Value Problem

Let C be a given curve in the (x, t)-plane described parametrically by the equations,

$$x = x_0(s), \qquad t = t_0(s); \qquad s \in I$$

where $x_0(s)$ and $t_0(s)$ are in $C^1(I)$. Let

$$\boldsymbol{\phi}(s) = \begin{bmatrix} \phi_1(s) \\ \phi_2(s) \\ \cdot \\ \cdot \\ \cdot \\ \phi_m(s) \end{bmatrix}$$

be a given vector function in $C^1(I)$. $\boldsymbol{\phi}(s)$ may be thought of as defining a vector function on the curve C. The initial value problem for the system (3.4) asks for a vector function $\mathbf{v} = \mathbf{v}(x, t)$ which is defined in a domain Ω of R^2 containing C, such that

(i) $\mathbf{v} = \mathbf{v}(x, t)$ is a solution of (3.4) in Ω,

(ii) On the curve C, $\mathbf{v}$ is equal to the given function $\boldsymbol{\phi}$, i.e.

(3.10) $$\mathbf{v}(x_0(s), t_0(s)) = \boldsymbol{\phi}(s), \; s \in I.$$

The curve C is called the *initial curve* of the problem while the function $\boldsymbol{\phi}$ is called the *initial data.* Equation (3.10) is called the *initial condition.*

We will now show that if the initial curve C is a characteristic curve, say C_i, of the system (3.4), then the initial value problem generally has no solution. This is because the ith equation of the system, which is an ordinary differential equation along C_i, imposes a condition on the initial data which are assigned on C_i. This condition is generally not satisfied, because the initial data are assigned to the initial curve without any reference to the system of partial differential equations. Indeed, since C_i is given by

$$x = x_i(t), \qquad t = t; \qquad t \in I$$

with t being the parameter s, the initial condition (3.10) is in this case,

(3.11) $$\mathbf{v}(x_i(t), t) = \boldsymbol{\phi}(t), \qquad t \in I.$$

The ith component of (3.11) is

(3.12) $$V_i(t) = \phi_i(t), \qquad t \in I.$$

Substituting (3.11) and (3.12) into equation (3.9), which is the ith equation of the system (3.4) along C_i, we obtain

(3.13) $$\frac{d\phi_i(t)}{dt} + c_i(x_i(t), t, \boldsymbol{\phi}(t)) = 0.$$

Equation (3.13) is a condition on the initial data $\boldsymbol{\phi}(t)$. In general, the assigned initial data will not satisfy this condition and consequently the initial value problem will not have a solution (no existence of solution). Incidentally, it can be shown that if, by accident or foresight, the assigned initial data do indeed satisfy condition (3.13), then the initial value problem has infinitely many solutions (no uniqueness of solution). An illustration of these possibilities is described in Problem 3.2.

Let us consider now the initial value problem for the system (3.4) with the initial curve being nowhere characteristic. For simplicity let us suppose that the initial curve is an interval (a, b) of the x-axis. The initial condition is

$$\mathbf{v}(x, 0) = \boldsymbol{\phi}(x), \qquad a < x < b. \tag{3.14}$$

It can be shown that if the coefficients $\lambda_i(x, t)$ and the initial data $\boldsymbol{\phi}(x)$ are C^1 while the functions $c_i(x, t, \mathbf{v})$ are C^2, then, in some domain of the (x, t)-plane containing the interval (a, b), the initial value problem (3.4), (3.14) has a unique solution which depends continuously on the initial data. In other words the problem is well-posed. The proof of this result, which can be found in the book of Petrovskii,[5] §10, uses the method of characteristics, described below, to transform the initial value problem (3.4), (3.14) to a system of integral equations, which in turn can be solved by the method of successive approximations.

The *method of characteristics* proceeds as follows. Let $(\bar{x}, \bar{t})$ be a fixed point with $a < \bar{x} < b$ and $\bar{t} > 0$. (The procedure when $\bar{t} < 0$ is the mirror image in the x-axis of what follows.) Through $(\bar{x}, \bar{t})$ pass m distinct characteristics $C_1, C_2, \ldots, C_m$, which are never parallel to the x-axis. If $\bar{t}$ is sufficiently small, all of these characteristics will intersect the x-axis at points in the interval (a, b) (see Problem 3.3). Let $(x_i^0, 0)$ be the point of intersection of the ith characteristic C_i with the x-axis (see Fig. 3.1), and let ℓ_i be the portion of C_i between the points $(x_i^0, 0)$ and $(\bar{x}, \bar{t})$. Remember that C_i is given by the solution $x = x_i(t)$ of (3.3) and ℓ_i is the portion of C_i for which $0 \le t \le \bar{t}$, so that

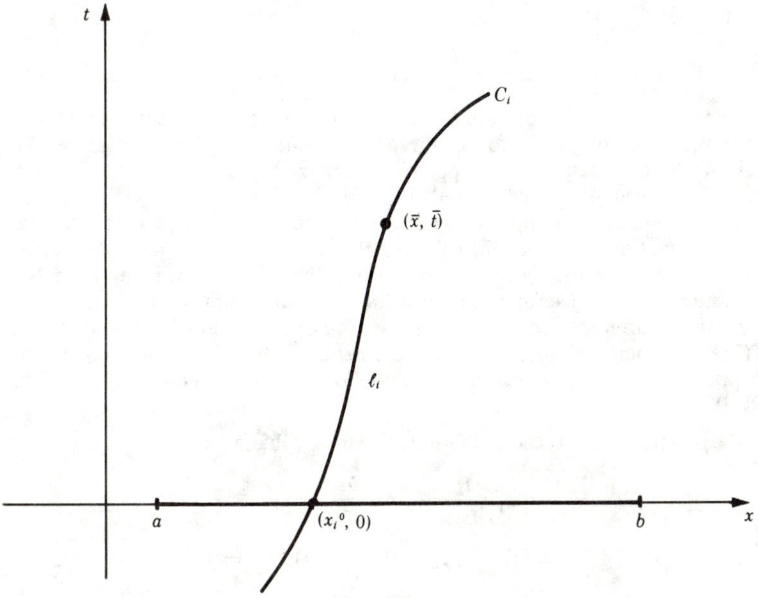

Fig. 3.1

(3.15) $$x_i(0) = x_i^0 \quad \text{and} \quad x_i(\bar{t}) = \bar{x}.$$

Along C_i the ith equation of system (3.4) is the ordinary differential equation (3.9). Integrating (3.9) along ℓ_i, which means integrating with respect to t from $t = 0$ to $t = \bar{t}$, we obtain

(3.16) $$V_i(\bar{t}) - V_i(0) + \int_0^{\bar{t}} c_i(x_i(t), t, \mathbf{v}(x_i(t), t))dt = 0.$$

From (3.7) and (3.15),

(3.17) $$V_i(\bar{t}) = v_i(x_i(\bar{t}), \bar{t}) = v_i(\bar{x}, \bar{t}),$$

and from the initial condition (3.14)

(3.18) $$V_i(0) = v_i(x_i(0), 0) = v_i(x_i^0, 0) = \phi_i(x_i^0).$$

Substituting (3.17) and (3.18) into (3.16) and remembering that the result is valid for every $i = 1, 2, \ldots, m$, we obtain the system of integral equations

(3.19) $$v_i(\bar{x}, \bar{t}) = \phi_i(x_i^0) - \int_0^{\bar{t}} c_i(x_i(t), t, \mathbf{v}(x_i(t), t))dt, \qquad i = 1, 2, \ldots, m.$$

Note that in (3.19) the points x_i^0 and the functions $x_i(t)$ depend on $(\bar{x}, \bar{t})$.

Briefly, the *method of successive approximations* for solving the system of integral equations (3.19) proceeds in the following manner. The first approximation $\mathbf{v}^{(0)}(x, t)$ is assumed to have components

(3.21) $$v_i^{(0)}(x, t) = \phi_i(x_i^0) \qquad i = 1, 2, \ldots, m.$$

To obtain the next approximation $\mathbf{v}^{(1)}(x, t)$, (3.21) is substituted into the right side of(3.19),

(3.22) $$v_i^{(1)}(\bar{x}, \bar{t}) = \phi_i(x_i^0) - \int_0^{\bar{t}} c_i(x_i(t), t, \mathbf{v}^{(0)}(x_i(t), t))dt, \; i = 1, 2, \ldots, m,$$

and the indicated integration is carried out. $\mathbf{v}^{(1)}(x, t)$ is then substituted into the right side of (3.19) to get the approximation $\mathbf{v}^{(2)}(x, t)$, etc. It can be shown that the sequence of approximations $\mathbf{v}^{(0)}(x, t)$, $\mathbf{v}^{(1)}(x, t)$, … , $\mathbf{v}^{(n)}(x, t)$, … , converges to the desired solution.

The computations involved in carrying out the method of successive approximations are too difficult and the method, except for some especially simple cases, has little practical value in actually computing the solution. A more useful method for finding approximations to the solution is a finite difference scheme which uses the ordinary differential equations (3.9) to compute approximate increments of the solution along linear approximations of the characteristics. For details see Petrovskii,[5] end of §10.

Example 3.2. Consider the initial value problem

(3.23) $$\frac{\partial v_1}{\partial t} + c_0 \frac{\partial v_1}{\partial x} = 0, \qquad \frac{\partial v_2}{\partial t} - c_0 \frac{\partial v_2}{\partial x} = 0,$$

(3.24) $$v_1(x, 0) = \phi_1(x), \qquad v_2(x, 0) = \phi_2(x); \qquad -\infty < x < \infty,$$

where ϕ_1 and ϕ_2 are given functions in $C^1(R^1)$. The corresponding system of integral equations (3.19) in this case is particularly simple:

$$v_1(\bar{x}, \bar{t}) = \phi_1(x_1^0), \qquad v_2(\bar{x}, \bar{t}) = \phi_2(x_2^0).$$

The characteristic curve C_1 corresponding to $\lambda_1 = c_0$ and passing through $(\bar{x}, \bar{t})$ is given by $x = \bar{x} + c_0(t - \bar{t})$, and its intersection with the x-axis is $x_1{}^0 = \bar{x} - c_0\bar{t}$ (see Fig. 3.2). The characteristic curve C_2 corresponding to $\lambda_2 = -c_0$ and passing through $(\bar{x}, \bar{t})$ is given by $x = \bar{x} - c_0(t - \bar{t})$, and its intersection with the x-axis is $x_2{}^0 = \bar{x} + c_0\bar{t}$. Therefore

$$v_1(\bar{x}, \bar{t}) = \phi_1(\bar{x} - c_0\bar{t}), \qquad v_2(\bar{x}, \bar{t}) = \phi_2(\bar{x} + c_0\bar{t}),$$

and after dropping the bars we obtain the desired solution of the initial value problem (3.23), (3.24),

$$v_1(x, t) = \phi_1(x - c_0t), \qquad v_2(x, t) = \phi_2(x + c_0t). \tag{3.25}$$

Example 3.3. In Section 1 of this chapter we described how an electrical transmission line is governed by the system (1.2). Let us consider now an infinite transmission line with positive constant electrical parameters C, G, R and L and suppose that initially the distribution of the current I and potential E in the line is known. The problem is to determine I and E at later time. We must solve the initial value problem for the system

$$\begin{cases} \dfrac{\partial I}{\partial t} + \dfrac{1}{L}\dfrac{\partial E}{\partial x} + \dfrac{R}{L} I = 0, \\ \dfrac{\partial E}{\partial t} + \dfrac{1}{C}\dfrac{\partial I}{\partial x} + \dfrac{G}{C} E = 0, \end{cases} \tag{3.26}$$

with initial conditions,

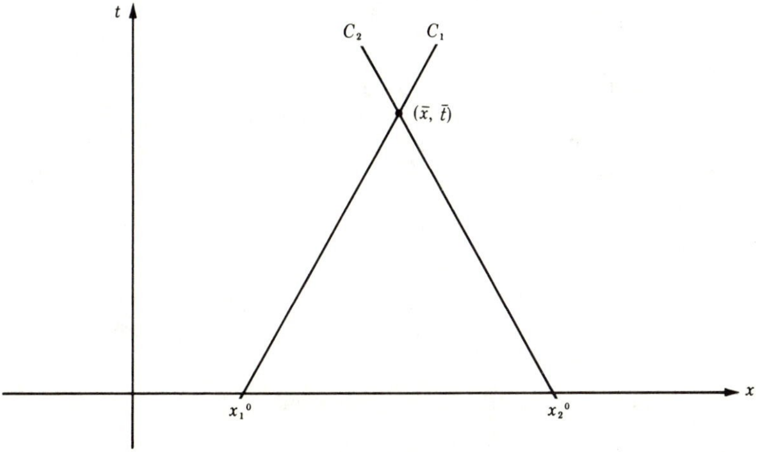

Fig. 3.2

(3.27) $\quad I(x, 0) = f(x), \qquad E(x, 0) = g(x); \qquad -\infty < x < \infty.$

The system (3.26) is hyperbolic in the whole (x, t)-plane with constant eigenvalues $\lambda_1 = 1/\sqrt{LC}$ and $\lambda_2 = -1/\sqrt{LC}$. Our first step is to transform (3.26) to its canonical form. In terms of the new variables v_1 and v_2 related to I and E by

$$(3.28)\quad \begin{bmatrix} I \\ E \end{bmatrix} = \begin{bmatrix} \sqrt{C} & \sqrt{C} \\ \sqrt{L} & -\sqrt{L} \end{bmatrix} \begin{bmatrix} v_1 \\ v_2 \end{bmatrix}, \qquad \begin{bmatrix} v_1 \\ v_2 \end{bmatrix} = \frac{1}{2\sqrt{LC}} \begin{bmatrix} \sqrt{L} & \sqrt{C} \\ \sqrt{L} & -\sqrt{C} \end{bmatrix} \begin{bmatrix} I \\ E \end{bmatrix},$$

the canonical form of (3.26) is

$$(3.29)\quad \frac{\partial}{\partial t}\begin{bmatrix} v_1 \\ v_2 \end{bmatrix} + \begin{bmatrix} \frac{1}{\sqrt{LC}} & 0 \\ 0 & -\frac{1}{\sqrt{LC}} \end{bmatrix} \frac{\partial}{\partial x}\begin{bmatrix} v_1 \\ v_2 \end{bmatrix} + \frac{1}{2LC}\begin{bmatrix} RC + LG & RC - LG \\ RC - LG & RC + LG \end{bmatrix}\begin{bmatrix} v_1 \\ v_2 \end{bmatrix} = \begin{bmatrix} 0 \\ 0 \end{bmatrix}$$

(Problem 3.7). From (3.27) and (3.28), the new unknowns must satisfy the initial condition

$$(3.30)\quad \begin{aligned} v_1(x, 0) &= \phi_1(x) \equiv \frac{1}{2\sqrt{LC}}[\sqrt{L}\, f(x) + \sqrt{C}\, g(x)] \\ v_2(x, 0) &= \phi_2(x) \equiv \frac{1}{2\sqrt{LC}}[\sqrt{L}\, f(x) - \sqrt{C}\, g(x)] \end{aligned}, \qquad -\infty < x < \infty.$$

Following the method of characteristics for solving the initial value problem (3.29), (3.30), let $(\bar{x}, \bar{t})$ be an arbitrary but fixed point in the upper-half (x, t)-plane. The characteristic curves C_1 and C_2 corresponding to $\lambda_1 = 1/\sqrt{LC}$ and $\lambda_2 = -1/\sqrt{LC}$ and passing through $(\bar{x}, \bar{t})$ are the lines,

$$(3.31)\quad \begin{cases} C_1: x = x_1(t) \equiv \bar{x} + \frac{1}{\sqrt{LC}}(t - \bar{t}), \\ C_2: x = x_2(t) \equiv \bar{x} - \frac{1}{\sqrt{LC}}(t - \bar{t}), \end{cases}$$

and these lines intersect the x-axis at the points

$$(3.32)\quad x_1^0 = \bar{x} - \frac{1}{\sqrt{LC}}\bar{t}, \qquad x_2^0 = \bar{x} + \frac{1}{\sqrt{LC}}\bar{t}.$$

(Again see Fig. 3.2.) The system of integral equations (3.19) in this case is

$$(3.33)\quad \begin{cases} v_1(\bar{x}, \bar{t}) = \phi_1\left(\bar{x} - \dfrac{1}{\sqrt{LC}}\bar{t}\right) - \dfrac{1}{2LC}\displaystyle\int_0^{\bar{t}} [(RC + LG)v_1(x_1(t), t) \\ \qquad\qquad + (RC - LG)v_2(x_1(t), t)]dt, \\ v_2(\bar{x}, \bar{t}) = \phi_2\left(\bar{x} + \dfrac{1}{\sqrt{LC}}\bar{t}\right) - \dfrac{1}{2LC}\displaystyle\int_0^{\bar{t}} [(RC - LG)v_1(x_2(t), t) \\ \qquad\qquad + (RC + LG)v_2(x_2(t), t)]dt, \end{cases}$$

where $x_1(t)$ and $x_2(t)$ are given by (3.31).

In general, in order to proceed any further with the computation of the solution, it is necessary to know the specific functional form of the initial data and employ a numerical approximation scheme. However, in the special case in which the electrical parameters of the transmission line satisfy the relation

$$(3.34)\qquad RC - LG = 0,$$

is is possible to obtain a general formula for the solution. This is due to the fact that when (3.34) holds, the canonical form (3.29) is separated in the sense that each equation involves only one of the unknowns v_1, v_2. These unknowns restricted on the characteristics (3.31) are

$$(3.35)\quad V_1(t) = v_1(\bar{x} + \frac{1}{\sqrt{LC}}(t - \bar{t}), t), \quad V_2(t) = v_2(\bar{x} - \frac{1}{\sqrt{LC}}(t - \bar{t}), t).$$

They satisfy the ordinary differential equations (3.9), which in this case are

$$(3.36)\qquad \frac{dV_1}{dt} + \frac{R}{L}V_1 = 0, \qquad \frac{dV_2}{dt} + \frac{R}{L}V_2 = 0,$$

and the initial conditions (3.18),

$$(3.37)\qquad V_1(0) = \phi_1(x_1^0), \qquad V_2(0) = \phi_2(x_2^0).$$

Solving the two initial value problems (3.36), (3.37), we obtain,

$$(3.38)\qquad V_1(t) = \phi_1(x_1^0)e^{-\frac{R}{L}t}, \qquad V_2(t) = \phi_2(x_2^0)e^{-\frac{R}{L}t}.$$

From (3.35),

$$(3.39)\qquad v_1(\bar{x}, \bar{t}) = V_1(\bar{t}), \qquad v_2(\bar{x}, \bar{t}) = V_2(\bar{t}).$$

Substituting (3.38) and (3.32) into (3.39) and finally dropping the bars we obtain the solution of the initial value problem (3.29), (3.30) when (3.34) holds,

$$(3.40)\qquad \begin{cases} v_1(x, t) = \phi_1\left(x - \dfrac{1}{\sqrt{LC}}t\right)e^{-\frac{R}{L}t} \\ v_2(x, t) = \phi_2\left(x + \dfrac{1}{\sqrt{LC}}t\right)e^{-\frac{R}{L}t}. \end{cases}$$

The solution of the original problem (3.26), (3.27) is obtained from (3.40) using (3.28) and (3.30),

$$
(3.41)\quad
\begin{cases}
I(x,t) = \left\{\dfrac{1}{2}\left[f\left(x - \dfrac{1}{\sqrt{LC}}t\right) + f\left(x + \dfrac{1}{\sqrt{LC}}t\right)\right] \right. \\
\qquad\qquad \left. + \dfrac{1}{2}\sqrt{\dfrac{C}{L}}\left[g\left(x - \dfrac{1}{\sqrt{LC}}t\right) - g\left(x + \dfrac{1}{\sqrt{LC}}t\right)\right]\right\} e^{-\frac{R}{L}t} \\
E(x,t) = \left\{\dfrac{1}{2}\sqrt{\dfrac{L}{C}}\left[f\left(x - \dfrac{1}{\sqrt{LC}}t\right) - f\left(x + \dfrac{1}{\sqrt{LC}}t\right)\right] \right. \\
\qquad\qquad \left. + \dfrac{1}{2}\left[g\left(x - \dfrac{1}{\sqrt{LC}}t\right) + g\left(x + \dfrac{1}{\sqrt{LC}}t\right)\right]\right\} e^{-\frac{R}{L}t}.
\end{cases}
$$

Note that the solution (3.41) shows that in the case $RC = LG$ the initial data propagate along the line without distortion other than exponential decay with increasing time. For this reason a line for which $RC = LG$ is called a *distortionless* line.

We close this section by pointing out certain aspects of the initial value problem for hyperbolic systems which are very similar to those for the wave equation. This is due to the fact that the wave equation is also hyperbolic. First, note that the solutions of the problems in Examples 3.2 and 3.3 are valid in the whole (x, t)-plane. In general, the solution of the initial value problem exists and can be determined in the whole (x, t)-plane if the system is linear, has constant coefficients and the initial condition is given on the whole x-axis. If the initial condition is given on a finite interval (a, b) of the x-axis, it is clear from the method of characteristics that the solution can be (uniquely) determined only in a "quadrilateral" domain bounded by characteristics through the endpoints a and b (see Problems 3.9 and 3.10). Next, Examples 3.2 and 3.3 also illustrate the following important fact: The value of the solution at a point (x, t) depends only on the values of the initial data over the part of the initial line cut off by the characteristics through (x, t). This part of the initial line is known as the *domain of dependence* of the solution. This concept and the dual concept of *range of influence* of the initial data were discussed in detail in connection with the wave equation (see also Problem 3.15).

Problems

3.1. (a) A direction defined by the non-zero vector (ξ_x, ξ_t) is said to be characteristic for the system (3.1) at the point (x, t) if

$$\det |\xi_t I + \xi_x A(x, t)| = 0.$$

Show that as a consequence of this definition, $\xi_x \neq 0$ and $-\xi_t/\xi_x$ is an eigenvalue of $A(x, t)$.

(b) A characteristic curve of system (3.1) is a curve such that at each of its points the normal vector to the curve defines a characteristic direction for (3.1) at that point. Use this definition to derive Definition 3.1.

3.2. (a) Consider the initial value problem for the system (2.16) with the initial condition

$$v_1(c_0 t, t) = t, \qquad v_2(c_0 t, t) = 0.$$

Show that this problem has no solution.

(b) Consider the initial value problem for the system (2.16) with the initial condition

$$v_1(c_0 t, t) = 1, \qquad v_2(c_0 t, t) = t^2.$$

Show that this problem has infinitely many solutions.

3.3. Consider the hyperbolic system

$$\frac{\partial v_1}{\partial t} + \frac{\partial v_1}{\partial x} + v_2 = 0$$

$$\frac{\partial v_2}{\partial t} - x^2 \frac{\partial v_2}{\partial x} + v_1 = 0.$$

This system is in canonical form with $\lambda_1 = 1$ and $\lambda_2 = -x^2$.

(a) Find the equation of the characteristic C_1 corresponding to λ_1 and passing through the point $(\bar{x}, \bar{t})$. Also find the point of intersection $(x_1{}^0, 0)$ of C_1 with the x-axis.

(b) Find the equation of the characteristic C_2 corresponding to λ_2 and passing through the point (i) $(\bar{x}, \bar{t}) = (1, 1)$, (ii) $(\bar{x}, \bar{t}) = (1, 1/2)$. Show that in case (i) C_2 does not intersect the x-axis, while in case (ii) C_2 intersects the x-axis. Find the point of intersection $(x_2{}^0, 0)$.

3.4. For the system of Problem 3.3, find the characteristic C_2 corresponding to λ_2 and passing through the point $(\bar{x}, \bar{t}) = (0, 1)$.

3.5. (a) Verify by direct substitution that (3.25) satisfies (3.23) and (3.24).

(b) If ϕ_1 has a jump discontinuity at $x = 0$, show that v_1, given by (3.25), has a jump discontinuity across the characteristic line $x = c_0 t$. Also describe the effects on the solution (3.25) of a jump discontinuity of ϕ_2 and of jump discontinuities in the derivatives $\phi_1{}'$ and $\phi_2{}'$. Conclude that discontinuities in the initial data or their derivatives propagate along corresponding characteristic curves. The solution (3.25) in this case is called a generalized solution.

3.6. Use the general solution (2.17) to obtain the solution (3.25) of the initial value problem (3.23), (3.24).

3.7. Derive the canonical form (3.29).

3.8. Verify by direct substitution that (3.41) satisfies (3.26) and (3.27).

3.9. Consider the initial value problem (3.23), (3.24) with the initial condition (3.24) given only on the interval $(0, L)$ of the x-axis rather than on the whole x-axis. Find the quadrilateral domain in which the solution can be determined.

3.10. Consider the initial value problem for the system of Problem 3.3 with the initial condition given on the interval $(0, 1)$ of the x-axis. Use the method of characteristics to show that in the upper half-plane the solution can be determined in the "triangular" domain bounded by the line $t = 0$, the line $x = t$ and the hyperbola $x = 1/(1 + t)$.

3.11. Consider the system (3.26) of Example 3.3 and assume that the unknowns $I(x, t)$ and $E(x, t)$ are of class C^2.

(a) Show that I and E must satisfy the telegraph equation (see Example 2.5, Chapter V),

$$I_{xx} - LCI_{tt} - (RC + LG)I_t - RGI = 0, \tag{3.42}$$

$$E_{xx} - LCE_{tt} - (RC + LG)E_t - RGE = 0. \tag{3.43}$$

(b) In the case of a distortionless line ($RC = LG$), show that $e^{\frac{R}{L}t}E$ and $e^{\frac{R}{L}t}I$ must satisfy the wave equation

$$u_{xx} - LCu_{tt} = 0.$$

Conclude that E and I must have the form

$$e^{-\frac{R}{L}t}\left[F_1\left(x - \frac{1}{\sqrt{LC}}t\right) + F_2\left(x + \frac{1}{\sqrt{LC}}t\right)\right] \tag{3.44}$$

where F_1 and F_2 are C^2 functions of one variable.

3.12. Consider the initial value problem (3.26), (3.27) of Example 3.3 with C^2 initial data and assume that the solution is C^2.

(a) Show that I must satisfy the initial value problem for the telegraph equation (3.42) with initial conditions

$$I(x, 0) = f(x), \qquad I_t(x, 0) = -\frac{R}{L}f(x) - \frac{1}{L}g'(x); \; -\infty < x < \infty,$$

while E must satisfy (3.43) with

$$E(x, 0) = g(x), \qquad E_t(x, 0) = -\frac{G}{C}g(x) - \frac{1}{C}f'(x); \; -\infty < x < \infty.$$

(b) In the case of a distortionless line, solve the initial value problems of part (a) to obtain the solution (3.41). [*Hint:* Use the form (3.44) of E and I.]

3.13. Consider a distortionless ($RC = LG$) semi-infinite transmission line occupying the positive x-axis and suppose that at the end $x = 0$ the current and potential are known functions of t for all t. In order to determine the current and potential in the line we must solve the initial value problem for the system (3.26) with initial condition given on the t-axis,

$$I(0, t) = f(t), \qquad E(0, t) = g(t); \qquad -\infty < t < \infty.$$

As in Problems 3.11, 3.12, reduce this problem to a pair of initial value problems for the telegraph equation and find the solution,

$$I(x, t) = \left\{\frac{1}{2}\left[e^{\frac{R}{L}(t-\sqrt{LC}x)} f(t - \sqrt{LC}\,x) + e^{\frac{R}{L}(t+\sqrt{LC}x)} f(t + \sqrt{LC}\,x)\right]\right.$$

$$\left. + \frac{1}{2}\sqrt{\frac{C}{L}}\left[e^{\frac{R}{L}(t-\sqrt{LC}x)} g(t - \sqrt{LC}\,x) - e^{\frac{R}{L}(t+\sqrt{LC}x)} g(t + \sqrt{LC}\,x)\right]\right\} e^{-\frac{R}{L}t}$$

$$E(x, t) = \left\{\frac{1}{2}\sqrt{\frac{L}{C}}\left[e^{\frac{R}{L}(t-\sqrt{LC}x)} f(t - \sqrt{LC}x) - e^{\frac{R}{L}(t+\sqrt{LC}x)} f(t + \sqrt{LC}\,x)\right]\right.$$

$$\left. + \frac{1}{2}\left[e^{\frac{R}{L}(t-\sqrt{LC}x)} g(t - \sqrt{LC}\,x) + e^{\frac{R}{L}(t+\sqrt{LC}x)} g(t + \sqrt{LC}\,x)\right]\right\} e^{-\frac{R}{L}t}.$$

[*Hint:* $I_x(0, t) = -[Gg(t) + Cg'(t)] = -Ce^{-\frac{R}{L}t}[g(t)e^{\frac{R}{L}t}]'$.]

3.14. Show that a C^1 solution of the transmission line system (3.26) with positive electrical parameters depending on x must satisfy the differential inequality,

$$(3.45) \qquad (IE)_x + \frac{1}{2}(LI^2 + CE^2)_t \leqq 0.$$

3.15. For a transmission line with L and C constant, use the energy method described in Section 3 of Chapter VIII to obtain the domain of dependence inequality,

$$\int_{\bar{x}-\frac{\bar{t}-T}{\sqrt{LC}}}^{\bar{x}+\frac{\bar{t}-T}{\sqrt{LC}}} (LI^2 + CE^2)\Big|_{t=T}\, dx \leqq \int_{\bar{x}-\frac{\bar{t}}{\sqrt{LC}}}^{\bar{x}+\frac{\bar{t}}{\sqrt{LC}}} (LI^2 + CE^2)\Big|_{t=0}\, dx$$

which is true for every $(\bar{x}, \bar{t})$ in the upper half-plane and for every T with $0 \leq T \leq \bar{t}$. [*Hint:* Start with (3.45) instead of (3.8) of Chapter VIII.] Prove uniqueness for the initial value problem and discuss domain of dependence and range of influence.

3.16. Consider a finite transmission line extending over $0 \leq x \leq a$, and having positive electrical parameters which are continuous functions of x.

(a) If the product IE vanishes at the end points $x = 0$ and $x = a$ for all $t \geq 0$, use (3.45) to obtain the energy inequality,

$$\int_0^a (LI^2 + CE^2)\Big|_{t=T}\, dx \leqq \int_0^a (LI^2 + CE^2)\Big|_{t=0}\, dx$$

which is valid for all $T \geqq 0$.

(b) Prove uniqueness of solution for the initial-boundary value problem, with boundary conditions

$$I(0, t) = \phi_1(t), \qquad E(0, t) = \phi_2(t);$$

$$I(a, t) = \psi_1(t), \qquad E(a, t) = \psi_2(t); \qquad t \geqq 0,$$

and initial condition

$$I(x, 0) = f(x), \qquad E(x, 0) = g(x); \qquad 0 \leq x \leq a.$$

4. Quasi-Linear Hyperbolic Systems

Let us consider the quasi-linear system of m equations in m unknowns and two independent variables which can be written in the matrix form,

$$\frac{\partial \mathbf{u}}{\partial t} + A(x, t, \mathbf{u})\frac{\partial \mathbf{u}}{\partial x} + \mathbf{b}(x, t, \mathbf{u}) = \mathbf{0}. \tag{4.1}$$

The system (4.1) is said to be *hyperbolic* in a domain Ω of the (x, t)-plane if for every (x, t) in Ω and for all possible solutions $\mathbf{u}$ of the system in Ω, the matrix $A(x, t, \mathbf{u})$ has m real and distinct eigenvalues $\lambda_1(x, t, \mathbf{u})$, $\lambda_2(x, t, \mathbf{u})$, $\ldots$, $\lambda_m(x, t, \mathbf{u})$. The characteristic curves of a quasi-linear system are defined in a manner analogous to that for a linear system, except that now a curve is said to be characteristic not only with respect to a system but also with respect to a particular solution of the system.

Definition 4.1. A curve C in the (x, t)-plane is said to be characteristic with respect to the system (4.1) and with respect to a solution $\mathbf{u}(x, t)$ of the system if C is given by $x = x(t)$, where $x(t)$ is a solution of the differential equation

$$\frac{dx}{dt} = \lambda(x, t, \mathbf{u}(x, t)) \tag{4.2}$$

with $\lambda(x, t, \mathbf{u})$ being an eigenvalue of the matrix $A(x, t, \mathbf{u})$.

The difference between linear and quasi-linear systems is fundamental. For a linear system, the characteristic curves can be determined solely from the coefficients of the system without any reference to a particular solution of the system. For a quasi-linear system, a solution of the system must be known in advance in order to determine the corresponding characteristic curves.

Example 4.1. Consider the quasi-linear system (1.4) of gas dynamics, the matrix form of which is

$$\frac{\partial}{\partial t}\begin{bmatrix} u \\ \rho \end{bmatrix} + \begin{bmatrix} u & c^2(\rho)/\rho \\ \rho & u \end{bmatrix}\frac{\partial}{\partial x}\begin{bmatrix} u \\ \rho \end{bmatrix} = \begin{bmatrix} 0 \\ 0 \end{bmatrix}. \tag{4.3}$$

The matrix

$$A(x, t, u, \rho) = \begin{bmatrix} u & c^2(\rho)/\rho \\ \rho & u \end{bmatrix}$$

has two real and distinct eigenvalues,

$$\lambda_+ = u + c(\rho), \qquad \lambda_- = u - c(\rho),$$

so that (4.3) is a hyperbolic system. The characteristic curves C_+ corresponding to λ_+ are solutions of

$$\frac{dx}{dt} = u + c(\rho) \tag{4.4}$$

while the characteristic curves C_- corresponding to λ_- are solutions of

$$\frac{dx}{dt} = u - c(\rho). \tag{4.5}$$

Equations (4.4) and (4.5) give the slopes of the characteristic curves but these slopes cannot actually be determined unless the values of the solution (u, ρ) are already known on the curves.

Just as in the case of linear systems, the characteristic curves of a quasi-linear system are exceptional for the initial value problem. Note that since the initial value problem assigns the values of the desired solution (initial data) along the initial curve C, it is possible to determine whether or not C is characteristic with respect to the system and with respect to the initial data. If C is characteristic, the initial value problem either has no solution or it has infinitely many solutions.

According to Definition 4.1, the characteristic curves of (4.1) are never parallel to the x-axis. In particular, the x-axis is not characteristic and we can consider the initial value problem for the system (4.1) with the initial condition prescribed along an interval (a, b) of the x-axis,

$$\mathbf{u}(x, 0) = \boldsymbol{\phi}(x), \qquad a < x < b. \tag{4.6}$$

It is shown in the book of Garabedian,[3] Chapter 4, §3, that under certain smoothness assumptions on the coefficients and initial data, the initial value problem (4.1), (4.6) is well-posed in the sense that in a domain of the (x, t)-plane containing the initial interval (a, b), the problem has a unique solution which depends continuously on the initial data. The proof consists of three steps. First, by introducing m additional variables the system (4.1) is reduced to a canonical form of $2m$ equations in $2m$ unknowns. The canonical form looks like the linear canonical form (3.4) except that now the diagonal matrix Λ is a function of x, t and of the unknowns. Next, by an extension of the method of characteristics described in Section 3, the initial value problem is transformed to a system of integral equations. Finally the system of integral equations is solved by the method of successive approximations.

Instead of going any further into the study of general quasi-linear systems, we will discuss in the next section the system (4.3) which arises in gas dynamics.

Problem

4.1. Show that the system (1.5) has three families of characteristics which must be solutions of the equations

$$\frac{dx}{dt} = u + c, \qquad \frac{dx}{dt} = u - c, \qquad \frac{dx}{dt} = u.$$

5. One-Dimensional Isentropic Flow of an Inviscid Gas. Simple Waves

In this section we will study in some detail the quasi-linear system (1.4) of two equations in two unknowns and two independent variables,

$$\text{(5.1)} \qquad \begin{cases} \dfrac{\partial u}{\partial t} + u\dfrac{\partial u}{\partial x} + \dfrac{c^2}{\rho}\dfrac{\partial \rho}{\partial x} = 0 \\[2ex] \dfrac{\partial \rho}{\partial t} + u\dfrac{\partial \rho}{\partial x} + \rho\dfrac{\partial u}{\partial x} = 0, \end{cases}$$

which, as we mentioned in Section 1, governs the one-dimensional isentropic flow of an inviscid gas. In (5.1), $u = u(x, t)$ and $\rho = \rho(x, t)$ are the velocity and density of the gas at position x and time t, while $c^2 = c^2(\rho)$ is assumed to be a known positive function of ρ. For example, for a perfect gas

$$\text{(5.2)} \qquad c^2(\rho) = A\gamma\rho^{\gamma-1},$$

where γ is the ratio of specific heats ($\gamma = 1.4$ for air), and A is a constant depending on the pressure and density of the gas at rest. The derivation of equation (5.1) can be found in most books on fluid mechanics such as Becker,[4] or in the book of Garabedian.[3]

As we saw in Example 4.1, the system (5.1) has two families of characteristics. The characteristics C_+ are solutions of

$$\text{(5.3)} \qquad \frac{dx}{dt} = u + c,$$

while the characteristics C_- are solutions of

$$\text{(5.4)} \qquad \frac{dx}{dt} = u - c.$$

We know that the characteristics are exceptional for the initial value problem. Also, across a characteristic the solution and its derivatives may have jump discontinuities. (Remember that when the solution is not C^1 we refer to it as a generalized solution; see Problem 3.5.) In fact, it can be shown that the characteristic curves are the only curves across which the derivatives of u and ρ can have jump discontinuities while u and ρ themselves remain continuous. Therefore, the characteristics can be viewed as describing the location of sound waves. Equations (5.3) and (5.4) then assert that the velocity dx/dt of a sound wave differs from the flow velocity u by $\pm c$. For this reason c is referred to as the *local speed of sound*.

It is clear from equations (5.3) and (5.4) that the characteristics depend on the particular solution (u, ρ) of (5.1) and can be determined only if that solution is already known. It would therefore seem hopeless to try to use the characteristics to solve the system (5.1). Nevertheless the characteristics can be used in an approximation scheme for finding the solution of the initial value problem for (5.1). This approximation scheme, which we describe below, is based on the fact that certain functions of u and ρ remain constant along characteristic curves. We obtain these functions in the next paragraph.

The system (5.1) takes a more symmetric form if the new unknown

$$\ell = \int_{\rho_0}^{\rho} \frac{c(\rho')}{\rho'}\, d\rho' \tag{5.5}$$

is used in place of ρ. In (5.5) ρ_0 is the density of the gas at rest. Since $d\ell/d\rho = c(\rho)/\rho > 0$, equation (5.5) can be solved for ρ in terms of ℓ. It is easy (Problem 5.1) to show that u and ℓ satisfy the system of equations,

$$\begin{cases} \dfrac{\partial u}{\partial t} + u\dfrac{\partial u}{\partial x} + c\dfrac{\partial \ell}{\partial x} = 0 \\[2ex] \dfrac{\partial \ell}{\partial t} + c\dfrac{\partial u}{\partial x} + u\dfrac{\partial \ell}{\partial x} = 0. \end{cases} \tag{5.6}$$

Now, adding and subtracting the equations in (5.6) we obtain

$$\frac{\partial}{\partial t}(\ell + u) + (u + c)\frac{\partial}{\partial x}(\ell + u) = 0$$

$$\frac{\partial}{\partial t}(\ell - u) + (u - c)\frac{\partial}{\partial x}(\ell - u) = 0,$$

from which it follows that the quantities

$$r = \frac{1}{2}(\ell + u), \qquad s = \frac{1}{2}(\ell - u) \tag{5.7}$$

satisfy the equations

$$\begin{cases} \dfrac{\partial r}{\partial t} + (u + c)\dfrac{\partial r}{\partial x} = 0 \\[2ex] \dfrac{\partial s}{\partial t} + (u - c)\dfrac{\partial s}{\partial x} = 0. \end{cases} \tag{5.8}$$

In view of equations (5.3) and (5.4) for the characteristics C_+ and C_-, the first of equations (5.8) states that the derivative of the quantity r along a characteristic C_+ is zero, while the second of equations (5.8) states that the derivative of s along a characteristic C_- is zero. It follows that *r is constant along the C_+ characteristics while s is constant along the C_- characteristics.* For this reason the quantities r and s are known as *Riemann invariants.*

Let us consider now the initial value problem for the system (5.1) with the initial condition given on an interval of the x-axis,

$$u(x, 0) = \phi_1(x), \qquad \rho(x, 0) = \phi_2(x); \qquad a < x < b, \tag{5.9}$$

where ϕ_1 and ϕ_2 are given C^1 functions. We know that the initial value problem (5.1), (5.9) has a unique solution in a domain of the (x, t) plane containing the interval (a, b). We will describe an approximation scheme for computing the values of the solution in this domain.

We note first that using the relations (5.5) and (5.7), the values of the Riemann invariants r and s can be computed from the values of u and ρ, and, conversely, u and ρ can be computed from r and s. In particular, from the initial condition (5.9), the values of r and s can be computed on the initial interval (a, b). The basic idea of the approximation scheme is the following: Let P and Q be two nearby points on the interval (a, b) (see Fig. 5.1) and suppose for a moment that we already know the characteristics. Let R be the point of intersection of the characteristic C_+ through P and the characteristic C_- through Q. Since r is constant on C_+ and s is constant on C_-, the values of r and s are known at R, since $r(R) = r(P)$ and $s(R) = s(Q)$. The only problem is that we do not really know the characteristics C_+ and C_-. However, from (5.3), (5.4) and the initial condition (5.9), we know the slopes of the characteristics at P and Q and therefore we can approximate C_+ by the line tangent to C_+ at P and C_- by the line tangent to C_- at Q. These lines intersect at R', and R' is close to R if P and Q are close to each other. Therefore, the values of r and s at R' are approximately equal to their values at R,

$$r(R') \cong r(R) = r(P), s(R') \cong s(R) = s(Q).$$

Suppose now that the closed interval AB is contained in (a, b). Then the solution can be determined in a "triangular region" with base AB by using the following procedure. Divide AB into n equal subintervals as shown in Figure 5.2. Through each point $P_0, P_1, \dots, P_{n-1}$ draw the straight lines approximating the characteristics C_+ and through $P_1, P_2, \dots, P_n$ draw the straight lines approximating the characteristics C_-. These lines intersect at the points $P_0', P_1', \dots, P'_{n-1}$, and the approximate values of the solution at these points can be computed according to the method described in the previous paragraph. The process is now repeated starting with the points $P_0', P_1', \dots, P'_{n-1}$, and eventually, approximate values of the solution are obtained on a grid of points of a "triangular" region with base AB. It can be shown that as $n \to \infty$, these approximate values converge to the actual

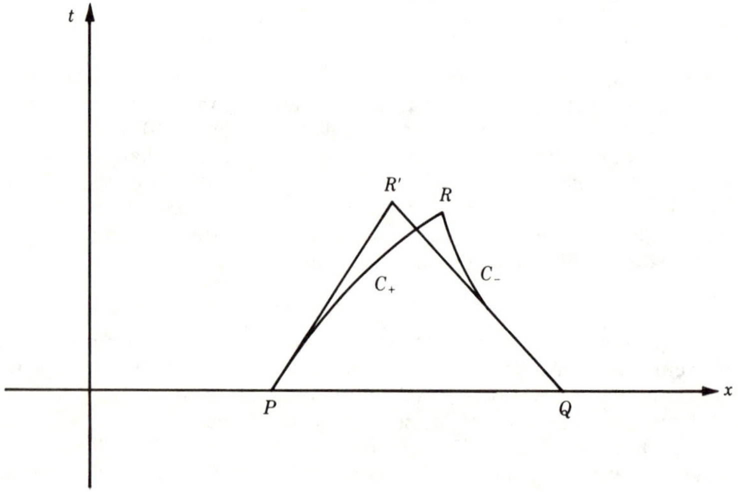

Fig. 5.1

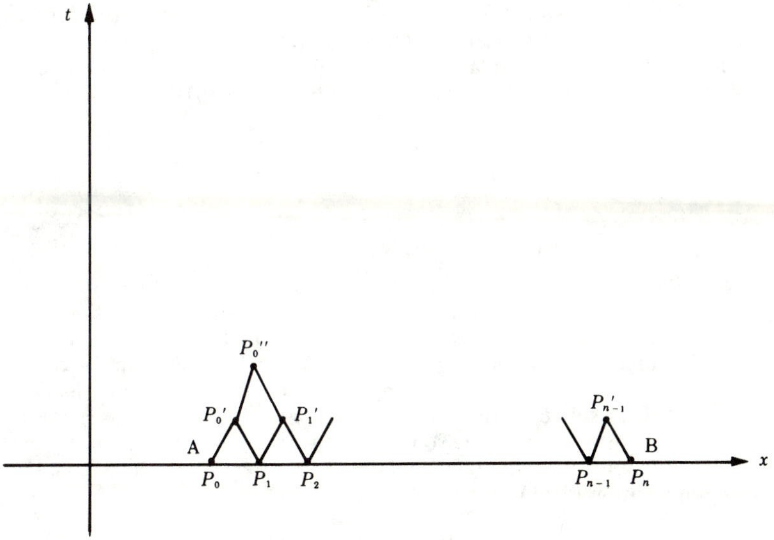

Fig. 5.2

values of the solution at points of the "triangular" region which are in the domain of existence of the solution.

We should emphasize here that by a solution of the initial value problem (5.1), (5.9) we mean a pair of C^1 functions u and ρ satisfying (5.1) and (5.9). Just as in the case of the conservation law (Section 5, Chapter III), which is a single quasi-linear equation, the nonlinearity of the system of equations (5.1) may cause the solution to develop discontinuities known as *shocks*. The above approximation scheme for the computation of the solution of the initial value problem is valid only in the part of the region which is free from shocks.

As a first step towards the understanding of the general nature of solutions of the system (5.1), it is useful to consider a *linear approximation* of that system. Actually, such an approximation gives a reasonable description of *small amplitude flows* such as one-dimensional acoustical waves. For these flows the values of u and ρ are not very different from their values $u = 0$ and $\rho = \rho_0$ at the rest state of the fluid. Assume that the deviations from rest, $u, \rho - \rho_0$ and $c - c_0$, and the derivatives of u and ρ are of the same order of magnitude which is small in comparison to ρ_0 and $c_0 = c(\rho_0)$. Neglecting products of small order terms in (5.1), we obtain the linear system

$$(5.10)\qquad \begin{cases} \dfrac{\partial u}{\partial t} + \dfrac{c_0^2}{\rho_0}\dfrac{\partial \rho}{\partial x} = 0 \\[2ex] \dfrac{\partial \rho}{\partial t} + \rho_0 \dfrac{\partial u}{\partial x} = 0. \end{cases}$$

(Note that (5.10) also follows from (5.1) if the coefficients of the derivatives of u and ρ are replaced by their values at rest.) With obvious changes in notation, the linear system (5.10) is system (2.6), the general solution of which is given by (2.18). The general solution of (5.10) is therefore

$$(5.11)\qquad \begin{aligned} u(x, t) &= \frac{c_0}{\sqrt{\rho_0}}\,[f(x - c_0 t) + g(x + c_0 t)], \\ \rho(x, t) &= \sqrt{\rho_0}\,[f(x - c_0 t) - g(x + c_0 t)], \end{aligned}$$

where f and g are arbitrary C^1 functions of one variable. If $g \equiv 0$, (5.11) represents a wave traveling undistorted in the positive x direction with speed c_0 and is called a *forward wave.* If $f \equiv 0$, (5.11) represents a wave traveling undistorted in the negative x direction with speed c_0 and is called a *backward wave.* Thus, c_0 is the speed of propagation of small disturbances, such as sound waves, and for this reason it is called the *speed of sound* in the fluid at rest. The solution of the initial value problem for the linear system (5.10) with initial condition (5.9) is easily obtained from the general solution (5.11),

$$(5.12)\qquad \begin{aligned} u(x, t) &= \frac{1}{2}[\phi_1(x - c_0 t) + \phi_1(x + c_0 t)] \\ &\qquad + \frac{c_0}{2\rho_0}[\phi_2(x - c_0 t) - \phi_2(x + c_0 t)], \\ \rho(x, t) &= \frac{1}{2}[\phi_2(x - c_0 t) + \phi_2(x + c_0 t)] \\ &\qquad + \frac{\rho_0}{2c_0}[\phi_1(x - c_0 t) - \phi_1(x + c_0 t)]. \end{aligned}$$

According to (5.12), an initial disturbance splits into forward and backward waves which travel undistorted for all t. This is due to the fact that nonlinearities have been completely neglected. Actually the nonlinearities of the original system (5.1) cause distortion of the traveling waves. This distortion builds up with time and may eventually lead to shocks. The usefulness of the linear approximation is thus limited to the study of very small disturbances over limited intervals of time.

The manner in which the nonlinearities of system (5.1) cause distortion of traveling waves is clearly demonstrated by a class of solutions of the system known as *simple waves.* These waves occur when the velocity u can be expressed as a function of the density ρ. Beginning with the assumption $u = u(\rho)$, we have

$$(5.13)\qquad \frac{\partial u}{\partial x} = \frac{du}{d\rho}\frac{\partial \rho}{\partial x}, \qquad \frac{\partial u}{\partial t} = \frac{du}{d\rho}\frac{\partial \rho}{\partial t}.$$

Multiplying the second equation in (5.1) by $du/d\rho$ and using (5.13) we obtain

$$\frac{\partial u}{\partial t} + u\frac{\partial u}{\partial x} + \rho\left(\frac{du}{d\rho}\right)^2 \frac{\partial \rho}{\partial x} = 0.$$

Comparing this equation with the first equation in (5.1) we find that

$$\left(\frac{du}{d\rho}\right)^2 = \frac{c^2(\rho)}{\rho^2},$$

or

$$\frac{du}{d\rho} = \pm \frac{c(\rho)}{\rho}.$$

After integration,

$$u = \pm \int_{\rho_0}^{\rho} \frac{c(\rho')}{\rho'} \, d\rho'$$

or

(5.14) $$u = \pm \ell,$$

where ℓ is the function of ρ defined by (5.5). Thus, if (u, ρ) is a solution of the system (5.1) such that u is a function of ρ, then this function of ρ must be one of the two functions ℓ or $-\ell$.

Since simple waves are solutions of system (5.1) for which $u = \pm\ell$, it is more convenient to work with the equivalent system (5.6). From (5.5), $d\ell/d\rho = c(\rho)/\rho > 0$, so that ρ can be expressed as a function of ℓ, $\rho = \rho(\ell)$. Therefore, system (5.6) can be written in a form that involves u and ℓ only,

(5.15) $$\begin{cases} \dfrac{\partial u}{\partial t} + u\dfrac{\partial u}{\partial x} + c(\rho(\ell))\dfrac{\partial \ell}{\partial x} = 0 \\[2ex] \dfrac{\partial \ell}{\partial t} + c(\rho(\ell))\dfrac{\partial u}{\partial x} + u\dfrac{\partial \ell}{\partial x} = 0. \end{cases}$$

In the first case, $u = \ell$, both equations in (5.15) reduce to the equation

$$\frac{\partial u}{\partial t} + [c(\rho(u)) + u]\frac{\partial u}{\partial x} = 0$$

or

(5.16) $$\frac{\partial u}{\partial t} + a(u)\frac{\partial u}{\partial x} = 0$$

where

(5.17) $$a(u) = c(\rho(u)) + u.$$

Equation (5.16) is a conservation law which we studied in Section 5 of Chapter III. Its general integral is

(5.18) $$u = F(x - a(u)t)$$

where F is an arbitrary C^1 function of a single variable. We conclude that *simple waves for which* $u = \ell$ *are forward waves* traveling with speed $a(u)$ which depends on u. In the second case, $u = -\ell$, both equations in (5.15) reduce to

$$\frac{\partial u}{\partial t} - [c(\rho(-u)) - u]\frac{\partial u}{\partial x} = 0$$

or

$$\frac{\partial u}{\partial t} - a(-u)\frac{\partial u}{\partial x} = 0 \tag{5.19}$$

the general integral of which is

$$u = G(x + a(-u)t). \tag{5.20}$$

Thus, *simple waves for which* $u = -\ell$ *are backward waves* traveling with speed $a(-u)$ which depends on u.

Simple waves are traveling waves whose profiles become distorted with time. This is due to the fact that the speed of propagation of each point of the u-profile depends on the value of u at that point. Any two points of the profile with different values of u travel with different speeds. The larger the amplitude of the wave, the more pronounced is the distortion. Writing (5.17) in the form

$$a(u) = c_0 + [c(\rho(u)) - c_0] + u \tag{5.21}$$

shows that for small amplitude flows the speed of propagation is nearly equal to the constant value c_0, since for such flows the last two terms in (5.21) are very small in comparison to c_0. It follows that over short intervals of time, simple waves of small amplitude behave almost like the undistorted traveling waves of the linear approximation theory. However, the last two terms in (5.21) always cause some distortion which builds up with time and may eventually lead to the formation of shocks. The following discussion of simple waves in a perfect gas will further illuminate this conclusion. At this point the reader should review Sections 5 and 6 of Chapter III.

For a perfect gas, $c(\rho)$ is given by (5.2) and the formulas for simple waves become more specific. Using (5.5) we find that

$$\ell(\rho) = \frac{2}{\gamma - 1}(c(\rho) - c_0). \tag{5.22}$$

Hence

$$c(\rho(\ell)) = c_0 + \frac{\gamma - 1}{2}\ell, \tag{5.23}$$

and from (5.17)

$$a(u) = c_0 + \frac{\gamma + 1}{2}u. \tag{5.24}$$

Therefore, forward simple waves ($u = \ell$) are defined implicitly by

$$u = F(x - [c_0 + \frac{\gamma + 1}{2}u]t), \tag{5.25}$$

and backward simple waves ($u = -\ell$) by

$$u = G(x + [c_0 - \frac{\gamma + 1}{2} u]t). \tag{5.26}$$

Let us discuss briefly a forward simple wave in a perfect gas. Such a wave is defined implicitly by equation (5.25) for some function F. Suppose that $u = u_0$ at the point (x_0, t_0) of the (x, t)-plane so that

$$u_0 = F(x_0 - [c_0 + \frac{\gamma + 1}{2} u_0]t_0).$$

Then $u = u_0$ at every point on the line

$$x - \left[c_0 + \frac{\gamma + 1}{2} u_0\right] t = x_0 - \left[c_0 + \frac{\gamma + 1}{2} u_0\right] t_0 \tag{5.27}$$

through (x_0, t_0). Indeed at any point (x, t) on this line

$$F\left(x - \left[c_0 + \frac{\gamma + 1}{2} u_0\right] t\right) = F\left(x_0 - \left[c_0 + \frac{\gamma + 1}{2} u_0\right] t_0\right) = u_0.$$

The line (5.27) is a C_+ characteristic. In general, for a forward simple wave in a perfect gas, the C_+ characteristics are straight lines on which u is constant and their inclinations are given by

$$\frac{dx}{dt} = c_0 + \frac{\gamma + 1}{2} u \tag{5.28}$$

(see Problem 5.5). Just as in Sections 5 and 6 of Chapter III, we can use this result to study forward simple waves and the development of shocks. Suppose for example that the profile of a forward traveling wave at $t = 0$ is given by

$$u(x, 0) = f(x), \qquad -\infty < x < \infty.$$

Then, for all t, u is defined implicitly by

$$u = f\left(x - \left[c_0 + \frac{\gamma + 1}{2} u\right] t\right).$$

If x_0 is any point on the x-axis and $u_0 = f(x_0)$, then $u = u_0$ along the line

$$x - \left[c_0 + \frac{\gamma + 1}{2} u_0\right] t = x_0 \tag{5.29}$$

through $(x_0, 0)$. If the initial velocity profile is nondecreasing with x (i.e., f is a nondecreasing function of x), then the lines (5.29) do not intersect in the upper half plane and hence no shocks develop for $t \geqq 0$. On the other hand, if the initial velocity profile is decreasing with x over any interval of the x-axis, some of the lines (5.29) will intersect in the upper half plane and a shock will develop at some positive time t.

We can give a more physical interpretation of the above explanation of the development of shocks. Since $u = \ell$, it follows from (5.5) that $du/d\rho = c(\rho)/\rho > 0$. Therefore, the larger the value of u at a point of the wave, the larger the value of the density ρ. In particular, $u > 0$ corresponds to $\rho > \rho_0$ (compression) while $u < 0$ corresponds to $\rho < \rho_0$ (rarefaction). Since the speed of propagation of a given point in the wave profile is given by (5.24) (or (5.28)), it follows that the larger the density at the point, the larger the speed is. Suppose now that initially two points A and B of the density profile are such that A is behind B and the density at A is larger than the density at B. Since A travels to the right faster than B, A will eventually reach B and the density profile between A and B will become vertical. This means that the density profile will develop a jump discontinuity which we call a shock.

Some applications of simple waves to the study of flows resulting from the motion of pistons in cylindrical tubes filled with a perfect gas can be found in Garabedian,[3] Chapter 14, Section 2 and in Landau and Lifshitz,[6] §94.

Problems

5.1. Derive the system of equations (5.6).

5.2. Derive the linear approximation (5.10) of system (5.1).

5.3. Show that the linear approximation (5.10) of system (5.1) leads to the one-dimensional wave equation of acoustics.

5.4. Derive formulas (5.22), (5.23) and (5.24) which are valid for a perfect gas.

5.5. For a forward simple wave in a perfect gas show that
(a) The Riemann invariants are $r = u$ and $s = 0$.
(b) The C_+ characteristics must satisfy equation (5.28)
(c) The C_+ characteristics are straight lines on which u is constant. [*Hint:* Use the fact that r must be constant on the C_+ characteristics.]

5.6. For a backward simple wave in a perfect gas formulate and prove results analogous to those of Problem 5.5.

5.7. Consider a perfect gas in a semi-infinite cylindrical pipe ($x > 0$) and suppose that at the end $x = 0$ of the pipe the velocity is known for all t, i.e., $u(0, t) = f(t)$. Show that in the resulting forward simple wave in the pipe, no shock will develop if $f(t)$ is a nonincreasing function of t, while a shock will always develop if $f(t)$ is increasing over any time interval.

5.8 In equation (5.19) introduce the new variables $x' = -x$ and $u' = -u$ to obtain the equation

$$\frac{\partial u'}{\partial t} + a(u')\frac{\partial u'}{\partial x'} = 0. \tag{5.30}$$

Compare (5.30) with (5.16) and conclude that backward simple waves are really forward simple waves in a different frame of reference. (Remember that u is the velocity in the direction of the positive x-axis.)

References for Chapter X

1. Churchill, R. V.: *Complex Variables and Applications,* New York: McGraw Hill Book Co., 1960.
2. Sokolnikoff, I. S., and Redheffer, R. M.: *Mathematics of Physics and Modern Engineering,* Ed. 2, New York: McGraw Hill Book Co., 1966.
3. Garabedian, P.: *Partial Differential Equations,* New York: John Wiley & Sons, Inc., 1964.
4. Becker, E.: *Gas Dynamics,* New York: Academic Press, 1968.
5. Petrovskii, I. G.: *Partial Differential Equations,* Philadelphia: W. B. Saunders Co., 1967.
6. Landau, L. D., and Lifshitz, E. M.: *Fluid Mechanics,* Elmsford, N. Y.: Pergamon Press Press, Inc., 1959.

Guide to further study

Chapter III

The theory of general nonlinear first order p.d.e.'s can be found in Garabedian, Chapter 2, Chester, Chapter 8 or John, Chapter I.

Chapter VI

§2. The equations of mathematical physics can also be derived by means of a variational procedure; see Courant and Hilbert, Volume I, Chapter IV, for a discussion of this technique. Tikhonov and Samarskii derive many equations by using techniques similar to those of this chapter.

Chapter VII

§1. For a definition and discussion of properties of generalized solutions of Laplace's equation see Garabedian, Chapter 8. The proof of our Theorem 1.1 follows from Theorem 11.2 and the results given in Garabedian. A proof of the real analyticity of solutions of general second order elliptic equations with analytic coefficients can be found in Garabedian, page 145.

§2. The form of Laplace's operator in general curvilinear coordinates can be found in Kellogg. Excellent treatments of spherical harmonics are contained in Müller or Hochstadt.

§3. Properties of the inversion mapping (3.13) are easily obtained by using analytic function techniques; see Kaplan, Chapter 7, §1.

§5. A proof of the maximum principle for solutions of a general class of elliptic equations with variable coefficients appears in Garabedian, Chapter 7, §1.

§6. Several different proofs for the existence of a solution of the Dirichlet problem are discussed in Garabedian, Chapters 7–10. These include conformal mapping techniques in the case of two independent variables, the Hilbert-space method of orthogonal projection and Dirichlet's principle, and the methods of potential theory and integral equations.

§8. Introductory discussions of topics in Fourier series such as convergence in the mean, the proof of Parseval's relation (8.34) for square-integrable functions, other modes of convergence and methods of summation are given in Tolstov. The nature of convergence of Fourier series near a jump discontinuity (Gibbs' phenomenon) is discussed and illustrated in Carslaw. A list of Fourier series expansions of selected functions can be found in Tolstov, page 147. Also, see Salvadori and Schwarz, page 372, for a useful collection of series.

§9. Discussions of properties of Green's functions and examples of these functions for certain boundary value problems can be found in Courant and Hilbert, Vol. I, Chapter V, and Weinberger, Chapter V. Stakgold, Vol. II, discusses several methods of determining Green's functions and gives examples. Weinberger, Chapter VIII, treats the construction of Green's functions by analytic function techniques in the case of two independent variables.

§11. The expansion of harmonic functions in terms of harmonic polynomials is covered in Hochstadt.

§16. The Green's function for the Neumann problem is discussed in Garabedian, Chapter 7.

Chapter VIII

§1. The reduced wave equation (1.25) is studied in great detail in Tikhonov and Samarskii, Chapter VII.

§3. The energy method extends naturally to a large class of problems and can be used as a basis for existence proofs; see Courant and Hilbert, Vol. II, page 656, and Garabedian, page 434.

§6. Baker and Copson contains an interesting discussion of Huygens' principle, and deals with several physical problems. This reference also contains a direct derivation of Kirchoff's formula (5.9). Generalized solutions of Cauchy problems are defined and briefly discussed in Petrovskii, Chapter 2, §9.

§8. The general Sturm-Liouville problem is treated in detail in Stakgold, Vol. I, and Titchmarsh, Part I. Both of these references contain examples of specific Sturm-Liouville problems and proofs of the associated eigenfunction expansion theorems.

§10. Titchmarsh, Part II, contains many examples of eigenfunction expansions associated with second order elliptic p.d.e.'s in two or more variables. In addition, properties of eigenfunctions and the distribution of eigenvalues are discussed for certain p.d.e.'s with variable coefficients.

Chapter IX

§1. The problem (1.1), (1.2), (1.3) with general non-zero boundary data f_1 and f_2 can be solved using Laplace transform methods; see Churchill's book on operational calculus.

§3. A verification of (3.5) appears in Tikhonov and Samarskii, p. 248. This book also contains a good selection of exercises.

§4. A verification that (4.3) gives the solution to the initial value problem (4.1), (4.2) can be found in Tikhonov and Samarskii, page 509, where it is assumed only that ϕ is piecewise continuous and bounded. Sobolev, Chapter XXII, contains a proof that (4.7) is actually a C^∞ solution.

Chapter X

A recent book on the mathematically theory of linear and nonlinear waves is Whitham.

The Method of Integral Transforms

Integral transforms provide an effective tool for solving many problems involving linear p.d.e.'s. The classical theory of the Fourier transform and its applications can be found in Sneddon. The Laplace, Mellin, Hankel and other transforms are also discussed in this book. Integral transform methods are also referred to as operational methods.

Potential Theory

A powerful method for the study of boundary value problems for Laplace's equation is the method of integral equations. This method is based on representing solutions in the form of integrals which are frequently encountered in physics and are known as potentials. The book of Gunter contains a detailed development of potential theory and its applications.

Additional Problems and Physical Applications

The book of Budak, Samarskii and Tikhonov consists of a large collection of problems of mathematical physics. Many of the problems are solved in detail. Another similar collection of problems can be found in Lebedev, Skalskaya and Uflyand. The books of Ames present an extensive survey of the theory and application of nonlinear p.d.e.'s that arise in engineering.

Numerical Methods

The theory and application of numerical methods for the solution of problems involving the Laplace, wave and heat equations can be found in the book of Forsythe and Wasow.

Equations of Mathematical Physics. (Advanced treatment)

A modern, more advanced treatment of the equations of Mathematical Physics using functional analysis and distributions (Dirac delta functions, etc.) can be found in the books of Stakgold and Vladimirov.

Modern Theory

In recent years, great advances in the general theory of linear p.d.e.'s have been achieved using the methods of functional analysis and in particular the theory of distributions. The study of this modern theory requires a higher level of mathematical sophistication than is necessary for the study of the present book. Treves' book "Basic Linear P.D.E.'s" is a good link between classical and modern theory. The books of Hormander and Treves ("Linear P.D.E.'s with Constant Coefficients") first develop the theory of distributions and then use it as the framework of the modern theory studied in their books. Friedman treats parabolic equations and Lions and Magenes study nonhomogeneous problems in the modern setting. The nonlinear theory can be found in Lions.

Bibliography for further study

Agmon, S.: *Lectures on Elliptic Boundary Value Problems,* Princeton, N. J.: Van Nostrand, 1965.

Ames, W. F.: *Nonlinear Partial Differential Equations in Engineering,* Vol. I, 1965; Vol. II 1970; New York: Academic Press.

Ames, W. F.: *Numerical Methods for Partial Differential Equations,* New York: Barnes & Noble Publications, 1969.

Baker, B. B., and Copson, E. T.: *The Mathematical Theory of Huygens' Principle,* Fair Lawn, N. J.: Oxford University Press, 1950.

Budak, B. M., Samarskii, A. A., and Tychonov, A. N.: *A Collection of Problems on Mathematical Physics,* Elmsford, N. Y.: Pergamon Press, 1964.

Carslaw, H. S.: *Introduction to the Theory of Fourier Series and Integrals,* Ed. 3, New York: Macmillan Co., 1930.

Chester, C. R.: *Techniques in Partial Differential Equations,* New York: McGraw-Hill Book Co., 1971.

Churchill, R. V.: *Modern Operational Mathematics in Engineering,* New York: McGraw-Hill Book Co., 1944.

Courant, R., and Hilbert, D.: *Methods of Mathematical Physics,* Vols. I and II, New York: Interscience Publishers, 1953.

Duff, G. F. D.: *Partial Differential Equations,* Toronto: University of Toronto Press, 1956.

Forsythe, G. E., and Wasow, W. R.: *Finite Difference Methods for Partial Differential Equations,* New York: John Wiley & Sons, 1960.

Friedman, A.: *Partial Differential Equations of Parabolic Type,* Englewood Cliffs, N. J.: Prentice Hall, 1964.

Garabedian, P. R.: *Partial Differential Equations,* New York: John Wiley & Sons, 1964.

Gunter, N. M.: *Potential Theory and its Applications to Basic Problems of Mathematical Physics,* New York: Frederick Ungar, 1967.

Hellwig, G.: *Partial Differential Equations,* New York, Blaisdell Publishing Co., 1964.

Hochstadt, H.: *The Functions of Mathematical Physics,* New York: Wiley-Interscience, 1971.

Hormander, L.: *Linear Partial Differential Operators,* New York: Springer-Verlag, 1963.

John, F.: *Partial Differential Equations,* Ed. 2, New York: Springer-Verlag, 1975.

Kaplan, W.: *Introduction to Analytic Functions,* Reading, Mass.: Addison-Wesley, 1966.

Kellogg, O. D.: *Foundations of Potential Theory,* New York: Dover Publications, Inc., 1953.

Lebedev, N. N., Skalskaya, I. P., and Uflyand, Y. S.: *Problems of Mathematical Physics,* Englewood Cliffs, N. J.: Prentice-Hall, 1965.

Lions, J. L.: *Quelques méthodes de résolution des problèmes aux limites non linéaires,* Paris: Dunod, 1969.

Lions, J. L., and Magenes, E.: *Non-homogeneous Boundary Value Problems and Applications,* New York: Springer-Verlag, 1972.

Mikhlin, S. G.: *Linear Equations of Mathematical Physics,* New York: Holt, Rine-

hart & Winston, Inc., 1967.
Miranda, C.: *Partial Differential Equations of Elliptic Type,* New York: Springer-Verlag, 1969.
Muller, C.: *Spherical Harmonics, Lecture Notes in Mathematics,* Vol. 17, New York: Springer-Verlag, 1966.
Petrovskii, I. G.: *Partial Differential Equations,* Philadelphia, W. B. Saunders Co., 1967.
Protter, M. H., and Weinberger, H. F.: *Maximum Principles in Differential Equations,* Englewood Cliffs, N. J.: Prentice Hall, Inc., 1967.
Salvadori, M. G., and Schwarz, R. J.: *Differential Equations in Engineering Problems,* Englewood Cliffs, N. J., Prentice Hall, Inc., 1965.
Smirnov, V. I.: *A Course of Higher Mathematics,* Vol. IV, Elmsford, N.Y.: Pergamon Press, 1964.
Smith, G. D.: *Numerical Solutions of Partial Differential Equations,* Fair Lawn, N. J.: Oxford University Press, 1965.
Sneddon, I.N.: *The Use of Integral Transforms,* New York: McGraw-Hill Book Co., 1972.
Sobolev, S. L.: *Partial Differential Equations of Mathematical Physics,* Elmsford, N. Y.: Pergamon Press, 1965.
Stakgold, I.: *Boundary Value Problems of Mathematical Physics,* Vols. I and II, New York: Macmillan Co., 1967.
Tikhonov, A. N., and Samarskii, A. A.: *Equations of Mathematical Physics,* Elmsford, N. Y.: Pergamon Press, 1963.
Titchmarsh, E. C.: *Eigenfunction Expansions Associated with Second-order Differential Equations,* Parts I & II, Ed. 2, New York: Oxford University Press, 1962.
Tolstov, G. P.: *Fourier Series,* Englewood Cliffs, N. J.: Prentice Hall, Inc., 1962.
Treves, F.: *Basic Linear Partial Differential Equations,* New York: Academic Press, 1975.
Treves, F.: *Linear Partial Differential Equations with Constant Coefficients,* New York: Gordon & Breach, 1966.
Vladimirov, V. S.: *Equations of Mathematical Physics,* New York: Marcel Dekker Inc., 1971.
Weinberger, H. F.: *A First Course in Partial Differential Equations,* Waltham, Mass.: Blaisdell Publishing Co., 1965.
Whitham, G. B. *Linear and Nonlinear Waves,* New York: John Wiley & Sons, 1974.

Answers to selected problems

Chapter I

2.3. $2x + 2y - z - 2 = 0$.

2.4. $x + 3y - 7 = 0$.

2.5. (a) $z = \sqrt{x^2 + y^2}$. (b) $z = -\sqrt{x^2 + y^2}$. (c) Not possible.

3.2. $(5, -4, -1)$ or any non-zero scalar multiple of this vector.

3.4. S_c: $2(x^2 - y^2 + 1) + z = 2$.

3.5. $x = \cos t$, $y = \sin t$, $z = -\cos t - \sin t$, $0 \le t \le 2\pi$.

4.1. $x(t) = e^{-P(t)}[c + \int^t e^{P(s)}q(s)ds]$.

4.2. (a) $x = t/2 + c/t$. (b) $x = 1 + ce^{-\frac{t^3}{3}}$.

4.3. (a) $x = t/2 + 3/(2t)$. (b) $x = 1 + e^{\frac{1-t^3}{3}}$.

4.4. (a) $x^2 = (c - 2t)^{-1}$. (b) $xt^{-1}e^{\frac{x^2}{2}} = c$.

4.5. $x = (1 - 2t)^{-\frac{1}{2}}$, $t < \frac{1}{2}$.

4.6. (a) $x_1 = c_1 \cos t + c_2 \sin t$, $x_2 = -c_1 \sin t + c_2 \cos t$.

(b) $x_1 = \cos t$, $x_2 = -\sin t$; $-\infty < t < \infty$.

4.7. (a) $x_1 = \dfrac{c_1c_2e^{c_1t}}{c_1 + c_2e^{c_1t}}$, $x_2 = \dfrac{c_1^2}{c_1 + c_2e^{c_1t}}$.

(b) $x_1 = \frac{2e^{2t}}{1 + e^{2t}}$, $x_2 = \frac{2}{1 + e^{2t}}$; $-\infty < t < \infty$.

Chapter II

1.4. (*b*) $x^4 + y^4 = c_1$, $z = c_2$.

2.2. $y = x$, $\frac{1}{2}xy - \arctan z = \frac{1}{2}$.

2.3. $\frac{x}{y} = c_1$, $\frac{1}{2}xy - \arctan z = c_2$; $y > 0$.

2.6. $u_2 = x + y + z$.

2.7. (a) $y + z = c_1$, $y - x/\log(y + z) = c_2$. (b) $(y - x)/xy = c_1$, $z/xy = c_2$.

2.8. (a) $x + y + z = c_1$, $x^2 + y^2 + z^2 = c_2$. (b) $x^2 + y^2 = c_1$, $(x^2 + y^2)y^2 + z^2 = c_2$.

3.1. (a) $u = F(z, y)$. (b) $u = F(z, x^2 + y^2)$. (c) $u = F(x, y^3 - z)$. (d) $u = F\left(\frac{x - y}{z}, \frac{x - y}{xy}\right)$.

3.2. (a) $u = F(y/x, (1/2)xy - \arctan z)$. (c) $u = F((x + z)/y, (x - y)^2 - z^2)$ (d) $u = F(x + y + z, xyz)$.

4.1. (a) $z = e^y \sin(x - y)$. (b) $z = xy$. (c) $x^2 + y^2 + z^2 - \frac{5}{9}(x + y + z)^2 = 0$.

4.2. (a) $F(x - y, ze^{-x}) = 0$, where F is any C^1 function satisfying $F(-1, 1) = 0$.

Chapter III

2.1. General integrals: (b) $F(z^2 - x^2, (x + y)^2 - (y + z)^2) = 0$. (f) $z = \tan[(1/2)xy + f(x/y)]$. (g) $F(x + y + z, xyz) = 0$. (h) $z^2 + y^2 = f(x)$.

2.4. $u_1 = yze^{-(y+z)}$

3.1. (a) $z = e^y \cos(x - y)$. (b) $z = \frac{xy}{xy + 2x - y}$.

(c) $z = \frac{x + y}{xy - 1}$. (e) $z = \frac{x^2 + y^2}{2} e^{\frac{x^2 - y^2}{2}}$. (g) $z = x\frac{3y^2 + 1}{3y^2 - 1}$.

4.1. (a) Unique solution. (b) Infinitely many solutions. (c) No solution.

4.2. Solutions exist only for linear f, $f(t) = ct$.

5.5. $z = x/(1 + y)$.

Chapter V

2.1. (a) $x_1 \pm 2(-x_2)^{\frac{1}{2}} = c, \quad x_2 < 0.$ (d) $x_1^2 + x_2^2 = c.$

2.2. (a) $x = \pm \frac{2}{5}(-y)^{\frac{5}{2}} + c, \quad y < 0.$ (b) $y = ce^{x^2}.$

6.1. (a) $u = xy - x^3/3 + (y - x^2/2)^2.$ (b) $u = xy - x^3/3 + \sin(y - x^2/2).$ (c) $u = xy - x^3/3 + y - x^2/2 + 5/6.$

7.7. (a) Elliptic in (x, y) plane. (b) Elliptic for $|x| < 1$, hyperbolic for $|x| > 1$, parabolic for $|x| = 1$. (c) Elliptic for $y > e^{-x}$, hyperbolic for $y < e^{-x}$, parabolic for $y = e^{-x}$.

Chapter VII

7.5. (b) $u(r, \theta) = \dfrac{-\log r}{\log 2} + \dfrac{1}{3}(4r - r^{-1}) \sin \theta.$

7.7. (a) $u(r, \theta) = \pi^2/3 + 4 \sum_{n=1}^{\infty} \dfrac{(-1)^n}{n^2} r^n \cos n\theta.$ (b) $u(r, \theta) = 2/\pi - (4/\pi) \sum_{n=1}^{\infty} \dfrac{1}{4n^2 - 1} r^{2n} \cos 2n\theta.$ (c) $u(r, \theta) = \pi^2/6 - 2 \sum_{n=1}^{\infty} \dfrac{1 + (-1)^n}{n^2} \cdot r^n \cos n\theta.$ (d) $u(r, \theta) = 4/\pi \sum_{n=1}^{\infty} \dfrac{1 - (-1)^n}{n^3} r^n \sin n\theta.$

8.14. Yes. Apply the Weierstrass M-test, using Mn^{-2}.

8.23. (a) $$\frac{L}{4} - \frac{2L}{\pi^2} \sum_{n=1}^{\infty} \frac{1}{(2n-1)^2} \cos \frac{(2n-1)2\pi x}{L} = \begin{cases} x, & 0 \le x < \dfrac{L}{2}, \\ L - x, & \dfrac{L}{2} \le x \le L. \end{cases}$$

(b) $$\frac{4}{\pi} \sum_{n=1}^{\infty} \frac{1}{n} \sin \frac{n\pi}{2} \sin \frac{n\pi a}{2L} \sin \frac{n\pi x}{L} = \begin{cases} 0, & 0 \le x < \dfrac{L-a}{2}, \\ 1, & \dfrac{L-a}{2} < x < \dfrac{L+a}{2}, \\ 0, & \dfrac{L+a}{2} < x \le L. \end{cases}$$

13.3. $G(\mathbf{r}', \mathbf{r}) = \dfrac{1}{4\pi} \sum_{k=-\infty}^{\infty} \left[\dfrac{1}{|\mathbf{r}' - \mathbf{r}_k|} - \dfrac{1}{|\mathbf{r}' - \mathbf{r}_k^*|}\right]$, where $\mathbf{r}_k = \mathbf{r} + (0, 0, 2k)$, $\mathbf{r}_k^* = \mathbf{r}^* + (0, 0, 2k)$; $k = 0, \pm 1, \pm 2, \ldots$ and $\mathbf{r}^*$ is symmetric to $\mathbf{r}$ with respect to the plane $z = 0$.

13.5. $G(\mathbf{r}', \mathbf{r}) = G_B(\mathbf{r}, \mathbf{r}) - G_B(\mathbf{r}', \mathbf{r}^*)$, where G_B is the Green's function for the full ball and $\mathbf{r}^*$ is symmetric to $\mathbf{r}$ with respect to the plane $z = 0$.

Chapter VIII

2.1. (a) $t = \ell/2$. (b) (i) $t = 2$, (ii) $t = 9$, (iii) $t = 4$.

2.4. (a) $u(\pi, t) = 0$ for $t = 0, \pi/2, 3\pi/2, 2\pi$, while $u(\pi, \pi) = 1/2$. $u(-\pi, t) = u(\pi, t)$ for $t \geq 0$. (b) $u(\pi, t) = 0$ for $t = 0$, $\pi/2$, $u(\pi, \pi) = 1/2$, $u(\pi, t) = 1$ for $t \geq 3\pi/2$.

4.1. $t = R$.

4.2. (a) $t = 4$. (b) $t = 9$. (c) $t = \sqrt{13} - 1$.

6.2. (a) For $n = 2$: $u(0, t) = t$ for $0 \leq t \leq 1$, $u(0, t) = t - \sqrt{t^2 - 1}$ for $t > 1$. For $n = 3$: $u(0, t) = t$ for $0 \leq t < 1$, $u(0, t) = 0$ for $t > 1$. (b) For $n = 2$: $u(0, t) = 1$ for $0 \leq t < 1$, $u(0, t) = 1 - t(t^2 - 1)^{-1/2}$ for $t > 1$. For $n = 3$: $u(0, t) = 1$ for $0 \leq t < 1$, $u(0, t) = 0$ for $t > 1$.

Chapter IX

2.1. $u(x, t) = 4U/\pi \sum_{n=1}^{\infty} \dfrac{1 - (-1)^n}{n^3} \sin nx \, e^{-n^2 t}$.

2.6. $u(x, t) = 100$.

2.7. $u(x, t) = aU/L + 4U/\pi \sum_{n=1}^{\infty} \dfrac{1}{n} \sin \dfrac{n\pi a}{2L} \cos \dfrac{n\pi}{2} \cos \dfrac{n\pi x}{L} e^{-\frac{n^2\pi^2}{L^2} t}$.

4.3. $u(x_1, x_2, x_3, t) = \dfrac{1}{2}(U_1 + U_2) + \dfrac{(U_1 - U_2)}{\sqrt{\pi}} \displaystyle\int_0^{\frac{x_3}{2\sqrt{t}}} e^{-\eta^2} \, d\eta$.

4.8. $u(r, t) = U_0$.

4.10. $u(r, t) = U_1 \left[1 - 2a/\pi \sum_{n=1}^{\infty} \dfrac{(-1)^{n+1}}{nr} \sin \dfrac{n\pi r}{a} e^{-\frac{n^2\pi^2}{a^2} t} \right]$.

4.11. (a) Extend the initial temperature distribution $\phi(x_1, x_2, x_3)$, $x_3 > 0$, to all of R^3 as an *odd* function of x_3 and apply (4.3). (b) Extend the initial temperature distribution $\phi(x_1, x_2, x_3)$, $x_3 > 0$, to all of R^3 as an *even* function of x_3 and apply (4.3).

4.14. Express the solution u as a superposition $u = u_1 + u_2$ where u_1 satisfies (4.14) and u_2 satisfies (4.19), (4.20), (4.21).

Chapter X

3.2. (a) The initial condition for v_1 is inconsistent with (2.17). (b) In (2.17), f can be any C^1 function satisfying $f(0) = 1$, while $g(x) = (x/c_0)^2$.

3.3. (a) $C_1: x = t + (\bar{x} - \bar{t})$, $x_1{}^0 = \bar{x} - \bar{t}$. (b) (i) $C_2: x = t^{-1}$; (ii) C_2: $x = \dfrac{2}{1 + 2t}$, $x_2{}^0 = 0$.

3.4. C_2: $x = 0$.

3.9. The quadrilateral has vertices $(0, 0)$, $(L/2, -L/2c_0)$, $(L, 0)$ and $(L/2, L/2c_0)$.

INDEX

A CATALOG OF SELECTED

DOVER BOOKS

IN SCIENCE AND MATHEMATICS

DE RE METALLICA, Georgius Agricola. The famous Hoover translation of greatest treatise on technological chemistry, engineering, geology, mining of early modern times (1556). All 289 original woodcuts. 638pp. 6¾ × 11.
60006-8 Pa. $18.95

SOME THEORY OF SAMPLING, William Edwards Deming. Analysis of the problems, theory and design of sampling techniques for social scientists, industrial managers and others who find statistics increasingly important in their work. 61 tables. 90 figures. xvii + 602pp. 5⅜ × 8½. 64684-X Pa. $15.95

THE VARIOUS AND INGENIOUS MACHINES OF AGOSTINO RAMELLI: A Classic Sixteenth-Century Illustrated Treatise on Technology, Agostino Ramelli. One of the most widely known and copied works on machinery in the 16th century. 194 detailed plates of water pumps, grain mills, cranes, more. 608pp. 9 × 12.
28180-9 Pa. $24.95

LINEAR PROGRAMMING AND ECONOMIC ANALYSIS, Robert Dorfman, Paul A. Samuelson and Robert M. Solow. First comprehensive treatment of linear programming in standard economic analysis. Game theory, modern welfare economics, Leontief input-output, more. 525pp. 5⅜ × 8½. 65491-5 Pa. $14.95

ELEMENTARY DECISION THEORY, Herman Chernoff and Lincoln E. Moses. Clear introduction to statistics and statistical theory covers data processing, probability and random variables, testing hypotheses, much more. Exercises. 364pp. 5⅜ × 8½. 65218-1 Pa. $9.95

THE COMPLEAT STRATEGYST: Being a Primer on the Theory of Games of Strategy, J.D. Williams. Highly entertaining classic describes, with many illustrated examples, how to select best strategies in conflict situations. Prefaces. Appendices. 268pp. 5⅜ × 8½. 25101-2 Pa. $7.95

MATHEMATICAL METHODS OF OPERATIONS RESEARCH, Thomas L. Saaty. Classic graduate-level text covers historical background, classical methods of forming models, optimization, game theory, probability, queueing theory, much more. Exercises. Bibliography. 448pp. 5⅜ × 8¼. 65703-5 Pa. $12.95

CONSTRUCTIONS AND COMBINATORIAL PROBLEMS IN DESIGN OF EXPERIMENTS, Damaraju Raghavarao. In-depth reference work examines orthogonal Latin squares, incomplete block designs, tactical configuration, partial geometry, much more. Abundant explanations, examples. 416pp. 5⅜ × 8¼.
65685-3 Pa. $10.95

THE ABSOLUTE DIFFERENTIAL CALCULUS (CALCULUS OF TENSORS), Tullio Levi-Civita. Great 20th-century mathematician's classic work on material necessary for mathematical grasp of theory of relativity. 452pp. 5⅜ × 8½.
63401-9 Pa. $9.95

VECTOR AND TENSOR ANALYSIS WITH APPLICATIONS, A.I. Borisenko and I.E. Tarapov. Concise introduction. Worked-out problems, solutions, exercises. 257pp. 5⅜ × 8¼. 63833-2 Pa. $7.95